T0221461

Admixtures for Concrete
Improvement of Properties

Other RILEM Proceedings available from Chapman and Hall

1 Adhesion between Polymers and Concrete. ISAP 86
  Aix-en-Provence, France, 1986
  Edited by H. R. Sasse

2 From Materials Science to Construction Materials Engineering
  Proceedings of the First International RILEM Congress
  Versailles, France, 1987
  Edited by J. C. Maso

3 Durability of Geotextiles
  St Rémy-lès-Chevreuse, France, 1986

4 Demolition and Reuse of Concrete and Masonry
  Tokyo, Japan, 1988
  Edited by Y. Kasai

5 Admixtures for Concrete
  Improvement of Properties
  Barcelona, Spain, 1990
  Edited by E. Vázquez

6 Analysis of Concrete Structures by Fracture Mechanics
  Abisko, Sweden, 1989
  Edited by L. Elfgren and S. P. Shah

Publisher's Note

This book has been produced from camera ready copy provided by the individual contributors. This method of production has allowed us to supply finished copies to the delegates at the Symposium.

Admixtures for Concrete
Improvement of Properties

Proceedings of the International Symposium held by RILEM (The International Union of Testing and Research Laboratories for Materials and Structures) and organized by RILEM Technical Committee TC-84AAC and Departmento de Ingenieria de la Construcción Escuela Técnica Superior de Ingenieros de Caminos, Canales y Puertos Universitat Politécnica de Catalunya, Spain.

Barcelona
May 14–17, 1990

EDITED BY
E. Vázquez

CRC Press
Taylor & Francis Group
Boca Raton London New York

CRC Press is an imprint of the
Taylor & Francis Group, an **informa** business
A TAYLOR & FRANCIS BOOK

CRC Press
Taylor & Francis Group
6000 Broken Sound Parkway NW, Suite 300
Boca Raton, FL 33487-2742

First issued in paperback 2019

© 1990 RILEM
CRC Press is an imprint of Taylor & Francis Group, an Informa business

ISBN-13: 978-0-412-37410-4 (hbk)
ISBN-13: 978-0-367-86345-6 (pbk)

This book contains information obtained from authentic and highly regarded sources. Reason-able efforts have been made to publish reliable data and information, but the author and publisher cannot assume responsibility for the validity of all materials or the consequences of their use. The authors and publishers have attempted to trace the copyright holders of all material reproduced in this publication and apologize to copyright holders if permission to publish in this form has not been obtained. If any copyright material has not been acknowledged please write and let us know so we may rectify in any future reprint.

Except as permitted under U.S. Copyright Law, no part of this book may be reprinted, reproduced, transmitted, or utilized in any form by any electronic, mechanical, or other means, now known or hereafter invented, including photocopying, microfilming, and recording, or in any information storage or retrieval system, without written permission from the publishers.

For permission to photocopy or use material electronically from this work, please access www.copyright.com (http://www.copyright.com/) or contact the Copyright Clearance Center, Inc. (CCC), 222 Rosewood Drive, Danvers, MA 01923, 978-750-8400. CCC is a not-for-profit organiza-tion that provides licenses and registration for a variety of users. For organizations that have been granted a photocopy license by the CCC, a separate system of payment has been arranged.

Trademark Notice: Product or corporate names may be trademarks or registered trademarks, and are used only for identification and explanation without intent to infringe.

British Library Cataloguing in Publication

Admixtures for concrete.
1. Materials: Concrete. Chemical mixtur‹
I. Vazquez, E.
620.136

Visit the Taylor & Francis Web site at
http://www.taylorandfrancis.com

and the CRC Press Web site at
http://www.crcpress.com

International Scientific Committee

Professor R. Rivera *Civil Engineering Institute, University of Nuevo León, San Nicolás de los Garza, Mexico (Chairman)*

Professor E. Vázquez *Department of Construction Engineering, Cataluña Technical University, Barcelona, Spain (Co-Chairman)*

Dr A-M. Paillère *Laboratoire Central des Ponts et Chaussées, Paris, France (Secretary)*

Professor M. S. Akman *Insaat Fakültesi, Technical University of Istambul, Turkey*

Dr F. Alou *Department of Materials, Swiss Federal Institute of Technology, Lausanne, Switzerland*

Dr M. Ben-Bassat *Technion City, Israel Institute of Technology, Haifa, Israel*

Dr S. Biagini *Modern Advanced Concrete, Treviso, Italy*

Dr J. Cabrera *Department of Civil Engineering, University of Leeds, UK*

Ing. E. A. Decker *Gifford Hill and Co., Dallas, Texas, USA*

Dr D. Dimic *Zavod za Raziskavo, Materiala in Konstrukcij, Ljubljana, Yugoslavia*

Professor F. Massazza *Central Chemical Laboratory, Italcementi, Bergamo, Italy*

Professor S. Nagataki *Department of Civil Engineering, Tokyo Institute of Technology, Japan*

Dr V. S. Ramachandran *Division of Building Research, National Research Council, Ottawa, Canada*

Contents

Preface   xiv

1 WORKABILITY

Superplasticizer and air-entraining agent demand in OPC mortars containing silica fume   1
A. DUREKOVIĆ and K. POPOVIĆ *Department for Concrete and Masonry Structures, Civil Engineering Institute, Zagreb, Yugoslavia*

Slump control and properties of concrete with a new superplasticizer. I: Laboratory studies and test methods   10
M. FUKUDA, T. MIZUNUMA, T. IZUMI and M. IIZUKA *Wakayama Research Laboratory, Kao Corporation, Wakayama, Japan*
M. HISAKA *Central Laboratory, Japan Testing Center for Construction Materials, Saitama, Japan*

Dynamic rheological properties of fresh concrete with chemical admixtures   20
S. KAKUTA *Department of Civil Engineering, Akashi Technological College, Akashi, Japan*
T. KOJIMA *Department of Civil Engineering, Ritsumeikan University, Kyoto, Japan*

Rheological studies on fresh concrete using admixtures   34
H. KIKUKAWA *Department of Civil Engineering, Meijo University, Japan*

High performance concrete with high filling capacity   51
K. OZAWA, K. MAEKAWA and H. OKAMURA *Department of Civil Engineering, University of Tokyo, Japan*

Influence du dosage et du mode d'introduction des superplastifiants sur le maintien de la maniabilité optimum des bètons à hautes performances avec et sans fumées de silice   63
(Influence of dosage and addition method of superplasticizers on the workability retention of high strength concrete with and without silica fume)
A. M. PAILLERE and J. J. SERRANO *Laboratoire Central des Ponts et Chaussées, Paris, France*
M. GRIMALDI *Laboratoire Régional des Ponts et Chaussées, Melun, France*

Influence des fluidifiants sur les caracteristiques rheologiques des pâtes de ciments fillerisés   80
(Influence of plasticizers on the rheological characteristics of cement pastes containing fillers)
R. SAADA, M. BARRIOULET and C. LEGRAND *Laboratoire Matériaux et Durabilité des Constructions, I.N.S.A.-U.P.S. de Toulouse, France*

Slump control and properties of concrete with a new super-plasticizer. II: High strength *in-situ* concrete work at Hikariga-oka Housing Project   94
C. YAMAKAWA *Construction Chemical Business, Kao Corporation, Japan*
K. KISHITANI *Department of Architecture, Nihon University, Japan*
I. FUKUSHI *Research Institute of Housing and Urban Development, Housing and Urban Development Corporation, Japan*
K. KUROHA *Technical Research Institute, Taisei Corporation, Japan*

2 SETTING

Effects of accelerating admixtures on cement hydration   106
B. E. I. ABDELRAZIG, D. G. BONNER and D. V. NOWELL *Divisions of Chemical Sciences and Civil Engineering, Hatfield Polytechnic, England*
J. M. DRANSFIELD and P. J. EGAN *Fosroc Technology, Birmingham, England*

Remedies to rapid setting in hot-weather concreting   120
M. S. EL-RAYYES *Department of Civil Engineering, University of Kuwait, Kuwait*

Mechanism of and solutions to rapid setting caused by addition of calcium lignosulfonate water reducer to cement with fluorogypsum as retarder   135
ZHANG GUANLUN and YE PING *Tongji University, Shanghai, China*

Comparative examinations of admixtures to cement   142
R. KRSTULOVIĆ, P. KROLO, A. ŽMIKIĆ, T. FERIĆ and J. PERIĆ *Laboratory for Inorganic Technology, University of Split, Yugoslavia*

3 STRENGTH

Study of the effectiveness of water-reducing additives on concrete with microsilica   156
J. C. ARTIGUES, J. CURADO and E. IGLESIAS *Concrete Admixture Research Department, TEXSA S.A., Barcelona, Spain*

The influence of superplasticizer type and dosage on the compressive
strength of Portland cement concrete in the presence of fly ash   168
    M. COLLEPARDI, S. MONOSI and M. PAURI *Department of
    Science of Materials and Earth, University of Ancona, Italy*
    S. BIAGINI and I. ALVERA *MAC-MBT, R&D Laboratories,
    Treviso, Italy*

Superplasticized silica fume high-strength concretes   175
    M. COLLEPARDI, G. MORICONI and M. PAURI *Department of
    Science of Materials and Earth, University of Ancona, Italy*
    S. BIAGINI and I. ALVERA *MAC-MBT, R&D Laboratories,
    Treviso, Italy*

Strength development of superplasticized plain and fly ash concretes   183
    M. K. GOPALAN and M. N. HAQUE *Department of Civil
    Engineering, ADFA, University of New South Wales, Australia*

A study on the use of a chloride-free accelerator   197
    S. POPOVICS *Department of Civil Engineering, Drexel University,
    Philadelphia, USA*

Strength and time-dependent strains of concrete with
superplasticizers   209
    E. N. SHCHERBAKOV *Research Institute for Transport Engineering,
    Moscow, USSR*
    Yu. V. ZAITSEV *Polytechnical Institute, Moscow, USSR*

4 DURABILITY

Inhibiting effect of nitrites on the corrosion of rebars embedded
in carbonated concrete   219
    C. ALONSO, M. ACHA and C. ANDRADE *Institute of Structures
    and Cement, 'Eduardo Torroja', CSIC, Madrid, Spain*

Impermeability and resistance to carbonation of concrete with
microsilica and water-reducing agents   229
    J. C. ARTIGUES, J. CURADO and E. IGLESIAS *Concrete
    Admixture Research Department, TEXSA S.A., Barcelona, Spain*

Mechanical properties and durability of superplasticized silica
fume mortars   241
    G. BARONIO *DISET, Politechnico of Milan, Italy*
    G. MANTEGAZZA and G. CARMINATI *Ruredil SpA,
    San Donato, Milan, Italy*

Calcium nitrite corrosion inhibitor in concrete   251
    N. S. BERKE *Construction Products Division, W. R. Grace
    and Company, Connecticut, USA*
    A. ROSENBERG *Washington Research Center, W. R. Grace
    and Company, Connecticut, USA*

The use of superplasticizers as steel corrosion reducers in reinforced concrete 269
M. COLLEPARDI, R. FRATESI and G. MORICONI
Department of Sciences of Materials and Earth, University of Ancona, Italy
S. BIAGINI *MAC-MBT, R&D Laboratories, Treviso, Italy*

Use of nitrite salt as corrosion inhibitor admixture in reinforced concrete structures immersed in sea-water 279
M. COLLEPARDI, R. FRATESI and G. MORICONI *Department of Sciences of Materials and Earth, University of Ancona, Italy*
L. COPPOLA *Enco, Engineering Concrete, Spresiano, Italy*
C. CORRADETTI *Snamprogetti S.p.A., Corrosion Protection Department, Fano, Italy*

Influence of plasticizers on corrosion of reinforcing bars in concrete 289
F. GOMÀ, J., VIVAR and J. MAURI *Laboratory of the Department of Construction, Polytechnic University of Barcelona, Spain*
J. M. COSTA and M. VILARRASA *Department of Physical Chemistry, University of Barcelona, Spain*

Effect of calcium nitrite and sodium molybdate on corrosion inhibition of steel in simulated concrete environment 299
B. B. HOPE *Department of Civil Engineering, Queen's University at Kingston, Ontario, Canada*
A. K. C. IP *Trow Consulting Engineers, Brampton, Ontario, Canada*

Influence des adjuvants sur la durabilité du bèton 307
(Influence of admixtures on the durability of concrete)
M. MAMILLAN *CEBTP, St-Rémy-Lès-Chevreuses, France*

Properties of super-durable mortars with admixtures 317
Y. OHAMA, K. DEMURA, Y. SATOH, K. TACHIBANA and T. ENDOH *College of Engineering, Nihon University, Japan*
Y. MIYAZAKI *Mitsui Petrochemical Industries Ltd, Japan*

Effets de superfluidifiants sur des bètons resistant aux sulfates 325
(Effects of superplasticizers on sulphate-resistant concretes)
Ph. SIMONIN and F. ALOU *Laboratoire des Matériaux de Construction, EPF, Lausanne, Switzerland*
M. ENDERLI *Bonnard & Gardel, Lausanne, Switzerland*

Microsilica based admixtures for concrete 346
P. J. SVENKERUD and P. FIDJESTØL *Elkem Materials a/s, Kristiansand, Norway*
J. C. ARTIGUES *TEXSA, Barcelona, Spain*

Admixtures for concrete under sea-water action 360
F. TAFLAN and I. FACAOARU *Building Research Institute, Bucharest, Romania*

Effect of zinc oxide admixture on corrosion inhibition of
ferrocement 375
C. TASHIRO, K. UEOKA and K. KOZAI *Faculty of Engineering, Yamaguchi University, Ube, Japan*
M. KONNO *Nippon Steel Corporation, Ooita, Japan*

Experimental study on the effectiveness of corrosion inhibitor in
reinforced concrete 382
F. TOMOSAWA *Department of Architecture, University of Tokyo, Japan*
Y. MASUDA *Building Research Institute, Ministry of Construction, Japan*
I. FUKUSHI *Housing and Urban Development Corporation, Japan*
M. TAKAKURA and T. HORI *Nissan Chemical Industries Ltd, Japan*

5 OTHER PROPERTIES

Permeability of the cement-aggregate interface: influence of the
type of cement, water/cement ratio and superplasticizer 392
U. COSTA, M. FACOETTI and F. MASSAZZA *Italcementi S.p.A., Bergamo, Italy*

Reduction of deformations with the use of concrete admixtures 402
H. CHARIF, J.-P. JACCOUD and F. ALOU *Department of Civil Engineering, Reinforced and Prestressed Concrete Institute, École Polytechnique Fédérale de Lausanne, Switzerland*

Influence of superplasticizers on the bleeding characteristics of
flowing concrete 429
E. G. F. CHORINSKY *Chemotechnik Abstatt GmbH, Abstatt, West Germany*

Une étude de l'influence du pourcentage de granulat sur la
resistance à la compression du bèton, lorque l'on utilise des
expansifs non-metalliques 433
(A study of the influence of aggregates on the compressive
strength of concrete containing non-metallic expansive agents)
S. B. DOS SANTOS and N. P. BARBOSA *Department Tecnologia da Construção Civil, Universidade Federal da Paraíba, Brèsil*

Admixing effect of high fineness slag on the properties of underwater concrete   440

    M. HARA and K. SATO *Steel Research Center, NKK Corporation, Japan*
    M. SAKAMOTO and T. HATSUZAKI *Technical Research Institute, Taisei Corporation, Japan*

Improved air entraining agents for use in concretes containing pulverised fuel ashes   449

    J. T. HOARTY and L. HODGKINSON *Cormix Construction Chemicals Limited, England*

Influence de la teneur en alcalins solubles du ciment sur la stabilité du réseau de bulles d'air en présence de superplastifiant   460

(Influence of soluble alkali content in cement on air void stability in the presence of superplasticizer)
    P. PLANTE *Fondatec Inc., Montreal, Canada*
    M. PIGEON *Department of Civil Engineering, Laval University, Quebec, Canada*

Improvement of drying shrinkage and shrinkage cracking of concrete by special surfactants   484

    M. SHOYA and S. SUGITA *Department of Civil Engineering, Hachinohe Institute of Technology, Hachinohe, Japan*
    T. SUGAWARA *Department of Civil Engineering, Hachinohe National College of Technology, Hachinohe, Japan*

6 TECHNOLOGY

Working with superplasticizers in concrete: a wide field application   496

    N. CILASON and A. H. S. ILERI *Sezai Turkes Feyzi Akkaya Construction Company Inc., Turkey*
    M. CHIRUZZI *Grace Italiana S.p.A. Construction Products, Italy*

Proprietés de certains bètons a adjuvants utilisés dans l'industrie des prefabriques en bèton   507

(Properties of concretes containing admixtures used in precast concrete manufacture)
    I. IONESCU and I. ISPAS *Research Institute for the Construction Materials Industry, Romania*

Use of magnesium calcium silicate admixture for instant-stripping concrete blocks   516

    K. KOHNO *Department of Civil Engineering, University of Tokushima, Japan*
    K. HORII *Department of Civil Engineering, Anan College of Technology, Japan*

Use of superplasticizers for concrete in underground construction work 524
F. MAÑA *School of Architecture, Technical University of Catalunya, Barcelona, Spain*

Improving quality with a new roller compacted dam (rcd) concrete admixture 533
M. TANABE, M. TAKADA and K. UMEZAWA *Nisso Master Builders Co. Ltd, Japan*

Improvement of bond strength of construction joints in inverted construction method with cellulose ether 540
Y. TAZAWA, K. MOTOHASHI and T. OHNO *Civil Engineering Department, Kajima Institute of Construction Technology, Japan*

Le Ca-acetate comme adjuvant pour mortiers et bètons 556
(Calcium acetate as an admixture for mortars and concretes)
G. USAI *Department of Chemical Engineering and Materials, University of Cagliari, Italy*

Method of determination of optimal quantity of cement in concrete, by using superplasticizers with cement-stone density is given in advance 569
T. R. VASOVIC and S. P. MANIC *Technical Division TKK Srpenica, Yugoslavia*

A study of some aspects of microair as air entraining admixture in flyash concrete 578
T. A. WEIGEL, J. P. MOHSEN and D. J. HAGERTY *Department of Civil Engineering, University of Louisville, USA*

Index 583

Preface

Chemical admixtures for concrete are being used more and more widely: for many applications they are used as a routine ingredient of mixes, and their technology is well understood and practised. Yet at the same time, new cementitious materials and new combinations of cementitious materials are being introduced, and admixtures are being used in combination to achieve particular properties. The need for a deeper understanding of the fundamental interactions between admixtures and the other components of concrete mixes is therefore increasing if technical and economic advantage is to be taken of the opportunities now available.

It is against this background that RILEM set up Technical Committee 84-AAC Applications of Admixtures for Concrete in 1985. Its terms of reference are to prepare a guide for the use of admixtures for concrete, which will be a state-of-the-art Report on the application of technology with special reference to the different types, brands and specifications of admixtures for concrete. This work is approaching completion at the time of writing. As a further aspect of the Committee's task, it was decided to organize an International Symposium 'Admixtures for Concrete Improvement of Properties'. This volume forms the Proceedings of the Symposium, which will be a valuable adjunct to the Committee's main task.

The contributions to the Symposium, and the topics discussed highlight the main current areas of technical development. Six main themes were selected for the Symposium: workability; setting; strength; durability; other properties; technology. These have attracted 50 papers covering many innovative topics as well as some which have been the subject of research and development for some time. For example, the combination of superplasticizing admixtures and silica fume is discussed in many papers, both for conventional and unusual applications of concrete. Continuing concern over the durability of concrete is reflected in the papers dealing with air-entraining admixtures and with corrosion-inhibiting admixtures.

We hope that the opportunity for exchange of information and ideas afforded by the Symposium and by these Proceedings will be of value to the concrete construction industry, and to all those who are involved in it.

The Committee have met on eight occasions prior to the Symposium at Barcelona, and they have reviewed all the abstracts and papers submitted. To all the researchers who have offered papers which were not finally accepted, RILEM TC 84-AAC extends its sincere appreciation.

I must also express, on behalf of RILEM, my appreciation to those organizations whose staff have been members of the Committee, for the facilities and support offered to all of them. I would extend special thanks to the Committee members themselves for their effort and hard work which has contributed greatly to the success of the Symposium.

Sincere thanks are due to the sponsorship and organization of the Departamento de Ingeniería de la Construccíon, de la Escuela Técnica Superior de Ingenieros de Caminos, Canales y Puertos de la Universitat Politécnica de Catalunya, headed by Professor Enric Vázquez Ramonich, who acted as Technical Secretary of the Symposium, and for the invaluable collaboration of the institutions acting as co-sponsors: University of Nuevo León, Monterrey, N.L. México; Ministerio de Obras Públicas y Urbanismo; Generalitat de Catalunya, Departamento Política Territorial y Obras Públicas; Ayuntamiento de Barcelona; Colegio Oficial de Ingenieros de Caminos, Canales y Puertos; Colegio Oficial de Arquitectos de Catalunya; Colegio Oficial de Arquitectos; Técnicos y Aparejadores de Catalunya; Colegio de Ingenieros Técnicos de Obras Públicas y Ayudantes de Obras Públicas; Asociación Nacional de Fabricantes de Hormigón Preparado; Agrupació de Fabricants de Ciment de Catalunya e Institut Catalá d'Enginyeria Civil.

I also wish to express my gratitude to UNESCO for the valuable financial support offered so that scientists from developing countries could attend.

Dr. Ing. Raymundo Rivera Villarreal
Chairman

1 WORKABILITY

SUPERPLASTICIZER AND AIR-ENTRAINING AGENT DEMAND IN OPC MORTARS CONTAINING SILICA FUME

A. DUREKOVIĆ and K. POPOVIĆ
Department for Concrete and Masonry Structures, Civil Engineering Institute,
Zagreb, Yugoslavia

Abstract
Blends of ordinary portland cement with different parts of silica fume
have been used for mixing of mortars in which the superplasticizer and
air-entraining agent demand was investigated.Mortars were prepared with
a water-binder ratio of 0.6 .The fresh mix consistancy (measured by flow
table test),the components blends proportions,micro-sieve analysis and
the particle size distribution (measured by sensing zone method) of anhydrous cement blends as well as the added HWR- and AEA-agent quantity
were considered.The results have shown that by the increase of the fine
particles share in anhydrous blends,i.e. their specific surface increase
rised also the admixtures demand in fresh mortars which are made with
such cements.
Key words: Silica fume blends,Superplasticizer,Aerant,Specific surface,
Particle size distribution.

1 Introduction

The researches into use of water-reducing agents (WRA or HWRA) and air-
entraining agents (AEA) on the cementitous composites containing condensed silica fume (CSF) pointed to some specific rules: the replacement of
cement by CSF led to an increased demand of AEA in concrete /Carette and
Malhotra 1983/;the CSF concrete needed a higher dosage of AEA than the
control concrete to reach a given air content /Virtanen 1983/,or for a
given dosage of an AEA an increase in CSF content resulted in reduced
air content in the fresh mixture /Virtanen 1985/;compared to the control
concrete the AEA-demand for CSF-concrete depends also on presence of WRA
- in the presence of WRA the AEA demand decreases /Sellewold 1987/.The
contradictory statements were reported in connection with stability of
the air content in fresh CSF-concretes with respect to vibration /Okken-
haug 1983,Aitcin and Vezina 1984/. It was also emphasized that the water-reducing agents had a much more pronounced effect on the CSF-concrete /Jahren 1983/.The retarding effect of lignin based admixtures is somewhat reduced when used in CSF-containing mixes /Sellewold 1987/ although
the lignosulphonate type,according to Sellewold and Radjy /1983/,appears
to be at least as efficient as the high range type of WRA when used in
CSF-containing concrete.
The basic mechanism of plasticizing effect on fresh cement mixes is explained by forming a temporarily stable double layer on cement particles

1

/Iler 1955,Kreijger 1980/.Since the formation of the double layer is also connected with the furface of particles,this paper focuses on the relationship between fineness characteristics of CSF-blends and the quantity of superplasticizer and air-entraining agent needed to realise a given mortars consistancy.

2 Experimental

Experimental data for HWRA- and AEA-demand obtained in this paper were taken from the mix proportions of mortars prepared with a water-binder ratio of 0.6 .The mortars were prepared using an ordinary portland cement (OPC),the CSF from ferrosilicon alloy production, quartz sand,a superplas-ticizer and air-entraining agent and distilled water.The OPC (type PC 45 B,according to Yugoslav standard JUS B.Cl.011) with the Blaine value of 3630 cm²/g,density of 3.06 g/cm³ had the following Bogue's composition: C_3S 50% ,C_2S 21% , C_3A 10% and C_4AF 8% .Condensed silica fume contained 86.9% of SiO_2 and had the bulk density of 0.22 g/cm³.The standard quartz sand with maximal grain size of 2 mm met requirements of Yugoslav standard JUS B.Cl.011. The used superplasticizer (solution with 20% dry substance) was of melamin-formaldehyde type.The surfactant (with 9% dry substance) added as the air-entraining agent was of the vinsol resin type The 1:5 water diluted solution was admixed to the mortars. The mortars consistancy measured by flow table test tried to keep uniform for all mixes at flow table diameter of 180 mm.Regarding the CSF-dosage there were 4 mixes containing 0%, 5%, 10%, and 15% of CSF replacing the same cement weight.Each of those mixes were in non-aerated and aerated form.
In order to obtain the fineness characteristics the anhydrous cement blends were investigated by micro-sieve analysis using micro mesh sieves (from "Feinoptik",Blankenburg,DDR) where the samples were dispersed in absolute methanol by supporting of ultra-sound agitation.The particle size distribution of anhydrous cement samples was determined by sensing zone method using "Coulter counter ZB" instrument and 5% LiCl solution in absolute methanol.Using the micro-sieve and Coulter counter results,the cumulative curves for particle size distribution of anhydrous blends were drawn.The specific areas were calculated on the basis of values read for 5μm fractions of particle size using from the Rosin nad Ramler's law /1933/ derived equation:

$$Sp = \frac{6}{\rho} \sum \frac{1}{x_i} \cdot w_i$$

where Sp is the specific surface per unit of mass, ρ is density, x_i is average size of particle fraction "i" which is calculated from equation $x_i = \left[\int_{x_{i,1}}^{x_{i,2}} x^3 \, dx / \int_{x_{i,1}}^{x_{i,2}} dx \right]$ .In calculation the Sp was increased for the value of 30 cm²/g (estimation based on previous research reported by Popović /1971/) as the corrections for the particle sizes in the range from 45 to 200 microns which were not included by measurement with Coulter counter.The shape factor of 0.5 determined for cement particles in earlier work /Popović 1971/ was also taken into specific surface calculations.

3 Results and discussion

Table 1 shows the characteristics of fresh mortars from which the admixtures demand was taken.

Table 1. Mix design characteristics of fresh mortars (water-binder ratio 0.6).

Mix sign	CSF content (g)	OPC content (g)	AEA/HWRA content (g)/(g)	Flow table diameter (mm)	Air content in mortar (%)
PC	0	2500	0/0	181	6.4
PC (AEA)	0	2500	30/0	182	13.6
5P	125	2375	0/18	177	5.2
5P (AEA)	125	2375	38/3	172	14.0
10P	250	2250	0/38	178	5.2
10P (AEA)	250	2250	46/25	172	13.7
15P	375	2125	0/75	180	3.8
15P (AEA)	375	2125	75/21	170	14.5

There is an evident increase in AEA and HWRA amount. The presence of silica fume in the composition of binders which are used for mortars increases the sieve fractions with the smallest particles as shown in table 2.

Table 2. Micro sieve analysis of anhydrous cements OPC and the blends with 5% (5P),10% (10P) and 15% (15P) of CSF.

Particle sizes x(microns)	PC	5P	10P	15P
	Percentage of particles in anhydrous cement			
x > 45	13.9	12.1	12.0	10.6
30 < x < 45	16.5	15.7	13.5	12.4
20 < x < 30	20.1	20.0	19.7	19.7
0 < x < 20	49.5	52.2	54.7	57.3

The addition of CSF which particles are smaller than those of OPC also cause a change in particle size distribution of anhydrous CSF-containing cement blends.This was proved as follows: For the particle range from 3.4 to 34 microns by sensing zone method obtained results and for particles of 40 and 45 microns obtained results by micro sieve analysis are used to make a cumulative particle size distribution curves which are afterwards evaluated in 5 microns fraction-step.The obtained 5 microns-step values were used for drawing the differential particle size distribution curves shown in Figure 1.

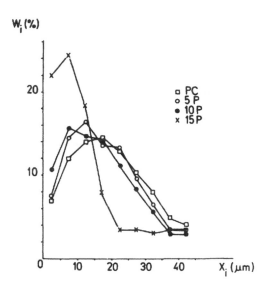

Figure 1.   Differential particle size distribution curves
            for anhydrous ordinary portland cement (PC sign)
            and the blends containing 5% (5P sign),10% (10P
            sign) and 15% (15P sign) of condensed silica fume.

Figure 1 clearly shows that the increased CSF content in anhydrous blends causes a shift of biggest weight portion in particle distribution to the smaller average size of particles.This fact has a consequences,as Figure 2 shows,for calculated specific surface of anhydrous blends which will be enlarged by increasing of CSF content.

    According to the basic principles of plasticizers action,that on the hydrating cement particles a double layer is formed it follows that CSF-containing cement dispersions with a larger portion of smaller particles, i.e. cement blends which have a larger specific surface will demand more HWRA to reach a chosen consistency when mixed with water.This is confirmed by correlations shown in Figure 3 and Figure 4.Namely,among the non--aerated mortars was PC-mix made whithout superplasticizer,so that the superplasticizer content added to the CSF-containing mortars could be directly correlated to the specific surface increase (Δ Sp) of cement blends used for mortars.The shown dependence of added HWRA and ΔSp of ce-

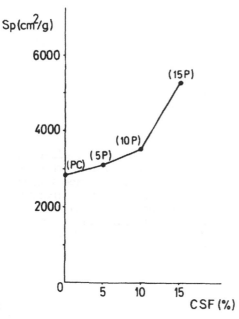

Figure 2. Calculated specific surfaces (Sp) of anhydrous ordinary portland cement (PC sign) and CSF-containing blends vs. added CSF content.

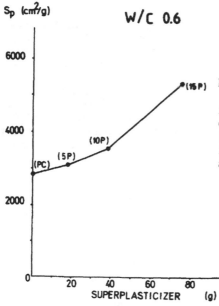

Figure 3. Superplasticizer demand for fresh mortars as function of the specific surfaces of used ordinary portland cement (PC sign) and the blends containing 5% (5P sign), 10% (10P sign) and 15% (15P sign) of CSF.

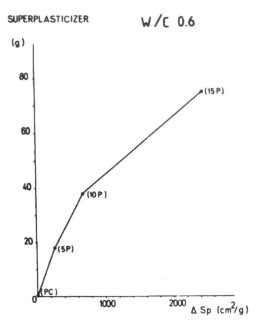

SUPERPLASTICIZER   W/C 0.6

Figure 4. Superplasticizer de-
         mand for fresh mortars
versus specific surface increase
(Δ Sp) of anhydrous ordinary portland cement (PC sign) and the
blends containing 5% (5P sign),
10% (10P sign) and 15% (15P sign)
of CSF.

ment blends is most likely also under influence of sterical hindrances
/Kreijger 1980/ which usually occur in such mixes.The CSF particles which
caused the increases of Sp are in the size range in which the floculation forces act.The emphasizing of the flocculation decreases the mix fluidity or,conversely ,it causes the increase of HWRA demand.
    In aerated mortars it was necessary to add a superplasticizer for
consistency correction.The superplasticizer of the melamin type realizes
strongly negatively charged zeta-potential /Collepardi et al.1981,Daimon
and Roy 1979/ as the used air-entraining agent also does.That means that
both admixtures originate the same kind of charge of the double layer and
thus mutually influence each other in their action.In spite of this ,Fi-
gure 5 clearly shows that increased addition of CSF increases the amount
of AEA,i.e. that AEA quantity per percentage of entrained air increases
with the increase of specific surfaces (ΔSp) of cement blends which are
used for mortars. The mechanism of this AEA-demand increase could also be
considered under speculation as reported by Ramachandran /1984/,that composites with bigg fine fractions tend to bind the water because of the
requirement that it coat their large surface areas.The water can then not
be a part of the air bubble generating and stabilizing process what resulted with a decrease of the air content in such mixes.Other to said,for a
given air content rise the needed AEA-amount.
Despite of difficulties by keeping a constant fresh mortars consistency
by changing the mix cohesiveness due to CSF addition Figure 6 shows that
AEA-demend results were obtained on the mixes of good mutual uniformity.

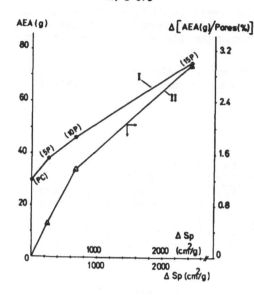

Figure 5. AEA-demand for fresh mortars expressed by correlation of total used AEA content vs. specific surface increases (ΔSp) and by correlation of the increase of AEA content per percentage of entrained air vs. s specific surface increases (ΔSp) of anhydrous cement blends.

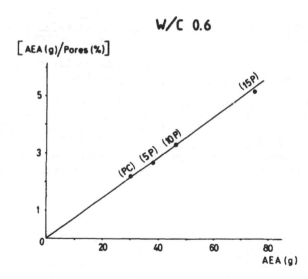

Figure 6. Relation between the total added AEA content (g) and the AEA demand needed to entrain one percent of air in the fresh mortars made with ordinary portland cement (PC sign) and the blends containing 5% (5P sign), 10% (10P sign) and 15% (15P sign) of CSF.

4 Conclusion

- The replacement of ordinary portland cement (OPC) by condensed silica fume (CSF) from ferrosilicon alloy production caused the change in the particle size distribution of anhydrous blends increasing the portion of smaller particles.
- The small particles increase was reflected on the specific surface enlargement of anhydrous OPC-CSF blends which consequently exhibited a higher HWRA and AEA demand.

References

Aitcin,P.C.,Vezina,D.(1984) Resistance to freezing and thawing of silica fume concrete.Cement,Concr.Aggreg.,CCAGDP.,6,1,38-42.

Carette,G.G.,Malhotra,V.M.(1983) Mechanical properties,durability and drying shrinkage of portland cement concrete incorporating silica fume. Cement,Concr.Aggreg.,CCAGDP., 5,1,3-13.

Collepardi,M.,Corradi,M.,Valente,M.(1981) Influence of polymerization of sulfonated naphtalene condensate and its interaction with cement.ACI SP-68 ,pp 485-498.

Daimon,M.,Roy,D.M.(1979) Rheological properties of mixes: II Zeta potential and preliminary viscosity studies.Cem.Concr.Res.,9,pp 103-110.

Iler,R.K.(1955) The Colloid Chemistry of Silica and Silicates. Cornell Univer.Press,Chapter III.

Jahren,P.(1983) Use of silica fume in concrete.Proceedings of the 1st International conference cn the use cf fly ash,silica fume,slag and other mineral by-products in concrete(ed.V.M.Malhotra),ACI SP-79,pp 625-642.

Kreijger,P.C.(1980) Plasticisers and dispersing admixtures.Proceedings of International congress on admixtures,Construction Press Ltd., London,pp 1-17.

Okkenhaug,K.(1983) Silikatsovets innvirkning pa luftens stabilitet i betong med L-stof og med L-stoff i kombinasjon med P-stoff.CBI Report 2-83,Swedish Cement and Concrete Research Institute at the Institute of Technology,Stokholm,pp 101-115.

Popović,K.(1971) Comparative determination of particle size and the influence of fineness on mechanical properties of cement (in croatio). Cement,No 4,pp 159-178.

Ramachandran,V.S.(1984) Concrete Admixtures Handbook.Noyes Public.,New Jersey,pp 280-281.

Rosin,P.,Rammler,E.,Sperling,H.(1933) The problems of coal ash grain sizes and their significance for the milling processes (in german).Berichte der Reichkohlenrates,Ber.C 52,Berlin.

Sellevold,E.J.(1987) Condensed silica fume in concrete: A world view. CANMET International Workshop "Condensed silica fume in concrete", Montreal,Paper No1,pp 1-77.

Sellewold,E.J,Radjy,F.F.(1983) Condensed silica fume (microsilica) in concrete:Water demand and strength development.Proceedings of the 1st Inter.Conf. on the use of fly ash,silica fume,slag and other mineral by-products in concrete (ed.V.M.Malhotra),ACI SP-79,pp 677-694.

Virtanen,J.(1983) Freeze-thaw resistance of concrete containing blast furnace slag,fly ash or condensed silica fume.Proceedings of the 1st Intern.Conf. on the use of fly ash,silica fume,slag and other mineral by-products in concrete (ed.V.M.Malhotra),ACI SP-79,pp 923-942.

Virtanen,J.(1985) Mineral by-products and freeze-thaw resistance of concrete.Dansk Betonforening,Publikation Nr 22:85,pp 231-254.

SLUMP CONTROL AND PROPERTIES OF CONCRETE WITH A NEW SUPERPLASTICIZER. I: LABORATORY STUDIES AND TEST METHODS

M. FUKUDA, T. MIZUNUMA, T. IZUMI and M. IIZUKA
Wakayama Research Laboratory, Kao Corporation, Wakayama, Japan
M. HISAKA
Central Laboratory, Japan Testing Center for Construction Materials, Saitama, Japan

Abstract
A new superplasticizer containing reactive polymer has
been studied as a new technology for slump control.
Fresh concrete is well known to lose its workability with
time. This phenomenon is called "Slump loss", particularly
high water reduced concrete has remarkable extent.
To find a new method of solving the slump loss problem, we
have developed a new superplasticizer (Mighty 2000) with
reactive polymer from various studies of dispersant.
This new superplasticizer enables the workability of fresh
concrete to be optionally controlled for a required working
time.
This report describes the technology and the method of
slump control by the newly developed superplasticizer.
Some properties of concrete with the new superplasticizer
are also reported.

Keywords: dispersant; dispersant precursor; hydrolysis;
reactive polymer; slump control; slump loss;
superplasticizer; workability.

1 Introduction

With the introduction of superplasticizer in the concrete
industries, the use of high strength concrete with low
water content became popular, and the durability of
concrete structures was largely improved.
Superplasticizers are generally known to be very effective
in improving the workability of regular concretes, but when
it is used to prepare high strength concrete of low water
content, slump loss becomes a serious problem.
Because the slump loss is caused by many complicated
factors, the method of preventing slump loss is not clear.

It is thought however to depend on physical and chemical factors. Regarding the physical factor, the primary cause is thought to be that particle numbers in unit volume increase due to the dispersion and apparent pulverization of cement particles. On the other hand, in the case of the chemical factor, consumption of dispersants due to the cement hydration reaction result in the deterioration of the dispersion function (see Fig.1).

Repeated dosage of a superplasticizer during the delivery to site, repeated again at the job site 1), and the use of granular superplasticizer 2) are not accepted widely due to the inconveniencies of changing the dosage procedure.

To find a new method of solving the slump loss problem, we tried many substances and found that a reactive polymer was very suitable for the purpose. So, by the application of this reactive polymer and superplasticizer together, we have developed a new superplasticizer (Mighty 2000), which enables the workability of fresh concrete to be optionally controlled for a required working time.

This report describes the technology and the method of slump control by the newly developed superplasticizer containing the reactive polymeric dispersant. Some properties of concrete with the new superplasticizer are also reported.

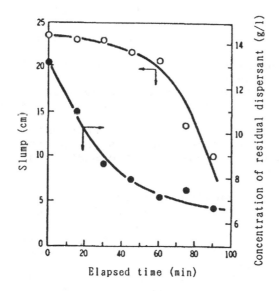

Fig. 1. Slump loss and concentration of residual dispersant with time

2. Mechanism of slump control with reactive polymer

Reactive polymer is gradually converted to dispersant by strong alkaline materials in concrete. The mechanisms of conversion of reactive polymer into soluble salt, are shown in Fig.2.3)

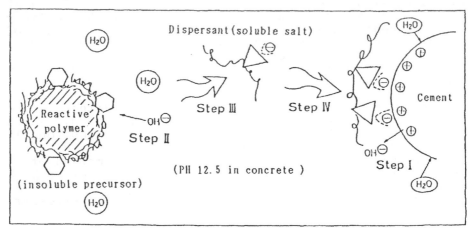

Fig. 2. Dispersing mechanism of reactive polymer

It is considered that these mechanisms consists of 4 steps as follows.
STEP 1 Generation of OH anion occurs by hydration with cement and water.
STEP 2 OH anions attack the reactive polymer.
STEP 3 The reactive polymer is converted into soluble dispersant.
STEP 4 The soluble dispersant is absorbed onto the surface of cement particles, enhances the surface potential energy and contributes to the stabilization of dispersion.

In optimizing the reactive polymer for slump control, it is necessary to study the factors influencing the rate of conversion. The conversion appears to be a heterogeneous reaction occuring at the interface between a liquid phase and a solid polymer particle because the reactive polymer is insoluble in water.
According to the unreacted core model, the rate of conversion of the reactive polymer. Xrp is expressed as follows;4),5)

$$X \, rp = 1 - (1 - \frac{b \cdot t \cdot K s \cdot C \, OH}{r \cdot \rho \, rp})^{3} \qquad (1)$$

where, t :Time, C OH:Concentration of alkalines,
       b :Constant, K s:Rate constant of conversion
       r :Initial size of reactive polymer
       ρ rp:Molar concentration of reactive polymer.

Ks is expressed as follows;

$$K s = K o \cdot E x p (- \Delta E / R T) \qquad (2)$$

where, Δ E :Activation energy, T :Absolute temperature,
       R :Gas constant,   K o:constant.

It is obvious from eqs.(1) and (2) that conversion of the
reactive polymer depends on the concentration of alkaline,
the temperature, and the particle size. Since both the
temperature and the alkaline concentration are the given
values depending on test conditions, the particle size of
the reactive polymer is the most important factor for
determining the rate of conversion.

3 Experiment

3.1 Materials
 The physical properties of materials used are shown in
Table 1.

Table 1. Materials

Cement	Ordinary portland cement (S. G=3. 16)
Fine agg.	River sand (S. G=2. 61, FM=2. 93)
Coarse agg.	Crushed Stone (S. G=2. 72, FM=6. 63, Gmax=20mm)

3.2 New superplasticizer (Mighty 2000)
 Our newly developed Mighty 2000 has two effective
functions for slump control and dispersant.
 As a slump control composition, dispersant precursor is
converted to the water soluble dispersant by reacting with
the strong alkaline materials in concrete.
 On the other hand, β -naphtalene sulfonate-formaldehyde
high condensate was used as a dispersing composition.
 Thus, Mighty 2000 greatly offers to reduce water content
and decrease slump loss.

3.3 Concrete mix designs
 The mix designs of concrete are shown in Table 2.
We set the initial slump at 18± 2 cm, and the air content
at 4.5± 0.5 %.

13

Table 2. Mix designs

Dosage of admixture (% wt. cement)			Unit weight (kg/m^3)				Water/cement ratio (%)	Obs. slump (cm)	Air content (%)
SP	RA	AE	Cement	Water	Fine agg.	Coarse agg.			
–	1.00	1.00*10-2	268	174	845	1000	65	18.3	4.5
1.50	–	6.00*10-3	282	169	844	999	60	18.7	4.6
1.50	–	5.50*10-3	307	169	823	999	55	19.3	4.5
1.50	–	5.50*10-3	328	164	812	995	50	17.7	5.0
1.50	–	5.00*10-3	367	165	784	1000	45	18.4	4.7
1.20	–	1.40*10-2	412	165	747	1000	40	18.7	4.5
1.60	–	1.40*10-2	472	165	698	1002	35	19.0	4.5
1.95	–	2.00*10-2	533	160	657	1000	30	18.2	4.7
3.50	–	3.50*10-2	618	154	555	1039	25	19.5	4.8

Notes: SP; New developed superplasticizer containing reactive polymer
RA; Water reducing admixture (lignosulfonate)
AE; Air-entraining agent

3.4 Testing method

The materials were poured into a slant drum mixer in the following sequence; gravel, half the amount of sand, cement, the rest of the sand and finally water.
This was mixed for 3 minutes. After the measurement of initial slump and air content, concretes were agitated at 4 rpm for 60 minutes and the slump measurements were carried out at 30 minute intervals. Samples for measurements of physical properties of hardened concrete were taken at the beginning.

3.5 Measurements of physical properties of concretes

Test for compressive strength, drying shrinkage, freezing-thawing resistance, and permeability were carried out according to ASTM C39, C157, C512, and DIN 1048 respectively.
The carbonation of concrete was observed by spraying (30°C , 60% RH, 5% CO_2) the phenolphthaleine indicator solution onto the cutout surface of the concrete samples and the depth of the uncoloured layer was reported.

4 Results

4.1 Properties of fresh concrete

Fig.3 shows the result of slump and air content variation with time. With the use of the new superplasticizer, it is possible to maintain the slump of concrete at the range of 18 ± 2 cm and the air content at the range of 4.5 ± 0.5 % within 60 minutes in all mix-designs.

Generally, it is known to increase the concentration of cement particles with decreasing water/cement ratio, and to reduce the slump largely due to the collision of cement particles with each other.

It is possible to control the slump by using the new superplasticizer in all mix-designs, this enables us to deal with high strength concrete easily, and the new super-plasticizer is expected to contribute to the increased use of high strength concrete.

4.2 Properties of hardened concrete

Fig.4 shows the results of compressive strength. In all mix-designs, there was a significant relationship between the cement/water ratio and the compressive strength. Compressive strength of 1026 kg/cm2 at 91 days was obtained using a water/cement ratio of 25%.

Fig.5 shows the relationship between the relative dynamic modulus of elasticity and the freezing-thawing cycle. The result of freezing-thawing resistance by using the new superplasticizer ensured durable concrete to 300 cycles.

Fig.6 shows the result of the carbonation of concrete. The rate of carbonation was restricted with the decrease of a water/cement ratio. Below a water/cement ratio of 35%, the carbonation was not seen after six months.

Fig.7 shows the result of permeability. The permeability coefficients decreased with the decrease of a water/cement ratio in each concrete sample under different dry conditions. Particularly, the permeability was largely improved below a water/cement ratio of 55%, and it was not seen below a water/cement ratio of 25%.

Fig.8 shows the result of drying shrinkage after six months. The amount of drying shrinkage decreased with the decrease of a water/cement ratio.

In the manufucture of concrete structures, using a decreased water/cement ratio as described, the durability of concrete is largely improved. Similarly in the case of high strength concretes by using the new superplasticizer, the durabilty is also improved.

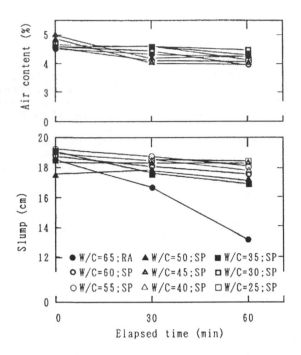

Notes: SP; New developed superplasticizer containing
reactive polymer
RA; Water reducing admixture (lignosulfonate)

Fig. 3. Slump and air content variation with time.

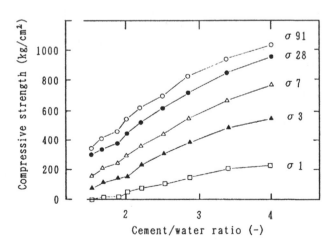

Fig. 4. Relationship between cement/water ratio and
compressive strength.

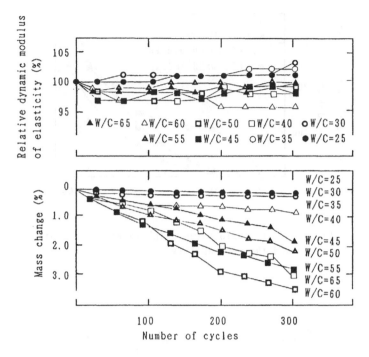

Fig. 5. Relationship between the freezing-thawing cycle
and the relative dynamic modulus of elasticity.

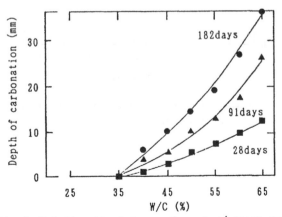

Fig. 6. Relationship between the water/cement ratio
and the carbonation.

17

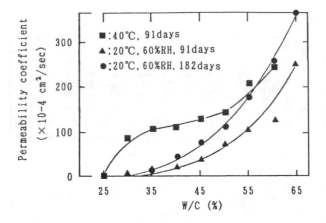

Fig. 7. Relationship between the water/cement ratio
and the permeability.

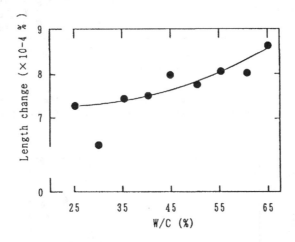

Fig. 8. Relationship between the water/cement ratio
and drying shrinkage after six months.

5 Summary

By using the new superplasticizer (Mighty 2000),
properties of water reduced concrete have been investigated
and the following information obtained.

(1) By using the reactive polymer, it is confirmed that the
    slump of fresh concrete can be optionally controlled in
    all mix-designs.
    It has been considered very difficult to maintain the
    workability of a low water/cement ratio, but the
    concrete containing the reactive polymer has been found
    to be very useful for maintaining the initial slump of
    ready mixed concrete.

(2) By using the new superplasticizer, it is confirmed that
    "High range water reduce and small slump loss" which
    are required of good concrete, are both achieved.
    With the increased density of concrete, the durability
    is largely improved due to the decreased water/cement
    ratio.

Recently, in Japan, the improvement of endurance in
towerblock concrete structures has progressed, and the
newly developed superplasticizer has been used for the
purpose of improvement of workability, high strength, and
slump control.
In the future, we want this new superplasticizer (Mighty
2000) to be used throughout the world, and to contribute
to the progress of concrete technology.

References
1) M.Iizuka, Y.Kazama, K.Hattori, CAJ Review of the 33rd
   General Meeting-Technical Session, P.239 (1979).
2) K.Hattori, E.Okada. T.Doi, and M.Sato, Cement and
   Concrete (Japan), No.443, P.10 (1984).
3) K.Kisitani, K.Kunikawa, M.Iizuka, and T.Mizunuma,
   Cement and Concrete (Japan), No.478, P.7 (1986)
4) O.Levenspiel, Chemical Reaction Engneering, 359,
   John Wiley and Sons, Inc.N.Y (1963).
5) A.Fujiu, H.Tanaka, M.Iizuka, CAJ Review of the 39th
   General Meeting-Technical Session, P.72 (1985).

DYNAMIC RHEOLOGICAL PROPERTIES OF FRESH CONCRETE WITH CHEMICAL ADMIXTURES

S. KAKUTA
Department of Civil Engineering, Akashi Technological College, Akashi, Japan
T. KOJIMA
Department of Civil Engineering, Ritsumeikan University, Kyoto, Japan

Abstract
A basic experimental investigation was carried out in order to obtain
an effective and rational method in consolidating fresh concrete with
chemical admixtures by using vibrator. The rheological properties of
fresh concrete during vibration were examined experimentally. Three
types of fresh concrete mix which contains a superplasticizer
(sulphonated naphthalene formaldehyde condensated base), an air
entraining agent(sodium abietate) and a segregation controlling
admixture (hydroxyethylcellulose base) were used in this study. A
rotating fan type rheometer and a pulse wave propagation rheometer
were used to obtain the dynamic rheological properties of fresh
concrete under vibration. Two types of vibrator, a table vibrator and
an internal vibrator were used in the experiment. To obtain the
properties of energy propagation produced by the vibrator, the wave
form which propagated in fresh concrete was measured.
The flow curves of concrete with admixtures were no longer the Bingham
relationship when it was vibrated,but they can be approximated as a
Newtonian or quasiviscous fluid. The effect of admixture under
vibration could be, therefore, estimated by rheological properties such
as coefficient of viscosity or non-Newtonian power number. There was
apparent difference in the wave spectrum of the above mixes.
Key words: Workability,Rheology,Superplasticizer,Vibration,Flow
curve,Dynamic viscosity,Wave propagation.

1 Introduction

The poor workmanship or technique in placing and compacting of fresh
concrete may become a reason for deterioration of hardened concrete.
It is important to control adequately the process of concrete work
before hardening. There is,however,only a few tools to estimate that
process. On the other hand,many types of chemical admixtures are being
used in concrete. Such concretes shown fluidity as compared with
normal concrete. Therefore,it is very difficult to estimate the
workability of fresh concrete with chemical admixtures by conventional
workability test.
This paper describes the rheological dynamic properties of fresh
concrete containing three types of admixtures,an air entraining
agent,an superplasticizer and so-called non-segregation-in-water
admixture(water soluble polymer). The effect of admixture on concrete

20

under vibration was estimated by the rheological parameters.

2 Experimental

2.1 Materials
Ordinary Portland cement,river sand for fine aggregate and crushed
hard sandstone for coarse aggregate were used in all mixes.
Three types of chemical admixtures were used. They are an anionic type
of air entraining agent,a superplasticizer based on naphthalene-
sulfonate and a non-segregation-in-water admixture containing a
mixture on aqueous cellulose system polymer. Details of materials used
in the experiment are summarized in Table 1.

Table 1. Materials

Material	Type
Cement	ordinary portland cement
Fine aggregate	river sand,specific gravity=2.58 F.M=2.41,water absorption=1.7%
Coarse aggregate	crushed stone,specific gravity=2.69 water absorption=0.6%,maximum size=20mm
Air-entraining agent	sodium abietate
Superplasticizer	sulfonated naphthalene formaldehyde condensate base
Non-segregation- in-water agent	hydroxyethylcellulose(H.E.C.) base (aqueous polymer)
Subsidiary agent	high condensate triazine type (superplasticizer)
AE water reducing agent	Lignosulfonate

2.2 Concrete and mortar mixes
Totally ten mixes for concrete and nine mixes for mortar were tested.
In all concrete mixes,water/cement ratio was 50% by weight,and sand
aggregate ratio was 48% by volume. Unit water content and type and
dosage of admixture are listed in Table 2. In mortar mixes,sand cement
ratio was 2.0 and water cement ratio was 40% in plain and SP mortar
and 40% and 50% in AE and Marine mortar. The type and dosage of
admixtures are also listed in Table 2.

Table 2. Mix proportions

Concrete and mortar	type
Plain concrete	W/C=50%,s/a=48%,W=190,200,210 and 218 kg/m^3 Slump............ 4, 18, 12 and 18 cm
AE concrete	W/C=50%,s/a=48%,dosage:0.06%/C by weight W=180 and 190kg/m^3
SP concrete	W/C=50%,s/a=48%,dosage:0.6%/C by weight W=180 and 190kg/m^3
Marine concrete	W/C=50%,s/a=48%,dosage:NSIW=0.6%/C by weight, subsidiary agent=2.0%/C by weight,AE W.R.A= 0.25%/C by weight,W=210 and 218 kg/m^3.
Plain mortar	W/C=40%,S/C=2.0 by weight
AE mortar	W/C=40%,50%,S/C=2.0 by weight,dosage:0.06%/C 0.03%/C by weight
SP mortar	W/C=40%,S/C=2.0 by weight,dosage:0.25%,0.50% and 0.75%/C by weight
Marine mortar	W/C=40%,50%,S/C=2.0 by weight,dosage:NSIW= 0.36%/C, subsidiary agent=2.0%/C by weight.

2.3 Dynamic rheometer

Fresh concrete,mortar and paste can be assumed a viscoelasticity materials. A dynamic rheometer for fresh concrete was tried to design in this experiment. Test set up is shown in Fig.1.

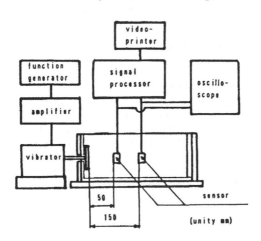

Fig.1. Pulse wave propagation rheometer

This rheometer was composed of the transmitting set,and the receiving
set. The transmitting set were comprised with a function generator,an
amplifier and a vibrator. The receiving set was comprised by the
sensors which resonance was 150kHz,signal processor(FFT analyzer) with
video-printer and an oscilloscope. The locations of two sensors were
5cm(S_1) and 15cm(S_2) from the vibrating plate respectively. The
container was made of a wood which dimensions were
20cm(height),30cm(transverse) and 45cm(longitudinal). To avoid the
effect of reflection from the wall of container,styrofoams were
attached to the bottom and all walls.
The transmitted wave and longitudinal waves which were detected by two
receivers is shown in Photo 1. In this picture,the frequency of
transmitting pulse wave was chosen to 500 Hz so that concrete and mortar
structure could not be failed and a relaxation by a compressive pulse
wave did not occur.
Dynamic modulus E' and dynamic viscosity η' were obtained from
following equations and Fig.2.

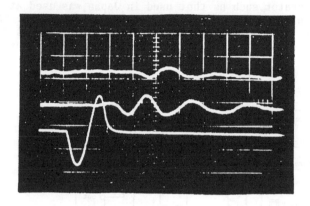

Photo 1. Propagated wave forms

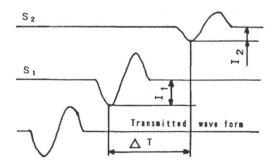

Fig.2. Schematic wave forms

$$E' = \frac{\rho V^2 \omega^2 (\omega^2 - \alpha^2 V^2)}{(\omega^2 + \alpha^2 V^2)^2}$$

$$\eta' = \frac{2 \rho \alpha \omega^2 V^3}{(\omega^2 + \alpha^2 V^2)^2}$$

$$\omega = 2 \pi f$$

$$\alpha = \ln(I_1 / I_2) / \Delta L \quad , \quad V = \Delta L / \Delta T$$

(2.1)

Where ρ is density of material, V is velocity of propagation wave and f is frequency of wave. α is a coefficient of attenuation calculated from pulse amplitude I_1, I_2 (see Fig.2) and propagation distance L. In the next step, the above measurement was carried out under vibration with sinusoidal wave motion. The frequency of vibrator was 200Hz. Measured propagated waves through concretes were processed to frequency spectrums by a signal processor.

2.4 Rotating fan type rheometer
A rotating fan type rheometer similar to that suggested by Tattersall was used. A table vibrator, such as that used in Japan, was used at the bottom of the container. The vibration frequency was 29.1Hz. The test set-up is shown in Fig.3. The accelerations of container at five locations were measured as shown in Fig.4 and measured mean values are summarized in Table 3.

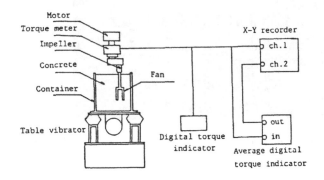

Fig.3. Rotating fan type rheometer

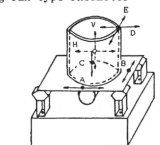

H: Measured horizontal acceleration of inner concrete
V: Measured vertical acceleration of inner cocrete

Fig.4. Locations and directions of acceleration gages

Table 3. Accelerations at each position of the container

Direction	Number of vibration (unit g)			
	1	2	3	4
A	4.33	4.50	4.00	5.33
B	0.67	1.67	1.33	1.00
C	2.00	2.50	2.67	3.17
D	7.50	9.67	10.2	11.5
E	0.67	1.67	1.50	2.17

2.5 Pore water pressure measurement
Locations of pore water pressure gages are shown in Fig.5. Two water pressure gages were set up in the middle of container at two levels of 7cm and 17cm from concrete surface.

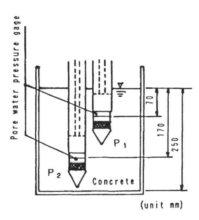

Fig.5. Locations of pore water pressure gages

3 Results and discussion

3.1 Dynamic rheological properties of mortar with chemical admixtures
The measured longitudinal wave velocity of fresh mortar with chemical admixtures are summarized in Fig.6. In this figure,AE1 and AE2 are AE mortars containing AE agent of 0.03% and 0.06% per cement content respectively,AE3 is a AE mortar with AE agent 0.03% per cement content and water cement ratio is 50%. In superplasticized mortars SP1,SP2 and SP3 dosages are 0.25%,0.50% and 0.75% per cement content respectively. Marine mortars contained 2.4kg/m^3 of non-segregation-in-water agent(NSIW) and water cement ratio of M1 and M2 are 40% and 50% respectively. Water cement ratio of mortars are 40% except of AE3 and

M1. Fig.6 shows the difference of wave velocity compared with three kind of admixtures and two or three dosages and two water cement ratios.

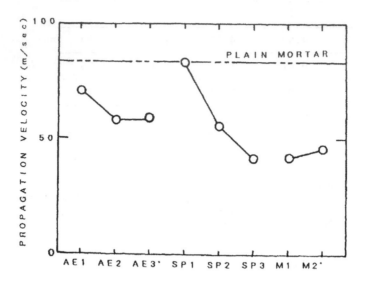

Fig.6. Longitudinal wave velocities of various mortars

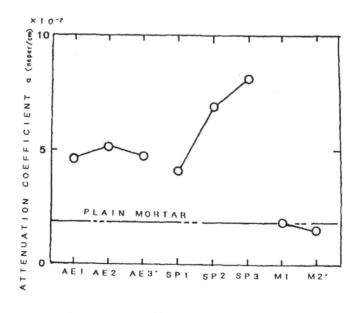

Fig.7. Coefficient of attenuations of various mortars

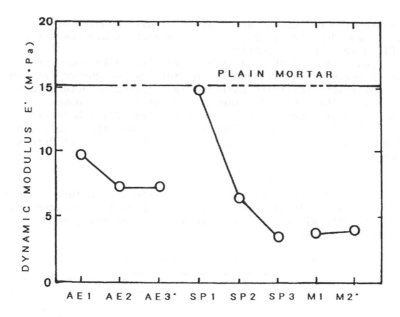

Fig.8. Dynamic modulus of various mortars

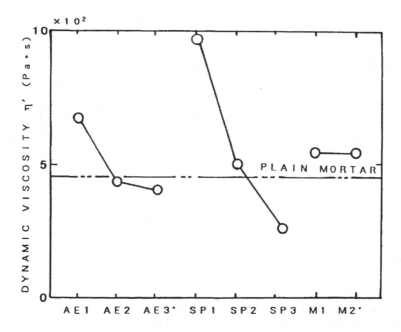

Fig.9. Dynamic viscosity of various mortars

The measured coefficient of attenuation,dynamic modulus of
elasticity and dynamic viscosity of fresh mortars are summarized in
Fig.7,Fig.8 and Fig.9 respectively.
It is obvious that the chemical admixture affects to the visco-
elasticity factors. In this experiment,increase of dosage lead to
decrease of visco-elasticity factors. That effect on water cement
ratio was less than that of dosage of admixtures. The dosage of
chemical admixture reduces the dynamic factors of fresh concrete and
increases the fluidity under vibration. The using the results,the
amount of dosage of chemical admixture will be able to be determined
in practice.

3.2 Wave spectrum of fresh mortar
The properties of wave propagation of concrete are influenced by many
factors,such as,acceleration of vibrator,propagation distance and
types of concrete mixes. Typical wave forms of concrete are
represented in Fig.10.

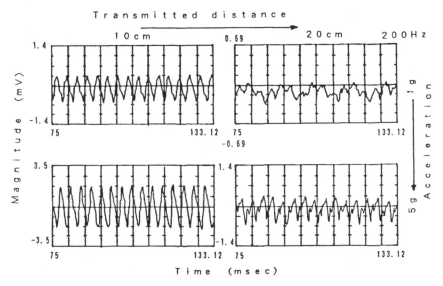

Fig.10. Typical wave forms under vibration

It is clear that the wave propagation was influenced by an
acceleration of vibrator and a propagation distance. Fig.11 shows the
frequency spectrum of SP2 mortar. Its frequency of vibration was 200Hz
and acceleration was 5g at vibrating plate. This figure shows the
propagation wave is consisted of the second and the third component of
resonance. Frequency spectrum of fresh mortar with chemical admixtures
are listed in Table 4. The magnitude of spectrum has correlation with
dynamic modulus of elasticity and dynamic viscosity. It is not
sure,how the ratio of dynamic factors affect to spectrum each other in
this point. However,it is clear on the effect of chemical admixture
were concerned with a frequency of vibration and an attenuation of
wave spectrum.

28

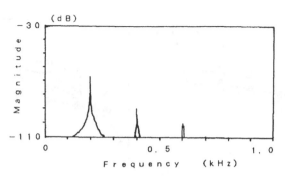

Fig.11. Frequency spectrum of superplasticized mortar under vibration

Table 4. Frequency spectrum of fresh mortar with chemical admixtures

Type of mortar	Dosage of admixture	Magnitude of frequency spectrum 100Hz (dB)		200Hz (dB)		Attenuation (dB/cm)	
	(%/C Wt.)	T_1	T_2	T_1	T_2	100Hz	200Hz
Plain	0	-78.0	-85.3	-66.7	-70.1	0.0934	0.0952
A E 2	0.06	-72.0	-82.0	-66.4	-67.5	0.0872	0.0948
S P 2	0.5	-74.2	-80.8	-69.6	-86.3	0.0918	0.0807
M 1	0.36	-79.6	-88.0	-70.8	-73.4	0.0905	0.0964

3.3 Flow behavior of fresh concrete under vibration
Fresh concrete is liquefied by vibration so that concrete behaves as a liquid within an effective range from vibrator. It is important to know the rheological properties of liquefied concrete because they give the capacity to hold aggregate particles and air bubbles under vibration. Rheological properties under vibration ,therefore,give a prediction for the resistance to segregation,compactability and fluidity. Fig.12 shows the flow curves of various concretes both under vibration and without vibration. The structural break down occurred by vibration so that the up-curve of flow curve almost agreed with the down curve. Therefore,only down-curves are shown in this figure. Although down-curves of concretes without vibration except marine concrete were straight lines,namely Bingham flow,flow curves under vibration became approximately a Newtonian flow or a quasi-viscous flow. A yield value disappeared by applying vibration and that an immediate drop in torque lead to liquefaction of concrete. In addition,it was clear that a fluidity changed with type of vibration. The effect of water content did not show for a fluidity of marine concrete due to its stickiness,in this experiment. Provided that properties of concrete under vibration can be estimated quantitatively from its flow diagram,it contributes to the development of effective vibratory consolidation technique. Usually,the rheological properties can be obtained from the flow diagram. Although many investigators support that concrete may be a Bingham fluid,concrete under vibration is still unknown.

29

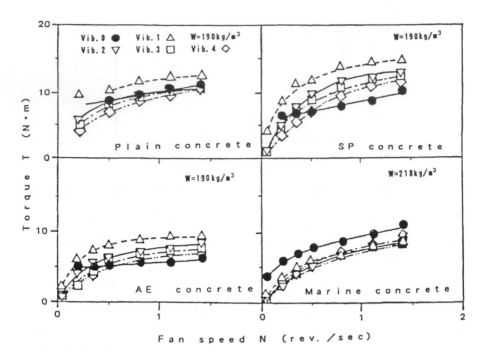

Fig.12. Flow curves of various concretes under vibration

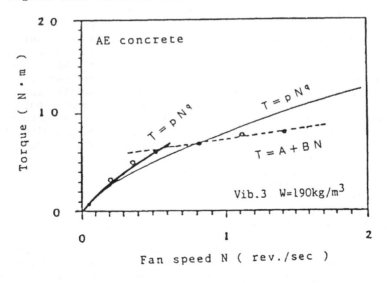

Fig.13. Analysis method for a flow curve under vibration

The exponential curves can be represented by a simple equation

$$T = p N^q \qquad (3.2)$$

where p(N.m/sec) is apparent coefficient of viscosity and q is apparent viscosity index. This equation gives an expression of non-Newtonian flow,namely pseudoplastic or quasiviscous flow. A typical flow diagram of AE concrete under vibration is shown in Fig.12. The flow curve under vibration could be represented by a power law equation as suggested by Tattersall,but a better agreement could be obtained when the curve was divided into two parts,and a straight line approximated to that for high shear rate range and exponential curve for low shear rate range by using consecutive four measured values as shown in Fig.13. Analytical results for various concrete obtained by above two method are summarized in Table 5.

Table 5. Analyzed results from flow curves

Types of concrete	Analysis	$T = p N^q$		$T = A + B N$	
		p	q	A	B
Plain	Full range	0.826	0.376	----	----
concrete	Divided range	11.08	0.473	9.007	0.906
S P	Full range	12.60	0.708	----	----
concrete	Divided range	15.89	0.831	9.176	2.421
A E	Full range	8.426	0.427	----	----
concrete	Divided range	10.34	0.736	5.793	1.933
Marine	Full range	9.556	0.608	----	----
concrete	Divided range	10.83	0.674	5.998	3.114

In practical case,concrete flows by small external forces such as self weight and mechanical weight like a surface vibrator.
Therefore,properties of fresh concrete can be obtained by using four point in lower shear rate.

3.4 Pore water pressure measurement
Stress of fresh concrete in a container is expressed by the following equation

$$p = p' + u \qquad (3.3)$$

where p is stress of fresh concrete in a container,p' is effective pressure and u is pore water pressure. Pore water pressure changes

when repeated shear stress is applied to fresh concrete. Fresh
concrete in the container may be assumed to be in undrained condition.
If a pore water pressure is increased by cyclic stress,then an
effective stress decreases. The liquefaction occurs simultaneously
when the effective stress becomes to zero. Typical relationships
between pore water pressure of concrete and vibration time are shown
in Fig.14.

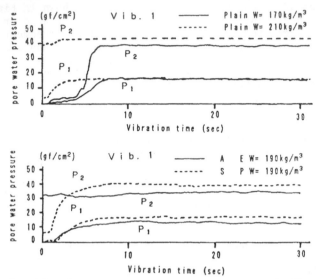

Fig.14. Relationships between pore water pressures and vibration times

Pore water pressures of various concretes were risen with
vibration time and became constant pressure(the liquid pressure)within
10 seconds. The excess pore water pressure of low slump raised slowly
with vibration time. In this case,a pore water pressure were raised as
the pore of concrete were filled with liquefied mortar. On the other
hand,the excess pore water of high slump concrete increased only a
little because this type concrete already liquefied without vibration.
Decrease of viscosity of concrete by vibration induce sedimentation of
particles and the pore water moves upward to surface of concrete.
Fig.14 shows the differences of water pressure between surface and
inner concrete indicating that the hydraulic gradient was occurred by
vibration. The compactability of concrete can be estimated by using
the coefficient of permeability calculated from following Darcy's
equation

$$v = k \ (d' \ / \ d_w) = e_v \ H/t \qquad\qquad (3.4)$$

where v is permeability speed of pore water,k is coefficient of
permeability,d' is a bulk density of solid particles in water,d_w is
the density of water,e_v is a vertical strain of concrete,H is a height
of concrete so that $e_v \ H$ is a amount of final sedimentation and t is
time when sedimentation is completed. Calculated results of
coefficient of permeability of concrete after 20 seconds are

32

summarized in Table 6. The permeability of plain concrete were influenced with water content and magnitude of vibration. There was remarkable difference of the coefficient in concrete with admixtures when compared with plain concrete. Especially,the value of marine concrete was negative.

Table 6. Coefficient of permeability of concrete with admixture

Type of concrete		Plain			A E	S P	Marine
Water content(kg/m^3)		170	180	200	190	190	210
Vibration Number	1	0.084	0.075	0.033	0.071	0.053	-0.043
	3	0.188	0.167	0.161	0.206	0.070	-0.023

(cm/sec)

4 Conclusions
(1) Properties of fresh mortar with chemical admixtures could be estimated by dynamic visco-elasticity factors.
Visco-elasticity factors are influenced by amount and types of admixture.
(2) Spectrum of the propagated wave through mortar with chemical admixtures were concerned with dynamic visco-elasticity factors.
(3) The flow curves under vibration showed Newtonian or quasi-viscous fluid behavior. An approach to evaluate the behavior of concrete with admixture under vibration by the non-Newtonian factor has been suggested.
(4) Flowing behavior under vibration had a considerable relation with the pore water pressure and the acceleration. The excess pore water pressure was observed when the concrete was vibrated and the measured values was affected by the type of admixture. Compactability of concrete with chemical admixture by vibration may be evaluated by a coefficient of permeability.

References

Tattersall,G.H. and Banfill,P.F.G.(1983) The Rheology of Fresh Concrete,Pitman Advanced Publishing Program,London.
ACI committee 309(1981) Behavior of Fresh Concrete During Vibration,ACI JOURNAL/January-February,36-53.
Tattersall,G.H. and Baker,P.H.(1988) The effect of vibration on the rheological properties of fresh concrete,Magazine of Concrete Research,**40**,143,79-89.
Kakuta,S. and Akashi,T.(1983) Study on Physical Properties of Cement Paste and Mortar at Early Age by Ultrasonic Pulse Method,Journal of the Society of Materials Science,**32**,353,175-181.
Kakuta,S. and Kojima,T.(1989) Effect of Chemical Admixtures on the Rheology of Fresh Concrete,Proceedings of Third CANMET/ACI conference on Superplasticizers and Other Chemical Admixtures in Concrete.

RHEOLOGICAL STUDIES ON FRESH CONCRETE USING ADMIXTURES

H. KIKUKAWA
Department of Civil Engineering, Meijo University, Japan

Abstract
Since admixtures for concrete are known to improve the properties of
fresh concrete and increase strength and durability of hardened concrete they are now widely used all over the world. However, rheological studies made on the influences of admixtures on fresh concrete
have been comparatively scarce up till the present. This paper describes rheological analyses of the properties of fresh concretes
containing admixtures. Particularly, it describes the improvement of
finishability of fresh concrete in a slab or pavement with admixtures,
and it also presents a method of estimating plastic viscosity of fresh
concrete containing a superplasticizer. This study will contribute
to the systematization of placeability of concrete, and through it,
it will also contribute to the conservation of energy resources.
Key words: Rheology, Fresh concrete, Viscosity equation, Plastic
viscosity, Yield value, Superplasticizer, Superplasticized concrete,
Finishability.

1   Introduction

Admixtures for concrete have the merits of improving various properties of fresh concrete and markedly increasing strength and durability
of hardened concrete, and they are, therefore, now used throughout
the world. Particularly, the advents of superplasticizers and admixtures for underwater concrete have contributed conspicuously to improved execution of concreting and safety of concrete structures.
However, it is felt that research works carried out with rheological
approaches to the effects of admixtures have been comparatively few
in number.
   Nowadays, improvement in placeability of concrete is being increasingly demanded, and it is of great importance for the intrinsic
nature of fresh concrete to be rheologically explained. Murata realized how important this was early on and pioneered research for improving placeability of concrete. The author is one of those who

were taught by him, and enjoyed the opportunity to contribute to the considerable development of studies on the rheology of fresh concrete. For example, it has become possible to estimate plastic viscosity employing a viscosity equation and to estimate yield value utilizing the correlation between slump and yield value.

This paper reports on methods of measuring rheological constants of fresh concrete and a method of estimating plastic viscosity using a viscosity equation, and in the aspect of application, the results of rheological studies on fresh concrete in case of using admixtures. In particular, improvement of the finishability of pavement and slab concrete, and the method of estimating the plastic viscosity of super-plasticized concrete using a viscosity equation were studied in the present research. Additionally, since yield value is also required besides plastic viscosity for the rheological analysis of fresh con-crete, a simplified method of estimation considering practical use was discussed. It is hoped that through the present study, a con-tribution can be made to systematization of the execution of concret-ing, and as a consequence, a contribution will be made to saving of labor in concreting.

2   Materials

The cement used was ordinary portland, the specific gravity of which was 3.15. Aggregates made to vary diversely with regard to kind, gradation, and particle diameter as shown in Table 1 were used. As for admixtures, two kinds were used: a material for improving

Table 1. Physical properties and gradations of aggregates.

		Classification		Specific gravity	Fineness modulus F.M.	Unit weight (kg/m³)	Solid volume (%)
Rg.		15~10mm. 10~5mm					
	1	Mix. 5:5		2.58	6.50	1582	61.3
	2	15~10mm		2.58	7.00	1538	59.6
	3	20~ 5mm		2.59	6.68	1558	60.2
	4	20~10mm		2.60	7.00	1476	56.8
	5		8:2 mixture	2.58	6.80	1550	60.1
	6	20~10mm	6:4 //	2.57	6.60	1549	60.3
	7	10~ 5mm	4:6 //	2.57	6.40	1549	60.3
	8		2:8 //	2.57	6.20	1548	60.2
Cs.	9	15~ 5mm		2.59	6.00	1580	61.0
	10	10~ 5mm		2.61	6.00	1473	56.4
Rs.	11	Under 5mm		2.58	2.88	1559	60.4
	12	under 5mm		2.56	2.43	1550	60.5

(Note) Rg: River gravel, Cs: Crushed stone, Rs: River sand

finishability and a superplasticizer. The material for improving
finishability was a polyol compound (Pozzolith No. 100) and the
superplasticizer a melamine sulfonate base compound (NP 20).

Moreover, for the purpose of preventing segregation during measurements of viscosity, a cellulose base water retention agent, NL 1850,
was used at a rate of 0.25% by weight of cement.

3   Method of measuring rheological constants

The particulars of the rotary internal cylinder viscometer used for
concrete were radius and length of internal cylinder 15 cm and 20 cm,
respectively, and radius of external cylinder 20 cm.

In order to eliminate absorption torque at the end of the internal
cylinder, a mortar layer of dry consistency (C:W:S = 1:0.4:3.5) was
laid between the bottom of the internal cylinder and the bottom of
the container, the concrete sample was then put in, and the flow
velocity distribution of the sample was measured by the multi-point
method, and thus the rheological constants were determined. In this
case, the ordinate and abscissa of a consistency curve are according
to the following equations:

$$\left. \begin{array}{l} \text{Ordinate} \quad V = \dfrac{2\dot{\theta}_i}{[1 - (\frac{r_i}{r_j})^2]} \\[2em] \text{Abscissa} \quad P = \dfrac{M}{2\pi r_i^2 h} \end{array} \right\} \tag{1}$$

where r_i radius at location 0.2 cm distant from internal cylinder
             wall (cm) (r_i = 15.2 cm)
        $\dot{\theta}_i$ angular velocity of sample at radius r_i (rad/s)
        h   length of internal cylinder (cm)
        M   torque (gf-cm)

As for the rheological constants, the plastic viscosity was determined as the inverse slope of the linear part of a consistency curve
according to Eq. (1), and the yield value was calculated by Eq. (2).

$$\tau_f = \frac{(\frac{r_i}{r_o})^2 - 1}{2\ell_n(\frac{r_i}{r_o})} \cdot \tau_a \tag{2}$$

where r_o outside radius of the region of sample flow (cm)
        τ_a intercept on abscissa by linear part of consistency
             curve (gf/cm^2)

4   Proposal of viscosity equation and estimation of plastic vis-
        cosity

Fairly complicated experiments are required for determining the plastic viscosity of fresh concrete. Therefore, it is desirable that the

plastic viscosity can be estimated by using a viscosity equation. Similarly to fresh cement paste and mortar, fresh concrete is a kind of high-concentration suspension, and relationships between solutes and solvents are considered to be as given in Table 2.

Table 2. Relationships between solvents and solutes.

Classification	Solvent	Solute
Cement paste	Water	Cement
Mortar	Cement paste	Fine aggregate
Concrete	Mortar	Coarse aggregate

In this research, as the result of various investigations, Roscoe's equation which considers high-concentration suspensions was adopted as the basic equation, and this was modified as follows so that it could be applied to cement paste, mortar, and concrete:

$$\eta_{re} = (1 - \frac{V}{C})^{-k} \tag{3}$$

where η_{re} relative viscosity of cement paste, mortar, or concrete
C absolute volume percentage of solute
V volume concentration of solute
k shape factor of solute

Eq. (3) was derived by considering the agglomerated condition of solute particles, but further, when the shape factor K of these solutes was investigated in detail, in the case of cement paste, K became the function of the volume concentration of cement particles and Blaine specific surface ϕ, and the exponential term of Eq. (3) is expressed by the following equation:

$$K = \alpha V^{\beta} \phi^{\gamma} \tag{4}$$

where α, β, γ experimental constants, with $\alpha = 1.00$, $\beta = -1.03$, and $\gamma = 0.08$.

Accordingly, Eq. (3) becomes as follows:

$$\eta_{re} = (1 - \frac{V}{C})^{-\alpha V^{\beta} \phi^{\gamma}} \tag{5}$$

However, as the results of various investigations, it was found that the effect of Blaine specific surface was small, and generally in the practical range of water-cement ratio (W/C = 0.45-0.85), K was the linear function of the volume concentration of cement particles. Accordingly, the following equation can be proposed:

$$\eta_{re} = (1 - \frac{V}{C})^{-(K_1 V + K_2)} \tag{6}$$

where K_1, K_2 experimental constants, and $K_1 = -15.6$, and $K_2 = 11.2$.

In the case of mortar or in the cases of concretes using coarse aggregates of maximum size 20 mm, it was shown that the shape factor K becomes the linear function of the fineness modulus of fine aggregate or coarse aggregate (see Figs 1 and 2).

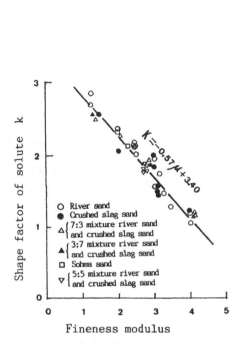

Fig. 1. Relationship between shape factor of solute and fineness modulus in fresh mortar. W/C=0.40∿0.60, s/c=0.2∿2.4

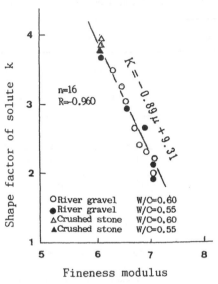

Fig. 2. Relationship between shape factor of solute and fineness modulus in fresh concrete. Max. size of coarse aggregate=20mm. sand–cement ratio of mortar(solvent) =1.68. Volumetric concentration of coarse aggregate =0.06∿0.33, W/C=0.55, 0.60.

Accordingly, the exponential term in Eq. (3) is expressed as follows:

$$K = a\mu + b \tag{7}$$

where μ fineness modulus of fine aggregate in case of mortar and of coarse aggregate in case of concrete
     a, b experimental constants

Therefore, the following equation can be proposed as the viscosity equation for mortar and concrete:

$$\eta_{re} = (1 - \frac{V}{C})^{-(a\mu + b)} \tag{8}$$

When the coefficients a and b were calculated on the basis of the experimental results, a = −0.57 and b = 3.40 were obtained in the case of mortar, and a = −0.89 and b = 9.31 in the case of concrete. Parts of the measured values of rheological constants for cement paste, mortar, and concrete are given in Tables 3 and 4.

Table 3. Plastic viscosity and yield value of cement paste.

W/C	1.00	0.85	0.75	0.70	0.65	0.60	0.55	0.50	0.45	0.40
Volume concentration	0.240	0.271	0.297	0.311	0.327	0.345	0.365	0.388	0.413	0.442
J-funnel flow time (s)	4.1	4.2	4.3	4.5	4.8	5.3	5.4	6.6	9.7	26.4
Plastic viscosity η_{pl} (P)	1.40	1.73	1.78	2.16	2.50	2.61	4.20	6.14	10.7	20.2
Yield value τ_f (gf/cm^2)	0.005	0.007	0.014	0.015	0.025	0.039	0.056	0.091	0.17	0.37

(Note) 1. Test values averages of 3 test results.
2. 1P=0.1Pa·s, 1gf/cm^2=98Pa.

The ratios of estimated values of plastic viscosity of cement paste (in the range of W/C = 0.40–1.00) determined by Eq. (5) to measured values were calculated in order to verify the effectiveness of the proposed viscosity equations, and as a result, with the number of samples 30, the ratios were 0.85 to 1.21, or 1.00 on average, and the coefficient of variation was 9.0%.

On the other hand, in the cases of mortar and concrete, the ratios of estimated values to measured values of the respective plastic viscosities calculated using Eq. (8) were 0.77 to 1.24, or 0.96 on average, and the coefficient of variation was 8.3% in the case of mortar with 312 samples. In the case of concrete, the ratios were 0.78 to 1.07, or 0.98 on average, and the coefficient of variation was 7.8% with the number of samples 116.

Accordingly, the plastic viscosities of cement paste, mortar, and concrete appear to be satisfactorily estimable using the proposed viscosity equations.

Table 4. Test results of mortar and concrete

Mortar

Classification	W/C	η_0	s/c	Volumetric concentration	Flow (mm)	Plastic viscosity η_{pl} (P)	Yield value τ_f (gf/cm²)
River sand SP**=2.58 F.M.=2.88	0.60	3.17	1.6 1.8 2.0	0.404 0.432 0.458	261 238 231	21.5 28.8 41.8	0.329 0.332 0.366
Crushed slag sand SP**=2.60 F.M.=3.04	0.50	5.70	1.4 1.6 1.8	0.394 0.426 0.455	253 243 233	63.7 105 197	0.300 0.350 0.400
Crushed slag sand 2.5~1.2mm SP**=2.53 F.M.=4.00	0.60	3.17	1.2 1.4 1.6 1.8	0.341 0.376 0.408 0.437	7.1* 276 257 244	12.7 18.5 28.5 52.1	0.085 0.150 0.250 0.300

Concrete

Classification	W/C	η_0	s/a	Volumetric concentration	Slump (cm)	Plastic viscosity η_{pl} (P)	Yield value τ_f (gf/cm²)
*** River gravel Specific gravity=2.59	0.55	48.1	0.52 0.49 0.44	0.280 0.310 0.350	20.0 19.5 16.5	400 617 863	1.13 1.25 1.47
Fineness modulus=6.68	0.60	32.7	0.55 0.51 0.40	0.250 0.280 0.380	23.5 21.5 16.5	232 307 154	0.870 0.990 1.23
Crushed stone *** 15~5mm Specific	0.55	48.1	0.63 0.57 0.55	0.200 0.240 0.260	23.0 20.5 19.0	227 326 404	0.830 1.11 1.17
gravity=2.59 Fineness modulus=6.00	0.60	32.7	0.77 0.68 0.62	0.110 0.160 0.200	27.0 25.0 24.0	85.5 119 137	0.450 0.510 0.600

(Note) 1. Test values of rheological constants averages of 2 test results.
2. Viscosity of water at 20 °C =1.002cP.
3. Estimated plastic viscosity of cement paste at W/C 0.55 =4.01 P.
4. η_0: Estimated plastic viscosity of solvent (cement paste or mortar) (P).
5. *: J-funnel flow time(s).
6. **: Specific gravity.
7. ***: Case of using fine aggregate No.12 in Table 1.

5   Basic properties of superplasticized concrete and estimation of plastic viscosity

Superplasticized concrete has been extensively used recently for the purpose of improving placeability and quality of concrete, but there has been little research done concerning its rheological properties.

The relationships between slump loss and rheological constants, the transitions in rheological constants and strength properties accompanying slump increments, and further, a method of estimating plastic viscosity of superplasticized concrete using a viscosity equation are discussed in this chapter. Two methods are conceivable for estimating plastic viscosity of superplasticized concrete. They are a method of estimating the plastic viscosity of superplasticized concrete from superplasticized cement paste, and a method of estimating using slump increment as the measure on the basis of the plastic viscosity of base concrete. The latter method was taken up in this study.

Table 5. Mix proportions of cement paste, mortar and concrete.

Series 1*

Classi-fication	W/C	s/c	s/a	Unit weight (kg/m³)					Slump (cm)	
				C	W	S	G	S p	B	S pc
	0.55	—	0.44	364	200	746	953	1.67 2.70 3.69 4.68	10	15 18 21 24
Concrete	0.55	—	0.44	373	205	737	942	0.86 1.71 2.57 3.73 5.60	12	15 18 21 24 27
	0.55	—	0.44	382	210	728	930	0.76 1.91 2.87 3.82	15	18 21 24 27

Series 2**

Classi-fication	W/C	s/c	s/a	Unit weight (kg/m³)					Slump (cm)	
				C	W	S	G	S p	B	S pc
Cement paste	0.50	—	—	1210	605	—	—	—	—	—
Mortar	0.50	1.49 1.44 1.39	—	708 718 728	354 359 364	1058 1036 1015	—	1.34~9.20 2.01~10.1 2.55~10.1	—	—
Concrete	0.50	—	0.42	450 460 470	225 230 235	637 664 655	925 913 901	0.84~5.85 1.29~6.45 1.64~6.54	10 12 15	12~24 15~27 18~27

(Note) 1.*: Mix proportions for investigating of basic physical properties.
2.**: Mix proportions for estimating plastic viscosity.
C: Cement, W: Water, S: Fine aggregate, G: Coarse aggregate,
Sp: Superplasticizer, B: Base concrete, Spc: Superplasticized concrete.

The superplasticizer was added after 15 min had elapsed from mixing of concrete, that is, so-called delayed addition was done. Base concrete was made to have slumps of 10, 12, and 15 cm, and by adjusting quantities of superplasticizer added, slump increments were made 3, 6, 9, 12, and 15 cm. The maximum value of slump was set at 27 cm. The mix proportions used are given in Table 5.

5.1 Relationships between slump losses and rheological constants

Examples of the results of experiments on changes with time in slump, yield value, and plastic viscosity of superplasticized concrete are shown in Fig. 3. The tendency for slump loss of superplasticized concrete to be larger than that of base concrete can be seen. This is the same as the tendency reported previously from various quarters. It is widely known, further, that superplasticized concrete of small slump loss has been developed as a result of more recent research.

Among rheological constants, plastic viscosity is not directly related to slump, and in case of a concrete having a large slump increment, the plastic viscosity hardly varies for about 50 min after mixing. However, it was found that hydration became more active with elapse of time, and accordingly, there is a gradual increase. On the other hand, as it has been said that yield value is closely related to slump, it was increased with elapse of time, and this tendency of increase was consistent independent of mix proportions.

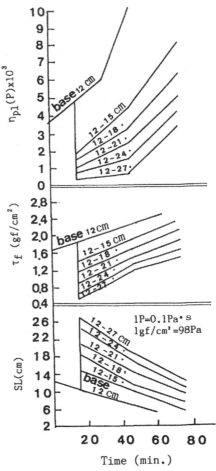

Fig. 3. Time-dependent change in rheological constant of superplasticized concrete.
(Slump of base concrete =12cm, t=20°C)

42

5.2 Slump increments and rheorogical constants of superplasticized
    concrete

The changes in rheological constants when slump was increased by
gradually increasing the dosage of superplasticizer are shown in Fig.
4. It was found that plastic viscosity fell rapidly at slump increment of around 3 cm after plasticizing, with gradual decrease thereafter as slump increment was increased.

On the other hand, the
yield value of superplas-
ticized concrete decreased
almost linearly as slump
increment became larger.
The rate of lowering of
yield value accompanying
slump increase was af-
fected by the consistency
of the base concrete also,
but within the scope of
the present experiments,
it was nearly constant in
every case. It is said
that yield value can be
estimated from the results
of slump tests, but it
can also be conjectured
from the results shown in
Fig. 4.

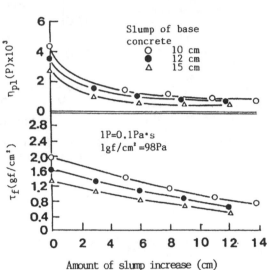

Fig. 4. Relationship between rheological
constant and amount of slump increase of
superplasticized concrete.
     Max. size of coarse aggregate=20mm,
     W/C=0.55, s/a=0.44, t=20°C.

5.3 Compressive strength
    of superplastized
    concrete

The results of experi-
ments in case of having
added superplasticizer
excessively to super-
plasticized concrete to examine strength properties are shown in Fig.
5. It is clear from this figure that lowering of strength occurred
when slump increment of the base concrete became large. It is considered that the main cause of this strength decrease was the segregation of concrete. The test pieces for compressive strength were
φ10 cm x 20 cm. Even when using such small test pieces the compressive strength ratio was about 75% in case of slump increment of a maximum of 17 cm over the base concrete, and in actual structures the rate
of decrease may be considered to be greater than this. Although a
summary statement cannot be made since the number of experiments conducted is still small, within the scope of the present experiments the
rate of decrease in compressive strength was low up to a slump increment of 10 cm. Accordingly, with regard to the dosage of superplasticizer, it appears better to limit it for a slump increment of not more
than about 10 cm. The Japan Society of Civil Engineers has prepared a
draft guideline for execution of superplasticized concrete placement
in which slump increments are limited to 10 cm with 5 to 8 cm as
standard.

5.4 Estimation of plastic viscosity of superplasticized mortar

Plastic viscosities of the cement paste and mortars in Table 5 in case of adding superplasticizer were measured and the results are given in Table 6. Furthermore, taking the measured values of plastic viscosity of cement paste in Table 6 as the viscosities of the solvent, the plastic viscosities of mortars were calculated using Eq. (8) and compared with the measured values. The results are also given in Table 6.

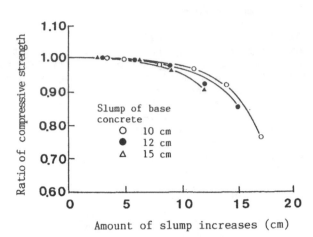

Fig. 5. Relationship between ratio of compressive strength and amount of slump increase of superplasticized concrete. Max. size of coarse aggregate=20mm, W/C=0.55, s/a=0.44, t=20°C.

Table 6. Accuracy of estimated plastic viscosity of superplasticized mortar. W/C=0.50, s/c=1.39∼ 1.49.

Classi-fication	s/c	Volume concentration	Dosage of super-plasticizer** (%)	Measured plastic viscosity of cement paste (P)	Plastic viscosity (P) Estimated A₂	Measured B₂	A₂/B₂
10*	1.49	0.415	0	6.10	47.0	47.0	1.00
			0.19	6.14	47.3	46.2	1.02
			0.46	5.66	43.6	44.6	0.98
			0.74	5.41	41.7	41.8	1.00
			1.02	5.19	40.0	38.4	1.04
			1.30	4.92	37.9	35.6	1.06
12*	1.44	0.406	0	6.10	43.3	43.0	1.01
			0.28	5.99	42.5	41.2	1.03
			0.56	5.56	39.5	39.8	0.99
			0.84	5.33	37.8	36.2	1.04
			1.12	5.10	36.2	34.2	1.06
			1.40	4.82	34.2	32.4	1.06
15*	1.39	0.398	0	6.10	40.4	40.6	1.00
			0.35	5.80	38.5	37.2	1.03
			0.70	5.35	35.5	34.0	1.04
			1.04	5.11	33.9	32.6	1.04
			1.39	4.84	32.1	29.5	1.09

(Note) 1. Test values averages of 2 test results.
2. *: Slump of base concrete.
3. **: Percentage by weight of cement.

The ratio of estimated values to measured values for superplasticized mortar were between 0.98 to 1.09, or 1.03 on average, within the scope of the present experiments, showing that the estimation could be carried out more or less satisfactorily.

5.5 Estimation of plastic viscosity of superplasticized concrete
The estimated values of plastic viscosity of superplasticized concretes calculated from superplasticized cement paste and then from mortar using Eq. (8) differed greatly from the measured values, and therefore, this method could not be used. Hence, in the method of estimating the plastic viscosity of superplasticized concrete it was considered to use slump increment as the measure. As shown in Fig. 6,

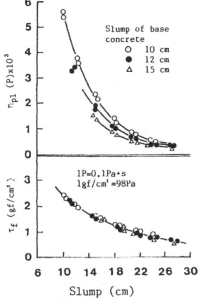 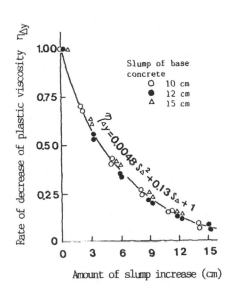

Fig. 6. Relationship between rheological constant and slump of superplasticized concrete. Max. size of coarse aggregate=20mm, W/C=0.50, s/a =0.42.

Fig. 7. Rate of decrease of plastic viscosity as a function of amount of slump increase. Max. size of coarse aggregate=20mm, W/C=0.50, s/a=0.42.

the plastic viscosity of superplasticized concrete is not directly related to slump like the yield value, but it was found that the rate of decrease in plastic viscosity is represented approximately by a quadratic function of slump increment, and the following equation holds true (see Fig. 7):

$$\eta_{\Delta y} = a_1 S_{\Delta}{}^2 + b_1 S_{\Delta} + C_1 \tag{9}$$

where $\eta_{\Delta y}$ rate of decrease in plastic viscosity

S_Δ slump increment (cm)

a_1, b_1, c_1 experimental constants, $a_1 = 0.0048$, $b_1 = 0.13$, and $c_1 = 1.0$

Accordingly, it is thought that after plastic viscosity of the base concrete has been estimated by Eq. (8), the plastic viscosity of superplasticized concrete can be estimated utilizing these relationships with slump increment as the measure. The relationship between value of plastic viscosity of superplasticized concrete estimated by this method and measured value is shown in Table 7. The ratios of estimated to experimental values were 0.81 to 0.98, and therefore, it appears that estimation can be done satisfactorily to a fair degree.

Table 7. Estimated plastic viscosity of superplasticized concrete.
(Value obtained using relation between rate of decrease of plastic viscosity and amount of slump increase)
Max. size of aggregate=20mm, W/C=0.50, s/a=0.42.

| Slump (cm) | Amount of slump increase (cm) | Dosage of super-plasti-cizer* (%) | Plastic viscosity (P) | | A_3/B_3 |
			Estimated A_3	Measured B_3	
15 (base)	0	0	1366**	1398	0.98
15—18	3	0.35	861	906	0.95
15—21	6	0.70	510	589	0.87
15—24	9	1.04	272	300	0.91
15—27	1 2	1.39	149	185	0.81

(Note) *: Percentage by weight of cement.
   **: Plastic viscosity of fresh concrete estimated using viscosity equation.

The abovementioned experiments were all cases of measurements immediately after adding superplasticizer, at 20 ± 2°C. In case of superplasticized concrete, there is slump loss with elapse of time, while plastic viscosity is also thought to become high. Therefore, it is necessary to investigate the effect of elapsed time also.

6   Examination of finishability of fresh concrete

Finishability is of importance for pavement concrete and slab concrete. The capacity for good or bad finish is expressed by the word "finishability," and it has been attempted from the past to express it quantitatively. However, there have been extremely few cases of the property having been measured specifically with numerical values. Since yield value among rheological constants is the stress at which concrete in contact with the wall of an internal cylinder starts to flow, concrete of small yield is thought can be easily finished with a trowel or other tool. Based on such a concept, the influence exerted

on the rheological constant was investigated using an admixture said
to be particularly effective for the improvement of finishability.

Parenthetically, the mix proportions of concrete were unchanged,
and comparative experiments were carried out on concretes with addition of a water-reducing agent for improving finishability (Pozzolith
No. 100) and plain concretes without addition. The results, as shown
in Fig. 8, were that in case of using the water-reducing agent for

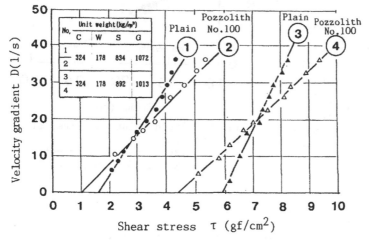

Fig. 8. Influence of admixture on consistency
curve of fresh concrete.
Max. size of coarse aggregate=20mm, W/C=0.50,
s/a=0.44, 0.47.

improving finishability, the yield value was small, the plastic viscosity was large, and these tendencies became more conspicuous as the
concrete became of stiffer consistency. These experiments are still
fragmentary and inadequate, but it is shown that finishability can be
evaluated to some extent by rheological constants, and they are
thought to provide an effective toe-hold to measurement of finishability.

7   Simplified method of estimating yield value

The slump of a concrete has a close relationship with its yield value,
and the relationship of the two has already been formulated. However,
it is difficult to directly estimate the yield value by the equation
for the relationship which takes into account the effect of friction
at the bottom of the slump cone. Therefore, it was decided to present
an empirical formula which expresses the relationship between slump
and yield value. That is, relationships between slump and yield value
were examined with 13 kinds of concrete in which maximum size of
coarse aggregate was 20 mm and water-cement ratios were 0.55 and 0.60.
The range of slump was from 13 to 25 cm and the relationship between
slump and yield value was determined as shown in Fig. 9.

It was found as a result that an exponential relationship exists between the two, and this is shown by the following equation:

$$\tau_f = A \log SL + B \qquad (10)$$

where τ_f yield value (gf/cm^2)
       SL   slump (cm)
     A, B   experimental constants,
          A = -4.83,
          B = 7.29

Yield values of the concretes were estimated using Eq. (10) and compared with measurements of yield values of 26 kinds of concrete having slumps of 13 to 25 cm, mixed separately with the

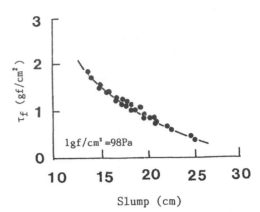

Fig. 9. Yield value vs. slump of fresh concrete.
Max. size of coarse aggregate =20mm, W/C=0.55, 0.60, s/a=0.32∿ 0.79.

objective of making confirmations. As a result, the ratios of estimated values to measured values were 0.88 to 1.14, or 1.03 on average, and the coefficient of variation was 7.8%. Therefore, it is considered that yield value can be estimated satisfactorily for practical purposes.

8   Conclusions

As a part of rationalization of concreting, investigations were made of the properties of fresh concrete in cases such as using admixtures. The principal results obtained within the scope of the present research are summarized below.

8.1 Measurements of viscosity were carried out for fresh cement paste, mortar, and concrete using a double-cylinder, rotary internal cylinder-type viscometer. From the results, based on Roscoe's viscosity equation for high-concentration suspensions, viscosity equations for fresh cement paste, mortar, and concrete were proposed according to the relationships between solutes and solvents (see Table 2 in main text).

8.2 It was found that the plastic viscosity in case of making the consistency of fresh concrete very soft by use of a superplasticizer could not be estimated by the viscosity equation mentioned in 8.1 above. Therefore, in the method of estimation, it was proposed to use slump increment as the measure. That is, after the plastic viscosity of base concrete has been estimated by the viscosity equation for concrete, the plastic viscosity after plasticizing can be determined by its rate of decrease. The rate of decrease in plastic viscosity of superplasticized concrete is given by the following equation:

$$\eta_{\Delta y} = 0.0048\ S_\Delta^2 + 0.13\ S_\Delta + 1$$

where $\eta_{\Delta y}$ rate of decrease in plastic viscosity of superplasticized concrete

        S_Δ slump increment (cm)

8.3 As a result of examining the strength properties of superplasticized concrete, it was found that when the slump increment of base concrete is large, lowering of strength occurs. It is considered that the main cause of this lowering of strength is the segregation of concrete. What can be said within the scope of the present experiments is that rate of lowering of compressive strength was low up to slump increment of about 10 cm. Accordingly, it appears advisable to limit the addition of superplasticizer so that the slump increment will be about 10 cm or under.

8.4 The effect in case of using an agent for improving finishability of pavement concrete was rheologically investigated. On plotting the results on a graph in which consistency curves had been drawn, it was clear that the yield value became small and plastic viscosity high in case of using the agent. Among rheological constants, yield value is the stress at which concrete in contact with the wall of the internal cylinder of a rotary viscometer starts to flow, and therefore, it may be considered that concrete having a small yield value can be easily finished with a tool such as a trowel. On the other hand, it is thought that when plastic viscosity becomes high, the capability of resisting segregation is strengthened. Accordingly, it is thought that finishability of concrete can be judged by using consistency curves. It was possible in this case to confirm that the agent for improving finishability was effective.

8.5 The yield value of fresh concrete is necessary for analyzing its behavior with the view of rationalizing concrete placement. It has been recognized that yield value is closely related to slump, and therefore, the relationship between yield value and slump was experimentally determined, and the following equation was presented as a simplified method of estimating yield value:

$$\tau_f = -4.83\ \log SL + 7.29$$

where τ_f yield value (gf/cm^2)
        SL slump (cm)

As the result of estimating yield value of concrete by this simplified equation, the ratios of estimated to measured values with 26 samples were 0.88 to 1.14, or 1.03 on average, and the coefficient of variation was 7.8%. It is thought, therefore, that the yield value of a concrete can be estimated more or less satisfactorily from the measured value of slump.

References

Murata, J. (1984) Flow determination of fresh concrete. <u>Materials and Structures</u>, 17, No. 98, 117-129.

Murata, J. and Kikukawa, H. (1985) The viscosity equation of mortar and concrete and its application. <u>Proceedings of 1985 Beijing International Symposium on Cement and Concrete</u>, 2, 232-242.

Murata, J. and Kikukawa, H. (1979) A proposal on method of measuring rheological constants of fresh concrete. <u>Trans. of JSCE</u>, 284, 117-126, (in Japanese).

Roscoe, R. (1952) The viscosity of suspension of rigid spheres. <u>British J. of Applied Physics</u>, 3, 267-269.

Kikukawa, H. (1987) Studies on the viscosity equation for fresh concrete and its applications, 238 pp. Doctoral thesis, Tokyo Metropolitan University, (in Japanese).

HIGH PERFORMANCE CONCRETE WITH HIGH FILLING CAPACITY

K. OZAWA, K. MAEKAWA and H. OKAMURA
Department of Civil Engineering, University of Tokyo, Japan

Abstract

The authors could find a mix of High Performance Concrete, which is defined as a concrete with high filling capacity. It can be filled into the all corners of formwork without using any vibrators.

The objective of this study is to investigate the role of chemical admixtures such·as superplasticizer and viscosity agent on the deformational and segregational behavior of fresh concrete, which is one of the most important studies for developing the concrete with high filling capacity. The optimum mix proportion of superplasticizer and viscosity agent was clarified for the concrete with high filling capacity. It was found that there exists the suitable viscosity of paste for improving not only the deformability but also the segregation resistance, which is highly dependent on the volume of frcc water in fresh concrete.

Key words: High Performance Concrete, Segregation, Filling capacity, Fresh concrete, Superplasticizer, Viscosity agent, Aggregates interaction, Free water

1 Introduction

The reliability of concrete structures is highly dependent on the construction works for concrete, such as transportation, placing and consolidation works. In case of placing concrete into the formwork where complicated reinforcing bars are fabricated and a shape of member is not simple, the honeycombing and segregation of concrete are often encountered due to the lack of consolidation or too much consolidation works. Then, in some cases, it has been supposed that such defects of concrete are caused by the short of skillfulness of construction workers.

We proposed the best way for improving the reliability of concrete structures, that is to develop and use the new concrete material with high filling capacity. It can be filled in the formwork without carrying out the vibrating works, even if it has the complicated sectional shape or complex arrangement of re-bars. The authors have studied for developing such kind of concrete and could find a mix of concrete with the combination of some commercial materials, including chemical admixtures(1989).

In this study, the authors investigated the role of chemical admixtures, such as superplasticizer and viscosity agent, on the deformational and segregational behavior of fresh concrete, which is one of the most important studies for creating the concrete with high filling capacity.

2 High Performance Concrete

2.1 Definition of High Performance Concrete
The authors call this type of concrete with high filling capacity High Performance Concrete and define the High Performance Concrete at three stages as follows.
The High Performance Concrete can fill into the every corner of the formwork without using any vibrators in fresh state. This is one of the most important properties for High Performance Concrete to realize dense concrete structures. In early age it has little initial defects such as cracks due to the heat generation of hydration, hardening and drying shrinkage. These cracks not only spoil the good appearance of concrete structures, but also trigger the corrosion of reinforcing bars. At the last stage, after hardened, it should have enough strength and sufficient resistance against the movement of potentially deteriorating factors such as oxygen, chloride ion and water especially in the surface of concrete structures. The High Performance Concrete with all these properties can realize the dense concrete and enhance the reliability of concrete structures, even if we have the poor structural design details or the poor construction works.

2.2 Properties of High Performance Concrete
Fig.1 shows the cross-section of the typical prestressed concrete bridge girder in Japan. It can be evaluated as a slimmed and rational section based on the structural design. However it can be seen that it is difficult to fill concrete especially into the bottom part of this section. There exists the complicated arrangement of sheaths and re-bars, and it has closed formwork. It also seems to be very difficult to insert vibrators into the bottom corners of the formwork.

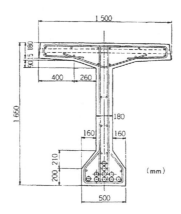

Fig.1. Cross-section of the typical prestressed concrete girder

Fig.2. High Performance Concrete filled into the formwork

The authors made the real size of this formwork in the bottom part and realized the arrangement of sheaths and re-bars. Then the High Performance Concrete was only poured into this section without using any vibrators. It can be seen that the High Performance Concrete is perfectly filled into the corners of the formwork as shown in Fig.2.

Some tests were carried out for the other properties after hardened and achieved good results as shown in Fig.3 and Fig.4. It can be seen in Fig.3, the results of drying shrinkage test, that the length reduction ratio of specimen of the High Performance Concrete is not greater than that of the conventional AE concrete which has about same 150 kg/m^3 of water content as the High Performance Concrete. Fig.4 shows the result of water permeability in the surface concrete by using the vacuum drying method(1989). The dewatering weight of the High Performance Concrete is much smaller than that of the conventional AE concrete which has 50% water to cement ratio. This result leads that the High Performance Concrete without using any vibrators can realize the dense concrete in the surface compared with the conventional concrete with the use of vibrators.

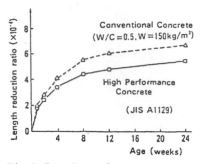

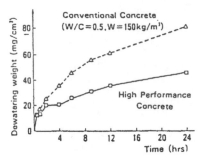

Fig.3. Results of
drying shrinkage test

Fig.4. Results of
dewatering test

We could get 277 kg/cm² at 7 days and 439 kg/cm² at 28 days for compressive cylinder strength of the High Performance Concrete under the water-cured condition at 20 degrees centigrade.

2.3 Mix proportion of High Performance Concrete
Mix proportion of this High Performance Concrete is given in Table 1. The properties of component materials used is shown in Table 2. Materials used are all commercial and the combination of these materials is one of the keys for the High Performance Concrete. It includes fine granulated blast-furnace slag, fly ash and expansive admixture as mineral admixtures and low slump-loss type of super-plasticizer and a small amount of viscosity agent are applied as chemical admixtures. It was confirmed that slump flow does little vary during 90 minutes after mixing, when adding this type of super-plasticizer. The mixer used was the forced mixing type of 100 litter in capacity.

Table 1. Mix proportion of the High Performance Concrete model

Water	OPC	Slag	Fly ash	Fine agg.	Coarse agg.	admix.	Slump flow	Air content
159	155	171	202	760	874	*)	57cm	2.0%

Maximum size of coarse aggregates: 20mm
*) 5544cc for superplasticizer and 20g for cellulose viscous agent

Table 2. Properties of materials used

	Specific gravity	Specific surface area		Specific gravity	F.M.	Absorption	solid volume
O.P.C.	3.15	3360 cm²/g	Fine agg.	2.62	2.59	1.58%	66.0%
Slag	2.90	5400 cm²/g	Coarse agg.	2.62	6.51	1.19%	65.4%
Fly ash	2.19	3010 cm²/g	(river)				

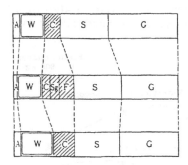

AE concrete
(w/c = 0.5, Slump = 17 cm.)

High performance concrete
(w/c + Sg + F = 0.3, Slump flow = 57 cm.)

Under water concrete
(w/c = 0.5, Slump flow = 45 cm.)

A : air, W : water, S : sand, G : gravel
C : ordinary portland cement, Sg : blast furnace slag
F : fly ash

Fig.5. Comparison of mix proportions in unit volume

Fig.5 illustrates the comparison of mix proportions between the conventional AE concrete, the conventional underwater concrete including a great amount of viscosity agent and the High Performance Concrete. It indicates that the High Performance Concrete has the smaller amount of coarse aggregates and the larger amount of powder materials, compared with the conventional AE concrete. On the other hand, in comparison with the conventional underwater concrete, the High Performance Concrete contains the smaller amount of water and the greater amount of powder materials.

3 Filling capacity of fresh concrete

Filling capacity of fresh concrete is governed by the workability as a material property and the boundary conditions. The workability of fresh concrete is governed by the deformability and the resistance to segregation. The authors carried out the experimental study for evaluating the filling capacity under the severe condition for the segregation of flowing concrete in order to clarify the influence of segregational behavior on the filling capacity of fresh concrete. This is one of the most important keys for developing the concrete with high filling capacity.

The setup of the experiment for evaluating the filling capacity is shown in Fig.6. Fresh concrete of about 35 litter in volume was poured into the formwork, the bottom of which meshes of reinforcement bars was attached. The formwork had a 30 X 30 cm section and meshes was fabricated with the D16 re-bars to make 5 cm clear distances between re-bars. This boundary condition is designed to simulate the rather severe condition such as a heavily reinforced area in the formwork. The volume of fresh concrete flowing through the meshes was measured to evaluate the filling capacity.

Concrete without adding any admixtures was used for this series of experiments. River sand (specific gravity 2.62, absorption capacity 2.08 %, fines modulus 2.59) and river gravel (specific gravity 2.62, absorption capacity 1.31 %, fines modulus 6.22) with maximum size of 15 mm were used. Materials were mixed by using the forced mixing type of mixer for 3 minutes.

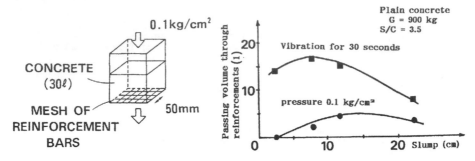

Fig.6. Setup for the
filling capacity test

Fig.7. Filling capacity for
plain concrete

Fig.7 indicates the relations between the measured volume of fresh concrete flowing through meshes and its slump value, which was controlled by changing the water content under the constant conditions of coarse aggregates content in 900 kg/m³ and sand to cement ratio by weight in 3.49. It can be seen that as for the concrete with the lower slump, the passing volume of concrete increases as the slump rises, that is, as the deformability increases. On the other hand, as for the concrete with the higher slump, the passing volume decreases though the deformability as a uniform material increases. Watching the concrete around meshes after a stop of flowing, it was found that the lower-slump concrete retained the uniformity of coarse aggregate and mortar, but that in case of the higher-slump concrete the water-discharged mortar remained around blocked coarse aggregates, which is called segregation of fresh concrete. Based on these results and observations, it can be concluded that filling capacity under severe boundary conditions is governed mainly by the deformability for the lower-slump concrete and mainly by the segregation resistance for the higher-slump concrete. Then we can get the maximum filling capacity as a result of the compound effect of deformability and segregation resistance.

Vibrators affect not only the deformability but also the segregation resistance of fresh concrete. Fig.7 shows that vibrations for 30 seconds betters the filling capacity especially for the lower-slump concrete. As for the lower-slump concrete, vibrations improve the deformability more than promote the segregation, which results in the good increase of the filling capacity. On the other hand, in case of the higher-slump concrete, the segregation is rather promoted than the deformability is bettered, which leads to the smaller increase of the filling capacity, compared with the lower-slump concrete. The value of slump at the maximum filling capacity is shifted to the lower by using vibrators, which means that the better vibrating works require the lower value of slump for the maximum filling capacity.

4 Free water in fresh concrete

It was clarified that the filling capacity of fresh concrete is controlled mainly by the deformability at the lower-slump and mainly by
the segregation resistance at the higher-slump. The authors take
free water as one of the governing factors of the deformability and
the segregation resistance. Free water is defined as the water which
is not retained by fine aggregates and powder materials as follows.

$$Wf = W - (Wp + Ws) \qquad\qquad (1)$$

where Wf gives a volume of free water, W is a total water content
and Wp and Ws indicates a retained water by powder materials and
sands respectively.

Water is one of the most deformable materials in the component
materials of fresh concrete. Then, we can easily guess the increase
of free water leads to the progress of deformability. How about the
relation between free water and the segregation resistance? The
segregation mechanism in the experiment for the filling capacity is
as follows. Coarse aggregates block flowing with the formation of
arch and mortar is likely to pass through the opening between
coarse aggregates, but fine aggregates also block. After that, only
paste or water can flow through the meshes.

IZUMI et.al.(1988) conducted the shear test of paste to investigate
the influence of paste on frictional resistance between solids as
shown in Fig.8. The relation between water to powder ratio by
volume and shear resistance under the constant normal stress is
given in Fig.9. The paste with the lower water to powder ratio, which
includes the smaller amount of free water, indicates the high intensity of shear resistance due to the powder particle-to-particle interactions such as collisions and frictions. On the other hand, the
shear resistance of the paste with the higher water to powder ratio
also increases, for part of too much free water is pushed away from
the gap of two plates by the normal force and the collisions and
frictions between two plates increase the shear resistance. Therefore, the minimum shear resistance is given under the condition that
there exists the appropriate volume of free water, where the inten- ·
sity of collisions and frictions between powder particles is not so
great and free water does not move out from paste under the normal
force.

In the experiment for the filling capacity, the similar situation
happens around meshes of re-bars as shown in Fig.10. Coarse aggregates are likely to approach each other at the position between
re-bars in flowing fresh concrete. As the concrete with the smaller
amount of free water is concerned, mortar can resist against the
flock of coarse aggregates. On the other hand, part of free water in
the mortar between coarse aggregates is easily moved away from the
concrete with the larger amount of free water, and coarse aggregates
can flock to form the arch when stresses due to collisions and frictions are increased between aggregate particles. It can be supposed
that vibrations release the water retained by sands and powder particles to increase the free water apparently.

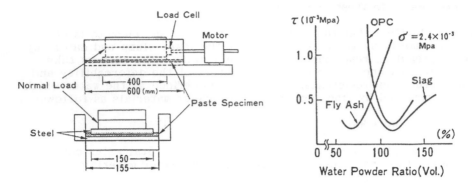

Fig.8. Apparatus for shear test
of paste between solids

Fig.9. Relation between shear
resistance and W/C by volume

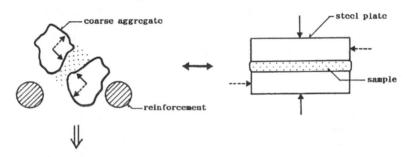

Fig.10. Situation of fresh mortar/paste
around re-bars and between steel plates

5 Role of chemical admixtures on the filling capacity

It was clarified that free water content in fresh concrete is one of
the governing factors for not only the deformability but also the
segregation resistance, and then for the filling capacity. The
authors investigated the effect of chemical admixtures such as superplasticizer and viscosity agents on the filling capacity from a
view point of free water content.

As concerned with the superplasticizer, it is supposed to release
the water retained by powder particles and fine aggregates due to
the transformation of flocculation structures of particles in water.
This process leads to increase the free water content in fresh concrete apparently. Fig.11 indicates the result of the filling capacity
test as explained in Section 3. for the concrete controlled by only
the volume of added superplasticizer without changing the mix
proportion of the other materials. The relation between the passing
volume of concrete and slump value is similar to that in Fig.7. This
result proves that the amount of free water in concrete controlled
by superplasticizer affects the filling capacity of fresh concrete.

58

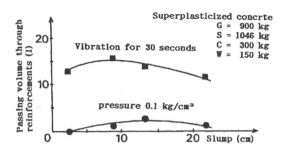

Fig.11. Filling capacity for superplasicized concrete

Viscosity agent is supposed to control the viscosity of water in concrete and to influence the deformability and segregation resistance. The relation between shear resistance and amount of cellulose viscous agent in the shear test of slag paste as explained in Section 4. is given in Fig.12. It describes that a small amount of viscosity agent lessens the shear resistance, but that the shear resistance increases by adding much volume of viscosity agent. The minimum shear resistance is denoted for about 0.2% in weight of viscosity agent to slag powder.

The paste with no viscosity agent corresponded to that with relatively large amount of free water. Therefore free water is moved away by the normal stress to increase the frictional resistance between solids due to the decrease of distance between solids. By adding the viscosity agent, free water is restrained to move and the frictional resistance between solids is controlled. However too much of viscosity agent represses the paste to deform as a liquid, which results in the total shear resistance will increase. The suitable amount of viscosity agent can improve the frictional resistance between solids 'and can decrease the total shear resistance effectively.

Furthermore, it was confirmed that there exists the two mechanism for shear resistance of paste between solids. The shear resistance at point A and B is the same in Fig.12. However they indicates the different behavior with the relation to the normal stress as shown in Fig.13. The shear resistance of paste at point A increased linearly according to the normal stress, which corresponds to the frictional mechanism. On the other hand, the shear resistance at point B denoted the constant value with the increase of normal stress, which is equivalent to the resisting mechanism as a viscous liquid. The shear resisting mechanism changed due to the increase of viscosity in water by adding the viscosity agent to the paste. From these results, the optimum amount of viscosity agent is supposed to be governed by the co-relation between the frictional mechanism as a solid and that as a viscous liquid. And it is expected not only to better the deformability due to the decrease of shear resistance, but also to improve the segregation resistance due to the unvaried shear resistance even with the increase of normal stress.

59

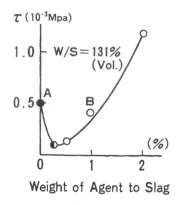

Weight of Agent to Slag

Fig.12. Relation between shear
resistance and weight of
viscosity agent

τ (10^{-3}Mpa)

σ' (10^{-3}Mpa)

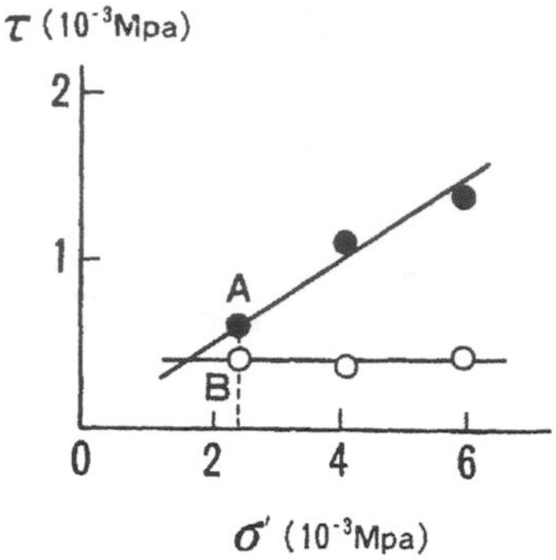

Fig.13. Relation between shear
resistance and normal
stress at point A and B

This expectation was proved by the filling capacity test as shown
in Fig.14, which describes the relation between the passing volume of
concrete and the amount of viscosity agent for the three kinds of
free water content in concrete. The volume of free water was controlled by the water to powder ratio and the different kinds of powder materials, and was computed according to the eq.(1). As for all
three cases of free water content, there existed the maximum filling
capacity, which proves that there exists the suitable amount of viscosity agent to each mix proportion. Furthermore it was found that
the optimum volume of viscosity agent increases according to the
amount of free water. It can be concluded that the optimum amount
of viscosity agent is highly dependent on the volume of free water.

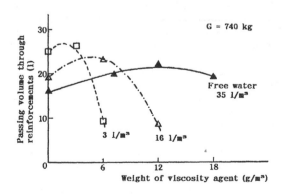

Fig.14. Relation between filling capacity and weight of viscosity agent

6 Concluding remarks

The authors could find a mix of High Performance Concrete, which is defined as a concrete with high filling capacity. From the experimental study, it was confirmed that the filling capacity is governed by the deformability and segregation resistance of fresh concrete. Then it was found that the amount of free water in concrete is one of the important factors for the deformational and segregational behavior of fresh concrete.

The role of chemical admixtures such as superplasticizer and viscosity agent on the deformational and segregation behavior of fresh concrete was investigated, which is one of the most important studies for developing the concrete with high filling capacity. The optimum mix proportion of superplasticizer and viscosity agent was clarified for the concrete with high filling capacity. It was found that there exists the suitable viscosity of paste for improving not only the deformability but also the segregation resistance, which is highly dependent on the free water content in fresh concrete.

Acknowledgment

The authors appreciate their help for experimental works by Mr.ENDOH,Y., Mr.TANGTERMSIRIKUL,S., Mr.NANAYAKKARA,A., Mr.FUNAI,K. and Mr.YAMAKOSHI,K. and express their gratitude to the Ministry of Education of Japan for partially providing the financial support by Grant-in-Aid No. 01750451.

References

OZAWA, K., MAEKAWA, K., KUNISHIMA, M. and OKAMURA, H. (1989) High Performance Concrete based on the durability design of concrete structures. Proc. of the Second East Asia-Pacific Conference on Structural Engineering and Construction, 1, 445-450.

OZAWA, K., MAEKAWA, K., and OKAMURA, H. (1989) Development of the High Performance Concrete. Proc. of the Japan Concrete Institute, 11-1, 699-704. (in Japanese)

KUNISHIMA, M. and OKAMURA,H. (1989) Durability design for concrete structures. Proc. of IABSE symposium on Durability of Structures, 57/2, 505-510

YAMANAKA, K., OZAWA, K. and KUNISHIMA, M. (1989) Quantitative evaluation of durability of concrete structures. Proc. of the Japan Concrete Institute, 11-1, 487-492. (in Japanese)

HASHIMOTO, C., MARUYAMA, K. and SHIMIZU, K. (1988) Study on visualization of blocking of fresh concrete flowing in pipe. Concrete Journal, 26-2, 119-127. (in Japanese)

OZAWA, K., NANAYAKKARA, A. and MAEKAWA, K. (1988) Flow and segregation of fresh concrete around bifurcation in pipe lines. Proc. of Third International Symposium on Liquid-Solid Flows, ASME FED-75, 139-144.

NANAYAKKARA, A., OZAWA, K. and MAEKAWA, K. (1988) Flow and segregation of fresh concrete in tapered pipes - Two-phase computational model -. <u>Proc. of Third International Symposium on Liquid-Solid Flows</u>, ASME, FED-75, 47-52.

IZUMI, T., MAEKAWA, K., OZAWA, K. and KUNISHIMA, M. (1988) Influence of paste on the frictional resistance between solids. <u>Proc. of the Japan Concrete Institute</u>, 10-2, 309-314. (in Japanese)

TSUJI, Y., NIWA, J., ITO, Y. and OKAMURA, H. (1987) Testing method for retained water of fine aggregate by utilizing centrifugal force. <u>Proc. of the Japan Society of Civil Engineers</u>, 384/V-7, 103-110. (in Japanese)

INFLUENCE DU DOSAGE ET DU MODE D'INTRODUCTION DES SUPERPLASTIFIANTS SUR LE MAINTIEN DE LA MANIABILITE OPTIMUM DES BETONS A HAUTES PERFORMANCES AVEC ET SANS FUMEES DE SILICE

(Influence of dosage and addition method of superplasticizers on the workability retention of high strength concrete with and without silica fume)

A. M. PAILLERE and J. J. SERRANO
Laboratoire Central des Ponts et Chaussées, Paris, France
M. GRIMALDI
Laboratoire Régional des Ponts et Chaussées, Melun, France

Résumé

A partir de matériaux, ciments et granulats bien définis, l'obtention des bétons à hautes performances avec ou sans fumées de silice, passe par des formulations à minimum de teneur en eau. Ceci peut être réalisé en utilisant des adjuvants à pouvoir réducteur d'eau très élevé tels que les superplastifiants à base de résine mélamine formaldehyde et naphtalène sulfonates. Cependant le maintien dans le temps des maniabilités élevées initiales (slump > à 20 cm) est généralement très limité.

Il est apparu ainsi que lorsque le dosage est en accord avec la réglementation française, c'est-à-dire à l'intérieur d'une plage comprise entre 0,2 et 0,8 % d'extrait sec, le maintien maximum de la maniabilité dans les bétons à faible rapport eau/ciment est obtenu grâce à l'introduction fractionnée de l'adjuvant, c'est-à-dire une partie dans l'eau de gâchage et une autre partie à la fin du malaxage ou avant la mise en oeuvre des bétons.

Dès que l'on aborde des dosages supérieurs, on constate, selon la nature des superplastifiants et du ciment, qu'il existe une plage de valeurs à l'intérieur de laquelle la maniabilité augmente avec le temps pour redevenir après 60 et parfois 135 min égale à la maniabilité initiale. Au-delà de cette plage, les dosages conduisent de nouveau à des raidissements rapides des bétons.

Il a été ainsi mis en évidence l'existence de trois plages de dosages influençant le maintien de la maniabilité du béton.

L'étude approfondie de ce phénomène donne une première approche basée sur l'adsorption ou le dosage saturant de l'adjuvant.

Mots clés : superplastifiant, béton hautes performances, maintien de la maniabilité, adsorption-désorption-fumée de silice.

1 Introduction

La formulation des bétons à hautes performances avec et
sans fumées de silice passe obligatoirement par
l'utilisation d'adjuvants appartenant à la famille des Superplastifiants Hauts Réducteurs d'Eau. Ces produits
conduisent à des teneurs en eau des bétons très faibles
(E/C de 0,28 à 0,35) et possédant des maniabilités élevées
(slump > 18 cm).

Toutefois, l'action fluidifiante des résines de synthèse,
que constitue cette famille d'adjuvants, est limitée dans
le temps et la perte de maniabilité est d'autant plus ra-
pide que la réduction d'eau est plus élevée.
Les conséquences de cette caractéristique étaient relativement limitées dans les bétons classiques, fermes ou plas-
tiques à slump de 4 à 10 cm, puisque leur introduction dans
le béton doit avoir lieu après la confection de celui-ci et
de préférence juste avant sa mise en oeuvre. C'est
d'ailleurs dans ces conditions que ces adjuvants ont le
maximum d'efficacité (Ramachandran 1984, Paillère et al.
1980).

Avec les bétons à hautes performances et à faible teneur en
eau et/ou avec fumées de silice, le problème s'est posé
différemment, d'une part, parce qu'il est impossible de fa-
briquer en centrale des bétons à slump nul, d'autre part,
parce que les fumées de silice exigent, lors du malaxage,
l'utilisation de superplastifiants.

Pour répondre à ces deux conditions, nous avions montré, au
cours de recherches antérieures (Paillère et "al." 1987),
l'intérêt d'introduire le dosage total en superplastifiant
en deux fractions, une fraction dans l'eau de gâchage,
l'adjuvant joue alors son rôle de plastifiant-réducteur
d'eau, la fraction restante à la fin du malaxage, voire
même avant la mise en oeuvre des bétons, l'adjuvant joue
alors son rôle fluidifiant.

Cependant, du point de vue du maintien de la maniabilité
dans le temps, il restait à connaître l'influence sur cette
méthode de la nature du ciment et de l'adjuvant. Par ailleurs, avec l'introduction des fumées de silice, les dosages en adjuvant ont été considérablement modifiés, ce qui
a rendu nécessaire la connaissance de leur optimum.

C'est ainsi que, toujours dans l'optique de l'obtention
d'une durée de maniabilité maximum, nous avons effectué la
présente recherche pour essayer de faire le point sur les
limites de la méthode d'introduction fractionnée et sur
l'influence du dosage en superplastifiants.

2 Conduite de la recherche

L'étude comprend deux parties, la première partie porte sur l'introduction fractionnée des superplastifiants. Elle ne concerne pas les fumées de silice et traite de l'influence de la nature du ciment et de l'adjuvant ainsi que la fraction introduite dans l'eau de gâchage.

Dans la seconde partie, on a abordé l'influence du dosage en superplastifiant et de la nature du ciment sur le maintien de la maniabilité des mortiers avec et sans fumées de silice.

2.1 Adjuvants
a) Nature

Dans l'ensemble de la recherche on a expérimenté quatre superplastifiants, dits de la première génération, c'est-à-dire sans effets retard.

Leur nature chimique de base était la suivante :

n° 1 et n° 2 résine mélamine formaldehyde (cation Na)
n° 3   résine naphtalène sulfoné (cation Na)
n° 4   résine naphtalène sulfoné (cation Ca)

La deuxième partie présente en outre les premiers résultats obtenus avec un superplastifiant, dit de la deuxième génération, (produit n° 5) à effet retard et à base de naphtalène sulfoné.

b) Dosages

Dans la première partie nous nous sommes limités à une seul dosage, 0,5 % d'extrait sec soit encore entre 1,5 et 2,5 % du produit commercialisé. Ce dosage respecte les règlements en vigueur c'est-à-dire :

. inférieur à 5 % du poids de ciment (définitions des adjuvants AFNOR P18-103.CEN - ISO-ASTM et RILEM

. compris dans les plages de dosages définies par les producteurs conformément aux spécifications de la norme NF 18-333

. correspondant à une moyenne de la plage de dosage des fluidifiants titulaires de la marque NF (France)

Dans la deuxième partie on a fait varier les pourcentages en extrait sec d'adjuvant au-delà des plages "normalisées" jusqu'aux limites conditionnées par le maintien maximum de la maniabilité, leur introduction a eu lieu juste avant la fin du malaxage.

2.2 Ciments

Le choix des ciments a été basé sur la teneur en C_3A. Les ciments C1, C2, C3,C4 ont été étudiés dans la première partie, leur teneur en C_3A varie de 1,8 à 11,9 %. Les ciments C4 et C5 ont été expérimentés dans la deuxième partie (tableau 1).

TABLEAU 1. CARACTERISTIQUES DES CIMENTS

CIMENTS	C1	C2	C3	C4	C5
CLASSES DE CIMENTS	CPA 55 PM	CPA HP	CPA 55 R	CPA HP	CPA 55 PM
C_3S	69,2	54,9	50,6	48,16	84,4
C_2S	17,7	21,1	26,2	26,3	5,99
C_3A	1,8	3,4	5,1	11,9	1,52
C_4AF	3,9	10,8	13,7	6	7,4

2.3 Rapport eau/ciment et maniabilité des mortiers étudiés

S'agissant de hautes performances, les superplastifiants ont été utilisés en tant que hauts réducteurs d'eau, c'est-à-dire que les formulations ont été déterminées à maniabilité constante. La maniabilité LCL (déterminée au maniabilimètre LCL norme française NF P18-452) fixée, a été de 15 ± 3 s, le rapport E/C a été défini en conséquence.

3 Première partie : Etude de l'introduction fractionnée de superplastifiant

3.1 Détermination de la teneur en eau minimale

La figure 1 résume les principaux résultats obtenus avec les quatre ciments et adjuvants étudiés.

Elle met en évidence que pour le même dosage de 0,5 % d'extrait sec de superplastifiant, on obtient, avec les 2 ciments à faible C_3A (1,8 et 3,4 %), les maniabilités souhaitées pour des teneurs en eau comprise entre 30 et 33 % du poids de ciment. Les adjuvants à base de naphtalène sulfonate conduisent aux plus faibles teneur en eau.

Lorsque le pourcentage de C_3A atteint des valeurs supérieures (5 à 12 %) l'efficacité des produits est moindre et les rapports E/C se situent entre 0,37 et 0,40. La différence entre les diverses natures chimiques des superplastifiants est moins marquée dans le cas de ces ciments.

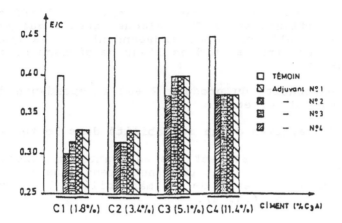

Figure 1   Teneur en E/C du mortier en fonction de la nature du ciment et de
l'adjuvant (dosé à 0,5 % d'extrait sec) pour une maniabilité constance,
égale à 15 ± 3 s.

3.2   Recherche du fractionnement optimal du superplasti-fiant

Nous avons étudié l'introduction en deux temps : d'abord
une partie du dosage total en superplastifiant dans 25 % de
l'eau de gâchage et ensuite l'autre partie directement
quelques secondes avant la fin du malaxage.

On a procédé, pour les superplastifiants 1-2-3-4 et les ciments C1, C2, C3, C4 à la recherche des fractionnements
optimaux de superplastifiant. Les rapports E/C ont été
déterminés pour chaque adjuvant afin d'obtenir une
maniabilité LCL de 15 ± 3 s immédiatement après malaxage.

On a caractérisé la double introduction par la proportion
P/S, avec P = Pourcentage de superplastifiant dans l'eau de
gâchage

S = pourcentage de superplastifiant introduit à la fin du
malaxage.

S + P étant le dosage total en superplastifiant.

La figure 2, montre que les maniabilités optimales (T_1 < 18
s) des mortiers après la dernière introduction d'adjuvant,
se situent aux alentours des proportions P/S comprises
entre 100/0 et 60/40 pour les produits à base de résine mélamine et les ciments C1, C2 et C3 et dans les valeurs comprises entre 0/100 et 60/40 pour les adjuvants à base de
naphtalène sulfonate et les ciments C2, C3 et C4.

Il est a noter que dans le cas du ciment C1, l'efficacité des superplastifiants est à T1 analogue, quelle que soit la proportion P/S, elle est donc indépendante de la quantité d'adjuvant introduite, soit dans l'eau de gâchage soit à la fin du malaxage.

De même, l'efficacité du produit 2 est indépendante de P/S et de la nature du ciment.

3.3 Etude du maintien de la maniabilité dans le temps

La différence de comportement des adjuvants en fonction de la nature du ciment apparait plus nettement dans l'étude de l'évolution dans le temps de la maniabilité des mortiers ainsi confectionnés (fig. 2).

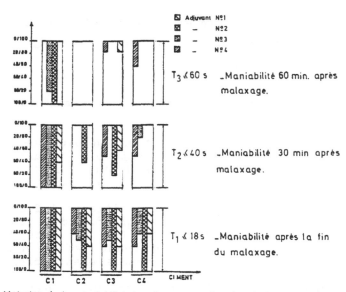

Figure 2 . Maintien de la maniabilité dans le temps en fonction de la nature du ciment et de l'adjuvant.

Si l'on admet comme maniabilités acceptables du mortier les temps d'écoulement LCL suivants :

. 30 min après fabrication T2 < 40 s
. 60 min après fabrication T3 < 60 s.

Il apparait, dans la figure synoptique 2, que, avec le ciment C4, 60 min après confection, seul le produit 4, utilisé pour des valeurs de P ≤ 40, permet d'obtenir un mortier à maniabilité T3 < 60 s. De même avec le ciment C3 et P < 20, on trouve les produits 1 et 4. Aucun produit ne donne une maniabilité T3 acceptable avec le ciment C2. Avec

le ciment C1, seul le produit 2 confère une maniabilité T3
< 60 s quelles que soient les proportions P/S ; il en est
de même avec le produit 3 à partir de P < 80.
Il est à noter qu'avec le ciment C1 à faible teneur en C_3A
et forte teneur en C_3S, le maintien de la maniabilité
jusqu'à 30 min peut être estimé satisfaisant pour
l'ensemble des fluidifiants.

4 Deuxième partie : recherche du maintien optimum de la maniabilité en fonction du dosage de superplastifiant
4.1 Evolution de la maniabilité dans le temps
4.1.1 Superplastifiants sans effet retard

On a étudié, en fonction du dosage en superplastifiant,
l'évolution de la maniabilité toutes les quinze minutes,
jusqu'à trois heures après confection des mortiers.
Les teneurs en eau des divers mortiers ont été déterminées
pour chaque dosage de façon à obtenir une maniabilité de 15
± 3 s.

4.1.1.1 Constatations principales
a) Première remarque

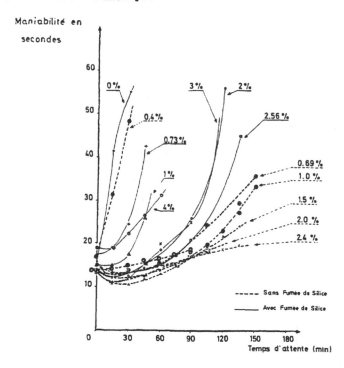

FIGURE 3.   MANIABILITE EN FONCTION DU TEMPS D'ATTENTE
ET DES DOSAGES EN SUPERPLASTIFIANT
CIMENT C.5 ET ADJUVANT N° 1 AVEC ET SANS FUMEE DE SILICE

Pour certains dosages en adjuvant, on constate l'accroissement, après confection, de la maniabilité en fonction du temps. La figure 3 donne un exemple de cette caractéristique pour des mortiers avec et sans fumées de silice.

Ce phénomène de refluidification est général pour tous les superplastifiants testés et passe par un optimum en dosage en adjuvant à partir duquel le maintien de la maniabilité commence à diminuer. Il est à noter toutefois que l'adjuvant n° 4 présente ce phénomène de façon très limitée.

TABLEAU 2

CIMENT	ADJUVANT N°	*F.S. %	ADJUVANT %	MANIABILITE EN SECONDES INITIALE	MAXIMALE	ADJUVANT %	MANIABILITE INITIALE (secondes)	MAINTIEN DE LA MANIABILITE Δt**(minutes)
5 1,52 % de C_3A	1	0	1,5	13,7	10,5 après 30min	2	16,5	75
		10	3	13	10,7 après 15min	2 à 3	14,1 à 13,8	30
	2	0	1,46	14,5	12,6 après 30min	2,24	19,3	75
		10	1,86 à 2,33	16 à 14,3	11,6 à 11,4 après 30min	1,86 à 2,33	11,6 à 11,4	60
	3	0	1,49	13,5	11,2 après 15min	1,49 à 1,96	13,5 à 16,3	75
		10	2,5	15	10 après 15min	3,2	14,7	75
	4	0	2,25	13,8	13,8 ***	-	13,8	0
		10	2,18	15	8,2 après 15min	2,18	15	30
4 11,9 % de C_3A	1	0	1,4	14,4	13,2 après 15min	1,47	16,9	60
		10	1,6	14	10,3 après 15min	1,6 à 1,8	14 à 16,8	45
	2	0	1,4	13,8	6,9 après 30min	1,4	13,8	135
		10	1,95	16,6	13,5 après 15min	1,95	16,6	30
	3	0	1,8	14,1	9,6 après 30min	1,8	14,1	120
		10	2,3	14,8	11,6 après 15min	2,3	14,8	45
	4	0	2	16,2	14,4 après 30min	2	16,2	45
		10	2	13	13 ***	-	-	0

* F.S.=FUMEE DE SILICE
** Δt =TEMPS MAXIMAL NECESSAIRE AU MORTIER FLUIDIFIE POUR REGAGNER LA MANIABILITE INITIALE
*** IMMEDIATEMENT APRES CONFECTION

Le tableau 2 et la figure 4 synthétisent les résultats des maniabilités maximales obtenues pour chaque dosage en superplastifiant et chaque ciment, ainsi que le temps Δt nécessaire pour que la maniabilité du mortier revienne à sa valeur initiale.

Il apparaît ainsi que :

. pour les mortiers sans fumées de silice la plus faible maniabilité obtenue et de 6,9 secondes avec le ciment n° 4 et 1,4 % d'adjuvant n° 2 à base de résine mélamine. Cette maniabilité est atteinte 30 min après confection et maintenue jusqu'à 45 min. Cette formulation de mortier met 135 minutes à regagner sa maniabilité initiale (tableau 2). Le maximum de maniabilité observé pour le ciment n° 5 est de 10,5 secondes avec 1,5 % d'adjuvant n° 1 et le maintien maximum de la maniabilité est de 75 minutes (tableau 2).

. pour les mortiers avec fumées de silice, paradoxalement l'adjuvant n° 4, dont les performances sont généralement faibles, conduit avec le ciment n° 5 à la maniabilité maximale observée (8,2 secondes). Toutefois le temps nécessaire pour regagner la maniabilité initiale n'est que de 30 min (tableau 2).

A cette exception près, on peut dire que les maniabilités maximales obtenues oscillent entre 10 et 10,7 secondes.

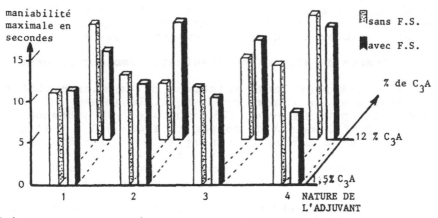

FIG 4. MANIABILITES MAXIMALES OBTENUES AVEC DES MORTIERS AVEC ET SANS FUMEES DE SILICE EN FONCTION DE LA NATURE DE L'ADJUVANT ET DU CIMENT

La fig. 4 montre que pour le ciment à faible C_3A la maniabilité du mortier avec F.S est toujours supérieure à celle du mortier sans F.S. Ce résultat est inversé dans le cas du ciment à forte teneur en C_3A et les adjuvants 2 et 3.

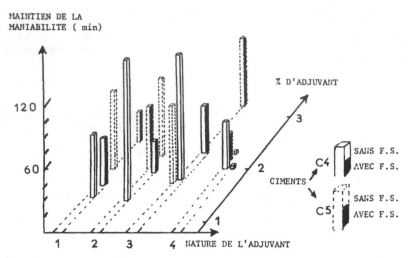

FIG 5 . MAINTIEN DE LA MANIABILITE EN FONCTION DE LA NATURE DE L'ADJUVANT ET DU CIMENT

L'examen de la figure 5 permet de dégager les remarques suivantes :

. Sans fumées de silice

Du point de vue du temps de maintien de la maniabilité, le ciment n° 4 à forte teneur en C_3A et les adjuvants n° 2 et 3 semblent les plus performants (fig. 5) 135 et 120 minutes respectivement. Ces valeurs sont obtenues pour les dosages les plus faibles des quatre adjuvants (fig. 5).

. Avec fumées de silice les résultats semblent inversés

En effet, les plus longs maintiens de la maniabilité sont constatés avec le ciment n° 5 à faible teneur en C_3A. La valeur la plus élevée (75 min) a été obtenue avec l'adjuvant n° 3 mais à un dosage très élevé 3,2 % (fig. 5). L'adjuvant n° 2 arrive 15 min après avec des dosages plus faibles : 1, 86 à 2,33. Il est à noter qu'avec ce ciment même l'adjuvant n° 4 a une bonne maniabilité avec un maintien de 30 min mais à un dosage de 2,18 %.

Le ciment à forte teneur en C_3A (n° 4) est moins performant en présence de fumée de silice. A signaler toutefois que son emploi avec l'adjuvant n° 1 permet un maintien de la maniabilité de 45 min et ceci pour les dosages en adjuvant les plus faibles : 1,6 et 1,8 % (fig. 5).

Que ce soit avec des ciments à faible ou forte teneur en C_3A, en présence de fumées de silice, les produits à base de résine mélamine semblent plus efficaces que les produits

à base de naphtalène sulfonates : maintien de la maniabilité satisfaisant pour de plus faibles dosages.

b) Deuxième remarque

L'étude de l'évolution de la maniabilité a fait apparaître
un point essentiel dans l'efficacité des superplastifiants:
c'est l'existence de 3 plages de dosages (figures 6 et 7).

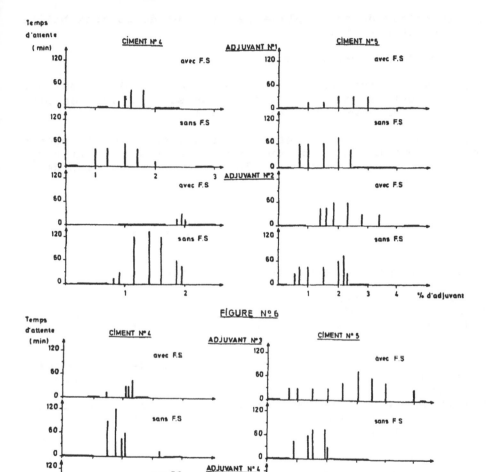

FIGURE N°6

FIGURE N°7 MAINTIEN DE LA MANIABILITÉ EN FONCTION DU DOSAGE EN ADJUVANT.

La plage de faibles dosages

Les pourcentages en adjuvant sont généralement inférieurs ou égaux à 1 % d'extrait sec, soit < 5 % de produit commercialisé. Cette plage prolonge légèrement les dosages fixés par le producteur, en accord avec les textes normatifs.

A l'intérieur de cette plage le maintien de la maniabilité est inférieur à 15 min.

La plage de dosages élevés

Ils peuvent atteindre des valeurs de 3 % et plus d'extrait sec, surtout en présence de fumée de silice, soit en produit commercialisé d'un ordre de 15 %.

C'est à l'intérieur de cette plage que se situe le phénomène de refluidification du mortier en fonction du temps d'attente et les valeurs les plus élevées de maintien de la maniabilité.

La plage de dosage très élevés supérieurs à 3 %

On observe des pertes de maniabilité pratiquement immédiates.

4.1.2 Superplastifiants avec effet retard

En absence de fumées de silice (fig. 8), on constate aussi l'existence des mêmes phénomènes précédemment cités : refluidification ou absence de maintien de la maniabilité selon les dosages.

Toutefois, il faut noter, lorsque l'on augmente le dosage pour obtenir la maniabilité fixée 15 ± 3 s et son maintien dans le temps, qu'un accroissement non négligeable de la teneur en eau est nécessaire alors qu'avec les adjuvants précédents l'augmentation du dosage était accompagnée d'une diminution du E/C.

Il y a par ailleurs (fig. 8) un déplacement des dosages vers les fortes valeurs (> à 2 %) ce qui, du fait de l'effet retard, prolonge de plus en plus le maintien de la maniabilité. Reste à connaître l'incidence sur le temps de prise et les résistances aux jeunes âges.

Avec fumées de silice, le comportement est similaire mais les dosages sont beaucoup plus élevés (> à 3 %) (fig. 9).

Une étude plus fine de rapport eau/ciment a fait apparaître de façon plus détaillée ce besoin en eau lié à l'augmentation du dosage. Ainsi, pour un pourcentage de 3,89 la maniabilité initiale de 15 secondes ne peut être

obtenue qu'avec 32 % d'eau alors que pour 1,75 % d'adjuvant
le E/C était de 0,28. Pour 4,67 %, le E/C atteint 0,34.

MANIABILITÉ EN FONCTION DU TEMPS
D'ATTENTE ET DES DOSAGES EN SUPERPLASTIFIANT
CIMENT Nº 4 ET ADJUVANT Nº 5

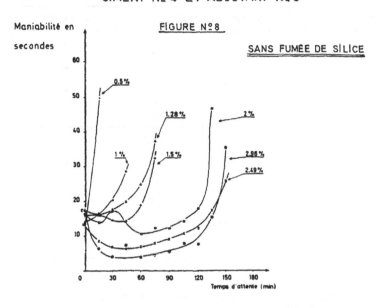

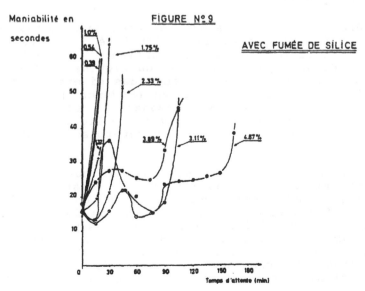

Enfin, pour terminer, il faut signaler que l'emploi de cet adjuvant est accompagné d'un entraînement d'air notable qui croît avec le pourcentage de superplastifiant. Ce phénomène est aussi constaté avec les fumées de silice (tableau 3).

TABLEAU 3 . POURCENTAGE D'AIR OCCLUS
ciment n° 4 - superplastifiant n° 5

TYPE DE MORTIER	% DE SUPERPLASTIFIANT	E/C	% D'AIR OCCLUS
SANS FUMEE DE SILICE	0	0,485	6
	0,5	0,387	7,5
	1	0,339	8
	1,5	0,332	11
	2	0,334	16,7
	2,5	0,368	36
	3	0,388	40
AVEC FUMEE DE SILICE	0	0,6	5,7
	0,39	0,492	6
	0,54	0,42	6,7
	1	0,318	6,8
	1,32	0,302	7,1
	1,75	0,282	7,6
	2,33	0,288	9
	3,11	0,302	18
	3,89	0,321	24
	4,67	0,342	32

4.2 Phénomènes plysico-chimiques accompagnant les plages de maniabilité

Une première approche de l'étude des phénomènes physicochimiques associés aux plages de maniabilité a été faite en mesurant l'adsorption de superplastifiants en fonction du temps et du dosage, par extraction du liquide interstitiel des divers mortiers étudiés. L'étude des divers isothermes d'adsorption a permis de mettre en évidence que, dans le cas des plages à faibles dosages, l'adsorption de l'adjuvant croît en fonction du temps. Le liquide interstitiel s'appauvrit de plus en plus en superplastifiant (fig. 10 a).

Dans le cas des plages à dosages élevés, correspondant à d'importants maintiens de maniabilité, on constate, pendant les premières minutes une adsorption puis, dans la période où il y a refluidification, une désorption de l'adjuvant. La solution interstitielle s'appauvrit en superplastifiant puis s'enrichit pour s'appauvrir de nouveau, alors que se produit le raidissement du mortier (fig. 10 b). Dans cette dernière phase où commence à intervenir la prise, le phénomène est plus complexe pour être appelé réadsorption.

POURCENTAGES D'ADSORTION
D'ADJUVANT EN FONCTION DU TEMPS.

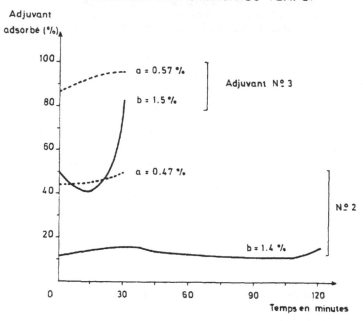

L'étude approfondie de ce phénomène d'adsorption-désorption est en cours, mais il semblerait qu'on soit en présence, dans les plages de refluidification, d'un déplacement de l'équilibre d'adsorption de la partie anionique des super-plastifiants.

Conclusions

Les formulations de bétons à hautes et très hautes performances conduisent à la recherche de la diminution la plus optimale de la teneur en eau, la tendance ou la tentation est donc de plus en plus grande d'augmenter les dosages en superplastifiants-hauts réducteurs d'eau afin d'abaisser au maximum le rapport eau/ciment des bétons.

Les résultats de l'étude du maintien de la maniabilité, montrent que la gamme de dosages en superplastifiants se partage de façon très nette en trois plages.

Une plage de faible pourcentages en adjuvants < à 1 % d'extrait sec, soit environ 5 % de produit commercialisé.

A l'intérieur de cette plage, le maintien de la maniabilité est faible, sa durée maximum étant de 15 minutes.

La méthode d'introduction fractionnée permet, en choisissant judicieusement, le couple superplastifiant-ciment d'obtenir des maniabilités instantanées élevées et pouvant se maintenir de façon satisfaisante dans le temps.

Les ciments à faible teneur en C_3A semblent donner les meilleurs résultats. A l'intérieur de cette plage on constate une adsorption progressive du superplastifiant.

Une plage de pourcentages élevés en adjuvants. Elle concerne des dosages en superplastifiants supérieurs à 1 % et pouvant atteindre jusqu'à 3 % et plus, en extrait sec, surtout en présence de fumées de silice. Ces pourcentages sont de l'ordre de 5 à 15 % de produit commercialisé.

On constate un phénomène de refluidification du matériau hydraulique en fonction du temps. Cette refluidification passe par un dosage optimal en adjuvant à partir duquel la durée de maniabilité décroît, elle est accompagnée d'une désorption de superplastifiant ou encore d'une augmentation de la concentration de ce produit dans la solution interstitielle du mortier.

Une plage de pourcentages très élevés en adjuvant. Les valeurs sont supérieures à 3 ou 4 % d'extrait sec. Dans cette plage, le maintien de la maniabilité est très faible.

Dans les limites de notre expérimentation, l'utilisation de superplastifiant à effet retard dans une plage de dosage autre que cette qui est normalisée, est délicate et peut conduire à des comportements contraires au but recherché.

L'augmentation des teneurs en eau et d'air entraîné impose une grande prudence dans le choix des pourcentages et de la nature de ces superplastifiants.

Bibliographie

Ramachandran V.S., 1984, Concrete Admixtures Handbook Properties, Science and Technology, Noyes Publications, N.J. USA

Paillère A.M., Briquet Ph., 1980, Influence of fluidifying synthetic resins on the rheology and deformation of cement pastes before and during the setting. Proceedings, International Congress on the chemistry of cement, Paris, V. 3, pp. 186-191

Paillère A.M., Buil M., Serrano J.J. de Larrard F., 1987, High performances concretes by optimization of the use of superplasticizers and steel fibres, International Association for Bridge and Structural Engineering, Symposium Paris-Versailles, September 2-4, 1987, Proceedings pp. 87-92

Buil M., Paillère A.M., Musikas N., 1989, Effect of Superplasticizers on the pozzolanic reactivity of silica fumes. Third International Conference on fly ash, silica fime pozzolans, Trondheim supplementary papers

Costa U., Manazza F., Barrilà A., 1982, Adsorption of superplasticizers on C_3S : changes in zeta potential and rheology pastes, Il Cemento 4/1982, pp. 323-336

Buil. M. et al., 1986, Plysicochemical Mechanism of the action of the naphtalene sulfonate based superplasticizers on silica fume concretes, ACI SP-91-46, pp. 959-971

INFLUENCE DES FLUIDIFIANTS SUR LES CARACTERISTIQUES RHEOLOGIQUES DES PATES DE CIMENTS FILLERISES

(Influence of plasticizers on the rheological characteristics of cement pastes containing fillers)

R. SAADA, M. BARRIOULET and C. LEGRAND
Laboratoire Matériaux et Durabilité des Constructions, I.N.S.A.-U.P.S. de Toulouse, France

Résumé
Le but de l'étude est de relier le comportement rhéologique des pâtes de ciment à leur structure. Pour ce faire, les méthodes expérimentales utilisées sont la viscosimétrie en cylindres coaxiaux pour la détermination du seuil de cisaillement et de la viscosité apparente des pâtes, des mesures sédimentométriques et des observations au Microscope Electronique à Balayage pour l'analyse de leur structure.
L'étude montre que les paramètres rhéologiques varient essentiellement en fonction de la surface spécifique de la phase solide et de l'état floculent de la pâte. Ce dernier peut être caractérisé à partir de mesures sédimentométriques. Le meilleur couplage filler-fluidifiant, d'un point de vue fluidification, en découle.
Mots clefs : Rhéologie, Ciments fillérisés, Fluidifiants, Sédimentométrie.

1 Introduction

L'arrangement géométrique des éléments solides d'un béton avant hydratation conditionne pour une grande part la microstructure du matériau durci et donc ses propriétés mécaniques et ses qualités de durabilité. Il dépend, pour un moyen de mise en place donné, de la façon dont est mis en oeuvre le matériau frais, donc de ses propriétés rhéologiques.

Par ailleurs, les éléments fins de la phase granulaire du béton, c'est à dire le ciment et les sables les plus fins ($< 100 \mu m$) présentent la plus grande partie de la surface solide en contact avec la phase liquide. Cette surface est en outre chargée électriquement. De ce fait, l'ensemble de la partie fine de la phase solide et de la phase liquide, désignée généralement par le terme de pâte interstitielle, joue un rôle fondamental dans la mise en oeuvre du béton frais.

Le comportement rhéologique de pâtes de ciment traditionnelles c'est à dire composées uniquement de ciment et d'eau a été largement étudié [1,2,3,4]. Il est conditionné par la structure plus ou moins floculée de la pâte. La difficulté expérimentale qu'il y a à réaliser l'étude directe de la structure de la pâte a conduit différents auteurs [1,2,5] à proposer des modèles basés sur des hypothèses issues des connaissances physico-chimiques et mécaniques des différents éléments constitutifs de la pâte.

Aujourd'hui l'utilisation quasi systématique d'ajouts minéraux et d'ajouts organiques dans la confection des bétons génère des modifications de comportement rhéologique qui ont essentiellement été étudiées en termes d'effet fluidifiant ou réducteur d'eau et de variation de ces propriétés dans le temps |6,7|. Par ailleurs, en amont, les mécanismes chimiques, notamment d'adsorption des fluidifiants qui conduisent à la fluidification des pâtes ont fait l'objet de nombreuses recherches depuis une dizaine d'années, le plus souvent sur des matériaux modèles (C_3S, CSH, $CaCO_3$...).

Notre étude consiste pour l'essentiel à relier les caractéristiques rhéologiques des pâtes de ciment à leur structure et se situe dans le domaine peu abordé séparant ces deux types de recherche.

Nous présentons dans cette communication les premiers résultats relatifs à des pâtes de ciments fillérisés ou non, avec ou sans fluidifiant. Le choix de travailler sur de telles compositions répond au double objectif d'étudier des pâtes de plus en plus utilisées dans la pratique et de diversifier les structures de pâtes à analyser, de façon à mettre en évidence les paramètres essentiels qui interviennent dans les mécanismes physiques qui régissent la mise en oeuvre des pâtes, mortiers et bétons.

L'étude rhéologique est effectuée à partir de mesures viscosimétriques réalisées dans un viscosimètre à cylindres coaxiaux.

L'obtention expérimentale d'informations précises permettant de décrire la structure de la pâte est délicate car d'une part le milieu pâteux se prête mal à une observation microscopique directe et d'autre part l'extrapolation à des suspensions concentrées telles que les pâtes de ciment, de mesures de type électrochimique (mobilité électrophorétique ou conductimétrie) susceptibles d'apporter des informations sur l'état électrique du milieu mais réalisables exclusivement sur des suspensions très diluées n'est pas satisfaisante.

Nous avons donc choisi de caractériser la structure des pâtes à partir de mesures de sédimentométrie réalisées sur des pâtes à concentrations en solide réelles, complétées d'observations au Microscope Electronique à Balayage d'échantillons traités par cryosublimation.

2 Méthodes expérimentales

2.1 Composition des mélanges

a) Ciments fillérisés

Les ciments fillérisés ont été fabriqués en laboratoire en mélangeant en proportions variables un ciment portland de classe haute performance dont les caractéristiques physico-chimiques sont données dans le tableau 1 à des fillers industriels calcaires ou siliceux chimiquement très purs dont les caractéristiques figurent dans le tableau 2.

Les mélanges obtenus sont caractérisés par le pourcentage pondéral de filler substituéau ciment dans une fourchette de 0 % à 30 %.

Afin d'étudier l'influence du paramètre surface spécifique, nous avons broyé le ciment de base de façon à obtenir de nouveaux ciments plus fins (figure 6).

Dénomination	masse volumique kg/m3	Surface spécifique du ciment de base m2/kg
CPA HP	3 100	337

CPA HP	SiO_2	Al_2O_3	Fe_2O_3	CaO	MgO	Na_2O	K_2O	SO_3	TiO_2	M_nO	Perte au feu
%	20	5.65	2.5	65	0.75	0.1	0.4	3.2	0.3	0.05	2

CPA HP	$CaSO_4$	C_4AF	C_3A	C_3S	C_2S
%	6.88	7.6	10.75	62.20	10.46

Tableau 1. Caractéristiques physico-chimiques du ciment

	fillers calcaires				fillers siliceux				
	c_1	c_2	c_3	c_4	s_1	s_2	s_3	s_4	s_5
surface spécifique moyenne (Blaine) m2/kg	300	650	1400	2000	250	300	450	600	1800
composition chimique	$CaCO_3$ supérieur à 99,75 %				SiO_2 supérieur à 99,6 %				
masse volumique kg/m3	2 700				2 650				

Tableau 2. Caractéristiques physico-chimiques des fillers

b) Fluidifiants

Deux produits ont été utilisés :
- un fluidifiant entrant dans la catégorie des condensats de mélamine sulfonée et de formaldéhyde,
- un fluidifiant du type condensat de naphtalène sulfoné et de formaldéhyde.

Tous deux se présentent sous forme liquide ; ils ont été introduits dans l'eau de gâchage et dosés à 0,8 % du poids de la phase solide des pâtes.

c) Mode opératoire
Le ciment et le filler ont été mélangés à sec dans un malaxeur de laboratoire pendant 3 mn. La durée du malaxage avec l'eau de gâchage
était de 3 mn. La teneur en eau des pâtes est caractérisée par le rapport E/C dans lequel E représente le poids de la phase liquide (eau
de gâchage + eau de dissolution du fluidifiant éventuellement) et C
le poids de la phase solide (ciment + filler éventuellement).

2.2 Mesures rhéologiques
L'étude rhéologique a été conduite à partir de mesures viscosimétriques réalisées dans un viscosimètre à cylindres coaxiaux en milieu
infini, 5 mn après le début de malaxage des phases solide et liquide.
   Nous caractériserons rhéologiquement les pâtes par le seuil de cisaillement de la pâte vierge τ_0 et par le coefficient de viscosité
apparente correspondant à un gradient de vitesse de $10s^{-1}$, déduit de
la courbe d'écoulement de la pâte pour laquelle le seuil de cisaille-
ment a été supprimé par vibrations |3|.

2.3 Mesures sédimentométriques
Nous présentons dans cette communication les résultats de mesures de
sédimentométrie simples réalisées sur les différents fillers mélangés
à un volume d'eau avec ou sans fluidifiant correspondant aux valeurs
des E/C des pâtes soumises à l'étude rhéologique. Ces mélanges sont
placés dans des tubes cylindriques à section circulaire de 30 mm de
diamètre et de 60 mm de hauteur. La hauteur de sédiment formé dans le
tube est mesurée en fonction du temps. Les courbes de la figure 1

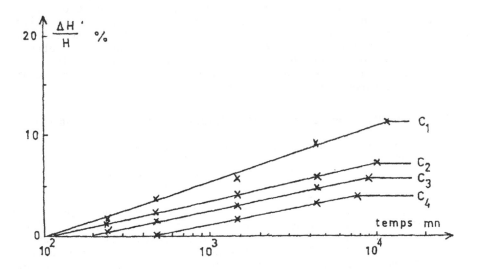

Fig. 1. Variation du tassement relatif du filler calcaire en fonction
du temps. E/C = 0,325 sans fluidifiant.

représentent la variation de tassement relatif de l'échantillon en fonction du temps qui peut être caractérisée par trois paramètres : le temps t_i d'initialisation de la sédimentation, la vitesse de sédimentation et le tassement relatif final $(\Delta H/H)_f$. Plus la structure de la pâte est floculée, plus le volume de sédiment est important, plus $(\Delta H/H)_f$ est faible et plus la vitesse de sédimentation est élevée.

Nous avons pu observer sur l'ensemble des mesures effectuées qu'il existe une bonne corrélation entre ces trois paramètres et nous avons choisi le tassement final $(\Delta H/H)_f$ comme unique paramètre caractéristique du degré de floculation de la pâte.

2.4 Observation au M.E.B.
L'observation en microscopie électronique à balayage de l'arrangement initial des grains anhydres de ciment et de filler dans la pâte est effectuée sur des échantillons soumis à une cryosublimation suivant une méthode mise au point dans notre laboratoire |8|. Nous nous bornerons à en rappeler le principe.

La congélation instantanée de l'eau de la pâte fraîche, réalisée en plongeant celle-ci dans l'azote liquide, permet d'éviter l'hydratation des grains de ciment. L'eau congelée subit ensuite une sublimation dans un lyophilisateur. L'échantillon lyophilisé est consolidé par imprégnation à la pression atmosphérique d'une résine très fluide. On réalise alors des sections d'éprouvettes qui sont polies à sec sur papiers abrasifs, puis métallisées au carbone.

L'observation s'effectue au microscope électronique à balayage en mode électrons rétrodiffusés, polarité inversée qui permet de différencier les constituants suivant leur numéro atomique : le ciment apparaît en noir, les fillers en gris, la résine en lieu et place de la phase fluide en blanc. On se rend compte que le filler se distingue d'autant plus du ciment que l'écart de masse volumique ciment-filler est important, soit dans notre étude pour le ciment fillérisé siliceux.

Les reproductions d'observations présentées nous permettront d'étayer qualitativement les interprétations des autres résultats expérimentaux obtenus, notamment pour traduire l'effet des deux fluidifiants sur le ciment fillérisé siliceux.

3 Relation entre caractéristiques rhéologiques et structure des pâtes

3.1 Influence de la surface spécifique de la phase solide
Les courbes des figures 2 et 3 montrent l'évolution, respectivement du seuil de cisaillement et de la viscosité lorsque varie la quantité de fillers calcaires substitués au ciment étudié pour des pâtes gâchées à E/C = cte = 0,325.

Une augmentation de la quantité de filler substitué entraîne une augmentation des valeurs des paramètres rhéologiques pour les fillers présentant une surface spécifique supérieure à celle du ciment et une diminution des valeurs de τ_0 et ν pour le filler de surface spécifique inférieure à celle du ciment. Nous avons retrouvé ce résultat, d'une part avec le filler siliceux, d'autre part lorsqu'on incorpore un fluidifiant (figures 4 et 5), confirmant que la surface spécifique de la phase solide est un paramètre fondamental dans l'évolution des caractéristiques rhéologiques du matériau.

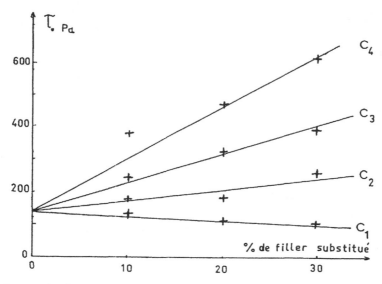

Fig. 2. Variation du seuil de cisaillement en fonction du pourcentage de filler calcaire substitué au ciment. E/C = 0,325 sans fluidifiant.

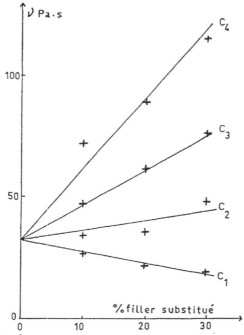

Fig. 3. Variation du coefficient de viscosité apparente en fonction du pourcentage de filler calcaire substitué au ciment. E/C = 0,325 sans fluidifiant.

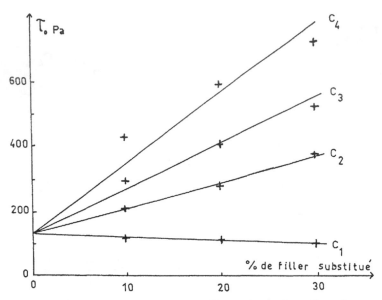

Fig. 4. Variation du seuil de cisaillement en fonction du pourcentage de filler calcaire substitué au ciment. E/C = 0,275 fluidifiant mélamine

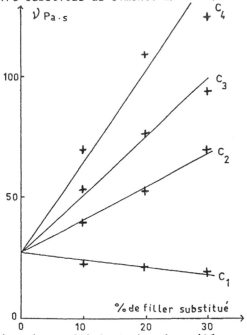

Fig. 5. Variation du coefficient de viscosité apparente en fonction du pourcentage de filler calcaire substitué au ciment. E/C = 0,275 fluidifiant mélamine.

A partir des courbes des figures 2 et 3 et de leurs homologues correspondant aux fillers siliceux, nous avons reporté sur les figures 6
et 7 les valeurs des paramètres rhéologiques en fonction de la surface
spécifique de la phase solide des différentes pâtes de ciments filléri-

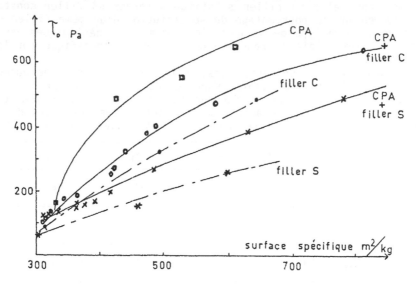

Fig. 6. Variation du seuil de cisaillement en fonction de la surface
spécifique de pâtes non fluidifiées - E/C = 0,325.

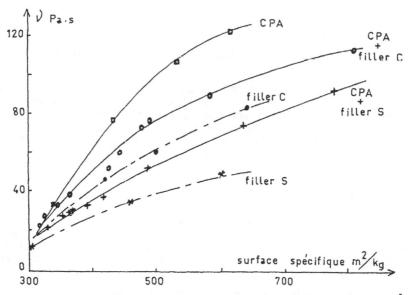

Fig. 7. Variation du coefficient de viscosité apparente ($\dot{\varepsilon} = 10s^{-1}$) en
fonction de la surface spécifique de pâtes non fluidifiées. E/C = 0,325

sés étudiés. On observe que pour chaque type de phase solide, indépendamment de la façon dont ont été composées les pâtes il existe une relation simple entre paramètres rhéologiques et surface spécifique.

Autrement dit, que l'on augmente la surface spécifique du mélange
en augmentant celle du filler substitué à teneur en filler constante
ou en augmentant le pourcentage de substitution d'un même filler plus
fin, pour un même type de mélange, les paramètres rhéologiques varient
suivant les mêmes lois en fonction de la surface spécifique totale de
la phase solide.

A dosage en eau constant, lorsque la surface spécifique augmente
il y a bien sûr neutralisation d'une quantité plus importante d'eau
résultant d'une plus grande surface de la phase solide et en même temps
une floculation plus importante comme le montrent les résultats des mesures sédimentométriques effectuées sur les fillers (tableau 3). Il

		c_1	c_2	c_3	c_4	s_1	s_2	s_3	s_4
	sans fluidifiant	11,4	7,6	6	4,4	14,5	12	8,5	6,5
E/C = 0,325	0,8 % mélamine	18	13,5	11	6	25,4	21,2	14,5	12,7
	0,8 % naphtalène	14,4	9,3	6,7	5	16	13,5	10,1	7,6
E/C = 0,275	0,8 % mélamine	15	8,5	6,8	3,4	23	19,5	14	10
	0,8 % naphtalène	8,5	6,8	4,2	3,3	10	8	6	4

Tableau 3. Tassement relatif final des fillers : $(\Delta H/H)_f$ en %

semble donc que, dans la gamme de substitution étudiée (< 30 %), la
diminution de "l'eau libre" due à l'augmentation de surface soit le
phénomène prépondérant influant sur la variation des paramètres rhéologiques lorsqu'il s'agit d'une même nature de mélange (ciment + filler
calcaire ou ciment + filler siliceux). Par contre, les différences
observées à surface spécifique constante pour les différents types de
pâtes (ciment, filler calcaire, filler siliceux, ciment + calcaire,
ciment + silice) résultent d'un état de floculation différent. C'est
ce que tendent à montrer les résultats de mesures sédimentométriques
effectuées sur les fillers calcaires et siliceux (tableau 3). On observe systématiquement un tassement plus important des fillers siliceux
qui en milieu aqueux présentent en surface des sites chargés électriquement négativement, ce qui induit des répulsions électrostatiques et
confère aux pâtes une structure peu floculée. Par contre les charges
de surface du filler calcaire dépendent de la concentration en ions
Ca^{2+} et Co_3^{2-} en solution. Le calcaire est moins électronégatif que le
filler siliceux ; Les forces d'attraction électriques induisent une

structure plus floculée.

Remarque
L'observation des courbes des figures 2 à 7 montre que le seuil de cisaillement et la viscosité des pâtes étudiées varient suivant des lois
similaires en fonction de la surface spécifique. Dans le but d'alléger
la présentation de cette communication nous ne donnerons par la suite
que les courbes relatives au seuil de cisaillement, celles représen-
tant la variation de la viscosité étant similaires.

3.2 Influence de l'état de floculation de la pâte
Les fluidifiants introduits lors du gâchage ont pour effet de modifier
l'état de floculation des pâtes.
    Les courbes des figures 8 et 9 montrent que l'allure de la varia-
tion du seuil de cisaillement en fonction de la surface spécifique
reste inchangée par rapport à celle obtenue pour une pâte sans fluidifiant. On constate une diminution de l'effet fluidifiant lorsque la
surface spécifique augmente, résultat observé de façon générale dans
la plupart des études réalisées sur ce sujet.
    Par ailleurs, on note pour chaque ciment fillérisé, un effet flui-
difiant plus important lorsqu'on utilise le fluidifiant à base de mélamine. Les mesures de sédimentométrie (tableau 3) indiquent que ce
fluidifiant induit une défloculation plus importante que celle obtenue
avec le fluidifiant à base de naphtalène. Les trois reproductions présentées illustrent l'observation effectuée au M.E.B. de la fluidification d'une même pâte de ciment fillérisé siliceux par les deux types
de fluidifiants.
    On note d'une façon générale une bonne répartition du filler dans
la masse de la pâte. Ces trois observations correspondent à trois pâtes présentant respectivement des seuils de cisaillement de 268 Pa,
541 Pa, 227 Pa.
    A teneur en eau égale l'effet fluidifiant plus prononcé du produit
à base de mélamine se traduit par une structure plus lâche (reproductions 2 et 3). Par ailleurs, bien que gâchée avec une quantité d'eau
plus importante, la structure de la pâte sans fluidifiant apparaît à
peu près identique à celle gâchée avec le produit à base de naphtalène.

    Les courbes de la figure 10 montrent l'influence du fluidifiant
à base de mélamine sur chacun des deux ciments fillérisés. On observe
que ce fluidifiant présente une plus grande efficacité sur le ciment
fillérisé siliceux que sur le ciment fillérisé calcaire. En effet, en
milieu aqueux la molécule de mélamine présente par rapport à celle de
naphtalène un supplément de charges positives dû à la libération de
groupements hydroxyles. L'adsorption des molécules de mélamine se
trouve de ce fait très importante sur les très nombreux sites négatifs
des fillers siliceux.
    Il est par ailleurs remarquable que les différentes courbes de la
figure 10 correspondant à des E/C différents et à des mélanges de nature différente avec ou sans fluidifiant se positionnent sur l'échelle
des τ_o dans le même ordre que les valeurs du tassement ($\Delta H/H)_f$ cor-
respondants (tableau 3).

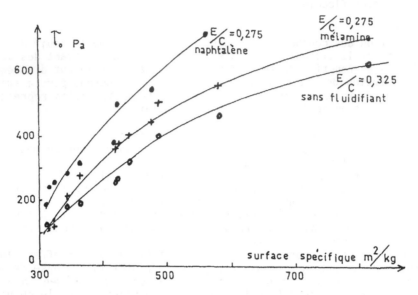

Fig. 8. Influence de la nature du fluidifiant sur le seuil de cisaillement des pâtes de ciment fillérisé calcaire.

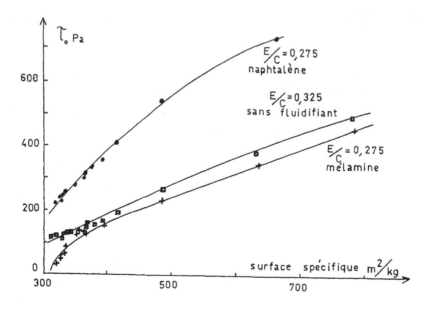

Fig. 9. Influence de la nature du fluidifiant sur le seuil de cisaillement des pâtes de ciment fillérisé siliceux.

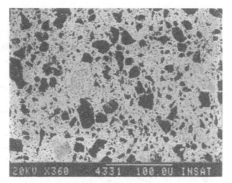

Repr. 1. E/C = 0,325
sans fluidifiant τ_0 = 268 Pa ; ν = 52 Pa.s

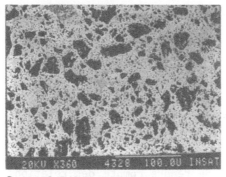

Repr. 2 E/C = 0,275
fluidifiant naphtalène τ_0 = 541 Pa ; ν = 98 Pa.s

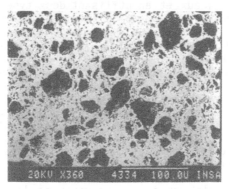

Repr. 3 E/C = 0,275
fluidifiant mélamine τ_0 = 227 Pa ; ν = 41 Pa.s

Reproductions d'observations au MEB correspondant à une pâte composée de ciment + 10 % de filler S_5 (surface spécifique 485 m^2/kg)

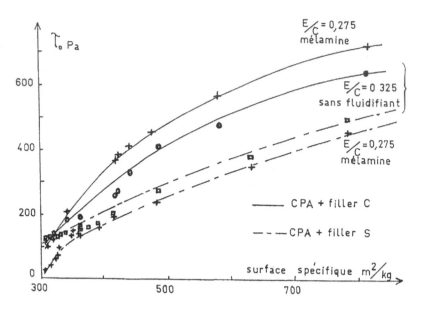

Fig. 10. Influence du fluidifiant à base de mélamine sur le seuil de cisaillement de pâtes de ciment fillérisé calcaire et siliceux.

4. Conclusions

Notre objectif était de rendre compte des relations existant entre le comportement rhéologique des pâtes et leur structure. Nous avons diversifié les pâtes en travaillant sur des ciments fillérisés avec des fillers calcaires et siliceux et en utilisant deux types de fluidifiants.

Etant conscients que l'observation réalisée dans le plan d'une section d'éprouvette ne constitue pas une méthode idéale pour rendre compte d'une structure tridimensionnelle floculée, nous avons utilisé, faute de mieux, des observations au M.E.B. aux fins de confirmation qualitative de résultats de mesures sédimentométriques qui paraissent bien appropriées pour caractériser l'état de floculation des pâtes.

Nous avons montré que les paramètres rhéologiques, seuil de cisaillement et viscosité varient dans le même sens en fonction de la surface spécifique de la phase solide et de l'état floculent du milieu. Nous avons observé que le fluidifiant à base de mélamine avait un effet fluidifiant plus prononcé sur les ciments fillérisés siliceux que sur les ciments fillérisés calcaires étudiés et était, à même dosage, plus efficace que le fluidifiant à base de naphtalène.

Le degré de floculation de la structure de la pâte dépend essentiellement de la nature et de la densité de charges électriques se trouvant soit originellement, soit par adsorption d'ions ou polymères en surface de grains dans un milieu aqueux à PH donné. Il convient donc, dans la pratique, d'être attentif aux compatibilités de couplage entre fluidifiantet phase solide, ciment et fillers éventuellement, ces der-

niers étant généralement imposés par l'environnement local, le choix
du fluidifiant à utiliser revêt une grande importance.

Références

|1| Powers, T.C. (1968) The properties of fresch concrete. Wiley,
New York.

|2| Bombled, J.P. (1973) Rhéologie des mortiers et des bétons frais,
influence du facteur ciment. Proceeding of RILEM, Leeds, Seminar.

|3| Legrand, C. (1972) Contribution à l'étude de la rhéologie du béton
frais, 1ère et 2ème partie. Matériaux et Constructions, 5, 275-295,
379-393.

|4| Barrioulet, M. et Legrand, C. (1977) Influence de la pâte intersti-
tielle sur l'aptitude à l'écoulement du béton frais, rôle joué par
l'eau retenue par les granulats. Matériaux et Constructions, 60,
365-373.

|5| Legrand, C. (1981) Etat floculent des pâtes de ciment avant prise
et ses conséquences sur le comportement rhéologique. Groupe Français de Rhéologie, Paris 1980. Editions anciens élèves E.N.P.C.,
129-136.

|6| Asagak, Roy, D.M. (1980) Propriétés rhéologiques des pâtes de ci-
ment : effects of superplasticizers on viscosity and yield stress.
Cement and Concrete Research, USA, vol. 10, 2, 287-295.

|7| Ramachandran, V.S. (1986) Les interactions d'adjuvants et d'addi-
tions dans le système ciment-eau. Il Cemento, Italie, vol. 83, 1,
13-38.

|8| Escadeillas, G. et Maso, J.C. (1990) Approach of the initial state
in cement paste mortar and concrete. Communication à la conférence
sur Advances in cementitious matérials, Gaithersburg 23-26 juin.

SLUMP CONTROL AND PROPERTIES OF CONCRETE WITH A NEW SUPERPLASTICIZER. II: HIGH STRENGTH *IN-SITU* CONCRETE WORK AT HIKARIGA-OKA HOUSING PROJECT

C. YAMAKAWA
Construction Chemical Business, Kao Corporation, Japan
K. KISHITANI
Department of Architecture, Nihon University, Japan
I. FUKUSHI
Research Institute of Housing and Urban Development, Housing and Urban
Development Corporation, Japan
K. KUROHA
Technical Research Institute, Taisei Corporation, Japan

ABSTRACT
In the Japanese construction field, there has been an increasing
demand for a ready-mixed concrete having a higher slump value and
ensuring high strength. One method of increasing the strength of
concrete is to use high range water reducing agents. However, high
range water reducing agents greatly reduce the slump value during
delivery and, therefore, are not suitable for practical application.
To solve this problem, a slump retentive high range water reducing
agent (hereafter referred to as "new superplasticizer") was developed.
The new superplasticizer has both high water reducing ability and
slump-value retaining ability provided by the slow-release of a
component reactive polymer. The present study confirmed the physical
properties of high strength concrete using a ready-mixed concrete of
high slump value prepared by adding the above admixture in an on-site
experiment.

Key words: high strength concrete, high-rise RC structure building,
slump retentive high range water reducing agent, new superplasticizer

1 Introduction

Conventionally, the design of RC buildings has been limited to six- or
seven-story buildings. However, recent developments in earthquake-
resistant designs have enabled planning and construction of high-rise
buildings of RC structure using high-strength concrete whose design
criteria for strength is 420 kgf/cm² or higher (mix designs: 500 to
600 kgf/cm²). In order to facilitate easy production, transportation,
and construction of such high-strength concrete, an admixture is
added, which ensures satisfactory workability and flowability for a
longer period at a lower water/cement ratio. By adding such an
admixture, high strength concrete can be handled in the same way as
usual concrete. Thus the admixture will increase the usage of high
strength concrete and further contribute to the spread of high-rise
RC-structure buildings.

Also, the use of high slump ready-mixed concrete will eliminate the need to use the flowing concrete method, which has the following disadvantages: quality control of flowing concrete is difficult; and, construction using flowing concrete causes noise at the worksite. In the present study, ready-mixed concrete, which uses the new superplasticizer that was developed to meet the above-mentioned requirements, was actually used in the construction of high-rise RC structure buildings; the physical properties of the high strength concrete were then determined. This is the first example in which the new superplasticizer has been actually used in concrete for a construction.

2 Test methods

2.1 Outline of the work
Table 1 shows the outline of the work and Fig. 1 shows the design criteria for strength. Construction was performed as follows: First, columns were cast with concrete up to the lower edge of the beams. A beam, which had been precast except for the upper section of the bar, and a slab were installed. After completion of the reinforcement, the panel zone and horizontal area were monolithically cast with concrete.

2.2 Outline of the test
Both when fresh and after hardening, the physical properties of the concrete were determined at mix designs of 540 kgf/cm², 430 kgf/cm², and 360 kgf/cm². Also, variation of quality was determined. Table 2 shows test parameters and test methods. Variation of quality and time-course changes of the slump value and air content were determined only in mix designs of 540 kgf/cm² and 430 kgf/cm². Variation in quality was evaluated based on the data obtained from tests repeated 49 times (seven tests per day over seven days). Each test parameter was determined using samples extracted from two agitator trucks on the same test day. Test results of each parameter were expressed as the mean value of the two determinations.

2.3 Materials and mix design
Table 3 lists materials used and Table 4 lists the mix designs. Concrete was prepared at mix design for required average strengths of 540 kgf/cm², 430 kgf/cm², and 360 kgf/cm², required slump of 18 cm, and air content of 4%. Compressive strength is that required on the 28th day.

2.4 Production and transportation of concrete
A batch of 2.5 m³ of concrete was mixed in a biaxial forced mixer. The new superplasticizer was weighed and added to the mixing water at the ready-mixed concrete plant. After mixing, two batches of concrete were loaded on an agitator truck and transported to the work site. Time for transportation was approximately 20 minutes.

Table 1. Outline of building

Name	GH Hikariga-oka B blok No.18
Owner	The Housing Corporation
Builder,Designer	Taisei Corporation
Use and Purpose	Apartment Building
Building area	1,676 m²
Total floor area	24,186 m²
Number of stories	25 stories with two basement floors
Height	Building height 75.0 m Maximum Building height 80.2 m
Structure	reinforced concrete

Table 2. Test Method and Parameters

Parameter	Method	Test age	Frequency (agitator truck)	Curing method
Slump	JIS A 1101		7 TRUCKS FOR EACH 7 DAYS	
Air content	JIS A 1128		"	
Compressive strength	JIS A 1108	28,91 days	"	Standard
Time course changes	Agitated in truck	90min.	2 TRUCKS FOR 1 DAYS	
Bleeding	JIS A 1123		"	
Setting time	ASTM C403		"	
Compressive strength	JIS A 1108	3,7,28,91 days	"	Standard
Young modulus	ASTM C469	28 days	"	"
Poisson ratio	ASTM C469	28 days	"	"
Drying shrinkage	JIS A 1129	until 26 days	"	"
Freez-thaw resistance	JIS A 6204	up to 300 cycles	"	"
Carbonation	30℃, 60%RH	until 6 months	"	"

Table 3. Materials

Cement	ordinary portland cement
Fine aggregate	sand s.g;2.56 absorption;1.71
Coarse aggregate	crushed stone s.g;2.68 absorption;0.85
Admixture	Mighty2000,2000H ; New superplasticizers

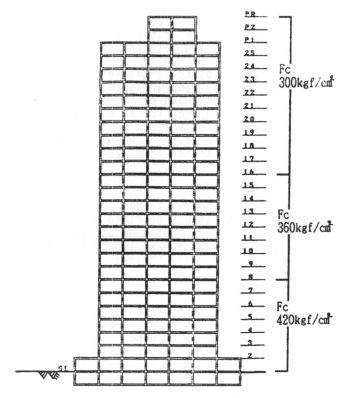

Fig. 1. Stories and design strength

Table 4. Mix design of concrete

Required compressive strength	W/C %	S/A %	Unit weight (Kg/m³)				Admixture and dosage
			Cement	Water	Sand	Gravel	
540 kgf/cm²	37.0	38.2	459	170	640	1077	Mighty 2000H C ×1.7%
430	44.5	41.7	382	170	725	1053	Mighty 2000H C ×1.6%
360	52.5	45.0	324	170	803	1021	Mighty 2000 C ×1.4%

3 Results

Tests of concrete of each mix design were performed according to the schedule of the work. Therefore, test conditions such as temperature and date differed with the mix design. Table 5 lists the results.

3.1 Physical property parameters of fresh concrete
3.1.1 Time-course change of slump value and air content
Although the water/cement ratios of each mix design were small, slump values were retained within 18 ± 2.5 cm for 90 minutes (Fig. 2). The results confirmed that the effect of the new superplasticizer, i.e. slow release of component reactive polymer, was achieved under actual working conditions. Air content tended to slightly lower as time passed.

3.2 Bleeding
As Fig. 3 shows, the amount of bleeding tended to decrease as the water/cement ratio lowered. The amount of bleeding of each mix design was 0.1 cm^3/cm^2 or less. In the mix design of 540 kgf/cm^2, the time required for the completion of bleeding was longer than those in other mix designs. This was because the test temperature was lower than those in the other tests.

3.3 Setting time
Fig. 4 shows the setting times of each mix design. For the same reason as in the bleeding test, the setting time tended to be longer in the mix design of 540 kgf/cm^2 than in other mix designs. The action mechanism of the new superplasticizer differs from that of the retarder; however, no abnormal setting delay was observed.

3.2 Physical property of hardened concrete
3.2.1 Compressive strength
As Table 5 shows, each mixing design achieved a respective required average strength on the 28th day: 600 kgf/cm^2 (mix design; 540 kgf/cm^2), 468 kgf/cm^2 (mix design; 430 kgf/cm^2), 397 kgf/cm^2 (mix design; 360 kgf/cm^2). The Young modulus was at a satisfactory level, and the Poisson ratio was approximately 0.21 in mix designs.

3.2.2 Drying shrinkage
As Fig. 5 shows, the drying shrinkage was approximately 6×10^{-4} on the 26th week in all mix designs. This is because the water content per unit was set at the same value in all mix designs.

3.2.3 Freeze/thaw resistance
As Fig. 6 shows, the relative dynamic modulus of elasticity after 300 cycles was approximately 90% in all mix designs.

3.2.4 Carbonation
As Fig. 7 shows, the rate of carbonation became slower as the water/cement ratio lowered. In the mix design of 540 kgf/cm^2, no carbonation was observed after six months.

Table 5.Test results of Measurement of Physical properties

Properties of fresh concrete					Properties of hardened concrete								
Mix Design kgf/cm²	Properties		Setting time (hr:min)	Bleeding	Compressive strength (kgf/cm²) days 3	7	28	91	Young modulus (kgf/cm²) 28th days	Poisson ratio 28th days	Drying shrinkage 26th weeks	Freeze&thaw resistance after 300cycles	Carbonation after 6 months
540	slump flow air content temperature of concrete temperature	16.5 cm 26.0 cm 3.4 % 16.0 ℃ 16.0 ℃	initial 8:03 final 10:45	amount 0.059cal/cm ratio 1.28 %	270	468	600	670	3.81 ×10⁵	0.22	5.91×10⁻⁴	89 %	0.0 mm
430	slump flow air content temperature of concrete temperature	19.0 30.0 4.5 26.0 23.5	initial 4:57 final 6:10	amount 0.062 ratio 1.41	259	363	468	509	3.22×10⁵	0.21	5.68×10⁻⁴	90 %	5.6 mm
360	slump flow air content temperature of concrete temperature	19.5 32.5 3.3 27.0 22.0	initial 5:50 final 7:07	amount 0.120 ratio 2.74	222	311	397	445	3.16×10⁵	0.21	6.42×10⁻⁴	91 %	8.2 mm

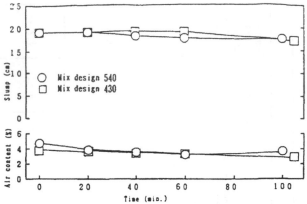

Fig. 2.Time-course changes in slump and air content

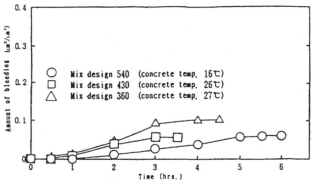

Fig. 3. Bleeding test

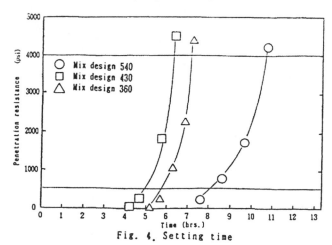

Fig. 4. Setting time

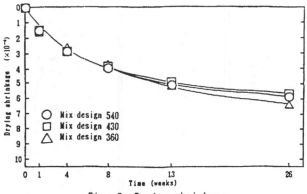

Fig. 5. Drying shrinkage

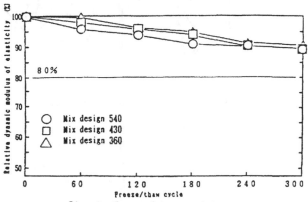

Fig. 6. Freeze/thaw resistance

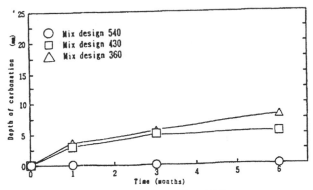

Fig. 7. Carbonation test

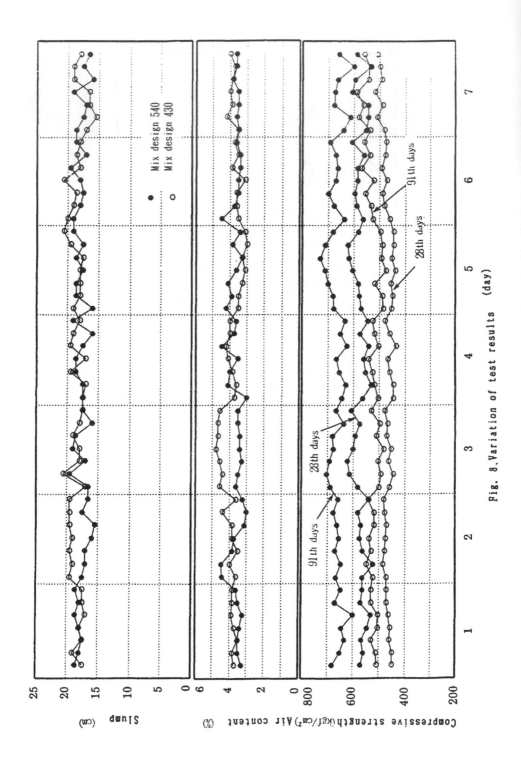

Fig. 8. Variation of test results (day)

Mix design 540
Mix design 430

Slump (cm)

Compressive strength(kgf/cm²) Air content (%)

28th days

91th days

91th days

28th days

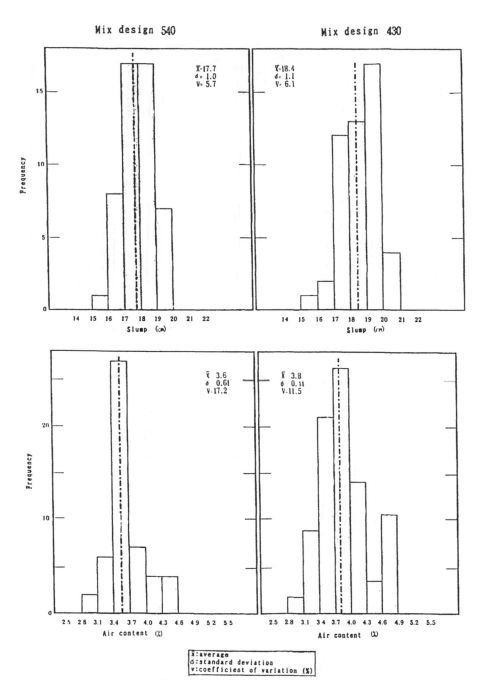

Fig. 9. Histogram of test results

103

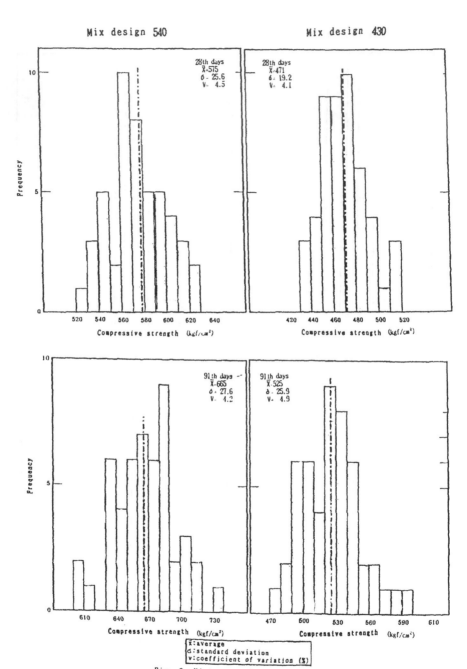

Fig. 9. Histogram of test results

3.3 Quality control

Fig. 8 shows the test results of each parameter on each test day. Fig. 9 shows a histogram of the test results.

3.3.1 Slump value and air content

Standard deviation of all measured slump values was approximately 1 cm, and 80% or more of the measured values were within 18 $\pm$ 1.5 cm, and all the measured values were within the acceptable range of 18 $\pm$ 2.5 cm. All measured values of air content were within the acceptable range of 4 $\pm$ 1%.

3.3.2 Compressive strength

Because of high strength concrete, mix design had been determined by assuming that variation of compressive strength was 10%. However, the results showed that the variation of compressive strength was approximately 4%.

4 Conclusion

(1) The effect of the new superplasticizer, i.e. slow release of component reactive polymer, was achieved in high slump ready-mixed concrete for high strength concrete used in actual work.

(2) High strength concrete prepared using the new superplasticizer was revealed to provide high durability and ensure the average required strength of each mix design. Variation in slump value, air content, and compressive strength was small. The new superplasticizer enables the handling of high strength concrete of high slump value in the same way as usual concrete. The new superplasticizer, which will eliminate the need to use the flowing concrete method, is expected to be widely used for the purpose of improving workability and strength.

Gratitude

We express our gratitude to Dr.Huminori Tomozawa,Professor of the department of technology in Tokyo University, Mr.Shigeyuki Nakajima,the Head of Hikariga-oka project of Taisei Corporation,and Mr,Kuniyuki Tomathuri,Maneger of Technical Research Institute of Taisei Corporation for their appropriate advice and help.

References

1)Kishitani,Kunikawa,Iizuka,Mizunuma : Cment concrete No.478,Dec.1986 P.7~14
2)Iizuka,Mizunuma : Cement concrete No.495,May.1988 P.8~15
3)Namiki,Yamamoto,Tomaturi,Kuroha : Summaries of Technical papers of Annual meeting, Architectural Institute of Japan Oct.1987 P.1240
4)Namiki,Tomaturi,Kuroha,Mizukami : Summaries of Technical papers of Annual meeting, Architectural Institute of Japan Oct.1988 P.1136

2 SETTING

EFFECTS OF ACCELERATING ADMIXTURES ON CEMENT HYDRATION

B. E. I. ABDELRAZIG, D. G. BONNER and D. V. NOWELL
Divisions of Chemical Sciences and Civil Engineering, Hatfield Polytechnic, England
J. M. DRANSFIELD and P. J. EGAN
Fosroc Technology, Birmingham, England

Abstract
Three commercially used accelerating materials, calcium chloride, calcium nitrate and sodium thiocyanate, were studied for their effects on the hydration of a particular ordinary Portland cement. Testing was carried out at equal weight percentage levels on cement pastes at a water:cement ratio of 0.50. Test procedures used were X-ray diffraction, differential scanning calorimetry, solution analysis, semi-adiabatic and semi-isothermal calorimetry. Results generally agreed with the known effects of the three admixtures in site practice with the calcium chloride being effective at all ages, the calcium nitrate at early ages and the sodium thiocyanate at later ages. There was also a good correspondence between the results obtained with the various test methods employed. Significant depressions of the levels of hydroxide and sulphate ions in solution in the first three hours were found, particularly with the calcium salts. Depression of the sulphate ion concentration was paralleled by the formation of gypsum in the early age X-ray diffraction and differential scanning calorimetry studies. The effects of the anions present in the admixtures appear to be significant in the later stages of hydration but no clear mechanisms were identified.
Key-words: Accelerators, Hydration, Solution Chemistry, Sulphate, Hydroxide, Portland Cement.

1 Introduction

Despite the amount of work carried out on the acceleration of cement hydration, the mechanisms by which acceleration is caused remain uncertain. Ramachandran[1] has listed twelve possible mechanisms which have been suggested to explain the effects of calcium chloride. The problems in producing a unified theory of the effects of acceleration are probably related to the wide variety of conditions and test methods used, which makes it difficult to compare one set of work with another. A further complication is that, for commercial reasons, a description or classification of an accelerator is often based upon its effects in speeding the stiffening, hardening and strength gain of concrete [2,3]. These factors are generally physical manifestations of the effects on the hydration of cement and are suitable as means of classification for commercial use but can cause confusion if the same physical effect is produced in chemically distinct ways by different materials. Classification of accelerators by their effect on cement hydration is more difficult as there are so many different possible aspects of hydration in general that may or may not be accelerated.

   Many chemicals have been used commercially as accelerators or otherwise as such [4,5]. In the work reported in this paper three chemicals used commercially as

106

have been selected and their effects on the hydration of a selected ordinary Portland cement studied with a view to forming some idea of which aspects are most affected. To obtain a broad picture of the effects occuring, a wide range of test methods has been used, following the hydration up to three days of age.

The admixtures chosen for this study were calcium chloride, calcium nitrate and sodium thiocyanate. Calcium chloride is still commercially the most widely used basis for accelerating admixtures because it is easily available and of low cost. It gives benefits in both early setting acceleration and strength gain. However, it is only suitable for use in concrete which does not contain embedded steel due to its effect in enhancing the corrosion of steel. There is still a need to produce acceleration of the hydration of cement without causing corrosion problems and therefore a need for chloride free accelerators. Calcium nitrate and sodium thiocyanate have both been used commercially but appear to have slightly differing properties [5]. Calcium nitrate gives greater effects on acceleration of concrete stiffening whilst sodium thiocyanate appears to be more effective on strength gain.

2 Test methods and sample preparation

2.1 Materials and mixing

Chemicals used were all of Analar grade. Each was used at a dosage of 1% by weight of anhydrous material on cement weight. This was chosen to represent the usual dosage method for admixtures as used commercially, however it gives differing dosages in molar quantities. Table 1 lists the dosages used by weight of cement and by weight of water for the chemicals and anions. An oxide and Bogue analysis of the cement used is given in Table 2. Except where otherwise noted, tests were carried out in cement pastes at a water:cement ratio of 0.50 using deionized water. Admixtures were predissolved in the water used.

Table 1. Admixture dosages (anhydrous)

Admixture	Dosage on cement				Dosage on water			
	(%)		(mmol/kg)		(g/litre)		(mmol/litre)	
	total	anion	total	anion	total	anion	total	anion
Calcium chloride	1.00	0.64	90.1	180.2	20.0	12.8	180.2	360.4
Calcium nitrate	1.00	0.76	60.9	121.9	20.0	15.2	121.9	243.8
Sodium thiocyanate	1.00	0.72	123.3	123.3	20.0	14.3	246.7	246.7

Table 2. Chemical analysis of OPC used

Constituent	CaO	SiO$_2$	Al$_2$O$_3$	Fe$_2$O$_3$	SO$_3$	MgO	Na$_2$O	K$_2$O	Free CaO
(% by weight)	64.9	20.8	4.6	2.9	2.6	1.2	0.1	0.6	1.0

Clinker Phase (Bogue)		C$_3$S	C$_2$S	C$_3$A	C$_4$AF
(% by weight)		63.6	11.7	7.3	8.8

2.2 Analysis of liquid phase

Previous work on the solution chemistry of Portland cement has tended to be at significantly higher water:cement ratios than those used in practice [6,7] although some work has been carried out at levels more typical of those used in actual construction [6] - only selected references have been included at the end of this paper. Work in this study was performed at the lower water:cement ratio level.

Initial mixing was for two minutes in a low speed mixer. The paste was then quickly poured into a sealed polythene container which was continually roll mixed at 22°C. Immediately before fluid extraction the sample was transferred to a glove box under a nitrogen atmosphere.

Fluids were extracted from the samples by using a Baroid type pressure filter using nitrogen gas to provide the pressure. Filters used were two Whatman No. 542 papers. Pore fluids removed by this procedure were collected and stored in polythene containers in a purged closed system. Extractions were made after 5, 20, 40, 60, 90, and 120 minutes of hydration. A final extraction was made after 150 or 180 minutes, depending on the state of stiffening of the sample. A pressure of 0.35 MPa was used in all cases, additional preliminary studies at 0.175 and 0.7 MPa showed no significant effect from changing the extraction pressure.

2.3 Examination of the solid phase

Initial mixing was for two minutes in a low speed mixer. Samples were then taken into stoppered polythene tubes. To ensure that no segregation of the mixed material occured, two stainless steel ball bearings were placed in each of the polythene tubes which were then tumble mixed at a very low speed for as long as possible. After setting, the tubes were sealed in plastic bags and immersed in a water tank. The tubes were stored at 22°C at all times.

Samples were taken for testing after hydration periods of 5 minutes, 1 hour, 3 hours, 6 hours, 24 hours and 3 days. Hydration was halted after the various times by grinding the samples in excess acetone. The residues were filtered and dried at 22°C in air pumped through silica gel and soda lime to remove moisture and carbon dioxide. This method was found to be very effective in stopping the hydration reaction and giving very low carbonation levels. It has been previously used in studies of the hydration of the major phases of Portland and aluminous cements and magnesia-phosphate cement. After drying the samples were transferred to stoppered glass tubes and stored under vaccuum in a desiccator.

X-ray diffraction (XRD) measurements were obtained using a Phillips PW-1050 diffractometer with Cu-Kα radiation, nickel filter and a scanning rate of 1° 2θ/ minute. Fully quantitative X-ray analysis was not possible. Differential scanning calorimetry (DSC) was performed on samples heated at 20°C per minute in a nitrogen atmosphere with a crimped aluminium crucible as reference.

2.4 Studies of heat of hydration evolution

Studies of the heat evolution from hydrating cement are usually performed under virtually isothermal conditions. This does not correspond to the environment of concrete when used in practice, where temperature rises during the hydration are common [8]. This temperature rise obviously will affect the kinetics of the hydration process. During the reported work tudies under both semi-isothermal (conduction calorimetry) and semi-adiabatic conditions were performed.

Samples for conduction calorimetry testing were cement pastes at a w/c ratio of 0.50. Mixing was carried out inside the calorimeter after reaching an equilibrium temperature at 25°C. The equipment used was an Oxford University design Calox calorimeter with a sample size of 18g. The test procedure produces a signal in millivolts from the equipment which can be converted, through a calibration procedure, to give a measure of the rate of heat evolution occuring.

Semi adiabatic tests were made using a sand cement mortar at a water:cement ratio of 0.50. The sand was added to reduce the temperature rise during the test to something approximating that to be expected under normal concreting conditions. The initial temperature of the test samples at mixing was 20°C. The calorimeter used was a silvered vacuum flask enclosed in foamed polystyrene. A Type T thermocouple [9] was embedded in the sample under test to record temperature rise. Although it is possible to carry out calibration procedures to allow calculation of actual heat evolution rates from the temperature changes this was not done in this study. Temperature rises during the tests corresponded approximately to those that would be expected from a concrete mix containing 300 kg/m³ ordinary Portland cement in large pours.

3 Test methods and results

3.1 Solution analysis
The solutions extracted under pressure filtration were analysed for the presence of cation and anion species using test methods as detailed in table 3. The methods used for the determination of levels of silicon and aluminium would not be expected to distinguish between the various possible oxyanions of these material. For convenience the terms Si and Al species have been used in this paper to represent the total amount detected, with no attempt being made to distinguish between the various possible contributing ions.

Table 3. Test methods used for identification of species in pore solution

Species	Test Procedure used
Li^+, Mg^{2+} and Fe^{3+}	Atomic absorption spectrophotometry (AAS)
Na^+ and Ca^{2+}	AAS and X-ray fluorescence (XRF)
Al species	XRF
K^+	Flame photometry and/or AAS
Si species	XRF and colorimetric using ammonium molybdate [10]
SO_4^{2-}	Turbidimetric methodwith barium chloride [11]
Cl^-	Colorimetric using ferric alum solution and alcoholic mercuric thiocyanate [12]
NO_3^-	UV absorption [13]
SCN^-	Colorimetric using acidic ferric nitrate [11]
OH^-	Titration

Concentrations of the various major species detected and the changes detected over the first three hours of hydration are shown on Figures 1 to 4. Other species tested for were only found at very low levels, which did not show any regular trends with time. These minor components of the pore solution being - Li^+, Fe^{3+}, Mg^{2+} and the Si and Al species. Results are shown on Table 4. Levels of these materials found in solution did not change significantly with time and no trends were apparent. The values in Table 4 represent the range of concentrations measured over the whole three hours of study for each component. In some cases no measurements at all were made of the Fe^{3+} and Mg^{2+} content due to the extremely low levels present, these instances are indicated in the table by asterisks.

109

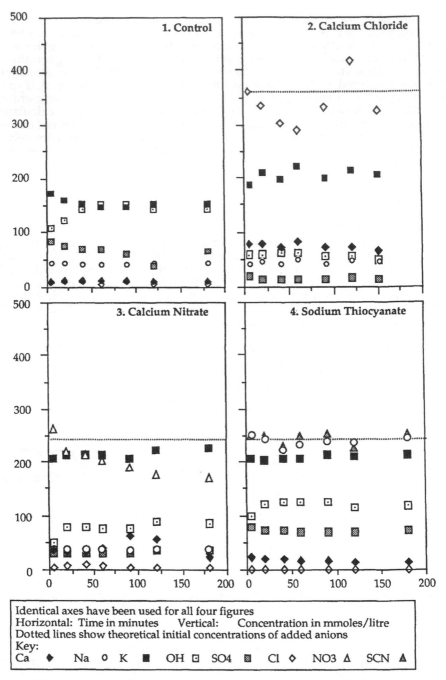

Identical axes have been used for all four figures
Horizontal: Time in minutes   Vertical: Concentration in mmoles/litre
Dotted lines show theoretical initial concentrations of added anions
Key:
Ca ◆   Na ○   K ■   OH ⊡   SO4 ▨   Cl ◇   NO3 △   SCN ▲

Figures 1 to 4: Major species detected in solution analysis

Table 4. Minor constituents of solutions analysed (asterisk indicates not determined)

| System studied | Range of concentration measured (mmoles/litre) | | | | |
	Si species	Li$^+$	Fe^{3+}	Mg^{2+}	Al species
Control	0.06 - 0.11	1.8 - 2.2	0.006 max	0.003 max	0.19 - 0.26
Calcium chloride	0.03 - 0.05	1.3 - 1.9	0.007 max	*	0.16 - 0.23
Calcium nitrate	0.17 - 0.36	1.2 - 1.7	0.007 max	*	0.22 - 0.33
Sodium thiocyanate	0.08 - 0.15	1.5 - 1.8	*	*	*

3.2 X-ray diffraction

Major peaks within the XRD spectrum were recorded and compared with previously identified reflections for the major phases present. In each case the strongest peak detected in the current work was used to provide a semi-quantitative measure of the amount of the particular phase present. As an aid to the analysis a set of descriptors was used, based on the height in scale units of the peaks; Table 5 summarises the semi-quantitative scale used.

Table 5.   Semi quantitative descriptors for XRD analysis

Descriptor	Abbreviation	Scale unit range
Off scale	vs	100 +
Strong	s	100 - 70
Medium	m	69 - 30
Weak	w	29 - 10
Very weak	vw	9 - 5
Trace	t	< 5

The peak chosen for estimation of calcium hydroxide was that corresponding to a d spacing of 3.48Å. This is not the reflection that is normally regarded as the main reflection for calcium hydroxide, which is at 2.63 Å, but was always the strongest peak detected. Similar results were found with Analar grade calcium hydroxide. Estimation of calcium hydroxide is known to be difficult due to orientation effects in the crystal plane reflections and it is believed that this is the reason in the current work for the low intensity of the normally accepted major reflection.

Table 6 summarises the changes in intensity of the main peaks detected, together with details of other, minor, peaks where found. Both scale values and descriptors are given and standard cement chemistry nomenclature for the has been used for the phases to which the reflections have been attributed. Compounds have been assigned to the minor peaks detected based on reflections previously reported in the literature. Clearly, because of the semi-quantitative nature of the work, small variations in the number of scale units in size of any particular peak should not be taken as significant. However, the results do allow an assessment of the change in phases present with time and between samples which can be taken together with results from other test methods to provide a fuller picture of the changes occuring.

A minor peak detected at a reflection corresponding to 7.56 Å was seen in many of the determinations. Gypsum and the complex carbonate phase C3A.CC.12H have both been reported to give a peak in this area. Other hydrations were carried out in a nitrogen atmosphere and also gave this reflection so it is believed to be unlikely that it is due to a carbonate phase. No gypsum was detected in XRD measurements made on the anhydrous cement, virtually all sulphate being found to be present as anhydrite with minor amounts of hemi-hydrate.

The intensity of the reflection used to quantify the level of anhydrite, at 3.48Å, is significantly higher in the samples from tests containing calcium nitrate and sodium thiocyanate, particularly so with the former. The reason for this is unclear. If only anhydrite is contributing to this peak, the implication would be that there were significantly higher levels of anhydrite present in the anhydrous cement in these cases. This seems unlikely as the cement used in these tests was all taken from the same drum of material. Another possibility is that there are other, as yet unidentified phases contributing to the reflection detected.

Table 6.   Semi quantitative analysis of reflections found in XRD examination

d value	C_2S (3.028Å)	$C\bar{S}$ (3.48Å)	CH (4.9Å)	$C_3A.3C\bar{S}.32H$ (9.72Å)	Others
i) Control					
5 minute	s (71)	w (19)	t (1)	t (3)	t, 7.56Å ($C\bar{S}$.2H)
1 hour	s (70)	w (19)	vw (5)	vw (5)	t, 7.56Å ($C\bar{S}$.2H)
3 hours	m (67)	w (10)	m (61)	vw (7)	t, 7.56Å ($C\bar{S}$.2H)
6 hours	m (54)	vw (5)	s (95)	vw (8)	
24 hours	m (40)	t (3)	vs	w (11)	
3 days	w (20)	t (2)	vs	w (11)	
ii) Calcium chloride					
5 minute	s (88)	w (12)	t (2)	vw (5)	t, 7.56Å ($C\bar{S}$.2H)
1 hour	s (77)	w (14)	t (4)	vw (6)	t, 7.56Å ($C\bar{S}$.2H)
3 hours	m (58)	w (18)	m (49)	vw (8)	t, 7.56Å ($C\bar{S}$.2H)
6 hours	m (55)	vw (4)	s (95)	vw (8)	
24 hours	m (35)	t (2)	vs	w (10)	
3 days	w (22)	t (2)	vs	w (11)	t, 7.9Å (C_4AH_{13} or solid solution)
iii) Calcium nitrate					
5 minute	s (75)	m (43)	t (1)	t (4)	t, 7.56Å ($C\bar{S}$.2H)
1 hour	s (77)	m (33)	vw (6)	vw (6)	t, 7.56Å ($C\bar{S}$.2H)
3 hours	s (78)	w (18)	m (48)	vw (9)	
6 hours	m (60)	w (10)	s (95)	w (10)	
24 hours	m (48)	t (3)	vs	w (14)	
3 days	m (36)	-	vs	w (15)	t, 8.59Å (solid solution)
iv) Sodium thiocyanate					
5 minute	s (75)	w (17)	t (3)	t (3)	t, 7.56Å ($C\bar{S}$.2H)
1 hour	s (71)	w (29)	t (4)	vw (6)	t, 7.56Å ($C\bar{S}$.2H)
3 hours	s (85)	w (20)	s (85)	vw (6)	t, 7.56Å ($C\bar{S}$.2H)
6 hours	m (65)	vw (6)	s (99)	vw (8)	
24 hours	m (28)	t (2)	vs	vw (9)	
3 days	w (20)	-	vs	w (10)	t, 8.85Å ($C_3A.C\bar{S}.12H$)

3.3 Differential scanning calorimetry

The energy absorbed in each endothermic reaction was computed for each run, values for the major peaks detected are given on Table 7. Standard cement chemistry notation has been used.

Table 7. Energy absorbed in DSC peaks (all values in J/g)

Temperature	$C_3A.3C\bar{S}.32H$ and CSH (90-130°C)	$C\bar{S}.2H$ (~145°C)	CH (450-490°C)	Others
i) Control				
5 minute	16.49	3.04	1.31	260°C, 0.47 J/g
1 hour	24.01	3.38	2.72	260°C, 0.70 J/g
3 hours	29.51	1.20	19.31	260°C, 0.50 J/g
6 hours	40.40	0.00	70.06	
24 hours	58.68	0.00	149.68	
3 days	81.60	0.00	172.85	
ii) Calcium chloride				
5 minute	20.30	6.36	5.41	290°C, 0.30 J/g
1 hour	26.98	8.43	18.82	
3 hours	48.25	2.03	55.08	
6 hours	46.34	0.00	97.59	
24 hours	81.27	0.00	168.59	
3 days	92.40	0.00	211.59	166°C, 0.90 J/g
iii) Calcium nitrate				
5 minute	26.10	6.87	2.38	260°C, 0.40 J/g
1 hour	47.28	1.79	9.30	260°C, 0.50 J/g
3 hours	69.45	0.00	46.32	
6 hours	80.70	0.00	82.46	
24 hours	92.46	0.00	144.35	
3 days	78.59	0.00	167.31	183°C, 0.75 J/g
iv) Sodium thiocyanate				
5 minute	19.83	1.30	1.13	260°C, 0.71 J/g
1 hour	26.18	3.32	3.97	260°C, minor
3 hours	31.17	1.07	20.96	260°C, minor
6 hours	33.52	0.00	85.82	
24 hours	68.07	0.00	166.16	
3 days	76.49	0.00	183.78	180°C, minor

The level of calcium hydroxide formed in a sample can be used as an estimate of the degree of hydration. Values for the percentage of calcium hydroxide present were calculated by the energy absorbed in the 480°C DSC peak in comparison with Analar calcium hydroxide. The results obtained may be a slight overestimate of the level present as an endothermic peak at around 445°C was noted in studies of the anhydrous cement and may overlap with the calcium hydroxide peak in the hydrated samples. The formation of other phases, particularly CSH gel and ettringite can also be used as an estimate of the total hydration. In this case it is not possible to produce an estimate of the total percentage of material present but the total absorbed energy can be used. This is obviously less accurate than the calcium hydroxide estimate as specific heat capacities vary from compound to compound. It does indicate any major changes, however. Calculated levels of calcium hydroxide are shown on Figure 5 and total energy absorbed on Figure 6.

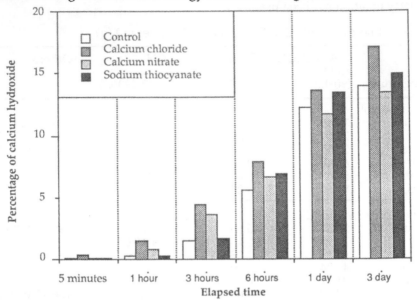

Figure 5:Percentage calcium hydroxide measured by DSC

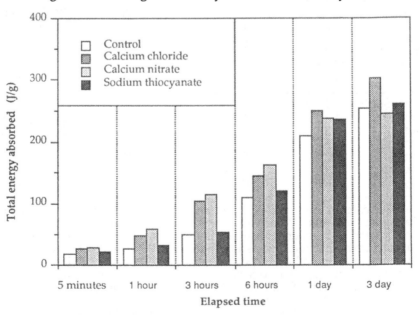

Figure 6. Total energy absorbed by hydration product phases

3.4 Calorimetry

Calibrated results for conduction calorimetry studies, giving the rate of heat evolution from the four systems, tested are given in Figure 7. Temperature rise values for the four systems as measured in the semi-adiabatic calorimetry tests are given in Figure 8. Values plotted are actual temperatures measured, starting temperatures were 20°C in each case.

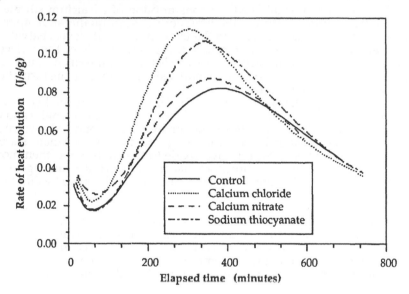

Figure 7. Heat evolution curves from conduction calorimetry

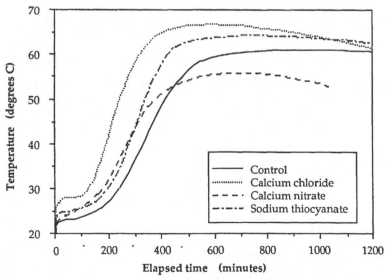

Figure 8. Temperature rise in semi-adiabatic calorimetry

4 Discussion

4.1 Examination of the solid phase

The two possible methods of studying the hydration of cement in the solid phase are to follow the consumption of the unhydrated phases or the production of hydration products.

X-ray diffraction analysis allowed the consumption of tricalcium silicate and anhydrite to be monitored. Within the limits of the semi-quantitative approach used there is little obvious difference in the tricalcium silicate levels between the samples examined. As previously discussed, the early levels of the reflections attributed to anhydrite, particularly in the samples containing calcium nitrate, do appear to vary. The four systems show generally similar trends in the formation of calcium hydroxide and ettringite.

Differential scanning calorimetry allowed the formation of calcium hydroxide to be followed and some estimate to be made of the formation of ettringite and CSH gel. Figure 5, showing the percentage of calcium hydroxide found, shows calcium chloride to give the greatest levels at each age. Sodium thiocyanate starts by showing similar amounts of calcium hydroxide to the control and then more at later ages. Calcium nitrate starts with more than the control but at later ages shows less. Similar trends were also found when the total energy absorbed by the reaction products was considered, as shown in Figure 6. In addition to the major phases formed, significant peaks were also seen at early ages at approximately 145°C, corresponding to XRD peaks found at 7.56Å. This has been attributed to the formation of gypsum in the first minutes of hydration as none was present in the anhydrous cement. This peak was found in all samples but was higher levels in the calcium chloride and calcium nitrate samples. Other minor peaks were found at early ages at around 260°C (290°C for calcium chloride), these have not been identified.

4.2 Solution analysis

Tests during the establishment of a procedure for the solution analysis indicated that there was very little difference in solution concentrations if differing pressures were used to extract the pore water. It was therefore assumed that the results obtained were indicative of the actual concentrations present in the whole of the pore water even though at most 45% of the initially added water was actually extracted.

The change in concentrations measured with time will obviously be affected to some extent by the fact that the water is steadily bound up in the hydration products formed. If no further material dissolved after the initial measurement the concentrations of all dissolved species would be expected to increase as water is combined, unless otherwise changed by solubility product effects. Continued hydration and precipitation of hydration product would further complicate the situation but it is clear that any decrease in concentration with time would represent material being lost from solution.

One point that should be made is that there was never an exact balance between the amounts of cations and anions measured in solution, although in some cases the balance was close enough to zero to be within experimental error. Totals generally showed an excess of anions.

The levels of cations detected in the pore solution removed from the control mixes (Figure 1) are relatively constant over the period examined. Calcium is present at approximately 12 to 14 mM/litre, sodium at 40 mM/litre and potassium at 150 mM/litre. The potassium is the only cation present which shows a change in concentration, dropping from an initial level of about 170mmole/litre over the first

116

hour. The levels of sodium and potassium measured in the first five minutes correspond to approximately 60% of the totals present in the cement as indicated by the oxide analysis. The level of calcium is at first sight very low, considering the available amounts of calcium in the hydrating cement. However, when solubility products are considered it is clear that the level of calcium must be suppressed by the presence of large quantities of sulphate and hydroxide ion. This is supported by the observation of gypsum in the early age results from the DSC and XRD tests when none was found in the anhydrous cement, implying its precipitation due to an initial supersaturation of the solution. When the anions are considered, the concentration of hydroxide detected shows a rise over the first hour from approximately 100 mM/litre to 150 mM/litre. The sulphate concentration started at approximately 85mM/litre and dropped to 65mM/litre over the whole period of examination. Levels of other materials are low and show no discernable trends. A possible interpretation of the changes seen is that they are initially caused by the dissolution of alkali sulphates and hydroxides, together with some contribution from calcium sulphate dissolution and cement phase hydration as insufficient sodium and potassium were detected to provide a balance to the sulphate and hydroxide ion levels. The increase in hydroxide concentration with time is probably indicative of the slowly continuing hydration as well as a slight effect due to the loss of water in chemical combination.

The addition of calcium chloride produces obvious changes in the solution chemistry of the system . The initial chloride ion concentration corresponds to the level of chloride added. However, the change in calcium level to approximately 80mM/litre, falling slightly with time, is well below the amount of calcium added in the admixture. This indicates that there must have been very rapid precipitation of calcium salts within the first five minutes of the hydration. This observation corresponds to the increased levels of gypsum detected in the early age DSC and XRD samples as compared to the control samples. The presence of the added calcium depresses the levels of sulphate and hydroxide significantly. The hydroxide reaches its maximum value more rapidly than in the control case. This is probably confirmation that the rise in the control case is due to continuing hydration as in the chloride case the maximum level is lower and therefore reached earlier. The level of sodium is hardly changed whilst that of potassium appears to be slightly higher. The only species that shows a significant change with time is the chloride ion, the level of which reduces over the first hour, increases and then decreases again. There is no obvious explanation for this effect, which was noticed in a number of samples. It may possibly be due to the change in free water in the system due to hydration, but that effect would be expected with the alkali metal concentrations as well. Other tests, using cement from the same batch and obtained at the same time but taken from a different drum, do not show the peak in chloride concentration at approximately two hours. These tests also show a lesser fall in the level of chloride over the whole period of measurements. Other workers [14,15] have suggested that the chloride binding capacities of Portland cement clinkers are related to cement fineness, it is possible that the variations found here are due to similar effects.

The addition of calcium nitrate had similar effects to those of calcium chloride although the sulphate and hydroxide ion levels were not depressed to as great an extent, probably due to lower molar quantities of calcium ion being added in the admixture. The initial level of nitrate detected in solution corresponds to the level added in the admixture and the decrease with time indicates that is in some manner becoming bound to the hydration products. The calcium concentration appears to go through a peak after about one hour but this does not correspond to any other changes.

117

Sodium thiocyanate shows similarities to and differences from the effects of the calcium salts. The level of hydroxide is again depressed from that of the control, despite there being no added calcium, although not to as great an extent as found in the other samples. The sulphate level, unlike the hydroxide, is hardly changed from that of the control. The calcium level is slightly higher than that of the control but it is unlikely that this is the cause of the reduction in hydroxide level as no calcium was added. It is more likely that the higher calcium level is possible becasue the hydroxide level has been reduced for other reasons. The sodium and thiocyanate ion levels correspond to those added in the admixture and do not change significantly with time.

The common effect of the accelerators added, whether or not calcium salts, is to depress the combined levels of sulphate and hydroxide ions in the solution. It is likely that this effect is related to the acceleration caused. The addition of accelerators had little effect on the level of minor constituents in solution.

4.3 Calorimetry measurements
Results from conduction calorimetry studies in many ways confirmed the results found in the other test methods. Calcium chloride showed the expected acceleration, both in the induction period and the main hydration peak. Calcium nitrate also showed an acceleration of the induction period but had only a small effect on the main hydration peak. Sodium thiocyanate showed little effect on the induction period but showed a significant increase in the main peak.

Semi-adiabatic calorimetry gave slightly contradictory results, possibly because all other test methods were effectively semi-isothermal due to the relatively small sample size. Sodium thiocyanate gave acceleration at all times instead of only after the first hours as was found in the other methods. Calcium nitrate did not show as great an effect in the early stages as calcium chloride, unlike in the conduction calorimetry, but was still accelerated over the control. Calcium nitrate did show a lower peak temperature than the control, possibly corresponding to the reduced reaction found in the solid phase analysis.

4.4 Overall comparisons
The calcium chloride clearly gave the greatest effect overall on the hydration, producing an acceleration or increase in all measures used. The effects noted on the solution chemistry of depressing the levels of hydroxide and sulphate in solution correspond to the detection of increased amounts of gypsum in the early periods by XRD and DSC.

Calcium nitrate gave similar early age effects to calcium chloride although not of as great an extent. One possible reason for this effect is the lower molar levels of calcium present but other factors may also be involved. Results after three hours indicate a reduction in reactivity of the cement, giving less reaction product and lower peak temperatures than the control specimens.

Sodium thiocyanate showed little effect in the first three hours on XRD, DSC or conduction calorimetry work, thereafter acceleration with respect to the control was found. Solution chemistry showed some depression of the levels of hydroxide and sulphate, and gypsum was noted in the XRD and DSC results, although only at the same levels as found in the control specimen. No clear change in thiocyanate ion level was found.

Experience in the use of the admixtures in the field suggests that calcium nitrate is most effective on accelerating the early period of hydration, reflected on site by the concrete stiffening. Sodium thiocyanate is more effective on the hardening processes which occur later. The results in this laboratory study appear to confirm this. Acceleration of stiffening appears to be due in some way to the depression of the sulphate and hydroxide levels in solution by high added amounts of calcium ions, leading to increased gypsum formation. Effects at later ages appear to be more related to the anion species present but no clear indications of processes involved were identified. Analysis of these effects is possibly complicated by the differing molar quantities of material present. Further work at equimolar levels has been completed and is in the process of being reported elsewhere.

References

1. Ramachandran, V.S. (1984) Concrete Admixtures Handbook. Noyes Publications
2. British Standards Institution (1982) BS 5075: Concrete Admixtures : Part1: Specification for accelerating admixtures, retarding admixtures and water-reducing admixtures
3. ASTM (1986) C 494: Standard specification for chemical admixtures for concrete
4. Skalny, J. and Young, J.F. (1980) Mechanisms of portland cement hydration, in Proceedings of the 7th International Congress on the Chemistry of Cement, Paris
5. Dransfield, J.M. and Egan, P.J. (1988) Accelerators, in Cement Admixtures; uses and applications 2nd edition (ed. P.C. Hewlett). Longman Group UK Limited.
6. Roberts, M.H. (1968) Effect of admixtures on the composition of the liquid phase and early reactions in Portland cement pastes. Building Research Station Current Papers 61/68
7. Thomas, N.L., Jameson, D. and Double, D.D. (1981) The effect of lead nitrate on the early hydration of Portland cement. Cement and Concrete Research, 11(1), 143-153
8. Fitzgibbon, M.E. (1974) Joint-free construction in rich concretes. Building Research and Practice. 2(3), 158 - 164
9. British Standards Institution (1981) BS 4937: International thermocouple reference tables. Part 5: Copper/copper nickel thermocouples. Type T
10. Vogel, A.I. (1978) Textbook of quantitative inorganic analysis.4th edition. Longmans.
11. Department of the Environment (1972). Analysis of raw, potable and waste waters. HMSO
12. Swain, J.S. (1956) Determination of low concentrations of chlorides. Chemistry and Industry. 20, 418 - 420.
13. Harries, R.C. (1981) Studies on the removal of nitrates from potable surface water by ion exchange. PhD thesis, Hatfield Polytechnic, Hatfield.
14. Byfors, K., Mansson, C. and Tritthart J. (1986) Cement and Concrete Research. 13, 760.
15. Blunk, G., Gunkel, P. and Smolszyk H.G. (1986) Proceedings of the 7th International Congress on the Chemistry of Cement, 4, 85 - 90.

REMEDIES TO RAPID SETTING IN HOT-WEATHER CONCRETING

M. S. EL-RAYYES
Department of Civil Engineering, University of Kuwait, Kuwait

Abstract
Concreting in very hot weather results in undesirable problems that could impair the properties and serviceability of concrete. This paper presents the results of a comprehensive experimental work aimed at evaluating the effect of ambient air temperatures on the times of initial and final setting of concrete having a wide range of water-cement ratios, and placed in extremely hot climates. Relationships between the setting behaviour and various influencing factors are investigated. Remedies intended to eliminate the adverse effects of rapid setting by selecting temperature - dependent optimal dosages of a water-reducing and set-retarding admixture are described. An expression is proposed for the admixture dosage necessary to prolong the setting time by a specified percentage when a concrete of any water-cement ratio is placed at a high environment temperature.
Key words: Admixtures, Concrete mixtures, Hot weather concreting, Setting times.

1   Introduction

As a result of a series of chemical reactions between the cement and water, a freshly mixed concrete undergoes a process of stiffening called setting. The setting process must not start too early to ensure that the freshly mixed concrete remains in a plastic condition for a period of time sufficient to permit satisfactory placing and compaction as well as proper finishing. Premature setting of concrete is also likely to result in the development of form-deflection cracks and cold joints in structural members.

It is customary to distinguish between the initial setting, signaling the start of disappearance of concrete plasticity, and the final setting, marking the complete stiffening of concrete. Of these two concepts, the initial setting has far greater significance. This is because the initial setting and the vibration limit, defined as the stiffening point when a newly placed concrete can no longer be made plastic by revibration, are synonymous. Thus, when hardening of a preceding concrete layer passes this point, it cannot be made mono-lothic with the next layer, and occurrence of cold joints and related undesirable consequences is imminent.

Hot-weather concreting involves some special problems arising both from the high temperature of the freshly mixed concrete and the high temperature of the air during placing and finishing. Higher than normal temperatures of fresh concrete result in excessive early hydration of cement. This leads to a rapid loss of workability, a rapid setting, and to a lower final strength (Mindess and Young 1981; Neville and Brooks 1987). Similarly, a high air-temperature associated with low relative humidity while concreting leads to a rapid evaporation of some of the mixing water causing yet a greater loss of workability, faster stiffening and extensive plastic shrinkage which may result in cracking (ACI Committee 305 1982).

In most regions of the Middle East, concreting during the long summer months almost always takes place at ambient temperatures far above the normal limits recommended. In such instances, the concrete construction process is most likely to be obstructed unless certain appropriate measures are taken to cope with the problems indicated above.

To effectively counteract the loss of workability and rapid setting caused by high temperatures, the use of a suitable set-retarding/ water-reducing admixture is a must. However, the effects of such an admixture on the properties of concrete are influenced by the mix proportions and in particular the water-cement ratio, the concrete and ambient temperatures, the admixture employed and its dosage, and the type and composition of the cement used (Popovics 1979; ACI committee 212 1981).

Experience shows that the composition of the admixture must be compatible with that of the cement (Schneider, and Shoenfelder 1972; Johnston 1987). Otherwise, the admixture could prolong the time of initial setting and/or reduce the amount of water needed for a given initial slump, but the concrete may harden faster even with a cement and an admixture that separately meet all specifications. For this reason, the admixture should be tested with the job materials under the actual climatic conditions in order to verify its compatibility with the concrete ingredients and its ability to produce the desired concrete properties under the prevailing conditions.

This research forms part of a wider test programme directed by the above considerations. The outcome of the overall study is discussed in two papers. The first (El-Rayyes 1987) addresses the way in which concrete workability is affected and remedied at Kuwait-like extreme hot-weather conditions. The present paper, on the other hand, is intended to quantitatively establish the influence of high climatic temperatures on the setting times, and how remedies, by using a suitable water-reducing and retarding admixture, can be optimally introduced at high ambient temperatures for different water-cement ratios.

2 Experimental program

2.1 Materials and sample preparation
A comprehensive experimental programme was undertaken to establish a fundamental understanding of the influece of various factors on the

setting times of concrete. An extensive number of concrete mixtures
were prepared in the laboratory under controlled conditions utilizing
the following materials:

(i) Locally-available crushed rock aggregates having quartz,
andesite and granite as the predominant minerals. Well—proportioned
coarse aggregate mixtures with size fractures of 9.5 mm, 12.5 mm,
and 19 mm were prepared. The fine aggregate has a fineness modulus
of 2.4.

(ii) Ordinary drinking water and a locally—available ordinary
Portland cement (Type I).

(iii) A non-entraining, set-retarding and water—reducing admixture of lignosulphonic base and conforming with type D of the ASTM
designation C494 (1984).

Concrete samples were mixed at water—cement ratios ranging from
0.40 to 0.65 in increment of 0.05. The admixture dosage, considered
as a part of the required amount of mixing water, was thoroughly
mixed with water before feeding it into the mixer within half a
minute after the start of water addition.

In addition to plain concrete, mixtures were also prepared with
different levels of admixture dosage. These were 200, 250, and 300
cm^3/50 kg of cement classified as normal, above normal, and high
dosage levels, respectively. Although each dosage was measured by
volume and not by weight, these dosages will be referred to herein
as 0.4, 0.5 and 0.6%, respectively, for simplicity. After the
completion of the mixing operations, mortar was sieved from each
concrete mixture and used to prepare a set of three test specimens.

The initial temperatures of the mortar specimens ranged essentially between 23 and 24°C. A small portion of the tested specimens
were stored at laboratory conditions, whereas the majority were
stored and tested outdoors at ambient temperatures ranging from 36 to
45°. However, within this temperature range, the major portion of
the testing program was undertaken at ambient temperatures of 40 to
42°C. During the testing program, outdoor specimens were kept in a
shielded enclosure to avoid direct exposure to the sun rays.

2.2 Determination of setting times

ASTM C 403-80 (1984) prescribes a technique for the determination of
the times of initial and final setting of concrete mixtures with
slump greater than zero. The method is based on testing by specified
needles the development of penetration resistance for a mortar sieved
from the concrete mixtures. According to this method, the times of
initial and final settings are defined as the elapsed time, after
initial contact of cement and water, required for the mortar specimens to reach a penetration resistance of 500 psi (3.5 MPa) and 4000
psi (27.6 MPa), respectively.

3. Presentation and discussion of results

The development of the penetration resistance with elapsed time after
mixing was determined for more than 120 concrete mortar specimens
having different water-cement ratios, admixture dosages and setting
temperatures. Figure 1 presents a typical setting curve illustrating

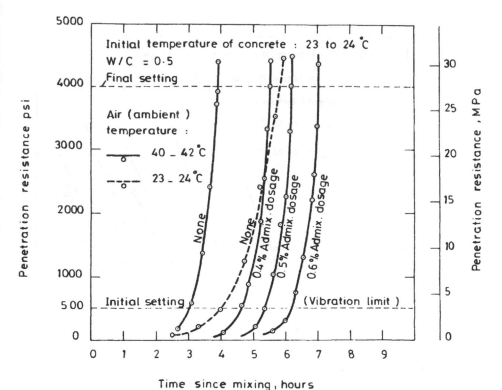

Fig. 1. Typical curves of standard penetration resistance and setting times for
concrete mortars with and without retard-
ing-reducing admixtures

the influence of the admixture dosage and ambient temperature on the
penetration resistance of a concrete mortar specimen at a water-
cement ratio of 0.5. At a given time after mixing, the penetration
resistance of specimens setting at high ambient temperatures (40-
42°C) increases with the decrease in the admixture dosage. Moreover,
the penetration resistance of plain concrete mixtures increases with
the increase of air temperature.

Times of initial and final settings of all concrete mortar specimens, corresponding to penetration resistance of 500 psi (3.5 MPa)
and 4000 psi (27.6 MPa), respectively, were determined from setting
curves, such as the ones illustrated in Fig. 1. Each setting time
represents an average of three tests on essentially identical
samples.

The results of the tests are plotted in Figs. 2 and 3, which
illustrate the influence of the water-cement ratio, ambient temperature, and admixture dosage on the initial and final setting times
of concrete mortars, respectively. Similar trends are exhibited in
Figs. 4 and 5. However, by plotting the setting times a log scale,
Figs. 4 and 5 clearly indicate that linear relationships exist bet-

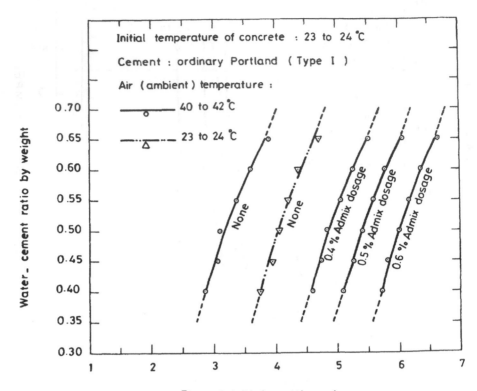

Fig. 2. Effect of water—cement ratio,
air temperature and dosage of admixture
on initial setting of concrete mortars.

ween the setting times and the water—cement ratio, in the range of
0.35 to 0.7, despite the variations in the admixture dosages and the
ambient air temperatures. For a given admixture dosage and specified
concrete and air temperatures, these linear relationships provide a
convenient means to predict the setting times of concrete of any
water—cement ratio merely by knowing the setting times for two water-
cement ratios. Moreover, on the basis of these linear relationships,
the original solid curves in Figs. 2 and 3 may be extrapolated to
water—cement ratios of 0.35 and 0.7 (dotted portions in Figs. 2 and
3), and can thus adequately serve for all the water—cement ratios
usually encountered in practice.

Analyses of Figs. 2 and 4 reveal the following regarding the
influence of various factors on the time of initial setting:

(1) An increase in the setting temperature from 23 to 41°C
reduces the time of initial set of plain concrete (i.e., without
retarding admixtures) from 25 to 17% for water—cement ratios betwen
0.35 and 0.7; the larger percentage of reduction occuring with the

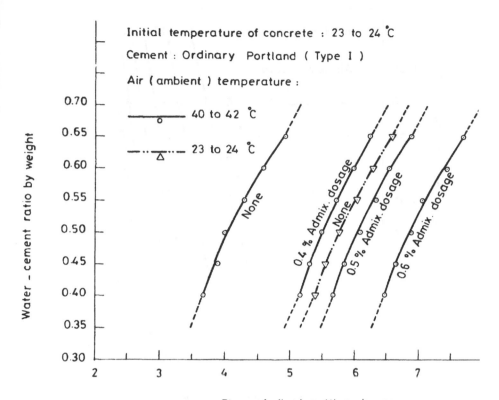

Fig. 3. Effect of water-cement ratio, air temperature and dosage of admixture on final setting of concrete mortars.

smaller water-cement ratio.

(2)   Addition of a retarding admixture prolongs the time of initial setting at 41°C compared to plain concrete setting at 23°C; the larger increase observed for the smaller water-cement ratio.   For example, for admixture dosages of 0.4, 0.5 and 0.6%, the increase in the initial set time at a water-cement ratio of 0.35 is 22, 37 and 54%, respectively, compared to 17, 29 and 40% at a water-cement ratio of 0.7.   However, when compared to plain concrete allowed to set at the same high temperature (41°C), the increase in the initial set time for the same dosage levels quoted above becomes 62, 81, and 104% at a water-cement ratio of 0.35, and 41, 54 and 68% at a water-cement ratio of 0.7.

Similar conclusions can be advanced from Figs. 3 and 5 regarding the time of final setting and its independence on variations in the water-cement   ratio,   admixture   dosage   and   setting   temperature. However, the addition of a retarding admixture has a lesser effect on the final set time compared to its effect on the initial set time.

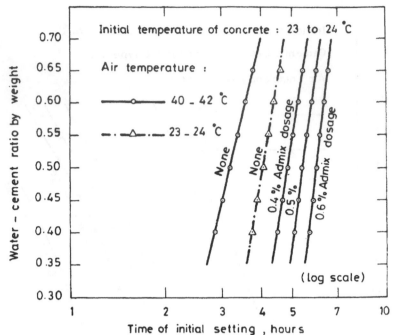

Fig. 4. Influence of water–cement ratio, air temperature and dosage of admixture on initial setting.

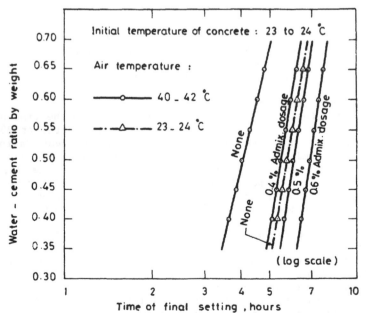

Fig. 5. Influence of water–cement ratio, air temperature and dosage of admixture on final setting.

This observation is clearly illustrated in Figs. 2 and 3, which show that, while an admixture dosage significantly less than 0.4% (about 0.27%) is capable of offsetting the reduction in the time of initial setting at a temperature of 41°C compared to that at normal temperature (22-24°C), a higher dosage (≅ 0.44%) is needed to achieve the same benefit for the time of final setting.

Making use of the curves plotted in Figs. 2, 3, 4 and 5, Fig. 6 is constructed to show the influence of the water-cement ratio and dosage level on the setting times at 41°C relative to those for plain concrete mortars at 23°C. For distinction purposes, a relative time of initial setting, in percent, is read on a solid line, whereas a relative time of final setting, in percent, is determined by a dashed line.

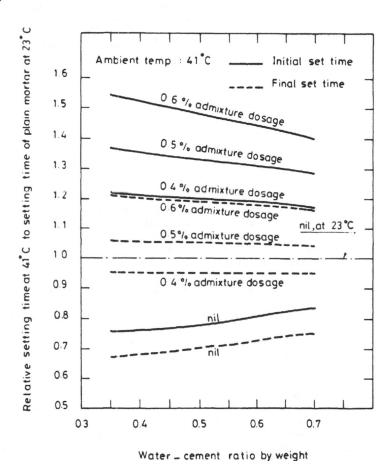

Fig. 6. Influence of water-cement ratio and admixture ture dosage level on the ratio of setting time at 41°C to setting time of plain concrete mortar at 23°C.

Table 1 gives penetration-test results concerning the times of initial and final setting for concrete mortars setting at climatic temperatures other than 41°C. Comparison of the results in Table 2 with those represented in Figs. 2 and 3 confirms the tendency for faster setting at higher ambient temperatures for concrete mixtures of similar proportions and having the same initial temperature.

Table 1. Times of initial and final setting of concrete mortars, with different admixture dosage levels and different water-cement ratios, and setting at various air temperatures.

Water-cement ratio	Dosage of set-retarding/water-reducing admixture							
	Nil		0.4%		0.5%		0.6%	
	I.S.T. hr^{min}	F.S.T. hr^{min}	I.S.T. hr^{min}	F.S.T. hr^{min}	I.S.T. hr^{min}	F.S.T. hr^{min}	I.S.T. hr^{min}	F.S.T. hr^{min}
0.40	2^{59}	3^{46}	4^{56}	5^{49}			5^{30}	6^{17}
	at 40°C		at 38°C				at 44°C	
0.50	2^{50}	3^{49}	4^{53}	5^{40}	5^{30}	6^{10}	6^{13}	7^{09}
	at 44°C		at 39°C		at 39°C		at 38°C	
0.60	3^{46}	4^{48}	4^{44}	5^{35}			6^{49}	8^{03}
	at 36°C		at 45°C				at 36°C	

Figure 7 shows that for different water-cement ratios and setting temperatures, a linear relationship exists between the penetration resistance of plain-concrete mortars and the elapsed time after mixing when plotted in a log-log system of coordinates. Results relating to a given setting temperature seem to form a family of parallel straight lines regardless of variations in the water-cement ratio. However, the common slope of each family of lines increases with the increase in the setting temperature. This indicates that the rate of stiffening, defined as the ratio between the final and initial setting times, decreases with the increase in the setting temperature.

Similar conclusions, for mortars with different admixture dosage levels but setting at a given temperature, can also be drawn from Fig. 8, which exhibits the straight-line approximation of the setting curves when plotted in the log-log scale. It can be noticed that the slope of the setting lines for concrete mortars with a retarding admixture is greater than that for plain mortars. This means that the addition of a retarding admixture reduces the rate of stiffening. It can also be seen that the setting curves for mortars with a retarding admixture and of a given water-cement ratio form approximately parallel straight lines in the log-log scale, indicating an equal rate of stiffening independent of the level of the admixture dosage; of course within the levels utilized in this study. Furthermore, the water-cement ratio does not seem to influence significantly the slope of the setting curves.

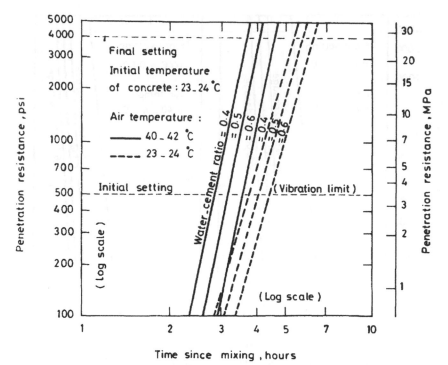

Fig. 7. Penetration resistance of mortars sieved from plain concretes of different water-cement ratios and setting at elevated and laboratory temperatures as a function of the time elapsed after mixing.

4 Generalization

From the foregoing discussion it is evident that the standard setting curves can be expressed mathematically as

$$p = c\, t^m \qquad\qquad (1)$$

where

p = penetration resistance,
t = time elapsed since mixing.

The parameters c and m are the intercept and slope, respectively, of the straight line forming the setting curve plotted in a log p – log t system. The c parameter is a function of the composition and type of cement, the water-cement ratio, the setting conditions and, to some extent, the admixture dosage. The m parameter represents the rate of stiffening of concrete and is a function of the setting temperature and retarding admixture dosage, but is independent to a great extent of the water-cement ratio.

Equation (1) makes it possible to establish a reliable method for the evaluation of the stiffening process. For instance, when two

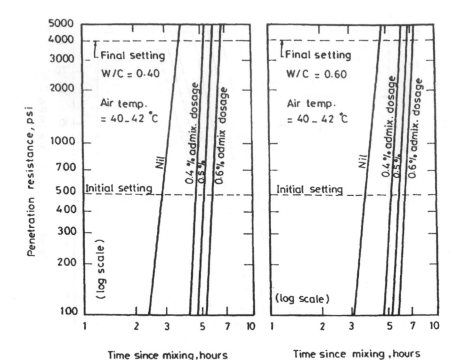

Fig. 8. Relationship between penetration resistance
of concrete mortars and the elapsed time after mixing, for different dosages of retarding admixture.

pairs of readings regarding the penetration resistance, p, and the
elapsed time, t, have been determined at an early stage while
carrying out the standard penetration test, the respective times of
initial and final setting can be directly estimated. Also, when the
slope, m, together with a single pair of the corresponding p- and t-
values are known, any other p and t- values can be predicted within
the limits of validity of Eq.(1).

The data presented in Table 2 are prepared with the aid of the
curves demonstrated in Figs. 2 and 4. For different water-cement
ratios, these values represent the time of initial setting, in percent, for concrete mortars with different admixture dosages relative
to that measured for plain concrete setting at 41°C. For a given
dosage level, the relative time of initial setting, averaged over all
the water-cement ratios utilized in this study, is plotted against
the dosage level in a semi-log system of coordinates in Fig. 9. It
can be seen that the resulting relationship is linear.

It has also been noted that the ratio of a specific relative time
of initial setting to the relevant average relative time of setting
varies only with the water-cement ratio and not by the dosage level.
Here, a coefficient β is introduced to give that ratio at a wide
range of water-cement ratios as shown in Table 3.

Table 2. Relative time of initial setting for concrete mortars with different water-cement ratios and various dosages of a set-retarding/water-reducing admixture

Water-cement ratio	Plain concrete set at 40-42°C	Retarded concrete set at 40-42°C Admixture dosage of		
		0.4%	0.5%	0.6%
	Relative time (%)	Relative time (%)	Relative time (%)	Relative time (%)
0.35	100	164	181	202
0.40	100	161	178	198
0.45	100	158	175	195
0.50	100	154	170	189
0.55	100	150	165	184
0.60	100	146	161	179
0.65	100	142	157	175
0.70	100	139	153	170

Table 3. Values of coefficient β for different values of water-cement ratio

Water-cement ratio	Coefficient β
0.35	1.08
0.40	1.06
0.45	1.04
0.50	1.01
0.55	0.99
0.60	0.96
0.65	0.94
0.70	0.91

By virtue of Fig. 9, and making use of Table 4, one can formulate
$$D = 2.233 \log_{10}(t_I/\beta) - 4.47 \qquad (2)$$
where

D   = admixture dosage in percent;

t_I = time of initial setting relative to that for plain concrete, in percent;

β = coefficient dependent on water-cement ratio and is given in Table 3.

As indicated in Fig. 9, Equation 2 can be applied satisfactorily within a reasonably wide range of dosage levels ranging from 0.3% to 0.7%.

Since we are interested in keeping fresh concrete, with normal water content and placed in an extremely hot climate, plastic enough to permit adequate compaction, and then alive enough to bond completely with the next layer, the importance of Eq. (2) should not be overlooked. Depending on the site location, placing conditions of

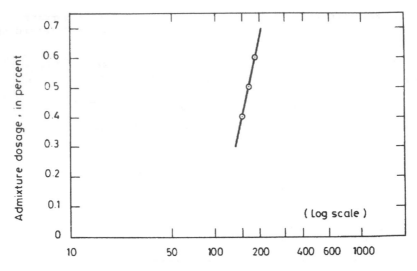

Average relative initial setting time , in percent

Fig. 9. Relationship between dosage of
retarding admixture and average relative
time of initial setting at 41°C.

the concrete and needs of the field engineer, it is of primary importance to know the percentage by which the setting time should be
extended or retarded. To prolong the time of setting when concreting
in hot weather, a suitable set retarding and water-reducing admixture
must be added to the concrete mixture. The appropriate dosage level
is decided upon according to the required percentage of retardation
in the time of setting. Evidently, this target can be achieved
simply and reliably by the application of Eq. (2).

Concrete technologists agree that ideal conditions for placing of
concrete occur when temperatures are 20–23°C, and the relative humidity is 50% or higher (Moore 1987). Thus, an inspection of Figs. 2
and 3 reveals that an average dosage of 0.27% is necessary to match
the initial time of setting of concrete placed at 41°C with that for
plain concrete placed at 23°C, whereas a dosage level of about 0.44%
is required to achieve the same benefit with respect to the final
time of setting.

In Kuwait and neighbouring regions, for instance, the hot months
of the year include May through October. During this period, high
environmental temperatures, 40–42°C on the average, prevail during
the day-time in the shade, but could occasionally rise up to 48°C and
fall to 36°C. Anyhow, the amount of admixture and the retardation
time should be limited to the minimum essential for proper execution
of the job (Weigler 1983; Johnston 1987).

In the light of the information and the guidelines stated above
and taking into consideration the results exhibited in Figs. 2 and 3
as well as in Table 1, and allowing for some rise in the initial temperature of concrete over 24°C as well as possible delays in
transporting and handling concrete between mixing and placing, the

132

following regime is proposed when concreting in hot climates:

(1) A dosage from 0.3% to 0.35% of lignosulfonate water-reducing admixture is recommended in May, June and October.

(2) A dosage from 0.4% to 0.45% is recommended in July, August and September.

(3) The higher the water-cement ratio, the greater the dosage within the above-specified limits.

5 Summary and conclusions

A high concrete temperature and a high air temperature result in an appreciable loss of workability, a rapid setting, and a lower final strength. A tangible decrease in the times of setting means a drastic decrease in the time available for furnishing adequate compaction, placement and finishing of the concrete.

To effectively counteract the loss of workability and rapid setting caused by high temperatures, it is necessary to use a suitable water-reducing and set-retarding admixture. Figures 2 through 6 give a detailed picture of the effects of the water-cement ratio, the dosage level of the admixture, and the ambient temperature on the times of setting of concrete mixtures. For plain concrete, a rise in the ambient temperature reduces the setting time; the smaller the water-cement ratio the larger the percentage of reduction. Compared to a plain concrete setting at 23°C air temperature, the addition of retarding admixture prolongs the time of setting; the relative prolongation increases with the increase of the dosage level but decreases with the increase of the water-cement ratio.

A linear relationship between the water-cement ratio and the times of setting, when plotted in a log scale, is maintained for all practical values of the water-cement ratio. The time of setting for a mixture of any water-cement ratio can henceforth be determined if only two setting times, corresponding to two water-cement ratios, are available.

In a log-log system of coordinates, the penetration resistance of concrete mortars follows a linear relationship with respect to the time elapsed after mixing regardless of the water-cement ratio, the setting temperature, or, within limits, the level of the admixture dosage.   On fixing the values of two of the preceding three variables; the resulting setting curves form a family of parallel straight lines.   The rate of stiffening of the concrete mortars decreases with the increase in the setting temperature, and also with the addition of admixture.

Based on a thorough analysis of the test results pertaining to this work and using lignosulfonate water-reducing admixtures, it has been argued and indicated that the optimum dosage level for concrete mixtures to be placed in extremely hot climates, with air temperatures from 36 to 48°C, is as follows:

(a)   0.3% to 0.35% in May, June and October.

(b)   0.4% to 0.45% in July, August and September.

The higher the water-cement ratio of the concrete, the greater the dosage level to be used within the above-specified limits.

References

ACI Committee 212 (1981) Admixtures for concrete (ACI 212.1R-81).
  Concrete International: Design and Construction, 3(5), 26-35.
ACI Committee 305 (1982) Hot weather concreting (ACI 305R-77).
  American Concrete Institute, Detroit.
ASTM Standards (1984) Concrete and mineral aggregates, vol. 04.02.
  American Society for Testing and Materials, Philadelphia.
El-Rayyes, M.S. (1987) Remedies to hot-weather concreting as related
  to loss of workability. J. University of Kuwait (Science), 14(2),
  252-260.
Johnston, C.D. (1987) Admixture-cement incompatibility: a case
  history. Concrete International: Design and Construction, 9(4),
  51-60.
Mindess, S. and Young, J. F. (1981) Concrete. Prentice-Hall,
  Englewood Cliffs, N.J.
Moore, W.C. (1987) Hot weather concreting: tips for builders.
  Concrete International: Design and Construction, 9(5), 63-64.
Neville, A.M. and Brooks, J.J. (1987) Concrete technology. Longman
  Scientific & Technical, Essex, England.
Popovics, S. (1979) Concrete-making materials. McGraw-Hill, New York.
Schneider, H. and Schoenfelder, R. (1972) Volume stability of
  concrete due to addition of setting retarders. Betonwerk und
  Fertigteil-Technik, 38(2), 97-104.
Weigler, H. 1983. Retarded concrete - technological problems and
  measures. Betonwerk und Fertigteil-Technick 49(6): 363-67.

MECHANISM OF AND SOLUTIONS TO RAPID SETTING CAUSED BY ADDITION OF CALCIUM LIGNOSULFONATE WATER REDUCER TO CEMENT WITH FLUOROGYPSUM AS RETARDER

ZHANG GUANLUN and YE PING
Tongji University, Shanghai, China

Abstract

CLS is a common water reducer for concrete, has some retarding effect on the setting of Portland cement, however, when added to fluorogypsum cement produced by Wusong Cement Manufacturer in Shanghai, it causes flash setting, which brings difficulties to concrete construction.

The principal reason for the aberrant setting is that CLS is absorbed at the surface of fluogypsum which has the great surface absorbed energy in the cement. The absorbed film of anionic surfactant restrains the solution of fluorogypsum, owing to the shortage of SO_4^{2-} in the liquid phase, sufficient AF_t can not form, therefore in a short time C_3A hydrates very quickly, the large amount of crystallites of aluminate hydrate interconnect and produce the flash setting.

It may decrease the surface absorbed energy of fluorogypsum particle to add a little quantity of CLS in one minute after adding mixing water, thus the absorbed film of CLS which covers the fluorogypsum particles will be weakened, more SO_4^{2-} will dissolve out and a certain amount of formed AF_t will control the normal setting of the cement paste effectively.

1 Introduction

Calcium Lignosulfonate (CLS) is a common water reducer for concrete. For the cement with gypsum as retarder (GC), it plays an important role in both water reducing and retarding, achieving good results not only in use but also in economic benefit. However, the addition of CLS to the cement with fluorogypsum as retarder (FC) usually causes rapid setting. Thus the inexpensive but substantial CLS

135

water reducer can not be used in the building engineering projects using FC.

From the experiment and analyses, the authors have put forward mechanism of rapid setting due to the addition of CLS water reducer to FC, furthermore, on the basis of this mechanism, worked out several selected solutions to rapid setting.

2 Experimental Results

2.1 Materials
Fluorogypsum and cement

Fluorogypsum is a by-product by neutralizing lime and waste from making freon. In order to change waste material into things of value and reduce cement production cost, Shanghai Wusong Cement Plant has substituted fluorogypsum for gypsum as cement rearder. Without CLS water reducer, the paste and concrete had good performance, but with this water reducer, rapid setting usually occurred in the paste and concrete made from the cement.

Sand, river sand ($M_k=2.87$)

Gravel, Size 5-25mm

Additives, CLS water reducer, AF air entraining high efficiency water reducer, Molasses water reducer, Sodium sulfate etc.

2.2 Results
2.2.1 Setting Times of Paste
A. Effect of Different Water Reducers on Setting Times. The effects of the type and dosage of additives on setting times are shown in Table 1.

The additives No 1, 4, 5, 6, 7 made the paste stay within the specified setting times whereas the addition amount of AF high efficiency water reducer was one time more than that of CLS water reducer. For the cost of water reducer in $1M^3$ concrete, the former exceeded the latter by 5 times. The slump loss of concrete with AF was greater than that with CLS. On condition that CLS water reducer and Na_2SO_4 were added as a combined admixture, and that the quantity of Na_2SO_4 added was over 0.35%, the paste had normal setting times. With increasing Na_2SO_4, initial setting times delayed, and the duration between initial and final setting times shortened. Separate addition is inconvenient to use.

Table 1. Effect of Various Additives on Setting Times

No	Type and dosage of additives	setting times (hr, min)		water reduction rate %
		initial	final	
1		6:00	8:53	0
2	0.25% CLS water reducer	0:08	0:10	10
3	0.15% molasses water reducer	0:07	0:09	10
4	0.50% high efficiency water reducer	6:30	8:58	14
5	0.25% CLS water reducer + 0.35% Na_2SO_4	6:08	6:35	10
6	0.25% CLS water reducer + 0.40% Na_2SO_4	6:45	7:25	10
7	0.25% CLS water reducer + 0.50% Na_2SO_4	7:45	8:34	10

B. Effect of Delayed Addition of CLS Water Reducer
If CLS water reducer was added after an initial mixing period of three minutes, setting times of the paste became normal. The more delayed the time of addition of CLS, the more noticeable the effect of retarding setting times was.

C. Effect of Small Dosage CLS and AF Water Reducer Admixture
After an initial mixing period of 1 minute, the addition of small dosage CLS and AF admixture made, the paste stay within the specified setting times. The concrete of high quality and low cost could be obtained.

2.2.2 Setting Times of Concrete
The experimental results of setting times of concrete are shown in Fig.1. It can be seen from Fig.1 that, of the concrete made from FC, the reference and the concrete with delayed addition of admixture were normal in initial and final setting times, but the concrete simultaneously admixed with 0.25% CLS water reducer started initial setting until 11 minutes .
In addition, according to test results, the slump loss of concrete with delayed addition of admixture was similar to that of reference concrete, its strength developed normally.

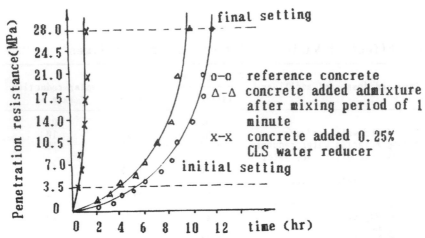

Fig. 1. The experimental results of concrete setting times

2.2.3 Measurement of SO_4^{2-} Concentration in Liquid Phase of Paste

Fig 2 shows the measurement results of SO_4^{2-} concentration in liquid phase of paste.

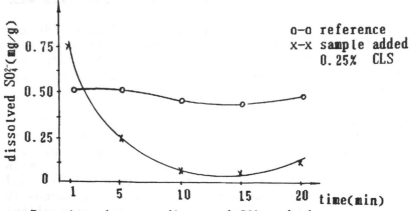

Fig. 2. Relations between dissolved SO_4^{2-} and time

It can be demonstrated that the dissolved SO_4^{2-} in the sample made from FC was regular, but in the sample added 0.25% CLS, the dissolved SO_4^{2-} quickly dropped the extremely low value.

2.2.4 Measurement of Physicochemical Properties and Hydrates
A. Adsorption Test
After three minutes, the concentration of residual CLS water reducer in the liquid phase of paste made from FC was measured only to be 12%.. However, under the same

condition, in the liquid phase of paste made from GC, it was 25%. This indicated that the adsorbability of CLS water reducer on FC was much stronger than that on GC.

B. Zeta Potential Test

The test results of the zeta potential in samples added CLS water reducer are given in Table 2. It shows that under the same condition, the zeta potentials after adsorbed water reducer on fluorogypsum and FC were much higher than those on gypsum and GC, and in FC, the potential of delayed addition of CLS was lower than that of simultaneous

Table 2. The test results of zeta potential

No	samples	amount of CLS added(%)	potential(ξ) mv
1	Fluorogypsum	0.25	13.71
2	FC	0.25	14.74
3	FC	0.25(delayed addition)	10.78
4	gypsum	0.25	6.60
5	GC	0.25	9.76

addition of CLS. It can be deduced from the zeta potential (ξ) variations that under the same condition, the adsorption potentials of CLS water reducer on fluorogypsum and FC were much higher those on gypsum and GC.

C. Measurement of Hydrates

It has been found by XRD that after admixed FC with CLS water reducer, during early hydration a very small proportion of AF_t formed, but it was obvious that there was calcium aluminate hydrates there. Conversely, in FC paste without CLS water reducer or with delayed addition of admixture, a noticeable amount of AF_t occurred but calcium aluminate hydrates were not detected in early hydration period.

3 Mechanism of and Solutions to Rapid Setting

3.1 Mechanism of rapid Setting

The analyses of the test results show that when admixed FC with 0.25% CLS water reducer by ordinary method, the reason for rapid setting in the paste and concrete is as follows,

Since CLS water reducer is swiftly adsorbed on the extremely strong adsorbent fluorogypsum in FC, the release of SO_4^{2-} in fluorogypsum is obstructed. Because of the insufficient SO_4^{2-} in liquid phase of paste, the amount of AF_t produced is not much enough effectively to restrain quick hydration of C_3A in cement. Then a considerable quantity of calcium aluminate hydrates form due to the fast hydration of C_3A so that rapid setting occurs. Under the condition of high content C_3A and low content SO_3 in cement, fine cement fineness, high temperature as well as high content F^-, the setting becomes faster.

3.2 Solutions to Rapid Setting
3.2.1 Selection of Suitable Additives for FC
Generally speaking, high efficiency water reducers, especially air entraining high efficiency water reducers, are more suitable for FC. But they are higher both in cost and in concrete slump loss. Therefore, there are certain difficulties in practice.

3.2.2 Addition of CLS and Na_2SO_4 Admixture
In general, simultaneous addition of 0.25% water reducer and over 0.30% Na_2SO_4 can prevent rapid setting while only addition of 0.25% CLS water reducer will cause rapid setting. However, this measure is inconvenient in the operation and Na_2SO_4 easily crystallizes. It is also difficult to put into effect.

3.2.3 Delayed Addition of Admixture
The admixture is added after an initial mixing period of about 1 minute. This method can avoid rapid setting that only addition of 0.25% CLS will result in Water reduction is desirable in a concrete mix, its strength develops normally, and the cost is lower. Now in Shanghai, this method has been adopted at several concrete mixing stations that use FC.

References

M.R.Rixom, Chemical admixtures for concrete
    E.& F. N. Spon, New york (1978).
S.M.Khalil & M.A.Ward, Mater. struct., 56, 57(1977)

W.C.Hansen, J.Mater, 5, 842(1970).
V.S.Ramachandran & R.F.Feldman, Mater. constr.(Paris) 5, 67(1972).
S. Diamond, J.Am, Ceram. soc.,54, 273(1971); 55, 177 (1972); 55, 405(1972); 56, 323(1973).
V. S. Ramachandran, R.F.Feldman, J. J. Beaudoin, CONCRETE SCIENCE, Treatise on current research, Heyden & son ltd (1981).

COMPARATIVE EXAMINATIONS OF ADMIXTURES TO CEMENT

R. KRSTULOVIĆ, P. KROLO, A. ŽMIKIĆ, T. FERIĆ and J. PERIĆ
Laboratory for Inorganic Technology, University of Split, Yugoslavia

Abstract
The aim of this paper has been to present the results of laboratory experiments regarding applicability of cement admixtures when comparative examination methods were used. Various commercial admixtures have been used, in different quantities. Special attention has been paid to hydration of cement in presence of admixtures. The calorimetrical, electrochemical, and spectrophotometrical methods have been used to indicate the changes in development of hydration in presence of admixtures to cement particles. In some cases, the physical and mechanical properties of the samples have been determined. The measurement results are presented graphically and in tables; a comparative analysis has been carried out. On the basis of data and results obtained, some conclusions have been reached regarding the influence of admixtures on the hydration mechanism and kinetics, as well as determination of the optimal quantity of admixture to the cement.
Key words: Cement, Admixture, Hydration process, Method, Electrochemical examination, Calorimetry, Spectrophotometry, Adsorption.

1 Introduction

Admixtures help to influence the development of the cement hydration process. They govern and control the cement hydration mechanism and contribute to new properties of mortar and concrete. Bearing in mind the requirements as regards application, workability and transportation, the ability to control the hydration process in preparation of cement mortars and concrete is of great interest. Because of this, the usage of admixtures has been increasing, although the theoretical basis for their usage and effect on cement hydration - due to their variety - has not yet been completely understood.

This paper presents an overview of our comparative examinations of effects of admixtures on cement hydration, where different methods were applied; the results obtained are also presented.

2 Experimental methods

2.1 The object of examinations
In order to explain the hydration process, examinations have been lately directed towards electrokinetic, thermokinetic, and adsorption properties of hydrating cement. A combination of methods has been applied: electrochemical, thermochemical, and spectrophotometrical methods, which provides for a simple and constant tracing of hydration process. These methods are especially useful in a study of effects of various parameters on the cement hydration process.

This paper presents the results obtained by these three techniques for tracing the effects of admixture and their applicability to cement/concrete.

2.2 Methods
The conductivity, electrokinetic, and electrodynamic potential were measured in order to determine the electrochemical properties acompanying the cement hydration process. The potentiometrical method associated to determination of electrodynamic potential (E) is often represented only by determination of pH-values. In order to obtain continuous measurements, metal indicator electrodes (Bi, Mo, etc.) are usually used instead of glass ones. These electrodes are particularly suitable for a study of effects of different admixtures on cement or concrete, where the curve $E = f(\tau)$ may directly lead to certain conclusions about the nature and effects of admixtures (accelerators, retarders, plastifiers, etc.) on cement hydration.

The second important feature of the electrochemical method is measuring the change in the electromotive force (EMF) in a galvanic cell consisting of two metal electrodes (lead and copper) immersed into a cement paste, mortar or concrete, which shows certain advantages over Vicat's method for determining the type of admixture, and may be used to determine the optimal condition for programmed cement or concrete setting for the quantity of admixture used.

A pH-meter linked to a recorder was used to measure the electro-dynamic potential (E). The Bi and Mo indicator electrodes were used, while the referent electrode was the standard calomel one; measurements were taken at 293K. The apparatus used in measuring the cement setting time consisted of a sample holder (Vicat's ring or a plastic vessel), a measuring electrode (Cu and Pb), a pH-meter, and a graphical recording device.

The second method applied was calorimetry. Techniques based on measuring thermal effects of chemical reactions are generally considered to be thermochemical methods. In cement studies, they are generally used to determine thermal effects accompanying the hydration process; different kinds of calorimeters are used mainly standardized. Thermochemical measurements were carried out by the thermos bottle method according to JUS B.C8.027.

The third technique used was spectrophotometry, to determine the adsorption properties of cement to admixture. The spectrophotometrical method is an analytical qualitative and quantitative method, used primarily in analysis of homogeneous solutions. Ultraviolet and visible light spectrophotometry provides for an analysis of hydrated cement solution composition in presence of admixtures by registering

143

the adsorption spectra of individual constituents or reaction products. This optical characterization in different reaction conditions makes it possible to trace the adsorption of admixtures in the cement-water-admixture system. Spectrophotometrical measurements of admixture adsorption in a solution provide for determining the mass of admixture in the solution, the maximum duration of adsorption, the adsorption isotherm, the cement surface coverage by admixture, and the optimal mass percent of the admixture for the cement in question. The Perkin Elmer 124 spectrophotometer with a recorder was used in quartz cuvettes, with the optical path p = 0.1 dm in the vawelenght range from 200 to 390 nm.

2.3 Samples
Industrial Portland cement was used in experiments either pure or with addition of slag (z) or puzzolana (p) and flying ash (lp). The markings were: PC-55, PC-20z-45, PC-20p-45, PC-20lp-55. Cement samples were examined and marked according to the Yugoslav standard (JUS). The admixtures used were commercial kinds of accelerators, retarders, plastifiers, and superplastifiers. Admixture types and markings are indicated in each figure and for every cement sample used in experiments.

3 Results

Measurements were carried out by the methods described, in order to prove that more exact examinations of cement hydration may be achieved as regards cement setting and hardening, especially when admixtures are present, and that they may be applied in practice.

Figures 1, 2, and 3 compare electrochemical and calorimetrical measurements. They show the curves of change in potential $E = f(T)$ when bismuth (Bi) and molybdenum (Mo) indicator electrodes are used. The setting time was determined by using the lead-copper (Pb-Cu) electrode in comparison to Vicat's needle indicated by a shaded rectangle in the figures. The curve of change in heat, dQ/dT with time was obtained from the calorimetrical measurements.

Comparative measurements (Figure 1) indicate the change in potential $E = f(T)$ and its maximum values to take place in the ascending part of the differential curve dQ/dT, i.e. that part of the curve characteristic of a period of rapid hydration. This period is characterized by intensive hydration reaction, by formation of hydration products, whose properties bring about the changes registered both by thermochemical and by electrochemical method. In the period of slower hydration (the descending part of the differential curve dQ/dT) the changes in potential measured by the Bi and Mo electrodes vary only slightly, and reach their final values at the end of this period. Figure 1 also shows the setting time measured by the electrochemical method and by Vicat's needle. Measurements shown indicate that the change in potential in the Pb-Cu electrode varies slightly until the beginning of the setting time determined by Vicat's needle (shaded area). The beginning of setting time is registered by an abrupt change in the Pb-Cu electrode potential, while the end of setting time is indicated by final values, i.e. slight changes in the

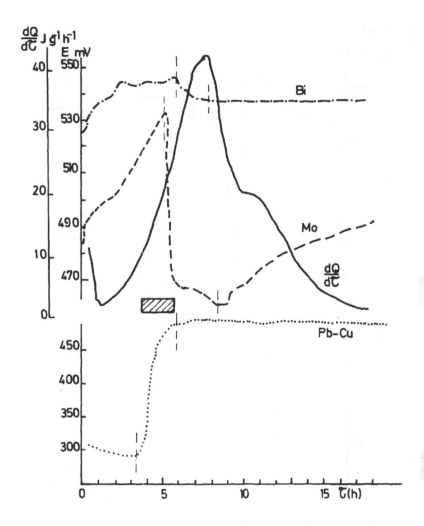

*Figure 1. A comparison of calorimetrical and electrochemical measure-
ments and setting time by Vicat's needle (the shaded rectangle)
for the PC-55 sample, W/C=0.3*

Pb-Cu electrode potential with time. Results obtained by the thermochemical method for samples with and without admixtures (Figures 2
and 3 – curves dQ/dτ) show a certain similarity. Differences which
may be observed are due to the intensity of effects registered and
their position in relation to the time axis, representing the hydration duration. The location of effects registered and the change in
intensity for samples containing admixtures as compared to samples
without admixtures are caused by the effect of admixture on cement
hydration. The admixtures applied do not influence the hydration
process mechanism, but they influence the period when intensive hydra-

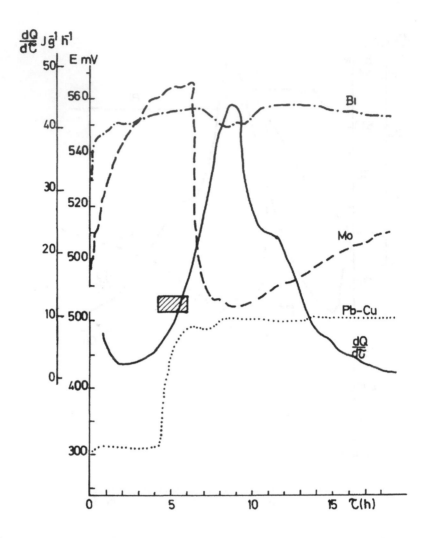

Figure 2. A comparison of calorimetrical and electrochemical measurements and setting time by Vicat's needle (the shaded rectangle) for the PC-55 sample, W/C=0.3, with 0.3% of the Delta Cementol admixture (accelerator plastifier)

tion occurs, and duration of individual hydration stages.

Admixtures with plastifying properties affect the induction period, which becomes longer if retarder type admixture is applied (e.g. Retardit).

The results obtained by thermochemical and electrochemical measurements show a good correspondence for the cement examined, which proves that these techniques may be justly used to trace the kinetic parameters influencing the cement hydration process. Measurements prove that admixtures regulate the setting time. They also prove the

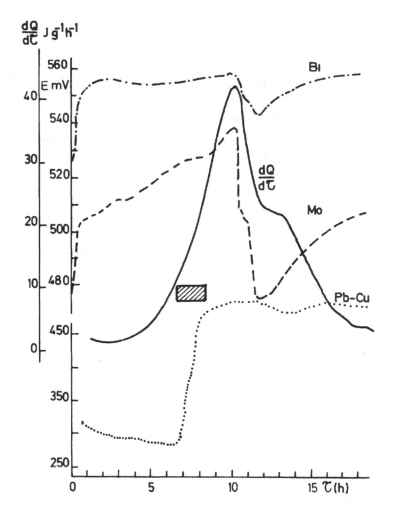

*Figure 3. A comparison of calorimetrical and electrochemical measure-
ments and setting time by Vicat's needle (the shaded rectangle) for
the PC-55 sample, W/C=0.3, with 0.6% of the Retardit admixture*

equivalence of the electrochemical and thermochemical methods.
    Determination of setting time by a metal indicator electrode
(Pb-Cu) was examined on the mortar and concrete samples made from the
same cement (PC-20p-45) with the Fluidal VX/OC (Ruredhil) admixture.
    Figure 4. compares the measurements of the setting time for cement
paste, mortar, and concrete with the same water/cement factor, and
for cement mortar with different quantities of admixtures.
    As shown by the results obtained, concrete, mortar, and cement
paste at the water/cement factor W/C=0.5 have the same setting time,
which makes it possible for the electrochemical method to be also
used in the systems where aggregates are present.

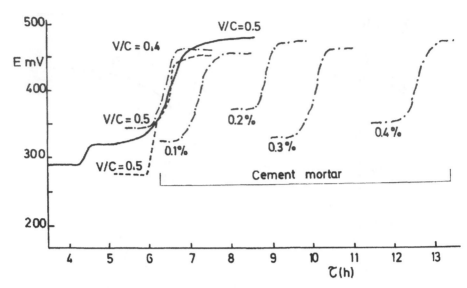

Figure 4. A comparison of setting time values obtained by the electrochemical method for cement paste, concrete and mortar samples, and the
effect of the Fluidal VX/OC (Ruredhil) admixture on mortar

The spectrophotometrical method was used to examine the effect of
admixtures on cement hydration in relation to adsorption of admixture
to cement surface. Three commercial superplastifiers were used: the
sulphonated melanine formaldehyde condensate (a1), and the sulphonated
naphthalene formaldehyde condensate of low molecular mass (a2) and
of high molecular mass (a3) with the industrial PC-55 cement with a
low content of aluminate and ferrite (c1) and with a high content of
aluminate and ferrite – PC-20lp-55 (c2). Measurements were taken in
the area of admixture mass percent ranging from 0.1% to 2.0% with
water/cement factors W/C=2 and W/C=1, at the temperature of 298K.
The adsorption of admixture to the cement suspension was traced
spectrophotometrically at λ max=220 nm and 228 nm. Prior to measuring
admixture adsorption to cement, the admixture absorption spectra were
determined in order to examine their stability in reaction conditions
in diluted solutions and in the first few minutes of initial hydration. Admixture adsorption to cement takes place within the first five
minutes of hydration; the admixture sensitivity in spectrophotometrical measurements ranges from $1x10^{-5}$ to $8x10^{-6}$ g/dm^3.
Measurements of admixture absorption at the corresponding vawe-
length, prior to and after adsorption determine the admixture mass
adsorbed to cement particles and the equilibrium mass concentration of
admixture in the solution. Figures 5 and 6 present the results
graphically.
The adsorption isotherms:

$$y = a \cdot \gamma / (1 + b \cdot \gamma)$$

were used to calculate linearized Langmuir's adsorption isotherms with
appropriate coefficients a and b, and cement surface coverage, θ :

148

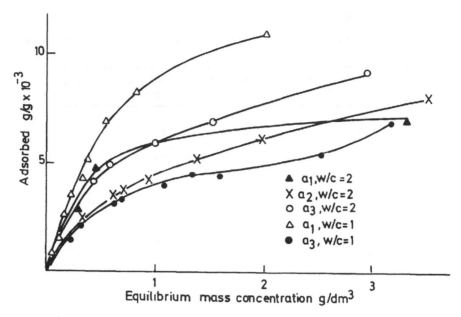

Figure 5. *Admixture adsorption to cement PC-55, c1, W/C=2 and W/C=1,
T = 298K*

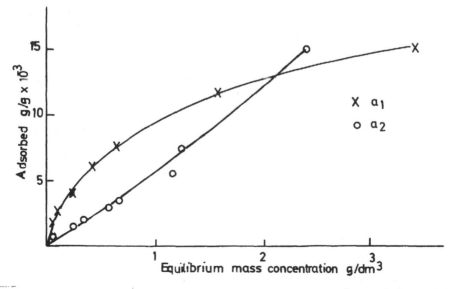

Figure 6. *Admixture adsorption to cement PC-55, c2, W/C=2, T=298K*

$$\Theta = b \cdot \Gamma / (1 + b \cdot \Gamma)$$

Table 1. Cement surface coverage by admixture at highest admixture
       mass percents

Admixture	Mass percent, c/c %	Cement	W/C	Surface coverage, Θ
a1	1.5	c1	2	.76
a2	1.0	c1	2	.67
	1.5	c1	2	.79
a3	1.0	c1	2	.69
	1.5	c1	2	.81
a1	1.5	c2	2	.49
	2.0	c2	2	.66
a2	1.5	c2	2	.14
	2.0	c2	2	.16
a1	1.5	c1	1	.89
a3	1.0	c1	1	.78

If adsorption isotherms are compared, it may be seen that cement
adsorbs the sulphonated melanine formaldehyde condensate best at
higher water/cement factors, when cement contains larger quantities of
$3CaOAl_2O_3$ and $3CaOAl_2O_3Fe_2O_3$. This difference in admixture adsorption
is ascribed to chemical compositions of admixture, solution, and
cement.

Beside SO_3 groups, the sulphonated melanine condensate contains
nitrogen, which may donate its free electron pair, thus increasing the
number of active locations able to adsorb, as opposed to the sulphonated naphthalene condensate. For that reason the c2 cement, richer in
aluminate and ferrite, adsorbs a higher quantity of the sulphonated
melanine condensate admixture. The higher the water/cement factor, the
more intense the hydrolysis and hydration, the lower the concentration
of ions leaving cement and going into the solution, the higher the
dispersion of cement particles, which all leads to a higher admixture
adsorption.

It should be noted that the admixture a2, although containing the
lowest quantity of polymeric molecules, adsorbs to cement c1 similarly
to the admixture a3, while the adsorption of admixture a2 to cement c2
increases linearly with the increase of the mass percent in the overall
work range from 0.1 - 2.0%. The results obtained are all the more
significant, as some authors claim that monomers and lower polymeric
fractions adsorb poorly.

All these effects may be studied in relation to the surface coverage calculated for given conditions. Although the admixture mass
adsorbed is higher with cement c2, its surface coverage is the lowest,
which is probably due to admixture adsorption, and also to the reaction
of admixture with hydrated and hydrolyzed constituents of aluminate
and ferrite.

Table 2 presents the values for non-adsorbed admixtures obtained by
approximation of admixture ratio prior to and after adsorption to the
null value of the mass percent.

Table 2. Non-adsorbed admixture

Admixture	Cement	W/C	Non-adsorbed admixture
a1	c1	2	.115
	c2	2	.145
a2	c1	2	.125
	c2	2	.245
a3	c1	2	.125
	c1	1	.145

The value of the segment on the ordinate corresponds to the non-adsorbed admixture present in the solution, and has been ascribed to presence of monomers in commercial admixtures in literature. The results obtained by these measurements prove that this ratio changes even for the same admixture in dependence on the water/cement factor and the adsorbent applied.

If this ratio really represents the non-adsorbed monomer present, it should be the same for all adsorptions examined. Besides, for admixture a2 there should be a higher quantity of non-adsorbed admixture than for admixture a3. As this is not true, this ratio is believed to represent the non-adsorbed admixture of all polymeric fractions, which means that not only polymers but lower fractions as well adsorb to cement.

In order to obtain a better explanation for admixture adsorption, beside "regular" adsorption in the cement-water-admixture system, examinations of "delayed" adsorption were carried out. The results obtained are presented in Figure 7 and Table 3.

Table 3. Langmuir's adsorption isotherm coefficients a and b, surface coverage, and non-adsorbed admixture for "regular" and "delayed" admixture adsorption to cement PC-55, W/C=2, T=298K

	Coefficients		W(a/c)%	θ	Non-adsorbed admixture
	a	b			
regular	.00803	1.410	1.5	.880	.185
	.00803	1.410	2.0	.911	.185
delayed	.00512	.888	1.5	.824	.234
	.00512	.888	2.0	.866	.234

A relatively small difference in admixture mass adsorbed and in surface coverage indicates that the adsorption takes place on non-hydrated and on hydrated cement samples.

In regular adsorption, dipolar molecules of water and admixture equally compete for cement surface. Although there is 100 - 2000 times more water present than admixture, the admixture adsorbs to cement surface, disperses the suspension, and thus makes possible further adsorption of both admixture and water molecules. Water molecules adsorbed now perform their function of cement hydration and hydrolysis.

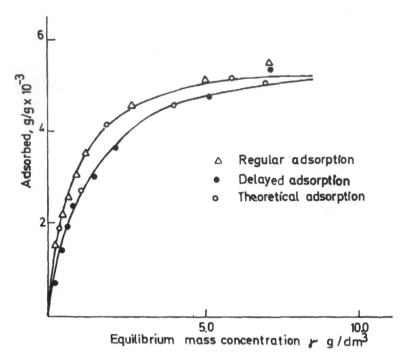

Figure 7. Admixture adsorption to cement PC-55, W/C=2, T=298K

In delayed adsorption, dipolar molecules of water existed alone for
2.5 minutes on cement surface and were adsorbed by chemisorption. The
admixture added after this initial adsorption competes for non-hydrated
and hydrated cement surface, so that processes of adsorption, hydration
and hydrolysis take place at the same time. The admixture, adsorbed by
its charge to SO_3-groups, displaces a part of water molecules already
adsorbed to cement surface. The admixture, being a polymeric molecule,
adsorbs according to Kondo's "flat" adsorption, and removes water
molecules from cement surface by its spherical disturbance. Because of
this, in regular adsorption there remains approximately 19% of non-
adsorbed admixture, while in the delayed adsorption this percent
increases to 23% approximately.

The effects of admixtures on bending and compressive strenght was
examined in the work range from 0.3 - 1.5% of admixture for the
constant water/cement factor W/C=0.4. Examinations were carried out
according to JUS standard, and the results were compared to the
standard for W/C=0.5 and W/C=0.4.

Figures 8 and 9 show some results obtained, while Table 4 shows a
greater number of results for all three admixtures examined.

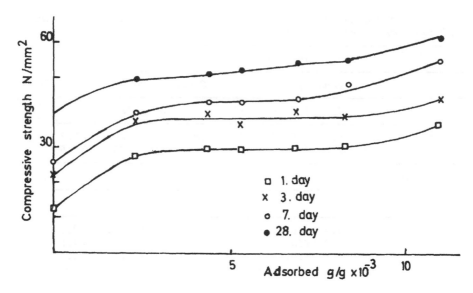

Figure 8. Dependence of compressive strenght on admixture a1 mass adsorbed for cement PC-55, c1, W/C=0.4, adsorption time 5 minutes, W/c1=2

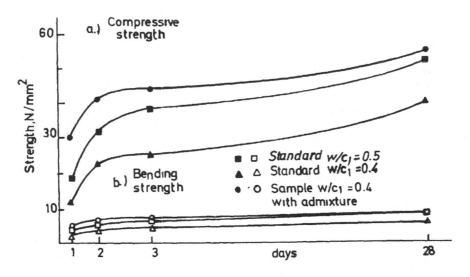

Figure 9. Dependence of compressive and bending strength on time for cement PC-55, c1, with 0.8% of admixture a1, W/c1=0.4, and standards for W/c1=0.4 and W/c1=0.5

Table 4. Dependence of cement compressive and bending strength on the admixture mass percent and setting time, W/C=0.4

	Strength, N/mm^2							
	Bending				Compressive			
days	1	3	7	28	1	3	7	28
standard W/C=0.5	4.1	5.7	6.8	7.7	18.7	32.3	38.6	51.9
standard W/C=0.4	2.5	3.8	4.7	5.7	12.2	22.4	25.2	39.5
mass percent %								
a1/c1								
0.3	4.5	6.6	5.3	6.9	27.5	38.0	40.0	49.5
0.8	5.3	7.0	7.1	7.7	30.1	41.4	43.9	54.9
1.5	6.0	8.0	8.0	8.9	37.2	48.9	55.1	62.5
a2/c2								
0.3	3.4	5.2	6.3	6.9	21.6	35.9	42.4	51.0
0.6	5.3	6.4	6.2	8.0	29.8	38.2	43.9	57.9
1.5	5.4	6.5	6.9	7.9	32.2	41.3	45.6	55.9
a3/c1								
0.3	4.7	6.3	6.3	6.4	24.1	37.1	44.1	62.9
0.6	5.9	6.4	7.3	7.3	31.8	43.8	49.1	54.8
1.5	6.8	7.6	8.4	8.8	35.2	47.9	54.3	64.2

Bending and compressive strength developed increases with the increase in the admixture mass percent and with the duration of strengthening. The analysis of results obtained indicates the highest increase to be achieved with 0.3% of admixture added and in the first day of strengthening. Further addition of admixture, until approximately 0.6% or 0.8%, further increases the strength, and afterwards assumes more constant values.

This increase in strength when the admixture mass increases corresponds to adsorption results. When adsorption increases, the dispersion of cement particles increases, and therefore the cement paste homogeny increases too. Homogeneous pastes yield higher strengths. On the other hand, this effect is counter-balanced by a slowing in the early stage in cement particle hydration when admixture adsorption increases. When admixture is added beyond the optimal mass concentration, dispersion does not increase, while strenth increases for only 12-20%. Because of this, strength increases are more prominent in the first day, approximately 70-200% in the overall work area, while in the 28th day, although strenth continues to increase, it is only for about 14-60%.

Strength measurements with admixtures of lower and higher molecular mass applied, do not indicate significant differences. This is an important result, as the examinations have so far led to a belief that only large polymeric parts of admixtures adsorb to cement and thus change its properties.

These mechanical results showing good correspondence to adsorption results, prove the optimal mass percent for admixture a1 to be approximately 0.8%, and for admixtures a2 and a3 approximately 0.6%, thereby reducing the quantity of necessary water by 20%.

References

Aschen, N (1966) Determining the setting time of cement paste, mortar and concrete with a copper-lead electrode, Mag. of Conc. Res. Vol 18, 56, 153

Burk, A.A, Gaidis, J.M. and Rosenberg, A.M. (1981) Adsorption of Naphthalene-based Superplasticizer on Different Cement, Res. Division of Inter. Conference on Superplast. in Conc. Ottawa, 485

Collepardi, M. et al. (1980) Influence of Sulfonated naphthalene on the fluidity of cement pastes, 7th Intern.Congr. on the Chem. of Cement, Paris, Vol. VI-20

Daimon, M. and Roy, D.M. (1978) Rheological Properties of Cement Mixes: 1. Methods, Preliminary Experiments, and Adsorption studies. Cem. Concr. Res. 8, 753

Krstulović, R., Krolo, P. (1984) A study of the kinetics and mechanism of the cement hydration process, Part I, Hem. Ind. Vol.38, 2, 36

Krstulović, R., Ferić, T., Krolo, P (1980) Study of kinetics of industrial cements hydration, 7th Inter. Symp. Chem. Cem., Vol II, II-153

Mtschedlow-Petrossian, O.P., Salop, G.A. (1966) Untersuchung des Erhärtungsvorganges durch pH-Wert-Messung, Silicattechnik, 17, 7, 205

Žmikić, A. and Krstulović, R. (1987) Spektrofotometrijska i konduktometrijsko praćenje utjecaja aditiva na hidrataciju cementa, Hem. Ind. Vol. 41. 2. 91.

3 STRENGTH

STUDY OF THE EFFECTIVENESS OF WATER-REDUCING ADDITIVES ON CONCRETE WITH MICROSILICA

J. C. ARTIGUES, J. CURADO and E. IGLESIAS
Concrete Admixture Research Department, TEXSA S.A., Barcelona, Spain

Abstract
This paper shows the influence that the use of different types of water-reducing admixtures in concrete with microsilica, exerts on two important characteristics of concrete: its mechanical strengths and its chemical resistance, reflected by the greater or lesser resistance against aggresive chemical attack. Results obtained show that the introduction of microsilica (MS) improves the resistances, but that when a water-reducing agent is also added, the increase in mechanical strengths is much greater. Thus, the carboxylic derivative-modified lignosulphonate based admixture is the one that originates a greater increase in resistances and strengths, as, besides its excellent water-reducing power, it provides a more stable structure in the concrete. The carboxylic polymer-based admixture is the best water-reducing agent (the best dispersing agent) and it also produces a strong increase in resistances. The presence of MS in the concrete produces a notable diminishing of the capillary porosity of same, thus hindering the penetration of aggresive fluids and resulting in a considerable increase in the resistance of the concrete against chemical attack. For the same reason as the diminishing of porosity (or increase in closeness), the admixtures producing a lesser water/cement ratio, are also those that improve attack resistance.
Key words: Microsilica, Admixtures, Mechanical Strength , Chemical Attack.

1 Introduction

Microsilica is a synthetic material made up of microscopic silica particles with a large specific surface ($20 \ m^2/g$).

When added to concrete, it provides two effects: as an ultrafine filler and a superactive pozzolana. These effects give high resistance and great durability to the concrete.

The high specific surface, ultrafine filler nature of microsilica will, in practice, demand a greater value of w/c ratio, for a specific consistency of the concrete; as this is not desirable, the joint use of Microsilica + Water-reducing agent is, in fact, the only viable way to make good use of the advantages of microsilica.

Therefore, the study of the most suitable water-reducing admixture to be combined with Microsilica, is very important.

As typical representatives of water-reducing admixtures, we have, in this study, used the following products:

* Modified lignosulphonates.
* New developments of the carboxylic polymer-based admixture.
* Melamine-Formol-Sulphonate (MFS) condensates.
* Naphtalene Sulphonate-Formol Aldehyde (NSF) condensates.

All these present different chemical structures, but the same physical-chemical behaviour in the cement paste: on disolving, they create "anions", with greater or lesser power-load density, which are absorbed into the surface of the cement particles as these are negatively loaded, they drive each other back, separate and disperse, thus diminishing the viscosity of the paste. For this reason, these admixtures can be classified as Anionic-type Dispersing Agents.

Other than these water-reducing admixtures, and solely as a simple comparison, we tried the use of a Vinsol resin-based (abietic acid derivative) aerating admixture.

2 Experimental methods

To expose the joint influence of Microsilica + admixtures, insofar as mechanical strength and chemical resistance of concrete is concerned, a series of experiences were carried out, based on the following parameters:

2.1 Materials used

(a) Microsilica (MS), Elkem Materials' type 920-D.

(b) Portland Cements: Two types of Portland cement were used: II-Z/35A or I/45 A.

(c) Admixtures:
       - A: Carboxylic polymer-modified lignosulphonate-based.
       - B: Carboxylic polymer-based.
       - C: Melamine-Formol Sulphonate (MFS) condensates.
       - D: Vinsol resin-based, combined with surface-active agents.
       - E: Naphtalene sulphonate condensed with formic aldehyde (NSF).

All these are Texsa S.A.-formulated products.

With these materials, and each case, concrete with the following cement aggregate composition was produced:

```
Cement                          15.90 %
Sand                            46.60 %
Crushed aggregate (5-14 mm)     20.50 %
     "         "    (14-25 mm)  17.00 %
```

_.2 Variables under study

Different batches of concrete using the indicated composition and abovementioned materials were manufactured, subject to the variation of the following parameters:

* Type of cement: II-Z/35 A and I/45 A.
* Two dosages of admixtures.
* Consistency: 8 cm and 17 cm measured as shrinkage in Abrams cone.
* Addition of microsilica: 10% of the cement weight.

The concretes resulting from the variation of these parameters are measured when fresh for several characteristics, such as consistency and entrained air, as well as two properties once hardened: mechanical strengths and chemical resistance.

2.3 Test Methods

(a) Fresh concrete: Production of the different batches of concrete, as well as the measuring of consistency and entrained air, were carried out according to ASTM C-494 standard.

(b) Hardened Concrete: Mechanical strengths and its durability under chemical attack are studied on the hardened concrete.

- Mechanical strengths: Compressive strength of the 15 x 30-cm test pieces is measured at the ages of 1 day, 7, 28 and 90 days, according to ASTM C-494 Standard.

- Resistance against chemical attack: This resistance is measured by the weight loss speed of the sample, either by means of dissolution of the component material of the concrete or breaking up of the parts of same, when submitted to different types of aggressive chemicals.

For the practical carrying out of these experiences, 5-cm thick discs are cut from the 15 x 30 test pieces, with a diameter of 15 cm, at the age of 28 days; three 15 x 5 discs are selected from the centre part of the test pieces, for each solution.

The discs are placed in the following aggresive solutions:

- Chlorhydric acid   1 %
- Acetic acid   5 %
- Lactic acid   1 %

These acid solutions are held at a temperature of $20 \pm 1^{o}C$, with weak recirculation of liquid and renewed every 6 days. After 30 days' exposure, the discs are withdrawn, washed, dried and weighed.

The relative weight loss is determined, average of every 3 discs from the same test piece, by means of a WEIGHT LOSS FACTOR (WLF) defined as follows:

WLF =   (Initial disc weight - Final Disc weight) / Initial disc weight)   x 100.

The chemical resistance test is carried out on 4 types of concrete, with II-Z/35 A and I/45 A cements, respectively, and both with and without Microsilica together with the studied admixtures.

3 Results

1) Mechanical strength
   (a) With II-Z/35 A Cement

Test nº	MS (%)	Admixture (%)	Air (%)	W/C ratio	Plasticity (cm)	Strength (Kg/cm²) days 1	7	28	90
1	–	–	2.2	0.469	9	94	184	228	246
2	10	–	1.8	0.516	8	94	221	296	326
3	10	A, 0.4	3.1	0.440	8.5	130	281	390	398
4	10	B, 0.7	2.8	0.438	8.5	128	268	373	379
5	10	C, 1.4	2.1	0.460	8	108	251	337	376
6	10	D, 0.03	4.5	0.491	8	85	191	286	291
7	10	E, 0.7	2	0.462	8	97	223	341	345
8	–	–	2.2	0.496	8	91	185	214	249
9	–	A, 0.4	3.5	0.442	8	122	236	278	311
10	–	B, 0.7	3.2	0.440	8	112	221	265	288
11	–	C, 1.4	2.6	0.447	8	112	205	256	301
12	–	D, 0.03	6.1	0.486	9	73	158	167	194
13	–	E, 0.7	2.7	0.460	8	97	190	223	261
14	–	–	1.9	0.577	17.5	77	160	193	243
15	10	–	1.2	0.650	17	76	201	286	290
16	10	A, 0.8	3.6	0.526	17	84	284	397	416
17	10	B, 1.2	2.6	0.525	17.5	119	264	382	430
18	10	C, 2.4	1.6	0.553	18	100	259	326	347
19	10	D, 0.06	8.5	0.616	17	43.5	131	179	250
20	10	E, 1.2	1.7	0.572	17	81	221	289	314
21	–	–	1.9	0.630	18.5	83	174	200	234
22	–	A, 0.8	4.2	0.542	19	77.5	223	278	324
23	–	B, 1.2	3.2	0.534	19	110	230	260	324
24	–	C, 2.4	2.6	0.540	20	100	209	251	280
25	–	D, 0.06	9	0.612	19	43.5	104	127	139
26	–	E, 1.2	2.1	0.533	19	97	195	223	273
27	–	–	2.2	0.460	8	97	205	231	254
28	10	–	1.9	0.507	8	104	227	302	339
29	10	A, 0.8	3.8	0.427	8	132	296	407	423
30	10	B, 1.2	3.3	0.422	8	150	287	393	407
31	10	C, 2.4	2.1	0.452	8	135	274	365	392
32	10	D, 0.06	6.4	0.490	8	98	185	274	284
33	10	E, 1.2	2	0.454	8	129	237	359	370

(b) With I/45 A cement

Test nº	MS (%)	Admixture (%)	Air (%)	W/C ratio	Plasticity (cm)	Strength (Kg/cm²) 1	7	28	90
34	–	–	2.4	0.440	8	161	257	313	325
35	10	–	1.7	0.507	9	181	302	413	418
36	10	A, 0.4	2.8	0.450	10	208	380	478	482
37	10	B, 0.7	2.5	0.433	8	207	345	442	450
38	10	C, 1.4	1.8	0.470	8	187	312	421	467
39	10	D, 0.03	4.2	0.482	8	163	287	367	368
40	10	E, 0.7	1.9	0.483	8.5	177	305	404	412
41	–	–	2.5	0.477	8	160	251	296	355
42	–	A, 0.4	3.2	0.447	9	203	325	377	417
43	–	B, 0.7	2.7	0.453	9.5	176	279	305	377
44	–	C, 1.4	2.7	0.450	9	198	281	351	382
45	–	D, 0.03	5.8	0.468	9	148	221	271	294
46	–	E, 0.7	2.9	0.450	8	167	265	305	343
47	–	–	2.3	0.566	17	150	224	259	316
48	10	–	1.4	0.651	18	154	286	353	384
49	10	A, 0.8	3	0.490	17	205	370	449	475
50	10	B, 1.2	2.5	0.552	17	203	343	434	448
51	10	C, 2.4	1.7	0.578	17	178	312	386	403
52	10	D, 0.06	7.5	0.640	17	116	211	251	320
53	10	E, 1.2	2	0.567	17	169	317	369	414
54	–	–	2.2	0.580	18	134	221	278	285
55	–	A, 0.8	3.4	0.538	18	174	281	339	390
56	–	B, 1.2	2.8	0.523	18	190	254	321	334
57	–	C, 2.4	2.6	0.527	18	169	281	314	352
58	–	D, 0.06	9	0.560	18	94	172	203	224
59	–	E, 1.2	3	0.523	18	153	233	281	350
60	–	–	2.3	0.459	9.5	140	243	292	385
61	10	–	1.9	0.512	9	158	311	403	405
62	10	A, 0.8	3	0.426	9	193	391	524	525
63	10	B, 1.2	2.8	0.412	9	218	389	502	518
64	10	C, 2.4	2.2	0.432	9	190	358	451	471
65	10	D, 0.06	6	0.495	9	135	249	331	360
66	10	E, 1.2	2.2	0.437	8.5	171	354	418	469

2) Chemical attack: Table of results

Test nº	Cement Type	MS (%)	Admixture (%)	Air (%)	W/C ratio	Weight loss factor HCl 1%	Acetic 5%	Lactic 1%
1	II-Z/35A	10	-	2.2	0.469	13.32	10.02	4.62
2	"	10	-	1.8	0.516	8.66	7.04	3.11
3	"	10	A, 0.4	3.1	0.440	7.42	6.12	2.56
4	"	10	B, 0.7	2.8	0.438	6.76	5.52	2.55
5	"	10	C, 1.4	2.1	0.460	7.21	5.84	2.51
6	"	10	D, 0.03	4.5	0.491	9.10	7.28	3.24
7	"	10	E, 0.7	2	0.462	7.51	6.20	2.70
8	II-Z/35A	-	-	2.2	0.496	14.85	12.06	5.16
9	"	-	A, 0.4	3.5	0.442	13.20	10.70	4.61
10	"	-	B, 0.7	3.2	0.440	12.75	10.36	4.41
11	"	-	C, 1.4	2.6	0.447	13.40	10.88	4.65
12	"	-	D, 0.03	6.1	0.486	15.23	12.29	5.30
13	"	-	E, 0.7	2.7	0.460	14.00	11.36	4.87
34	I/45A	-	-	2.4	0.440	11.10	9.03	3.87
35	"	10	-	1.7	0.507	7.05	5.73	2.41
36	"	10	A, 0.4	2.8	0.450	6.63	5.27	2.28
37	"	10	B, 0.7	2.5	0.433	6.14	4.96	2.09
38	"	10	C, 1.4	1.8	0.470	6.84	5.64	2.39
39	"	10	D, 0.03	4.2	0.482	7.26	5.91	2.54
40	"	10	E, 0.7	1.9	0.483	6.92	5.61	2.41
41	I/45A	-	-	2.5	0.477	12.20	9.91	4.26
42	"	-	A, 0.4	3.2	0.447	10.88	8.86	3.81
43	"	-	B, 0.7	2.7	0.453	11.46	9.33	4.01
44	"	-	C, 1.4	2.7	0.450	11.02	8.95	3.84
45	"	-	D, 0.03	5.8	0.468	13.15	10.70	4.60
46	"	-	E, 0.7	2.9	0.450	11.22	9,12	3.89

4 Conclusions

Analysis of the data obtained, reflected in these experimental results, enables us to extract the following conclusions for each of the properties object of our study.

4.1 On Fresh Concrete
- It is evident that the most energic water-reducing agent is type B (carboxylic polymer-based), for both medium consistency (8-9 cm) and fluid consistency (17-19 cm) concrete, with both types of cement.

The next admixture, insofar as effectiveness is concerned, is the modified lignosulphonate-based admixture (A).
- The addition of MS causes a reduction of entrained air in the

concrete; this is due to a more uniform occupation of the empty spaces by the small MS particles.

MS, therefore considerably reduces capillary porosity.

- Admixture D is withdrawn from this general good behaviour of water-reducing admixtures, owing to it being a typical aerating admixture for attaining levels of air of 4-9%, even at doses as low as 0.03%.

4.2 On hardened concrete

(a) Mechanical strengths: Admixture A, in spite of not being the highest water-reducing agent, is the one that affords greatest compressive strengths; this is due to the chemical composition of the admixture, which creates a more stable structure in the hardened matrix of the concrete.

Admixture B, carboxylic polymer-based and especially at high consistency, also gives good results.

C and D, MFS and NSF unmodified condensates cause a reduction in the water/cement ratio, although not as high as A and B, and therefore give lesser strengths than A and B.

The results make it evident that MS, without admixture, improves the strengths considerably (20-40%), in relation to the control concrete (see first 2 graphs in figures CS-1 and CS-2).

- The addition of any water-reducing (all except D), improves the strength of the concrete even more, when the latter already has MS. These concretes, with both MS + water-reducing agent, present increases in strength, as opposed to the control sample, of 70-100% in the case of reducing agents A and B (see figs. CS-1 and CS-2).

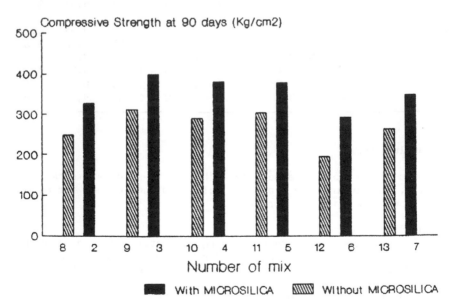

Figure CS1. Commpressive strength
vs admixtures. II-Z/35 A cement.
Slump 8. Low dose.

- There is no daubt that reducing agents such as A and B, given their chemical "anionic" nature, more potent and originating in solution anions with a very high negative power-load density, give a better dispersing effect less water/cement and, therefore, better mechanical strengths (see figs. CS-3 and CS-4).

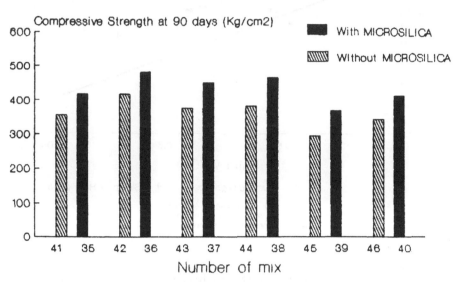

Figure CS2. Compressive strength
vs admixtures. 1/45 A cement.
Slump 8. Low dose.

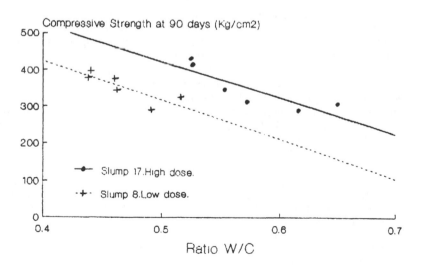

Figure CS3. Compressive strength
vs ratio W/C.II-Z/35 A cement
with microsilica and admixtures.

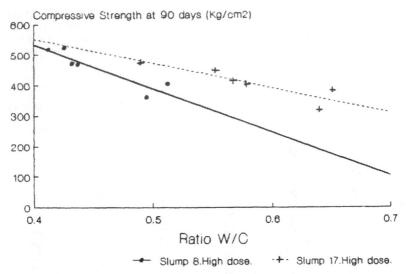

Figure CS4. Compressive strength
vs ratio W/C. I/45 A cement
with microsilica and admixtures.

- Logically, higher strengths are attained using I/45 A, at the
lower consistency (8 cm) and with the higher dose of admixtures (for
lesser w/c). But the improvement offered by MS or better still, by
MS + Admixture, is valid for both cements and for the two
consistencies.
(b) Chemical resistance: The order of aggresivity of the 3 acids
under study, for the 4 types of concrete, is from greater to lesser:
HCl 1% Acetic 5% Lactic 1%, with quite a difference especially in the
lactic acid versus the other two (see figs. CA-1 and CA-2).
- Note that when the cement used in the concrete is I/45 A, it
resists the attack of any of the 3 acids better than when II-Z/35 A
is used(compare figure CA-1 with CA-2).
- The weight loss factor (WLF) is 50-70% loss, when MS is added to
the concrete; this is valid in general, both for HCl and Acetic and
lactic acids. We observe the same principle when comparing the
batches of concrete with Admixtures + MS, with the same type of
cement,with those without MS (see figures CA-3 and CA-4).

- The effect of the different admixtures is less intense than that
of MS, although it is also evident. The best water-reducing agents,
A and B, are those that further increase the attack resistence, very
probably as they originate a lower w/c ratio and, as a result,
greater closeness. In this sense, it can be observed that the
concrete with the addition of D, being more porous, is the most prone
to attack, even more so than the actual control sample, provided MS
is added, which counteracts the effect by reducing porosity.

164

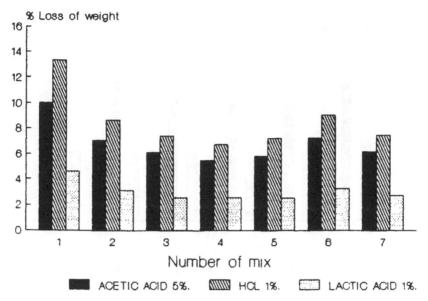

Figure CA1. Chemical attack.
Microsilica and admixtures.
II-Z/35 A cement.

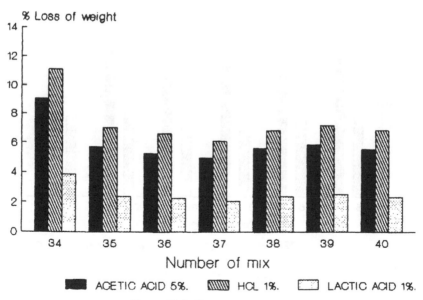

Figure CA2. Chemical attack.
Microsilica and admixtures.
I/45 A cement.

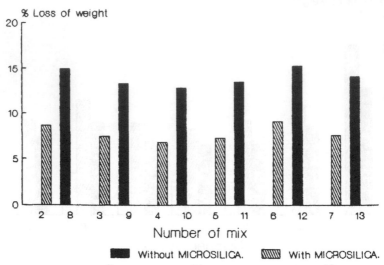

Figure CA3. Chemical attack HCl 1%.
Microsilica and admixtures.
II-Z/35 A cement.

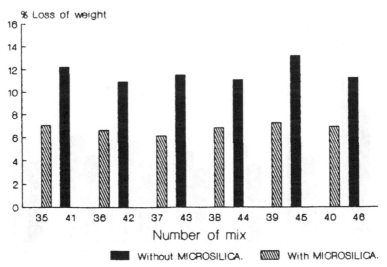

Figure CA4. Chemical attack HCl 1%.
Microsilica and admixtures.
I/45 A cement.

References

Aitcin, P.C., Editor, (1983) <u>Condensed Silica Fume,</u> Department of
    Civil Engineering, University of Sherbrooke.
Carette, G.G. and Malhotra, U.M. (Summer 1983),Mechanical Properties,
    Durability and Drying Shrinkage of Portland Cement Concrete
    Incorporating Silica Fume, <u>American Society for Testing Materials,</u>
    <u>Vol. 5, Nº 1</u>
Gjørv, O.E. (aug. 193) Durability of concrete containing condensed
    Silica Fume, <u>American Concrete Institute, Special Publication,</u>
    <u>SP-79, Vol. II.</u>
Hjorth, L. (1982), <u>Microsilica in Concrete,</u> Nordic Concrete
    Federation, Nordic Concrete Research, Publication Nº 1.
Mehta, P.K. and Gjørv, O.E. (1982), <u>Properties of Portland Cement</u>
    <u>Concrete containing Fly Ash and Condensed Silica Fume,</u> Report STF
    65 A 82030, Cement and Concrete Research Institute, The Norwegian
    Institute of Technology, Trondheim, Norway.
Okkenhang, K. and Gjørv O.E. (Aug. 24, 1982), <u>Influence of Condensed</u>
    <u>Silica Fume on the Ari Void System in Concrete, Report STF 65 A</u>
    <u>82044,</u> Cement and Concrete Research Insitute, The Norwegian
    Institute of Technology, Trondeheim, Norway.
Regourd, M., Mortureux, B. and Hornain, H. (Aug. 1983). Use of
    Condensed Silica Fume as Filler in Blended Cements, <u>American</u>
    <u>Concrete Institute, Special Publication SP-79, Vol. II.</u>
Sellevold, E.J., Bager, D.H. and Klitgaard Jensen, K. (1982) <u>Silica</u>
    <u>Fume Cement Pastes: Hydration and Pore Structure,</u> Report Nº
    8.610, Division of Building Materials; The Norwegian Institute of
    Technology, Trondheim, Norway.
Sellevold, E.J. and Radjy, F.F. (Aug. 1983), Condensed Silica Fume
    (Microsilica) in Concrete: Water Demand and Strength Development,
    <u>American Concrete Institute, Special Publication SP-79, Vol. II.</u>
Skrastins, J.I. and Zoldners, N.G. (Aug. 1983), Ready Mixed Concrete
    Incorporating Silica Fume, <u>American Concrete Institute, Special</u>
    <u>Publication SP-79, Vol. II.</u>
Tratteberg, A. (1978); Silica fume as a pozzolanic material, <u>Il</u>
    <u>Cemento,</u> Vol. 75.

THE INFLUENCE OF SUPERPLASTICIZER TYPE AND DOSAGE ON THE COMPRESSIVE STRENGTH OF PORTLAND CEMENT CONCRETE IN THE PRESENCE OF FLY ASH

M. COLLEPARDI, S. MONOSI and M. PAURI
Department of Science of Materials and Earth, University of Ancona, Italy
S. BIAGINI and I. ALVERA
MAC-MBT, R&D Laboratories, Treviso, Italy

Abstract
The effect of superplasticizer type - Sulphonated Naphtalene Polymer SNP, or Sulphonated Melamine Polymer SMP - and of its dosage rate (2 or 4% by weight of cement) on compressive strength of concrete mixes containing fly ash (15% by weight of cement), has been studied. The results of the present paper indicate that the combined addition of superplasticizer and fly ash increases concrete strength at a given workability. The highest strength increase occurs in fly ash concrete mixes when type III Portland cement and SMP superplasticizer type (at a 4% dosage rate) are used.
Key words: Fly ash, Superplasticizer, Naphtalene sulphonated polymer, Melamine sulphonated polymer.

1 Introduction

The purpose of the present work is to optimize the compressive strength in fly ash concrete mixes by varying the Portland cement type (I or III according to ASTM), the superplasticizer type SNP or SMP (Sulphonated Naphtalene or Sulphonated Melamine Polymers), and the superplasticizer dosage rate (1-3).

In general, when used in form of a 40% aqueous solution, the dosage rate of SNP or SMP superplasticizers is about 1% by weight of cement (1). Higher dosage rates such as 2 to 4% could cause a retarding effect on cement hydration at early ages.

2 Experimental

Two cement factors (300 and 400 kg/m^3) for both type I and type III Portland cement have been used in reference concrete mixes without fly ash addition (Table 1): superplasticizers (in form of 40% aqueous solution of SNP

168

or SMP) at a dosage rate of 2 and 4% by weight of cement have been used to reduce mixing water at a given slump (220 +/- 10 mm) with respect to plain mixes (no superplasticizer addition).

Fly ash (15% by weight of cement) has been used to replace cement (Table 2) or as an additional component without any cement reduction (Table 3).

In general the air volume in the fresh concrete mix appears to be increased by the addition of superplasticizer: the increase is negligible when SMP is used and more remarkable in the presence of SNP particularly at 4% dosage rate.

All the concrete mixes have been cured at 20°C and compressive strength on cubic specimens (100 mm) has been determined at 1, 7 and 28 days.

3   Compressive strength of type I Portland cement concrete mixes

Figure 1 shows the effect of 2 and 4% SNP or SMP superplasticizer additions on compressive strength of concrete mixes containing 300 kg/m^3 of type I Portland cement, in the absence (Fig. 1A) or in the presence of fly ash (Fig. 1B & C); fly ash has been used in two different ways: to replace 15% by weight of cement (Fig. 1B) or as an additional component (15% by weight) without any cement reduction (Fig. 1C).

Figure 2 shows the same parameters as those of Fig. 1, the only difference being the cement factor which is 400 kg/m^3 instead of 300 kg/m^3.

The SMP superplasticizer performs a little better than the SNP one, independently of the curing time, the fly ash presence, the way of fly ash utilization (Fig. 1 and 2).

When fly ash is used as an additional component (Fig. 1C and 2C) the compressive strength is increased in comparison with that of the reference mixes (Fig. 1A and 2A), whereas the opposite is true when fly ash replaces the same amount of Portland cement particularly at early ages (Fig. 1B and Fig. 2B).

As far as the superplasticizer dosage effect is concerned, SMP and SNP behave in a similar manner: by increasing the superplasticizer dosage from 2 to 4%, concrete compressive strength at early ages (1 day) is reduced; at later ages compressive strength does not depend significantly on the superplasticizer dosage. In particular at 28 days there is a small strength reduction in the leaner cement mixes (Fig.

Table 1. Composition of concretes without fly ash at a given slump (220 +/- 10 mm)

Cement type	Cement factor (kg/m^3)	Superplasticizer type	Superplasticizer dosage (% w.c.)	W/C ratio	Air volume (%)
I	300	-	0	0.73	1.0
I	400	-	0	0.55	1.4
I	300	SNP	2	0.57	6.9
I	400	SNP	2	0.37	7.3
I	300	SNP	4	0.49	9.2
I	400	SNP	4	0.32	5.7
I	300	SMP	2	0.57	1.9
I	400	SMP	2	0.39	2.5
I	300	SMP	4	0.49	1.8
I	400	SMP	4	0.34	1.9
III	300	-	0	0.69	1.4
III	400	-	0	0.57	1.0
III	300	SNP	2	0.46	6.7
III	400	SNP	2	0.35	4.8
III	300	SNP	4	0.43	5.5
III	400	SNP	4	0.32	4.0
III	300	SMP	2	0.46	2.6
III	400	SMP	2	0.35	2.6
III	300	SMP	4	0.43	2.6
III	400	SMP	4	0.32	2.4

Table 2. Composition of concretes with fly ash replacing the cement content (15% by weight) at a given slump (220 +/- 10 mm).

Cement type	Cement factor (kg/m^3)	Fly ash (kg/m^3)	Superplasticizer type	Superplasticizer dosage (% w.c.)	W/C ratio	Air volume (%)
I	255	45	-	0	0.82	1.6
I	340	60	-	0	0.61	1.8
I	255	45	SNP	2	0.69	6.0
I	340	60	SNP	2	0.51	4.6
I	255	45	SNP	4	0.59	7.4
I	340	60	SNP	4	0.44	4.1
I	255	45	SMP	2	0.68	1.8
I	340	60	SMP	2	0.53	2.0
I	255	45	SMP	4	0.61	1.7
I	340	60	SMP	4	0.44	2.3
III	255	45	-	0	0.83	1.6
III	340	60	-	0	0.63	1.7
III	255	45	SNP	2	0.69	2.5
III	340	60	SNP	2	0.54	2.3
III	255	45	SNP	4	0.47	5.2
III	340	60	SNP	4	0.36	5.0
III	255	45	SMP	2	0.78	1.3
III	340	60	SMP	2	0.55	1.4
III	255	45	SMP	4	0.50	1.8
III	340	60	SMP	4	0.36	2.5

Table 3. Composition of concretes with fly ash as additional component (15% by weight of cement) at a given slump (220 +/- 10 mm).

Cement type	Cement factor (kg/m^3)	Fly ash (kg/m^3)	Superplasticizer type	Superplasticizer dosage (% w.c.)	W/C ratio	Air volume (%)
I	300	45	-	0	0.69	1.7
I	400	60	-	0	0.50	1.9
I	300	45	SNP	2	0.57	2.4
I	400	60	SNP	2	0.39	3.6
I	300	45	SNP	4	0.49	4.6
I	400	60	SNP	4	0.34	3.6
I	300	45	SMP	2	0.57	1.9
I	400	60	SMP	2	0.39	2.3
I	300	45	SMP	4	0.52	2.3
I	400	60	SMP	4	0.36	2.0
III	300	45	-	0	0.71	1.6
III	400	60	-	0	0.60	1.8
III	300	45	SNP	2	0.57	2.5
III	400	60	SNP	2	0.44	3.4
III	300	45	SNP	4	0.40	4.2
III	400	60	SNP	4	0.31	4.8
III	300	45	SMP	2	0.62	1.8
III	400	60	SMP	2	0.46	1.5
III	300	45	SMP	4	0.40	2.4
III	400	60	SMP	4	0.31	2.2

1) and a slight strength increase in the richer ones in the presence of fly ash (Fig. 2B & C).

4 Compressive strength of type III Portland cement concrete mixes

Figures 3 and 4 show the effect of 2 and 4% superplasticizer additions, with or without fly ash, on compressive strength of concrete mixes containing respectively 300 and 400 kg/m^3 of type III Portland cement.

By using type III Portland cement, the change from 2 to 4% in the SMP superplasticizer dosage significantly increases the compressive strength of concrete mixes containing fly ash (Fig. 3B & C and Fig. 4B & C), whereas the same change does not affect remarkably compressive strengths of concrete mixes without fly ash (Fig. 3A and 3B). The performance of 4% SMP superplasticizer in fly ash - type III Portland cement concrete mixes is so effective that a 28 day compressive strength of 80 MPa can be attained even with a relatively low cement factor such as 255 kg/m^3 (Fig. 3B). This could be particularly advantageous when high strength is required in lean cement mixes for massive concrete structures. The SNP superplasticizer too performs very well in fly ash - type III Portland cement concretes when 4% dosage of superplasticizer is used, even if to a

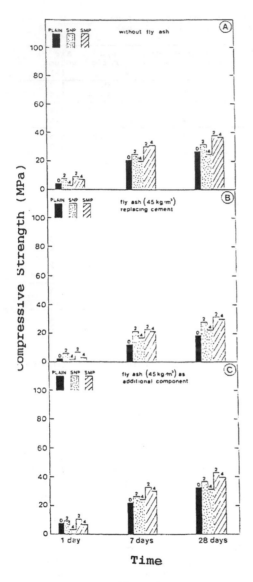

Figure 1

Compressive strength of concretes with 300 kg/m³ of type I Portland cement without fly ash (A), with fly ash to replace 15% by weight of cement (B), or as 15% additional component (C).

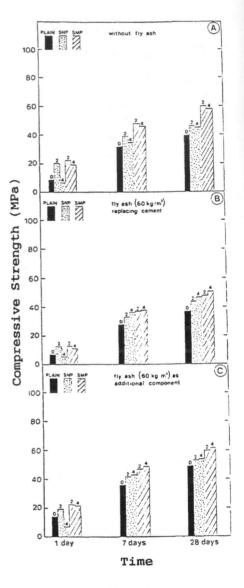

Figure 2

Compressive strength of concretes with 400 kg/m³ of type I Portland cement without fly ash (A), with fly ash to replace 15% by weight of cement (B), or as 15% additional component (C).

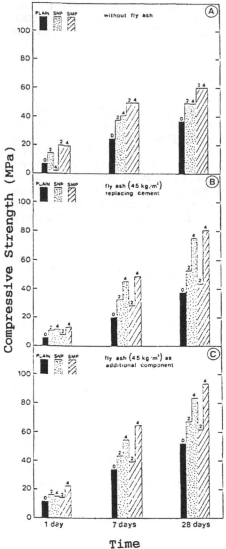

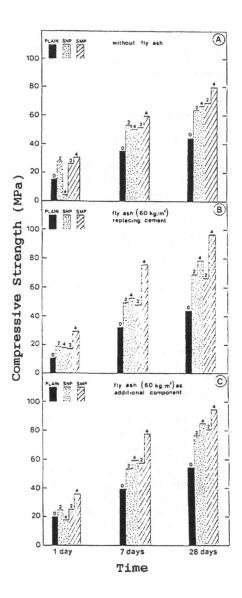

Figure 3

Compressive strength of
concretes with 300 kg/m^3
of type III Portland
cement without fly ash
(A), with fly ash to
replace 15% by weight of
cement (B), or as 15%
additional component (C).

Figure 4

Compressive strength of
concretes with 400 kg/m^3
of type III Portland
cement without fly ash
(A), with fly ash to
replace 15% by weight of
cement (B), or as 15%
additional component (C).

slightly lower extent than the SMP one. At 2% dosage, no significant difference has been found between SMP and SNP superplasticizer types in fly ash - type III Portland cement mixes.

5 Conclusions

The combined addition of superplasticizer used to reduce mixing water, and fly ash used to replace cement or as an additional component (15% by weight of cement) is particularly effective in concrete mixes containing type III Portland cement. An increase in the superplasticizer dosage, such as from 2 to 4%, could be advantageously used in fly ash concrete mixes even with a relatively low type III cement factor, such as 255 kg/m^3 in order to obtain high strength level (80 MPa) for massive concrete structures.

When a less reactive Portland cement is used, such as type I, the increase from 2 to 4% in the superplasticizer dosage does not cause a significant change in the compressive strength of fly ash concrete mixes.

In the absence of fly ash, a change from 2 to 4% in the superplasticizer dosage, does not appear to give any additional strength increase independently of the Portland cement type.

The SMP superplasticizer appears to be in general more effective than the SNP one for the strength increase, except for fly ash - type III Portland cement mixes with 2% superplasticizer dosage rate. At a dosage rate of 4%, the SMP superplasticizer is more effective than the SNP one, independently of the cement type and fly ash presence: this could be ascribed to the difference in the air volume between SMP and SNP concrete mixes, when a very high dosage rate, such as 4%, is used.

6 References

(1) Ramachandran, V.S. and Malhotra, V.M. (1984) "Superplasticizers", in Concrete Admixtures Handbook, Editor V.S. Ramachandran, Noyes Publication, Park Ridge, New Jersey, USA, pp. 211-268.
(2) Mehta, P.K. (1984) "Mineral Admixtures", in Concrete Admixtures Handbook, Editor V.S. Ramachandran, Noyes Publication, Park Ridge, New Jersey, USA, pp.303-333.
(3) Malhotra, V.M. (1986) Editor, Proceedings of International Conference on "Fly ash, silica fume, slag and natural pozzolans in concrete", ACI SP-91, Detroit, Michigan, USA.

SUPERPLASTICIZED SILICA FUME HIGH-STRENGTH CONCRETES

M. COLLEPARDI, G. MORICONI and M. PAURI
Department of Science of Materials and Earth, University of Ancona, Italy
S. BIAGINI and I. ALVERA
MAC-MBT, R&D Laboratories, Treviso, Italy

Abstract
The effect of combined additions of silica fume and superplasticizer on concrete compressive strength has been studied by taking into account the following parameters: a) type and dosage rate of superplasticizer; b) type and content of Portland cement; c) way of silica fume utilization (as additional component or as cement replacement). In the presence of silica fume, for both type I and type III Portland cement the Melamine Sulphonated Polymer superplasticizer performs better than the Naphtalene Sulphonated Polymer one, particularly when a high dosage such as 4% is used. A change from 2 to 4% superplasticizer dosage rate in general does not modify or reduce compressive strength in the absence of silica fume, whereas significantly increases compressive strength in the presence of silica fume.
Key words: Silica fume, Superplasticizer, Naphtalene sulphonated polymer, Melamine sulphonated polymer.

1 Introduction

The combined addition of silica fume and superplasticizer is generally used to produce high-strength concretes (1-3). The purpose of the present work is to examine the effect of the following parameters on the concrete compressive strength:
- type of superplasticizer: Sulphonated Naphtalene Polymer (SNP) or Sulphonated Melamine Polymer (SMP);
- dosage of superplasticizer in form of a 40% polymer aqueous solution: 2 or 4% by weight of cement;
- type of Portland cement: type I or III according to ASTM;
- cement content: 300 or 400 kg/m^3 in the reference mixes without silica fume;
- way of silica fume utilization: as additional component (15% by weight of cement) without any cement reduction or as cement replacement (15%).

2 Experimental

Table 1 shows the concrete composition of plain mixes (in the absence of silica fume) with or without SNP or SMP superplasticizers (2 or 4%) by using type I or type III Portland cement.

Table 2 indicates the composition of the same concrete mixes as those shown in Table 1, the only difference being a replacement of 15% by weight of cement by silica fume.

Table 3 shows the composition of the same mixes as those shown in Table 1, the only difference being the addition of silica fume (15% by weight of cement) without any reduction in the cement factor.

All the concrete mixes have been manufactured at a given slump (220 +/- 10 mm) and cured at 20°C.

Compressive strength measurements on cubic specimens (100 mm) have been carried out at 1, 7 and 28 days and their values are shown in Fig. 1-4.

3 Effect of silica fume and superplasticizer on strength of concretes containing type I Portland cement

Figures 1 and 2 show the effect of superplasticizer addition, with or without silica fume, on the compressive strength of concretes containing 300 and 400 kg/m^3 respectively of type I Portland cement. Figures 1A and 2A indicate the behaviour of plain or superplasticized concretes both in the absence of silica fume. Silica fume has been used to replace 15% by weight of cement (Fig. 1B and 2B) or as additional component without any reduction in the cement factor (Fig. 1C and 2C).

In the absence of silica fume, early and later compressive strength are reduced when the superplasticizer dosage rate is increased from 2 to 4%. Such a reduction is in general negligible and becomes more significant for the 1 day strength when 4% of SNP superplasticizer is used (Fig. 1A and 2A). Moreover, compressive strengths of SMP superplasticized concretes appear to be higher than those of SNP treated concretes particularly at 4% superplasticizer dosage rate (Fig. 1A and 2A).

In the presence of silica fume, the change from 2 to 4% of SNP or SMP superplasticizer dosage rate increases remarkably the compressive strength, and this appears to be more effective for the SMP superplasticizer than for the SNP one (Fig. 1B & C and 2B & C).

Table 1.   Composition of concretes without silica fume at a given slump
(220 +/- 10 mm)

Cement type	Cement factor (kg/m³)	Superplasticizer type	Superplasticizer dosage (% w.c.)	W/C ratio	Air volume (%)
I	300	-	0	0.73	1.0
I	400	-	0	0.55	1.4
I	300	SNP	2	0.57	6.9
I	400	SNP	2	0.37	7.3
I	300	SNP	4	0.49	9.2
I	400	SNP	4	0.32	5.7
I	300	SMP	2	0.57	1.9
I	400	SMP	2	0.39	2.5
I	300	SMP	4	0.49	1.8
I	400	SMP	4	0.34	1.9
III	300	-	0	0.69	1.4
III	400	-	0	0.57	1.0
III	300	SNP	2	0.46	6.7
III	400	SNP	2	0.35	4.8
III	300	SNP	4	0.43	5.5
III	400	SNP	4	0.32	4.0
III	300	SMP	2	0.46	2.6
III	400	SMP	2	0.35	2.6
III	300	SMP	4	0.43	2.6
III	400	SMP	4	0.32	2.4

Table 2.   Composition of concretes with silica fume replacing the cement content
(15% by weight) at a given slump (220 +/- mm).

Cement type	Cement factor (kg/m³)	Silica fume (kg/m³)	Superplasticizer type	Superplasticizer dosage (% w.c.)	W/C ratio	Air volume (%)
I	255	45	-	0	0.88	1.0
I	340	60	-	0	0.62	1.1
I	255	45	SNP	2	0.76	1.8
I	340	60	SNP	2	0.50	2.9
I	255	45	SNP	4	0.51	6.0
I	340	60	SNP	4	0.35	5.1
I	255	45	SMP	2	0.79	1.1
I	340	60	SMP	2	0.51	1.5
I	255	45	SMP	4	0.51	1.8
I	340	60	SMP	4	0.35	2.2
III	255	45	-	0	0.84	1.1
III	340	60	-	0	0.61	1.3
III	255	45	SNP	2	0.67	2.6
III	340	60	SNP	2	0.50	4.0
III	255	45	SNP	4	0.44	5.4
III	340	60	SNP	4	0.34	4.0
III	255	45	SMP	2	0.73	1.3
III	340	60	SMP	2	0.56	1.4
III	255	45	SMP	4	0.45	2.3
III	340	60	SMP	4	0.36	2.4

Table 3. Composition of concretes with silica fume as additional component
(15% by weight of cement) at a given slump (220 +/- 10 mm).

Cement type	Cement factor (kg/m^3)	Silica fume (kg/m^3)	Superplasticizer type	Superplasticizer dosage (% w.c.)	W/C ratio	Air volume (%)
I	300	45	–	0	0.76	1.1
I	400	60	–	0	0.49	1.2
I	300	45	SNP	2	0.61	2.2
I	400	60	SNP	2	0.38	3.3
I	300	45	SNP	4	0.43	4.5
I	400	60	SNP	4	0.29	4.2
I	300	45	SMP	2	0.66	2.0
I	400	60	SMP	2	0.39	2.2
I	300	45	SMP	4	0.43	2.4
I	400	60	SMP	4	0.29	2.3
III	300	45	–	0	0.69	1.1
III	400	60	–	0	0.53	1.3
III	300	45	SNP	2	0.55	2.4
III	400	60	SNP	2	0.43	3.4
III	300	45	SNP	4	0.39	4.0
III	400	60	SNP	4	0.32	4.3
III	300	45	SMP	2	0.59	2.0
III	400	60	SMP	2	0.44	1.9
III	300	45	SMP	4	0.39	2.2
III	400	60	SMP	4	0.32	2.4

Silica fume in superplasticized concretes when used as an additional component (Fig. 1B and 2B) is of course more effective than when used as cement replacement (Fig. 1C and 2C).

The combined addition of silica fume (45 kg/m^3) and superplasticizer (4% SMP) is so effective that even with a significantly low cement factor, such as 255 kg/m^3, the 28 day compressive strength is much higher than that of a richer mix (cement factor 400 kg/m^3) containing or silica fume either superplasticizer. For instance the 28 day compressive strength is about 65 MPa in the leaner cement mix with the combined addition of silica fume and 4% SMP superplasticizer (Fig. 1B), and less than 60 MPa in the richer cement mix containing only 4% SMP (Fig. 2A); even lower (less than 40 MPa) is the compressive strength of the richer cement mix containing only silica fume (Fig. 2C).

On the other hand, silica fume as additional component or cement replacement does not cause any strength increase in the absence of superplasticizer (Fig. 1 and 2). Really in plain mixes silica fume reduces strength particularly at earlier ages (1 day). This is due to the fact that the required mixing water at a given slump increases when silica fume is used without superplasticizer (Tables 1-3). At later ages (7-28 days) the strength reduction caused by the presence of silica fume in plain concrete mixes, becomes gradually lower because of the slow pozzolanic

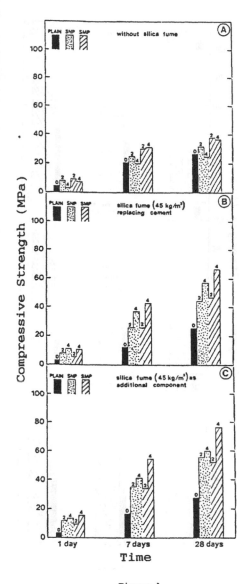

Figure 1

Compressive strength of concretes with 300 kg/m^3 of type I Portland cement without silica fume (A), with silica fume to replace 15% by weight of cement (B), or as 15% additional component (C).

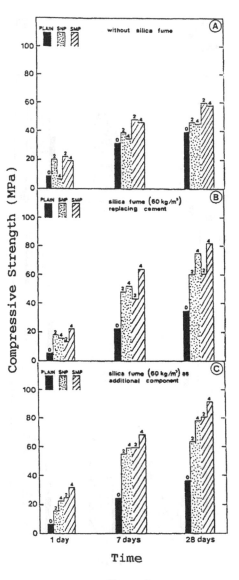

Figure 2

Compressive strength of concretes with 400 kg/m^3 of type I Portland cement without silica fume (A), with silica fume to replace 15% by weight of cement (B), or as 15% additional component (C).

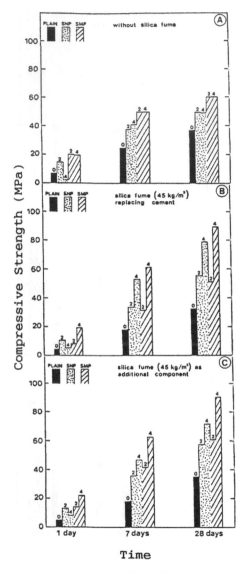

Figure 3

Compressive strength of concretes with 300 kg/m^3 of type III Portland cement without silica fume (A), with silica fume to replace 15% by weight of cement (B), or as 15% additional component (C).

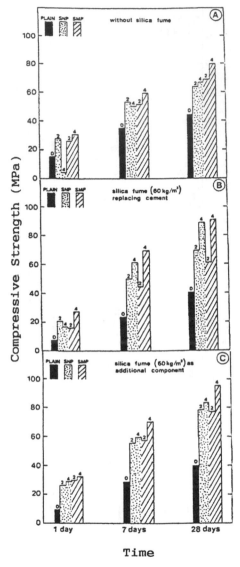

Figure 4

Compressive strength of concretes with 400 kg/m^3 of type III Portland cement without silica fume (A), with silica fume to replace 15% by weight of cement (B), or as 15% additional component (C).

reaction between silica fume and $Ca(OH)_2$ (produced by the cement hydration) which partly compensates the higher w/c ratio caused by the presence of silica fume.

4   Effect of silica fume and superplasticizer on strength of concrete containing type III Portland cement

Figures 3 and 4 show the effect of superplasticizer on compressive strength of concretes containing 300 and 400 kg/m³ respectively of type III Portland cement with or without silica fume.

No substantial difference has been found in the strength change caused by superplasticizer and/or silica fume additions between concretes containing type I Portland cement (Fig. 1 and 2) on one hand, and those produced by using type III Portland cement (Fig. 3 and 4) on the other hand.

Of course type III Portland cement concrete mixes appear to be stronger than the corresponding concretes containing type I Portland cement. For instance, at the cement factor of 255 kg/m³, the 28 day compressive strength of the concrete containing 45 kg/m³ of silica fume and 4% of SMP superplasticizer is 65 or 90 MPa when type I or type III respectively Portland cement is used (Fig. 1B and Fig. 3B).

5   Conclusions

Independently of the type of Portland cement used (I or III), in the absence of silica fume the compressive strength of SMP superplasticized concrete appears to be higher than that of SNP treated concretes; in the presence of silica fume the SMP superplasticizer performs significantly better than the SNP one only at 4% dosage, whereas no substantial difference has been found in general between the performances of SMP and SNP at the 2% superplasticizer dosage.

The change in the superplasticizer dosage from 2 to 4% causes a remarkable increase in the compressive strength of both type I and type III Portland cement concretes only in the presence of silica fume; the effect of this change is more remarkable for the SMP superplasticizer than for the SNP one.

6 References

(1) Ramachandran, V.S. and Malhotra, V.M. (1984) "Superplasticizers", in <u>Concrete Admixtures Handbook</u>, Editor V.S. Ramachandran, Noyes Publication, Park Ridge, New Jersey, USA, pp. 211-268.

(2) Mehta, P.K. (1984) "Mineral Admixtures", in <u>Concrete Admixtures Handbook</u>, Editor V.S. Ramachandran, Noyes Publication, Park Ridge, New Jersey, USA, pp. 303-333.

(3) Malhotra, V.M. (1986) Editor, <u>Proceedings of International Conference on "Fly ash, silica fume, slag and natural pozzolans in concrete"</u>, ACI SP-91, Detroit, Michigan, USA.

STRENGTH DEVELOPMENT OF SUPER-PLASTICIZED PLAIN AND FLY ASH CONCRETES

M. K. GOPALAN and M. N. HAQUE
Department of Civil Engineering, ADFA, University of New South Wales, Australia

Abstract
Compressive strength and water penetration of four mixes of structural grade plain concrete with cement content ranging from 250 to 400 kg/m^3 and fly ash concrete with fly ash cementitious ratio of 0.32 are reported. These mixes were redesigned by adding a proprietary superplasticiser at a recommended maximum dosage of 1.2 litres per 100 kg of cementitious material, maintaining similar workability. The strength development in three curing regimes was monitored. The results suggest that lower grade mixes of fly ash concrete benefited the most when proper curing was provided. However, the strength development of plain concretes was better than the companion fly ash mixes when limited initial curing was provided. The higher strength development of the superplasticised concrete was the maximum for mixes with high water cementitious ratios.
Key words:Strength development, Superplasticiser, Fly ash, Curing regime, Water penetration, Size of specimen

Introduction

Superplasticisers are widely used in modern concrete industry as a dispersant to provide fluidity and cohesion during handling. The mechanism of the dispersive action is due to the adsorption of the admixture into cement particles causing them to be mutually repulsive because of its anionic nature (CCA Report 1976, Massazza et al 1981). The dispersive action allows for more surface area of cement to be in contact with water. Once the plasticising action of the superplasticiser subsides, hydration of the cement can take place more rapidly, resulting in higher early strengths.

According to Malhotra (1981) the superplasticisers in concrete can be used in three possible ways:

a) to design high strength concrete by reducing the water content of the mix;

b) to make concrete with reduced cement content by maintaining constant water cement ratio;

c) to provide flowing concrete for easy pumping without segregation.

The flowing concrete has other advantages as it is self levelling and self compacting. Malhotra and Malanka (1979) found that the strength of cylinders cast without vibrating was comparable to the vibrated. The superplasticisers are mainly used to produce flowing concrete and their application as a water reducing agent has not achieved the same popularity (Aignesberger and Kern 1981) . The impact of superplasticisers was localised in Australia until 1983 (Mielenz 1984).

This study was, therefore, directed to produce concrete with reduced water content. The economic merit of using a superplasticiser results from the early higher strength. This increase in strength was attributed to the reduced water content (Perenchio et al 1979). Johnston et al (1979) found that the superplasticised concrete developed higher compressive strength at early ages than the reference concrete even when the water- cement ratio was kept constant. This extra strength was found to diminish with with age. Gu Dezhen et al (1982) reported contrary results and argued that the superplasticised concrete had less water for hydration. Mor and Mehta (1984) studied this phenomenon and concluded that the superplasticised concrete developed 19 percent more hydration than the reference concrete at 28 days. Rixom (1975) reported that there was no significant increase in the strength of the superplasticised concrete up to 28 days. Similar results were published by Ryan and Munn (1979). Carrasquillo and Carrasquillo (1988) observed that concrete samples cured under ambient conditions, with or without a curing compound, yielded strength similar to those stored under standard curing.

Lane and Best (1979) found that the superplasticisers were compatible with fly ash and produced no detrimental effects. Mukherjee et al (1982) reported that high strength concrete, up to 63 MPa at 28 days, containing large percentage of fly ash can be produced when a superplasticiser was used. They also observed that the properties of water reduced superplasticised concrete were superior to the reference fly ash mix. Samarin and Ryan (1975) suggested that the flowing concrete required extra fine material and fly ash was ideal for this purpose. He also reported that the slump loss was unaffected by the variation of cement fly ash ratios.

Pomeroy (1972) found that the curing conditions influenced the strength of the specimen through size dependency. Smaller size specimen needed careful curing when the water-cement ratio was high. For the wet cured samples the strength decreased significantly, as the cube size was increased. Carrasquillo and Carrasquillo (1988) observed that the smaller size specimen registered lower strength for superplasticised concrete.

Curing

Curing is important in enhancing the strength and quality of the concrete. Aitcin and Riad (1988) observed that the 28 day compressive strength of specimens cured under standard condition gave a fair representation of actual strength when water cementitious ratio was below 0.3. Kunze (1989) found that special attention was required for curing if specified design strengths were to be achieved. Beresford (1978) indicated that curing was less important to high strength concrete than to normal strength concrete. Gopalan and Haque (1987) concluded that proper curing was more important to the strength development of fly ash concrete than plain concrete.

Aim of Experiment

The aim of the investigation was to evaluate the strength development of superplasticised plain and fly ash concretes of different grades under adequate, inadequate and no curing conditions. The water content of the superplasticised concrete was reduced to maintain identical workability. The strength development of these concretes cured in the above regimes was studied. The relative quality of the concretes with and without the superplasticiser was assessed by testing for water penetration at 91 days.

Details of the Experiment

The materials used were 20 and 10 mm maximum size crushed gravel, river sand, Type A portland cement, bituminous fly ash (ASTM Class F) and a proprietary sulphonated naphthalene formaldehyde condensate superplasticiser which conformed with the ASTM C494 requirements for type G admixture. Chemical composition and physical properties of the fly ash are given in Table 1.

TABLE 1.Chemical and Physical Properties of Flyash

Chemical:

compound	percent
silica	52.70
alumina	26.50
calcium oxide	4.26
iron oxide	7.44
magnesium oxide	1.40
sodium and potassium oxide	2.13
sulphur trioxide	0.45
loss on ignition	1.80
others	3.32

Physical:

bulk density	$2.40 \ t/m^3$
residue on 45μm sieve	7.9%

Test Details

Four mixes of structural grade plain concrete with a cement content of 250, 300,350 and 400 kg/m^3 (hereafter referred to as G1,G2,G3 and G4 respectively) were initially designed. Another four fly ash mixes of similar grade were designed using the mix design techniques (Gopalan and Haque 1985) at a fly ash cementitious ratio of 0.32. These eight mixes were recast by incorporating the superplasticiser at a maximum recommended dosage of 1.2 litres per 100 kg of cementitious material. The slump of these mixes were maintained at a constant value of 50 ±20 mm. The cylinders from each batch were cured in the following regimes:

    a) fog curing at $23 \pm 2°C; 95 \pm 3\% R.H.$ (F cured);

    b) controlled environment room at $23 \pm 2°C; 50 \pm 3\% R.H.$ (D cured);

    c) fog curing for 7 days and then in the controlled environment room FD cured).

Each mix was designated by the grade, fly ash cementitious ratio and the type of the mix. For example, G1-32-S identifies a mix of grade 1 with a fly ash cementitious ratio of 0.32 and containing the superplasticiser. Similarly, G1-32-N represents the same mix without the superplasticiser. The concrete without fly ash is referred to as plain concrete and that without superplasticiser as normal concrete. The mix parameters and some of the properties of the fresh concrete are given in Table 2

Compressive strength testing was done on 100 mm diameter cylinders at 28 and 91 days. The testing for size effects on compressive strength was done at 28 and 91 days on FD cured samples only as this curing regime was assumed to be more realistic to the on site cured concrete. For all concretes, three specimens from each curing regime were tested. The results are given in Tables 3 and 4.

Results and Discussion

Figures 1 to 4 show the variation of strength with curing condition for each grade of the superplasticised concrete. In order to make the comparisons of the results more realistic, these figures are plotted after adjusting the 28 day strength of the normal concrete to same values. From those figures, it is evident that the curing conditions influence the strength development significantly. As expected the strength developed is the least for the D cured concrete. The strength of the lower grade mixes did not improve with age for inadequate curing. The fly ash concrete showed significantly higher strength at 91 days for the fog cured condition. However, for higher grade concretes, this increase in long term strength for fly ash concrete is not evident. Thus, the addition of a superplasticiser is more beneficial to the strength development of lean fly ash mixes when proper curing is provided.

    The 28 day strength of the FD cured mixes is found to be similar to that of the the F cured. The nominal increase in the 28 day strength of the FD cured cylinders over the fog cured, may be due to the changes in the overall moisture

TABLE 2. Properties of Fresh Concrete.

Mix	C. Agg.	F. Agg.	Cem.(c)	Ash(f)	Water(w)+ Sup. Pl.(s)	(w+s)/ (c+f)	Slump
	(kg/m^3)	(kg/m^3)	(kg/m^3)	(kg/m^3)	(l/m^3)		(mm)
G1-0-S	1308	673	246	0	172	0.70	30
G1-32-S	1273	686	185	89	157	0.57	35
G2-0-S	1274	685	307	0	160	0.52	35
G2-32-S	1219	656	229	111	164	0.48	30
G3-0-S	1251	673	361	0	157	0.43	30
G3-32-S	1162	625	278	135	171	0.41	70
G4-0-S	1196	645	400	0	176	0.44	35
G4-32-S	1136	611	344	166	150	0.29	70
G1-0-N	1271	685	249	0	180	0.72	50
G1-32-N	1244	671	182	88	175	0.65	30
G2-0-N	1232	667	298	0	186	0.62	30
G2-32-N	1207	648	227	109	173	0.51	40
G3-0-N	1212	651	351	0	183	0.52	30
G3-32-N	1154	620	276	133	177	0.43	45
G4-0-N	1174	633	394	0	191	0.48	50
G4-32-N	1114	600	337	163	166	0.33	30

content of the hardened concrete (popovics 1986). In this curing condition, plain concrete developed higher strength than the fly ash concrete for all grades of concrete. The fly ash concretes have suffered considerable loss of strength at 91 days when compared to the fog cured. There is no significant improvement in strength between 28 and 91 days. This shows that continued adequate curing is essential for the long term strength development of fly ash concrete. The plain concretes are not affected as much as the fly ash concrete. Thus the initial curing is more beneficial to the strength development of plain concretes than fly ash concretes.

The strength of the D cured concrete is always lower than the samples cured in the other two regimes. The figures show that the grade of the concrete is not very important for this curing condition. The 28 and 91 day strengths are practically the same. However, there is a significant difference in the loss of strength between plain and fly ash concretes. The average loss of strength due to lack of curing is 30 and 45% at 28 days and 38 and 55% at 91 days for the plain and fly ash concretes respectively. Thus the fly ash concrete containing a superplasticiser suffered about 50% more loss of strength due to lack of curing.

TABLE 3. Properties of Normal Concrete.

Mix	strength (MPa)				Water	
	28 day			91 day	Depth	
	F	D		F	(mm)	
	100x200	150x300	75x150	100x200	FD	D
G1-0-N	25.6	20.3	14.7	26.6	21.0	50.0
G1-32-N	22.6	16.5	10.7	32.5	29.5	50.0
G2-0-N	37.0	27.1	19.6	39.5	17.0	35.5
G2-32-N	35.7	25.1	21.8	42.1	21.2	37.0
G3-0-N	45.6	35.2	24.9	49.5	14.5	27.0
G3-32-N	41.9	30.1	24.2	55.4	16.0	36.0
G4-0-N	50.5	40.4	32.3	54.9	11.0	28.5
G4-32-N	50.2	33.5	27.8	61.3	15.5	33.0

TABLE 4. Properties of Superplasticised Concrete.

Mix	strength (MPa)						Water Depth(mm)	
	28 day			91 day				
	F	FD	D	F	FD	D	FD	D
G1-0-S	35.6	44.0	20.6	37.5	41.9	20.3	13.5	34.5
G1-32-S	33.6	37.0	20.8	45.4	36.1	19.9	16.0	30.0
G2-0-S	50.7	58.0	32.3	56.2	59.0	32.7	11.0	24.0
G2-32-S	41.6	44.3	21.2	56.7	45.8	21.3	14.5	30.5
G3-0-S	54.7	60.9	41.5	60.8	63.9	41.5	12.5	16.0
G3-32-S	51.5	52.4	29.5	58.2	52.9	29.2	13.0	26.0
G4-0-S	61.1	63.0	49.3	76.9	73.6	52.4	11.5	15.0
G4-32-S	58.3	58.2	27.4	75.6	60.4	36.4	15.0	27.5

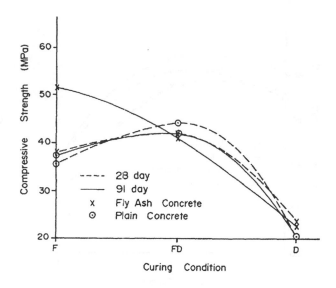

Fig. 1. Variation of Strength with
Curing Condition. – G1

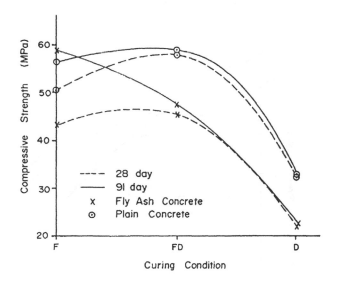

Fig. 2. Variation of Strength with
Curing Condition. – G2

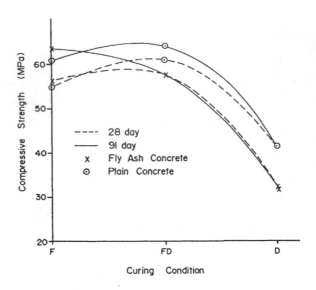

Fig. 3. Variation of Strength with
Curing Condition. - G3

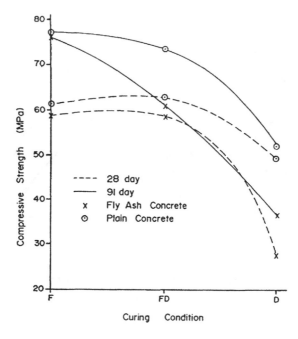

Fig. 4. Variation of Strength with
Curing Condition. - G4

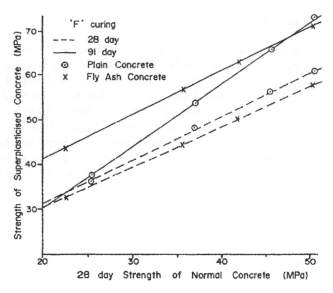

Fig. 5 . Regression of Strength of Superplasticised
Concrete on 28 day Strength of
Normal Concrete.

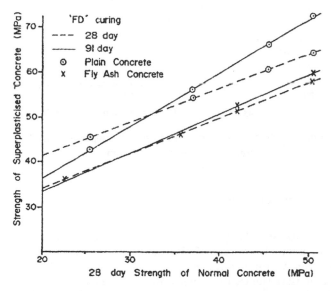

Fig. 6. Regression of Strength of Superplasticised
Concrete on 28 day Strength of
Normal Concrete.

Strength Development with Grade

The strength of the superplasticised concrete is found to correlate well with the 28 day strength of the normal concrete by a linear regression; the average correlation being above 90%. Figures 5,6 and 7 were plotted using the calculated strengths as given by the regression equation. This is done to eliminate the random errors in the strength measurement. It is evident from the figure 5 that the long term strength of the rich,superplasticised, plain concrete is higher than that of the fly ash mix. At 28 days, the plain concrete is only marginally superior than the fly ash mix; properties improving for richer mixes. For leaner mixes the strength development is quite different. The 91 day strength of the fly ash concrete is higher than that of the plain concrete.

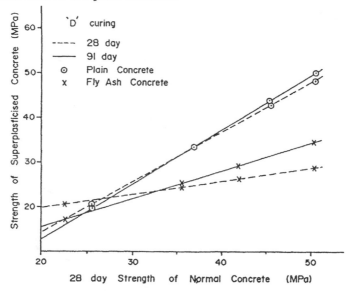

Fig. 7. Regression of Strength of Superplasticised Concrete on 28 day Strength of Normal Concrete.

Increase in Strength of Superplasticised Concrete

As reported by Johnston(1979) the superplasticised concrete developed higher strength than normal concrete of similar water cement ratio. The percentage increase in strength for various grades of F cured concrete are given in Table 5. It is evident that the extra strength due to the superplasticiser is highest for the mix with higher water cementitious ratio. This is true for both plain and fly ash concretes. It is also obvious that the extra strength was developed at 91 days. Statistically, there is no significant difference in extra strengths at 28 and 91 days.

TABLE 5. Increase in Strength of Superplasticised Concrete (%)

Mix	28 day	91 day
G1-0-S	27.8	27.1
G1-32-S	24.6	18.8
G2-0-S	8.6	13.8
G2-32-S	17.7	15.6
G3-0-S	5.7	11.7
G3-32-S	13.0	13.6
G4-0-S	2.2	23.8
G4-32-S	4.7	9.9

Effect of Specimen Size

The strength variations due to the size of the test samples for the superplasticised concrete was investigated for the FD cured samples. The results are tabulated in Table 6. The results are analysed by statistical methods; paired t test; for three size combinations, namely 100 & 150, 100 & 75 and 150 & 75 and are given in Table 7. It is evident that the size of the specimen did not have any significant effect on the measured strength at 91 days. However, at 28 days the results are dependent on the size. Table 8 shows the two-sample t test to see if the fly ash concrete samples behaved differently to the plain concrete. It shows that there is no difference in the strength at 28 days, but the strength at 91 days is significantly different.

TABLE 7. Sample Size and Strength

Age	Size of Sample	't' Calculated	Prob. of a larger 't'(%)
	150x75	2.42	4.7
28 day	150x100	5.58	1
	100x75	0.16	50
	150x75	0.41	50
91 day	150x100	1.22	27.5
	100x75	1.18	29.1

TABLE 6. Variation of Strength due to Size of Sample

Mix	strength (MPa)					
	28 day			91 day		
	150x300	75x150	100x200	150x300	75x150	100x200
G1-0-S	40.8	44.2	44.0	40.8	40.9	41.9
G1-32-S	36.7	38.0	37.0	38.8	35.3	36.1
G2-0-S	54.5	51.7	58.0	53.3	58.5	59.0
G2-32-S	38.3	41.9	44.3	46.1	39.9	45.8
G3-0-S	56.1	64.9	60.9	55.5	61.5	63.9
G3-32-S	46.8	51.4	52.4	54.0	52.7	52.9
G4-0-S	55.3	68.3	63.0	72.5	75.1	73.6
G4-32-S	54.3	55.7	58.2	59.9	61.8	60.4

TABLE 8. Type of Concrete and Strength

Age	Size of Sample	't' Calculated	Prob. of a larger 't'(%)
28 day	150x75	0.81	44.7
	150x100	0.51	50
	100x75	0.74	¿48.8
91 day	150x75	2.64	4.1
	150x100	2.6	4.2
	100x75	0.44	50

Water Penetration

The quality of the concretes, as influenced by the FD and D curing is investigated by comparing the water penetration at 91 days. The depth of water penetrated into the samples, immersed in a constant head of water for two hours, was measured by splitting the cylinders. The average depth of water penetration for the FD and D curing is 15.8 and 31.3 respectively. The results show significant increase in the water penetration for inadequately cured concrete. The average depth of water for FD cured normal and superplasticised concrete is 18.2 and 13.3 respectively. The corresponding values for the D cured samples are 37.2 and 25.4. This shows that there is significant reduction in water penetration for superplasticised concrete for both curing regimes. It must be noted that the superplasticised concrete developed higher strength than the normal mix.

Conclusions

The addition of a superplasticiser was more beneficial to lower grade fly ash concrete than the higher grade mixes under adequate curing conditions. Continued adequate curing was found to be essential for the long term strength development of fly ash concrete. When the concrete was initially cured for 7 days only, the strength development of the plain concrete was higher than the corresponding fly ash mix. Under D curing the fly ash concrete suffered 50% extra loss of strength than similar plain concrete.

The superplasticised concrete developed higher strength than the normal concrete of same water cementitious ratio. This extra strength was found to depend on the grade of the concrete; the maximum gain in strength was for mixes with higher water cementitious ratio. It is also shown that the extra strength at 28 days was not significantly different from that at 91 days.

The size of the specimen did not influence the strength at 91 days, but the results at 28 days were statistically different.

The penetration of water into the FD cured cylinders was significantly lower than that of the D cured ones. However, there was no significant difference between the depth of water penetration of the fly ash and plain concretes.

References

Aignesberger, A. and Kern, A. (1981) Use of melamine based superplasticiser as a water reducer. ACI SP 68, pp61-80.

Aitcin, P.C. and Riad, N. (1988) Curing temperature and very high strength concrete. Concrete International, vol. 10, no. 10, pp 69-72.

Beresford, F.D. (1978) Myths of curing. Concrete Institute of Australia, CIA news, vol. 14, no. 3, pp 12-15.

Carrasquillo, P.M. and Carrasquillo, R.L. (1988) Evaluation of the use of current concrete practice in the production of high strength concrete. ACI materials J., vol. 85, no. 1, pp 49-54.

CAA/CCA Report (1976) Superplasticing admixtures in concrete. Joint working party report no.45.030. Cement Admixtures Assoc./Cement and Concrete Assoc. London.

Gopalan, M.K. and Haque, M.N. (1985) Design of fly ash concrete. Cement and Concrete Research, vol. 15, pp 694-702.

Gopalan, M.K. and Haque, M.N. (1987) Effect of curing regime on the peoperties of fly ash concrete. ACI Materials J., vol. 84, no. 1, pp 14-19.

Gu Dezhen, Xiong Dayu and Lu Zhang (1982) Model of mechanism for naphthalene series water reducing agents. ACI J., vol. 79, no. 5, pp 378.

Johnston, C.D., Gamble, B.R. and Malhotra, V.M. (1979) Effect of superplasticisers on properties of fresh and hardened concrete. Proc. TRB Symposium, Trans. Res. Rec. no. 720.

Kunze, W.E. (1989) High strength concrete in United States. xiv Biennial Conference, Concrete Inst. of Australia.

Lane, R.O. and Best, J.F. (1979) Laboratory studies on the effects of superplasticisers on the engineering properties of plain and fly ash concrete. ACI SP 62, pp 193-207.

Malhotra, V.M. (1981) Superplasticisers: their effect on fresh and hardened concrete. Concrete International, vol. 3, no. 5, pp 66-81.

Malhotra, V.M. and Malanka, D. (1979) Performance of superplasticisers in concrete: Laboratory Investigation. ACI SP 62, pp 209-243.

Massazza, F., Costa, U. and Barrila A. (1981) Adsorption of superplasticisers on calcium aluminate monosulphate hydrate. ACI SP 68, pp 499-514.

Mielenz, R.C. (1984) History of chemical admixtures for concrete. Concrete International, pp 40-53.

Mor, A. and Mehta, P.K. (1984) Effect of superplasticising admixtures on cement hydration. Cement and Concrete Research, vol. 14, pp 754-756.

Mukherjee, P.K., Loughborough, M.T. and Malhotra V.M. (1982) Development of high strength concrete incorporating a large percentage of fly ash and superplasticisers. Cement Concrete and Aggregates, vol. 4, no. 2, pp 81-86.

Perenchio, W.F., Whiting, D.A. and Kantro D.L. (1979) Water reduction, slump loss and entrained air void systems as influenced by superplasticisers. ACI SP 62 pp 137-155.

Pomeroy, C.D. (1972) Effect of curing conditions and cube size on the crushing strength of concrete. Cement and Concrete Assoc. of Australia, Tech. Report.

Popovics, S. (1986) Effect of curing method and final moisture condition on compressive strength of concrete. ACI J. vol. 83, no. 4, pp 650-657.

Rixom, M.R. (1975) New type of concrete admixtures. Workshop on the use of chemical admixtures in concrete, The University of New South Wales, Australia.

Ryan, W.G. and Munn, R.L. (1979) Some recent experiences in Australia with superplasticising admixtures. ACI SP 62, pp 123-136.

Samarin, A. and Ryan W.G. (1975) Experience in the use of admixtures in concrete containing cement and fly ash. Workshop on the use of chemical admixtures in concrete, The University of New South Wales, Australia.

A STUDY ON THE USE OF A CHLORIDE-FREE ACCELERATOR

S. POPOVICS
Department of Civil Engineering, Drexel University, Philadelphia, USA

Abstract

The strength accelerating effects of a chloride-free admixture is investigated with portland cements, flyashes, and with and without other chemical admixtures. The influence of curing temperature on the effectiveness of the accelerator is also tested.

The accelerator is a water-soluble organic material belonging to the carboxylic acid group. The results show that considerable strength increases are produced in portland cement mortars and concretes by the use of this accelerator. These strength increases were particularly large with curing at elevated temperatures. This makes the accelerator particularly suitable for steam-cured concrete producing not only high early strengths but also high strengths at the age of 28 days and later.

In addition to strengths, times of setting as well as the mechanism of the acceleration are also tested.

Key words: Accelerator, Chloride-free, Compressive strength, Concrete, Flyash, Mortar, Portland cement, Time of setting.

1 Introduction

Since calcium chloride is not permitted in effective quantities in reinforced concrete, the importance of chloride-free accelerators increased. Despite the need, the published literature on this topic is very little. One can hardly find information about the strength increasing effects of chloride-free accelerators even in relatively recent books on chemical admixtures (1,2), perhaps because there are hardly any such admixtures. Therefore, laboratory strength tests were performed with various combinations of an accelerating admixture to establish its effectiveness in the presence of various other admixtures. The common characteristic of these combinations is that they are either chloride free.

The accelerator to be discussed is a water-soluble organic material* belonging to the carboxylic acid group. It will be referred to in this paper as "the accelerator." It will be shown that this accelerator is applicable in a wide variety of cementitious compositions including portland cement, non-hydraulic cements such as expoxies, finely divided mineral admixtures, such as fly ash, water-reducing admixtures, or any combinations of these. The quantity of theaccelerator used in a mixture will be given as percent of the cement weight.

Since many admixtures may affect more than one property of a concrete, sometimes adversely, the optimum use of any admixture, including this one, may require careful attention to the type and amount of cement used, type and amount of other admixtures present, and other characteristics of the concrete. Up to this point, the available evidence shows that this chloride-free accelerator increases significantly the strengths of a wide range of cementitious compositions.

2 Effects of the accelerator on the time of setting

After the dormant period a fresh cement paste starts stiffening, less and less plasticity can be observed, and finally all the plasticity is gone and the paste becomes brittle, although it is still without any sizable strength. This stiffening process is called *setting* and is the result of a series of reactions between the cement and water. It is customary to talk about *initial setting*, which is basically the beginning of the stiffening, and *final setting*, which is marked by the disappearance of plasticity.

The kinetics of the setting process is influenced by many factors, among which the temperature of the paste and chemical admixtures are probably the most important. The setting process should not start too early because the freshly mixed concrete should remain in plastic condition for a sufficient period to permit satisfactory compaction and finishing after transportation and placing. On the other hand, too long setting period is also undesirable because this can produce excessive bleeding in the placed concrete and will cause a useless delay in the strength development. The elimination of the too long initial setting time is so important that products are available on the market that shorten significantly the setting times at the expense of the strength development at later ages.

To see the accelerating affect of the admixture on the times of setting, experiments were performed. The tested cements were portland cements that complied with the specifications of ASTM C 150. TheVicat method specified in ASTM C 191 was used. The test results are summarized in Table 1. The data show that the accelerator reduces both the initial and the final setting times to about half when the paste as well as curing temperature are around 20°C.

---

*U. S. Patent No. 4,419,138.

Under certain circumatances, such as construction in cold weather, this is quite useful, especially if this is followed by increased strength development.

Table 1. Setting times of two portland cements with and without the accelerator

Cement	Accelerator	
	0%	2%
Type I		
initial setting	3 hr 25 min	1 hr 50 min
final setting	5 " 25 "	2 " 55 "
Type III		
initial setting	2 hr 45 min	1 hr 37 min
final setting	3 " 35 "	2 " 28 "

3 Experiments with mortars

The compatibility of the accelerator with polymers was tested in a test series with portland cement mortar. A Type I (ordinary) portland cement was used with a locally available concrete sand. Four polymers were used in emulsion form, namely three epoxies and a latex. Some of the obtained compressive test results are presented in Figure 1. It can be seen that the specimens containing the accelerator (the A specimens) produced significantly higher strengths under wet curing than the specimens without the accelerator (the NA specimens.) Similar strength increases were produced by the accelerator under dry curing conditions (3) which is in accordance with the observations by Rezansoff and Corbett with concretes cpntaining calcium chloride. (4)

Figure 2 demonstrates the compatibility of the accelerator with fly ash.(5) A Type I portland cement was combined with a locally available concrete sand and one of two commercially available Class F fly ashes with or without l.5% accelerator. The aggregate-cementitious material ratio was 3, and the liquid-cementitious materials ratio 0.5 by weight, in each series. Other details of the mortar composition are presented in the figure. The consistency of the fresh mortars was plastic; 2-in. (50-mm) cubes were prepared and tested essentially according to ASTM C 109. The specimens were cured in a fog room at standard temperature. (6) In each and every test series the specimens with the accelerator provided higher compressive strengths than the comparable controls without the accelerator.

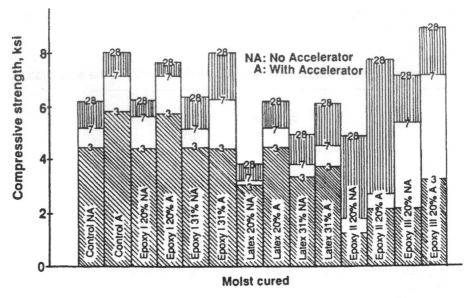

Fig. 1. Effect of various polymer modifications as well as accelerator on compressive strength, II (curing: continuously in fog room; 1 ksi = 6.90 MPa) (3)

In Figure 3 the effects of 3% accelerator are presented on the compressive strength of Ottawa sand mortars along with that of 5% calcium formate and compared to strengths of mortars without any admixture (Control A and B). The 2-in. (50-mm) cubes were prepared and tested essentially accord- ing to ASTM C109 except that some of the specimens were steam cured. The cementitious material consisted of 80% Type I portland cement and 20% natural pozzolan, by weight. The aggregate-cementitious materials ratio was 2.75 and the water-cementitious materials ratio 0.54, by weight, in every mixture. One can see from the figure not only the excellent strength increasing effects of the accelerator after water curing at room temperature but also after steam curing. Note particu-cularly the sizeable strength increase of the steam-cured specimens with ac-celerator after 7 days. The poor performance of the calcium formate here might be attributed to the presence of pozzolan.

Figure 4 illustrates the effect of the quantity of accelerator on the strength increases. It shows for a Type III that the magnitude of the relative strength increase may also be a function of the age as well as the quantity of the acce-lerator used. In this figure the compressive strengths are presented as per-centage of the strength of the comparable mortar without admixture at the same age. The sand-cement ratio was 3 and the water-cement ratio 0.5, by weight, in every mixture. The 2-in. (50-mm) cubes were cured in a fog room at standard

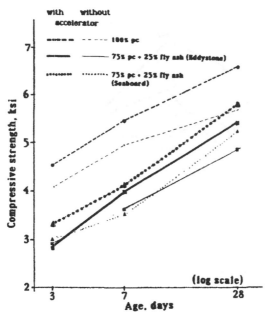

Fig. 2 --Effect of 1.5% accelerator on the compressive strength of
various portland cement-fly ash mortars.(5) Cement: Type I
1 ksi = 6.89 MPa

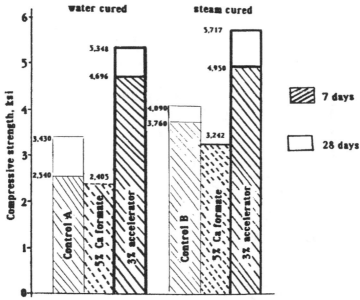

Fig. 3 --Effect of the accelerator on the strength of water- and
steam-cured Ottawa-sand mortars.(5)
Cementitious material: 80% Type I cement + 20% volcanic ash
Thin lines: without accelerator   Thick lines: with 3% accelerator
1 ksi = 6.89 MPa

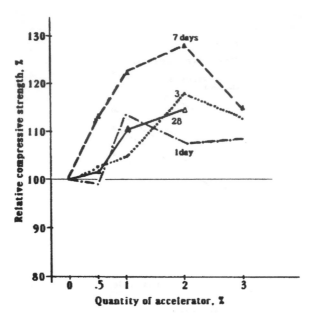

Fig. **4** –Effect of the quantity of accelerator on the relative
strengths of mortars at various ages.**(5)**
Cement: Type III--the strength without admixture is 100% at the same age

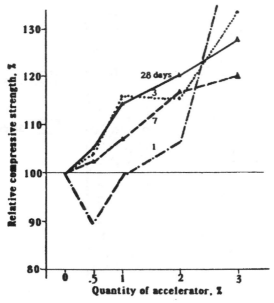

Fig. **5** –Effect of the quantity of accelerator in the presence of
a retarder on the relative strengths of mortars at various ages.**(5)**
Cement: Type III--the strength without admixture is 100% at the same age

temperature up to six months. The mortars represented in Figure 5 are similar to those in Figure 4 except that these mixtures contain also a set retarding admixture. Here again the strengths usually increase with the increase of the quantity of the accelerator.

Another variable investigated was the type of portland cement as it affected the strength increases produced by the accelerator. It appears from Figure 6 that the Type V cement used is clearly the least compatible with the accelerator, and that the largest strength increases are produced by Type III cement with 3% accelerator. Similar experiments repeated with an additional set retarding admixture show the same trend .

The results obtained seem to suggest that the accelerator follows the pattern of the strength-increasing effects of calcium formate which was found an effective accelerator by Gebler (7) when the ratio of C_3A to SO_3 was greater than 4 and the SO_3 content was low.

The effect of delayed addition of the accelerator on mortar strength increases was also investigated. It was found that 2% accelerator works better with delayed addition. (Fig. 7) In this figure relative strengths are plotted again 100% being the corresponding   strength at the same age but w2ithout the accelerator. In other words, higher strength increases were obtained regularly when first the dry components of the mortar were mixed with a portion of the mixing water and the accelerator was added subsequently with the rest of the mixing water, than when the accelerator was added to the dry components with the mixing water. The same trend was noticed with a Type III cement with and without an additional set-retarding ad- mixture.

In brief, the overwhelming majority of the mortar tests presented demonstrates considerable strength increases produced by the accelerator under a wide variety of circumstances.

4 Experiments with Concrete

Figure 8 shows the strength results of four concrete series. Two of these contain 2% accelerator, the other two did not (control). Type II cement, natural quartz sand and crushed stone of 1/2-in. (12.5-mm) maximum particle size were used. The cement-fine aggregate-coarse aggregate ratio was 1 : 2.75 : 2.0, and the water-cement ratio 0.52, by weight, in every mixture. Some of the cube specimens were cured in a fog room at 73°F (23°C), the rest at 140°F (60°C). The results indicate considerable strength increases produced by the accelerator, especially at the higher curing temperature, which is in accordance with the findings shown in Fig. 3.

Figure 9 shows the compatibility of the accelerator with a high-range water-reducing admixture (superplasticizer). Here the compressive strengths of steam-cured concretes are plotted. A Type III (high-early strength) portland cement, natural quartz sand and a crushed stone of 12.5 mm (1/2 in) maximum particle size were used. The cement, fine aggregate, coarse aggregate ratio

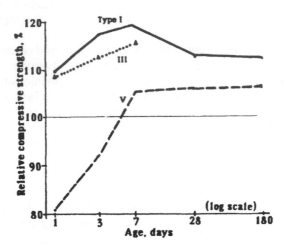

Fig. **6** --Effect of cement type on the relative increases of mortar strengths produced by 3% accelerator at various ages--the strength without accelerator is 100% at the same age.(5)

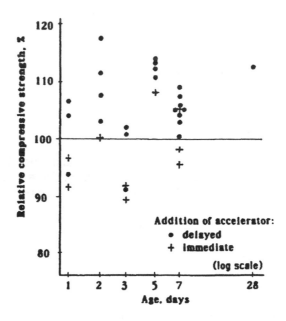

Fig. 7--Effect of the time of addition of 2% accelerator on relative mortar strengths at various ages--the strength without accelerator is 100% at the same age.(5)
    Cement: Type I

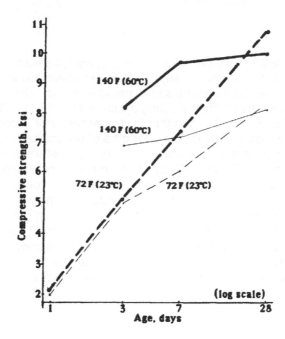

Fig. 8 --Effect of accelerator on the compressive strength of
concrete at normal and elevated temperatures.(5)
        Cement.: Type II
        Thin lines: without accelerator
        Thick lines: with 2% accelerator
        1 ksi = 6.89 MPa

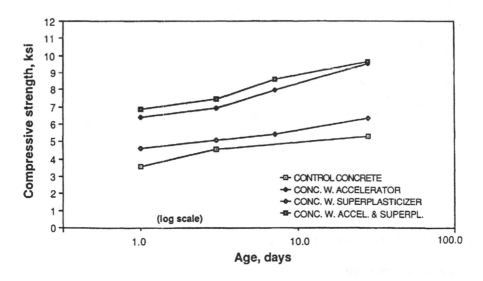

Figure 9. Comparison of the strengths of steam-cured concretes with
and without the accelerator at various ages. (8) Cement: Type III.
1 ksi = 6.90 MPa.

was 1 : 1.69 : 2.33 by weight. The water-cement ratio was 0.52 by weight, except in the mixtures containing superplasticizer where w/c = 0.46. The consistency of all concretes was the same, namely plastic. The cube specimens were steam-cured by traditional curing cycles with 77⁰C (170⁰F) maximum temperature. (8) Figure 9 shows that (a) the accelerator is compatible with the superplasticizer used; and (b) the strengths of the mixtures with the accelerator are typically more than double of the strengths of the control mixture, especially at the age of 28 days. This is much more than the strength increases produced by the superplasticizer through the reduction of the water- cement ratio. In other words, the use of the accelerator seems a practical tool for the production of steame-cured concrete of high final strength. (8)

The relationship between the compressive strengths of the control concrete and those of the concretes with the accelerator can be seen in Figure 10. This shows that the greater the strength of the control, the more effective the accelerator becomes. It was also noticed that the accelerator increased the 28- day strengths more than the 1-day strengths.

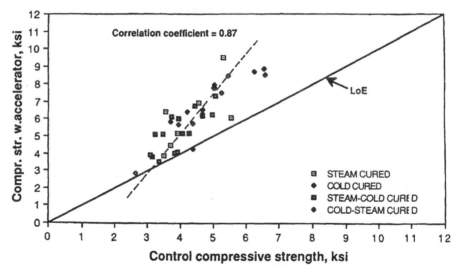

Figure 10. Demonstration of the strength-increasing effects of the accelerator on concretes subjected to different curing methods. (8) Cement: Type III.   LoE: line of equality.   1 ksi=6.9 MPa.

5 Mechanism of strength acceleration

The mechanism of strength increase was investigated through checking the effect of the accelerator on the hydration of the cement by silylation. This method examines the structure of the silicate anions in hardened cement pastes that were cured under wet condition usually until 28 days then kept in dry air.

The examinations are made by a chemical method utilizing the *principle of structure retention*. This consists of the end-blocking of the reactive hydroxil groups of the silicic acid liberated from the parent silicate structure by a non-reactive trimethyl silyl, $(CH_3)3Si$-, group in order to prevent the spontaneous polycondensation of the silicic acid. The end-blocked trimethyl silyl esters of the silicate anions are both hydrolytically and thermally stable and the volatile ones can be separated and analyzed by gas-liquid partition chromatography. (9)

The silicates in an unhydrated portland cement are in the form of mono-silicate. Due to hydration,, the share of monosilicate structure in the hardening paste decreases and simultaneously short silicate polymers, so-called oligo-mers, are formed. Unfortunately, only the oligomers of n = 2 to 5 Si atoms in the anion can be determined by gas chromatography. These limited results show no difference between the hydration products in the pastes with the accelerator and those without. That is, no new kind of hydration products were developed by the accelerator. One may surmise that perhaps the accelerator acted as a catalyst on hydration of the silicates in the cement.

6 Conclusions

The experimental data presented clearly show that the accelerator reduces the times of setting and increases the strength of a wide variety of cementitious compositions. Specifically, the accelerator has the tendency to

reduce the standard setting times to about half;

be compatible with a group of epoxies and superplasticizers;

produce greater strength increases with Type III (high-early-strength) portland cement than with Type I, or especially with Type V cements;

yield higher early as well as higher late concrete strengths than the control concrete;

produce greater strength increases at high curing temperatures than at normal or low temperatures. Thus, it is an excellent accelerator to make steam-cured concrete of high late-strength;

create greater strength increases with delayed addition.

References

1. Rixom, M.R., Editor, Concrete admixtures: use and application, Construction Press, London, 1977.

2. Rixom, M.R., Chemical Admixtures for Concrete, E. & F.N. Spon Ltd., London, 1978.

3. Popovics, S., "Strength Losses of Polymer-Modified Concretes Under Wet Conditions," Polymer Modified Concrete, SP-99, American Concrete Institute, Detroit, 1987. pp. 165-189.

4. Rezansoff, T. and Corbett, J. R., "Influence of Accelerating Admixtures on Strength Development of Concrete under Wet and Dry Curing," ACI Materials Journal, Vol. 85, No. 6, November - December 1988. pp. 519 - 528.

5. Popovics, S., "Strength Increasing Effects of a Chloride-Free Accelerator," Corrosion, Concrete, and Chlorides, F. W. Gibson, Editor, SP-102, American Concrete Institute, Detroit, 1987. pp. 79-106.

6. Popovics, S.,"Improved Utilization of Fly Ash in Concrete through a Chloride-Free Accelerator," in ASTM Cement, Concrete and Aggregates, CCAGDP, Vol. 7, No. 1, Summer 1985, pp. 49-51.

7. Gebler, S., "Evaluation of Calcium Formate and Sodium Formate as Accelerating Admixtures for Portland Cement Concrete," ACI Journal, Proceedings Vol. 80, Sept.-Oct. 1983, pp. 439-444.

8. Popovics, S., "Effects of a chloride-free accelerator on concrete strength," Second International Symposium on Concrete in Developing Countries, Session 8: Mineral and Fibre Admixtures to Concrete, Bombay, India, Jan. 3-8, 1988. pp. 24-36.

9. Tamas, F.D., Sarkar, A.K., and Roy, D.M., "Effect of Variables Upon the Silylation Products of Hydrated Cements" Hydraulic Cement Pastes: Their Structure and Properties, Proceedings of a conference held at University of Sheffield, 8-9 April, 1976. Cement and Concrete Association, Wexham Springs, Slough, 1976, pp. 55-72.

STRENGTH AND TIME-DEPENDENT STRAINS OF CONCRETE WITH SUPERPLASTICIZERS

E. N. SHCHERBAKOV
Research Institute for Transport Engineering, Moscow, USSR
Yu. V. ZAITSEV
Polytechnical Institute, Moscow, USSR

Abstract
Experimental results on strength, drying shrinkage and
creep of concrete with superplasticizer are presented. It
is shown that the effect of superplasticizers on strength
and time-dependent strains of concrete is complicated and
depends on the content of superplasticizer in concrete mix
and other variables. The multiplicative models for
prediction of shrinkage and specific creep strains of
normal-weight concrete without superplasticizer are
obtained. The results of a multiple regression analysis
(based on 280 experimental data) for concrete with
superplasticizer are given. It is proved that the
proposed models allow to predict the time-dependent strain
of concrete with high accuracy independently of admixture
addition.
Key words: Strength, Drying shrinkage, Creep, Mathematical
models, Prediction, Superplasticizers effect.

1 Introduction

Widely spread superplasticizers application for reinforced
concrete structures, prestressed concrete ones among them,
raise a problem of concrete strain parameters prediction,
modified due to such admixtures addition, while designing
reinforced concrete stuctures. We have every reason to
assume that plasticizers appreciable effect on concrete
strength and (or) concrete mix water content as a result
of its addition would influence noticeably the concrete
time-dependent strain parameters and therefore change the
basic performance characteristics of the structures
designed. The results of many investigations devoted to
this problem solution have been lately published by
Nagataki and Yonekura (1978), Johnston, Gamble and
Mahlhotra (1979), Brooks,Wainwright and Neville (1981),
(1983), Dhir and Yap (1983) et al.
    However in most cases the studies suffer from grave
methodological shortcomings, as there has been made an
attempt to obtain some average ratio of time-dependent

strain of concrete with superplasticizer and plain
concrete without plasticizer on the basis of experiments,
neither aim nor specific conditions for admixture use
having been taken into account, and admixture own
influence and the accompanying effects followed
(alteration of concrete mix water content or strength of
concrete) having been not separated. As the result the
assessments obtained during the experiments differ
significantly not only by magnitude but by sign as well.
The latter complicates their application in design
practice and may lead to erroneous prediction of strains.

2   Experimental results

As it can be seen in Fig. 1, presented in author's paper
(1989) the relative reduction of water content in concrete
mix Wp/W (Wp and W − initial water content in mix with
superplasticizer and without it) by a given dosage of
USSR-produced superplasticizer S-3 (in this case 0,5% by
cement weight) depends on the cement content of concrete
mix. Accordingly the strength of hardened concrete is
altered in a sophisticated manner. As it is shown in Fig.
1 there is no tendency for any kind of mix flow effect.
But one should take into consideration that water reducing
effect for concrete mix at a given cement content also
depends on cement properties, dosage and type of
superplasticizer, which is not presented by the data of
Fig. 1. And finally, it should be noted that admixture
addition frequently produces air− entraining effect (up to
4% by volume) and influences concrete strength gain.
Undoubtedly the latter exerts additional influence on the
magnitude time−dependent strains of concrete.

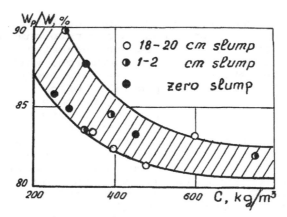

Fig. 1.   The effect of superplasticizer S-3 on
          reduction in water content of concrete
          mix W_p /W at a given constant mix flow.

The stated above facts show, that the effect of
superplasticizers on concrete creep and shrinkage appears
to be a sophisticated and ambiguous phenomenon.   To give a
proper estimation of this effect one should first of all
take into account the alteration in concrete mix water
content and (or) strength, induced by admixture addition
(depending on superplasticizer application purpose).   All
these effects being properly taken into account, an
attempt can be made to separate admixture own effect with
regard to its chemical composition and dosage in concrete
mix.

Fig. 2 shows the results of processing of test data on
concrete drying shrinkage.   Superplasticizers addition
was carried out under conditions of conservation of
contstant mix composition, aggregate grading being
different.   Superplasticizeres of 4 types (according to
British Cement and Concrete Association), which dosage
varied within the wide range (P = 0...5 %),were used.
Test duration was 420 days, relative humidity of
environment was 55-60 %.

Ratio of drying shrinkage measured values for concrete
with superplasticizer and corresponding plain concrete
$\varepsilon_{sp}/\varepsilon_s$ are plotted in Fig. 2.

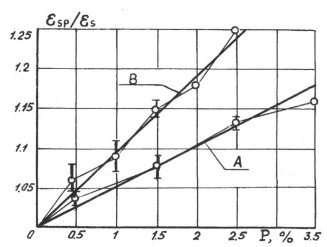

Fig. 2.   The effect of superplasticizers dosage
          on concrete shrinkage prepared from
          constant proportion mixes.   Author's plot
          from test data by Dhir and Yap (1983).

Fig. 2 indicates that the ratio $\varepsilon_{sp}/\varepsilon_s$ under
consideration, other being equal conditions, is a
statistically distributed value, the expected value of

which regularly increases approximately in proportion to
dosage of admixture in the mix, coefficient of residual
variation remaining within 4-5 % range.

Certain effect of admixture chemical composition (type)
can be observed as well. In the given case
naphtalene-formaldehyde superplasticizes (straight line
regression B), which are similar to admixture S-3, exert
great influence on shrinkage growth, as compared to
melamine ones (straight line A).

The effect of other plasticizing admixtures, modified
lignosulfonates (USSR-produced analog is LSTM-2) among
them, approximates straight line B.

Physical and chemical character of superplasticizers
own effect remains unknown. Fig. 3 shows experimentally
obtained relationships. It is seen that the addition of
admixture S-3 into concrete effects slightly the moisture
volume, evaporating from concrete (Fig.3a), as well as the
relationships between shrinkage and moisture loss (Fig.
3b) under conditions of the experiments conducted.

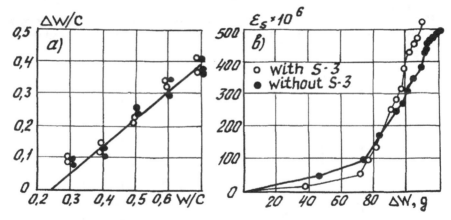

Fig. 3.   The effect of superplasticizer S-3 on the moisture
           loss from concrete (a) and relationship between
           shrinkage of concrete and moisture loss (b).

However Fig. 3b shows some increase of linear coefficient
of shrinkage strain for concrete with superplasticizer.
Superplasticizers addition into concrete mix affects large
and small pores (capillaries) ratio in hardened cement
paste and increases concrete strain reaction on
evaporation of similar moisture volume.

Some other explanations are not excluded, which may be
associated with alteration of binding energy of liquid
phase with hydration solids or diffusion characteristics
of a system as a whole.

3 Results of multiple regression analysis

3.1 Subjects

Complex assessment of all the mentioned predictors effect on time-dependent strains of concretes with superplasticizer has been conducted by means of statistical analysis of test results published in various countries. A great amount of original data, containing 280 average results of deformations measurements in the course of 700 specimen testing, has been proccessed. The tests conducted involved 16 kinds of superplasticizers, which were introduced into reference concrete mix in up to 3.5 % by weight of cement both with the purpose of getting flowing mix (mix proportion being constant) and water-reduced mix (mix flow being constant). Most data have been obtained by application of superplasticizer S-3. Multiple stepwise regression analysis has been conducted by the procedure stated by the author (1983).

The main objective of the analysis was substantiation of mathematical models for prediction ultimate values of long-time strains, equally suitable both for concrete with superplasticizer and without it, including all possible variants for plasticizisers effect in concretes.

3.2 Shrinkage strain

Some of the test data sample, designed for statistical analysis of shrinkage strain, involved N = 194 results. As a response the ratio shrinkage strain of concrete with superplasticizing admixture and that of plain concrete (without admixture) independently of superplastizcizer introduction purpose has been considered.

In accordance with the multiplicative mathematical model, presented in the author's paper (1988) for test data sample (input data matrix N = 522 x 18) predicted ultimate value of shrinkage strain for normal-weight concrete without superplasticizer

$$\hat{\varepsilon}_s(\infty, t_w) = k_s \, W^{3/2} \left(\frac{0.012 + 1.12M}{0.06 + M} \right) \left(\frac{98 - \theta}{28} \right)^{1/3} \times$$

$$\times \left(\frac{7.2 + 0.9t_w}{6.5 + t_w} \right) \left(\frac{8.5 + 0.9D}{6.5 + D} \right) \tag{1}$$

where $\varepsilon_s(\infty, t_w)$ - predicted mean ultimate (for $t \to \infty$) value of relative shrinkage strain of concrete drying from age t_w, days; W - initial water content of concrete (mixing water) in litres per 1000 l of mix; M - open for drying specific surface of specimen (surface : volume ratio), θ - relative humidity of environment, %; D - maximum size of aggregate, mm; $k_s = 0.128 \times 10^{-6}$.

Dimnesionless coefficient value estimation k_s in eq. (1) corresponds to expected value of ultimate shrinkage strain on the axis of concrete specimen, fabricated from Portland cement with optimal content of SO_3 and moist-cured before drying. Variability of chemical and mineralogical characteristics of industrial produced Portland cements taken into account value of coefficient k_s in eq. (1) is to equal $k_s = 0.145 \times 10^{-6}$. For this case a 95 % probability level of shrinkage prediction (independently of individual characteristics of cements) will be achieved. For present test data sample (N = 522) the predictive model (1) posesses satisfactory accuracy; unexplained variation is only about 9 %. Superplasticizer not exerting its own effect on shrinkage magnitude, the condition

$$\varepsilon_{sp}/\varepsilon_s \cong \left(\frac{Wp}{W}\right)^{3/2}$$

should be satisfied according eq.(1). The offered suggestion check for sample N = 194 has yielded relationship

$$\varepsilon_{sp}/\varepsilon_s = 1,14 \left(\frac{Wp}{W}\right)^{3/2} \qquad (2)$$

which provides predictive error, characterized by residual variation coefficient $\delta = 10.8$ %. According to eq. (2) for sample N = 194, shrinkage of concrete with superplasticizer exceeds on the average by 14 % that of plain concrete of the same nominal composition (Wp = W).

More detailed statistical analysis showed that the indicated relationship (2) can not be considered in a general case as a constant one. The statistical significant effect of superplasticizer dosage P in mix (% by the cement weight) has been found to depend on Wp/W ratio. The model obtained for those effects estimation

$$\varepsilon_{sp}/\varepsilon_s = \left(2 + 0,09\ P - \frac{Wp}{W} \right) \left(\frac{Wp}{W}\right)^{3/2} \qquad (3)$$

allows in comparison with eq. (2) to increase the prediction reliability ($\delta = 8,05$ % versus $\delta = 10,8$ %). According to eq. (3) for the case of mix without admixture (when P = 0, $W_p = W$) $\varepsilon_{sp} = \varepsilon_s$. As follows from eq. (3) numerical values of the first multiplier in brackets may differ greatly from mean value 1,14 in eq. (2), depending on admixture concentration in a mix and its use purpose.

Speaking about concretes made from mixes equal

proportion (Wp = W), shrinkage increase due to superplasticizer addition is calculated by eq. (3) as equal to 9 % per every percent of admixture content in mix. For usual concentrations of superplasticizers (0.5...0.6 %) predicted shrinkage strain increase is only 5 %.

In case of equal flow concrete mixes use (Wp < W) shrinkage of concrete with admixture is on the contrary less, as compared to that of concrete without admixture, by 9 % (estimation is based on the analysis for N = 56 results). The results obtained shows that effect of superplasticizer addition depends in a complex manner on specific conditions for admixture use (its concentration in mix, water-reducing effect et al.).

In this connection experimental revision should be made of the author's assumption (1989) that admixture concenration relative not to cement weight but to water phase volume is considered to be the measure for superplasticizer own effect on concrete shrinkage.

In any case it is evident that there is no possibility for determination of mean ratio between shrinkage of concrete with superplasticizer and without it, which would be equally valid for all the cases. The latter is well illustrated by the model (3).

The effect of other factors, involved in the model (1), on the shrinkage being assumed to depend slightly on admixture presence, one has every reason to extend the model (1) to shrihkage prediction of concrete with superplasticizer and without it. For this purpose according to (3) the following expression for coefficient k_s in model (1) should be accepted

$$k_s = 0.145 \left[2 + 0.09\ P - \frac{Wp}{W} \right] \times 10^{-6} \qquad (4)$$

3.3 Specific creep strain
Statistical ahalysis of superplasticizing admixture effect on creep strain of normal-weight concrete has been conducted for test data sample N = 80. The analysis supported the suggestion that concrete creep alteration induced by superplasticizers addition is satisfactorily predicted by model obtained in the above-mentioned author's paper (1988) for concrete without plasticizer (input data matrix N = 631 x 22) for both water and steam-cured concretes):

$$\hat{C}_{cr}(\infty, t_o) = k_c \frac{W}{\Delta + R_\tau} \left[\frac{0.16 + 1.48M}{0.35 + M} \right] \left(\frac{135 - \theta}{65} \right) \qquad (5)$$

where $C_{cr}(\infty, to)$, MPa^{-1} — predicted mean ultimate (for t → ∞) value of specific creep strain of concrete, loaded in age to, days; R_τ – cubic strength of concrete at age of loading, MPa; $\Delta = 5$ MPa; $k_c = 20.2 \times 10^{-6}$.

The estimation of value of dimensionless coefficient k_c in model (5) corresponds to the expected value of ultimate specific creep of concrete specimen made on ordinary Portland cement (Type 1), crushed granite and quart sand and moist cured until age of loading. For steam-cured concrete $k_c = 18.2 \times 10^{-6}$. For test data sample N = 631 prediction precision of specific creep by the model (5) is characterized by unexplained variation about 11 %.

Specific creep of concrete with superplasticizer as well as that of concrete without admixtrure according to predictive model (5) is princpally determined by its effect on alteration of strength and (or) initial mix water content in concrete. At the same time creep of superplasticized concrete features are the following:

(a) specific creep of concrete loaded in water saturated environment is higher of the average by 40 % compared to that for concrete without admixture;

(b) scatter of specific creep prediction for these conditions of loading is also higher as compared to test results obtained in atmospheric humidity;

(c) dosage of superplasticizer in a mix exerts its influence on ultimate specific creep value similar to its effect on shrinkage strain.

The latter has been illustrated by Fig. 4, plotted by the results of statistical analysis both creep and shrinkage test data.

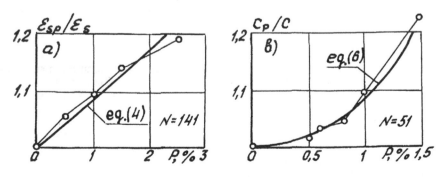

Fig. 4.   The effect of superplasticizer dosage in concrete mix on shrinkage (a) and creep (b) of concrete.

Fig. 4b show superplasticizers own effect on specific creep (mix composition and strength of concrete being equal).   According to statistical estimation plotted in

Fig. 4b the following expression for coefficient k_c in model (5) should be accepted

$$k_c = 20.2 \left[1 + 0.9 \, (P)^{2.25} \right] \times 10^{-6}. \qquad (6)$$

The plot in Fig. 4a for shrinkage strain corresponds to eq. (4) for Wp = W.

The data illustrated by Fig. 4 and eq. (4) and (6), are worthy of detailed study in the course of further accumulation of test data. Table 1 presents statistical estimation of the results of prediction of ultimate specific creep strain on the model (5) basis, accounting for stated features of superplasticizer effect on these strains.

Table 1. Results of statistical analysis for creep data.

Test conditions	Sample (N)	Statistics for experimental to predicted value ratios	
		mean value	variation coeff. δ (%)
in air (θ= 50...70 %)	55	1.002	17.0
in water	25	1.026	25.6
as a whole	80	1.010	20.0

Table 1 illustrated, that the model (5), designed for specific creep prediction with superplasticizer dosage correction according to eq. (6), provides as accurate estimation of these concrete strain as that for concrete without admixture (according author's data δ = 22,4 %) . Bias in estimate in all cases is absent in fact. The conclusion is valid for wide-ranging type of superplasticized concretes (including steam-cured concrete) with strength at age of loading varied within range $R_{\tau} \cong 6,5...136$ MPa.

References

Nagataki, Sh. and Yonekura, A. (1978) Studies of the volume changes of high-strength concrete with superplasticizer. J. of Jap. PCEA, 20, 26-33.

Johnston, C.D. , Gamble, B.R. and Malhotra V.M. (1979) Effects of superplasticizers on properties of fresh and hardened concrete. Transp. Res. Rec. 5, 1-17.

Brooks, J. J., Wainwright, P. J. and Neville, A. M. (1981)
Time-dependent behaviour of high-early strength
concrete, containing of superplasticizers, in
Development in Use of Superplasticizers, ACI Spec.
Publ. SP-68, Detroit, pp 81-100.

Brooks, J. J. and Wainwright, P. J. (1983) Properties of
ultra-high strength concrete, containing
superplasticizer. Mag. of Concr. Res.,35, 205-213.

Dhir, R. K. and Yap, A. W. F. (1983) Superplasticizer
high-workability concrete: some properties in the fresh
and hardened states. Mag. of Concr. Res.,35, 214-228.

Shcherbakov, E. N. and Zaitsev, Ju. V. (1976) Voraussage der
Festigkeits- und Verformungseigenschaften von Beton
under Dauerlast. Cem. and Concr. Res. 6, 515-528.

Shcherbakov, E. N. (1983) Prediction of physical and
mechanical properties of concrete as composite material,
in Studies of Strain, Strength and Durability of

Concrete For Transport Structures, Res. Inst. for

Transp. Eng., Moscow, pp 5-21 (in Russian).

Shcherbakov, E. N. (1989) Developments in improvement of
codes for concrete structures design, in Design

Problems of RC Structures Quality and Material Savings,

Res.Inst. for Transp.Eng.,Moscow, pp 4-19 (in Russian).

Shcherbakov, E. N. (1989) On prediction of time-dependent
strain of concrete with superplasticizers, in Design

Problem of RC Structures Quality and Materiail Savings,

Res. Inst. for Transp. Eng., Moscow, pp 51-60 (in
Russian).

4 DURABILITY

INHIBITING EFFECT OF NITRITES ON THE CORROSION OF REBARS EMBEDDED IN CARBONATED CONCRETE

C. ALONSO, M. ACHA and C. ANDRADE
Institute of Structures and Cement, 'Eduardo Torroja', CSIC, Madrid, Spain

Abstract
Concrete carbonation induces a pH drop of the pore solution down to values around the neutrality. When the carbonation front reaches the rebar, the steel is depassivated and its corrosion starts at a rate which depends on the humidity inside the concrete pores.
In order to prevent or avoid this corrosion, several methods have been developed which, in general, need to be foreseen during the design or the construction. In the present paper, results on the inhibiting effect of nitrites, added as admixtures during the mix, on the rebar corrosion in carbonated concrete, are presented.
Nitrites have been tested in concrete as inhibitors of the chloride attack, but their effect against carbonation is poorly known.
The tests have been carried out in mortar. The results show that nitrites passivate the steel in carbonated media, regardless the solution pH value or the cement type. When the inhibiting action is not complete, at least the reduction of the corroding area is very high. The inhibiting effect was not detected when chlorides were added to the mixing before carbonation.
Key Words: Concrete, carbonation, rebars corrosion, Nitrites.

1 Introduction

The employment of inhibitor admixtures to prevent the corrosion of reinforcements when aggressives are present in the concrete has been a common practice for several years. It was the Soviet Union (Ratinov, V.B. (1972) and Akinova, K.M. and Ivanov, F.M. (1976)) that pioneered the use of these admixtures since their extremly cold climate requires the implementation of antifreeze admixtures such as $CaCl_2$.
Among the different chemical substances tested as inhibitors of rebars corrosion in concrete, Sodium or Calcium Nitrites are those which have also shown the best physico - chemical compatibility with concrete (Rehm, G. and Rauen, A. (1969) and Berke, M.S., Grace, W.R. and Cº (1985)).
Nitrites have the advantage of being able to operate in alkaline and neutral media which allows their use in conditions similar to those of concrete polluted with chlorides (alkaline nature) as well as

in carbonated concrete (neutral conditions).

This admixture has proved to induce an inhibitor effect against the action of chlorides (Alonso, C. and Andrade, C. (1983) and Gouda, V.K. and Sayed, S.M. (1973) and Treadaway, K.W. and Russell, A.D. (1968) although the durability of its effectiveness still remains controversial in spite of the excellent study previously carried out (Lundsquist, J.T., Rosenberg, A.M. and Gaidis, J.M. (1979) and Rosenberg, A.M. and Gaidis, J.M. (1979). However no references have been found on their ability to protects rebars from corrosion due to the concrete carbonation.

In the present paper the inhibiting action of NO_2^- in synthetic solutions simulating the pore concrete one and carbonated mortars is studied by means of electrochemical techniques for corrosion control.

2 Experimental Methods

2.1 Solution tested

Different synthetic solutions were prepared with distilled water and $Ca(OH)_2$ to reach a saturated state. Amounts of 0 and 0.05 M $Ca(NO_2)_2$ were also employed.

The composition of the solutions tested were:

1) $Ca(OH)_2$ sat
2) $Ca(OH)_2$ sat + 0.1 M NaOH + 0.1 M KOH
3) $Ca(OH)_2$ sat + 0.5 M NaOH

4) $Ca(OH)_2$ sat + 0.05 M $Ca(NO_2)_2$
5) $Ca(OH)_2$ sat 0.3 M KOH + 0.05 M $Ca(NO_2)_2$
6) $Ca(OH)_2$ sat + 0.5 M KOH + 0.05 M $Ca(NO_2)_2$

These solutions were carbonated bubbling CO_2 through them and simultaneously the decay of the pH until a constant value was recorded.

2.2 Mortar specimens

Mortar specimens of 2 x 5.5 x 8 cm in size were fabricated with six different types of cement: OPC, SRPC, OPC + 30% of Fly Ash, Fly Ash C, Slag C and Pozz. C. The water/cement and cement/sand ratios were: 0.5 and 1/3 respectively.

Some of them were prepared with chlorides (2% $CaCl_2$) and without admixtures and also it was added in the mix different amounts of $NaNO_2$ (0, 2 and 3% in weight of cement).

During the curing period, the specimens were held at 90% R.H. at 20 ±
2ºC for 28 days. Then the specimens were hastly carbonated in a
chamber saturated with CO_2 and 60% humidity. Finally the carbonated
specimens were submitted to dry and wet periods (100%, 50% R.H. and
partial immersion).

2.3 Techniques for corrosion control

Both the mortar specimens and the solutions had two similar steel bars
embedded, as duplicate working electrodes in which the corrosion
process was tracked, and a graphite bar used as counter electrode.
Corrosion potential was determined using a calomel reference
electrode. Linear polarization technique was also employed. The Rp was
obtained from the slope of a potentiodynamic ramp carried out in the
anodic direction from –10 to + 10 mV around the corrosion potential at
a sweep rate of 10 mV/min. The corrosion intensity was determined from
Stern and Geary's equation $i(\mu A/cm^2) = B/Rp$. As constant B it was
used a value of 26 mV when the specimenswere actively corroding and 52
mV when they were passive.

3 Results

3.1 Carbonated solutions

Figures 1 and 2 illustrate the corrosion potential and the corrosion
intensity measurements of the rebars immersed in the solutions before
and after carbonation.

The solutions saturated with $Ca(OH)_2$ and having different alkaline
additions (figure 1) show that the rebars before carbonation have a
Ecorr between –200 and –300 mV and a Icorr ranging between 1 to 3
$\mu A/cm^2$. After carbonation the Ecorr decreases to a value lower than
–600 mV and Icorr of 20-30 $\mu A/cm^2$. After five days of carbonation the
Ecorr becomes similar except for the solution with $Ca(OH)_2$sat + 0.5 M
KOH where Ecorr increases till values of –300 mV, this, perhaps is,
due to the high amount of $CO_3^=$ and HCO_3^- present in the solution after
carbonation; anyway the corrosion intensity still remains high.

In presence of $Ca(NO_2)_2$ is shown in figure 2 the corrosion potential
that before carbonation was around –200 mV, after carbonation
increases and stabilizes at $\sim$ –50 mV, while the corrosion rate reaches
a value between 0.1 – 0.2 $\mu A/cm^2$. When the specimens were removed from
the solutions it was not found any corrosion product in those rebars
immersed in solutions with NO_2^-.

3.2 Carbonated mortars

After a carbonation process the mortar specimens were submitted to wet
and dry periods at different enviromental humidities (50% R.H., 100%
R.H. and partial immersion).

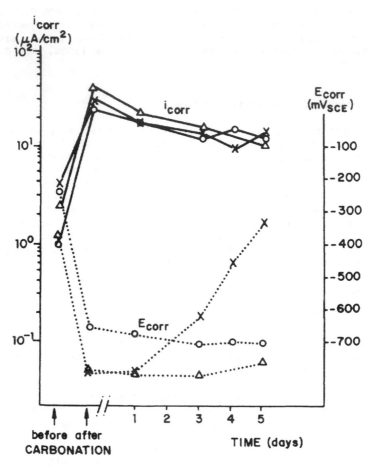

FIGURE 1

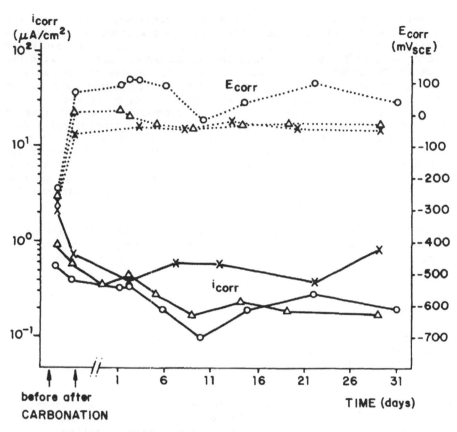

FIGURE 2

The evolution of the corrosion intensity (Icorr) was traced over 400 days for the six cement types tested, with and without $CaCl_2$ and with and without $NaNO_2$. In figures 3, 4 and 5 is represented the evolution of Icorr in time for bars embedded in carbonated mortars made with OPC and 0% or 3% of $NaNO_2$ and 3% $NaNO_2$ + 2% $CaCl_2$. It may be deduced that the presence of NO_2^- significantly reduce the corrosion rate due to carbonation, mainly in atmospheres with high humidity content.

In figures 6 and 7 is summarized the electrochemical weight loss of the rebars for all the conditions studied. The electrochemical losses were calculated integrating the Icorr – time curves as that of figures 3, 4 and 5 and through Faraday's law. From figure 6 the inhibiting effect of nitrites is clearly deduced. The use of 2 or 3% of $NaNO_2$ show a similar behaviour although the higher the proportion of nitrite is the lower the corrosion rate becomes and, consequently, the smaller the metal losses are.

When chlorides and carbonation are acting together, the amount of nitrites here used nearly do not affect the corrosion rate of the rebars, as consequence similar electrochemical weight loss can be seen in figure 7.

4 Discussion

The use of nitrites as rebar corrosion inhibitor has been mainly studied to prevent chloride attack. In the present work a new possibility for the use of nitrites is presented in relation with the carbonation attack.

Although the carbonation of the concrete cover is a slow process, the increasing amount of relatively old structures showing rebars corrosion due to this phenomenon has become a worring problem.

The aim of the present study was to explore the ability of nitrites to avoid this type of possible danger for the rebars and try to extend service life of concrete structures. Although the results presented here were carried out in simulated pore concrete solutions and in mortars in laboratory conditions they try to contribute to a better understanding of the rebars corrosion process.

Solution tests allow to confirm the inhibiting action of nitrites in carbonated conditions simulating the pore chemistry of a carbonated concrete.

Regarding mortar studies nitrites have shown to reduce and, in most of the cases, completely avoid corrosion caused by the concrete carbonation. It has been detected that the type of cement has not a significant influence on the corrosion rate after carbonation especially in presence of NO_2^-.

On the other hand the presence of nitrites seems to enhance their

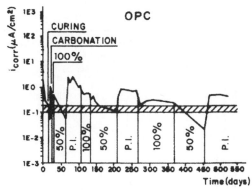

FIGURE 3

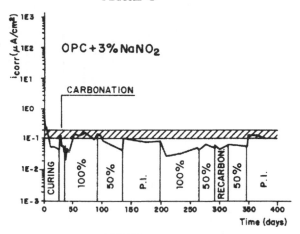

FIGURE 4

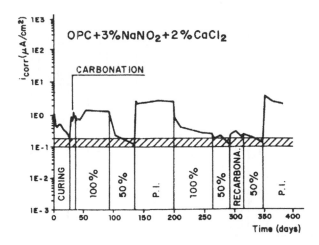

FIGURE 5

225

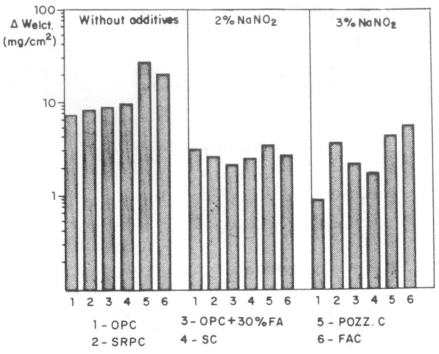

1 - OPC   3 - OPC+30%FA   5 - POZZ. C
2 - SRPC   4 - SC   6 - FAC

FIGURE 6

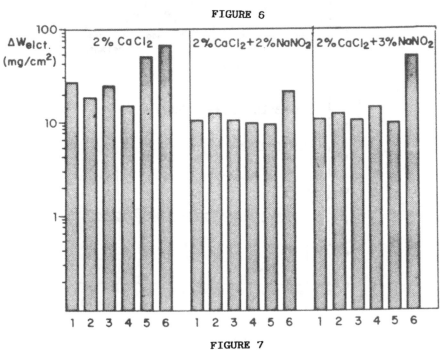

FIGURE 7

action when the concrete is wet. This turns out to be a favourable effect since wet concrete is riskiest condition for rebar corrosion. Unfortunately when chlorides added in the mix act together with carbonation the nitrites are not able to avoid the attack. However the presence of nitrites always reduce corrosion.

Concerning the optimum proportion of nitrites to use in case of carbonation attack, it may be deduced from this work that 3% has proven efficient enough for preventing corrosion. So that the higher the proportion of nitrites used which do not disturb the concrete properties, the higher the protection level results. (González, J.A., Algaba, S. and Andrade, C.(1980) and Alonso, C. and Andrade, C. (in press)).

5 Conclusions

- Nitrites added in the mix (3% by weight of cement) protects reinforcements for further corrosion due to concrete carbonation.

- Nitrites enhance their inhibiting action in wet concrete.

- When the proportion of nitrites is not high enough to completely avoid rebars corrosion, an increase of the attack is not found. Always the presence of nitrites reduce the attacked area (weight loss).

Acknowledgments

The authors are indebted to W.R. GRACE and Co. and to Spanish CAICYT of the Ministry of Education and Science for financial support of this research.

References

Stern, M. and Geary, A.L.- "Electrochemical polarization I: A theoretical analisys of the slope of polarization curves". J. Electroch Soc. 104 (1957) 56.

Treadaway, K.W. and Russell, A.D. - "Inhibition of the corrosion of steel in concrete". Highways and Public Works, 36 (1968) 19.

Rehm, G. and Rauen, A. - "Electrochemical studies on corrosion inhibition of steel in concrete". Colloque RILEM "Durability" Prague (1969).

Ratinov, V.B. - "Concrete and mortars with anti-corrosive admixture for repair of reinforcing structures". Beton i Zhelezobeton, 7 (1972) 22.

Gouda, V.K. and Sayed, S.M. - "Corrosion behaviour of steel in solutions containing mixed inhibite an aggressive ions". Corr. Sc. 13 (1973) 841.

Akinova, K.M. and Ivanov, F.M. - "Anti-Corrosive protection of reinforcement in aggressive environment by inhibitors". Beton i Zhelozobeton, 2 (1976) 38.

Rosenberg, A.M. and Gaidis, J.M. - "The mechanism of nitrite inhibition of chloride attack of reinforcing steel in alkaline aqueous environments". Materials perform. Nov. (1979) 45.

González, J.A., Algaba, S. and Andrade, C. - "Corrosion of reinforcing bars in carbonated concrete". Brt. Corros. J. 15 nº 3 (1980) 135.

Alonso , C. and Andrade, C. - "Efecto inhibidor del $NaNO_2$ en la corrosión de armaduras provocada por amasado de mortero con agua de mar". Rev. Ib. Am. Corros. y protec. (1983) 141.

Berke, N.S., Grace W.R. and Co. - "The effects of calcium nitrite and mix design on the corrosion resistance of steel in concrete". Congress Corrosion 85. Boston, March (1985) Paper nº 273.

Alonso, C. and Andrade, C. - "The effect of nitrite as a corrosion inhibitor in contaminated and chloride free carbonated mortar". Journal ACI (in press).

Lundsquist, J.T., Rosenberg, A.M. and Gaidis, J.M. - "Calcium nitrites an inhibitor of rebar corrosion in chloride containing concrete" Materials Performance, 18 (3) (1979) 36.

IMPERMEABILITY AND RESISTANCE TO CARBONATION OF CONCRETE WITH MICRO-SILICA AND WATER-REDUCING AGENTS

J. C. ARTIGUES, J. CURADO and E. IGLESIAS
Concrete Admixture Research Department, TEXSA S.A., Barcelona, Spain

Abstract
The use of microsilica in concrete is under great development, owing to the excellent properties it confers to the concrete on a durability level. This paper shows that to attain optimization of the properties of concrete filled with microsilica, the use of water-reducing admixtures with a high effectiveness is required, and, within these last, different behaviours have been observed depending on their chemical nature. This study discusses the combined effects of the microsilica and water-reducing admixtures in two fundamental properties of the concrete, in relation to its durability: its waterproofness and its carbonation resistance. The different types of admixtures used have shown differing behaviour, not only in their effectiveness as far as reducing the water/cement ratio is concerned, but also in the durability and compressive strength of the concrete, depending on their chemical constitution. Modifiel lignosulphonate and carboxylic polymer-based admixtures show greater effectiveness than these based on naphtalene sulphonate-formaldehyde condensate and sulphonated melamina-formaldehyde condensate.
Key words: Microsilica, Durability, Carbonation, Waterproofness, Water-reducing admixtures, Water/cement ratio, Compressive strength.

1 Introduction

Microsilica is a by-product obtained from the reduction of quartz with coal in electric arc furnaces, in the production of silicon and ferrosilicon alloys. Its SiO_2 content exceeds 90% and its appearance is that of a greyish coloured, ultrafine $(0.1 \mu m)$ filler.

The basic properties of microsilica are its pozzolanic activity and its compacting capacity as an ultrafine filler.

When microsilica is added to concrete, it reacts with the calcic silicates, which tend to close the capillary pores reducing the carbonation effect, according to the reaction:

$$Ca(OH)_2 + CO_2 \rightarrow CaCO_3 + H_2O$$

The improvement in the carbonation resistance, optimum compacting of the concrete and greater waterproofness obtained with the use of microsilica, will depend directly on the water/cement ratio used.

229

Therefore, and owing to the extreme fineness and large specific surface of microsilica, in order to obtain maximum performance of the microsilica, the use of water-reducing admixtures is necessary, in specific quantities to ensure good dipersability of the cement and the microsilica in the concrete, without sacrificing its workability or its optimum water content.

According to different authors, there are various factors that influence carbonation speed: type of cement, environmental conditions, water/cement ratio and microstructure of the concrete. In this paper, we have added a new factor: the influence of the combination of microsilica with water-reducing admixtures of diverse chemical nature.

2 Experimental method

The concrete used has the following composition:

Cement	15.9 %
Siliceous sand (0-3 mm)	46.6 %
Crushed aggregate (5-14 mm)	20.5 %
" " (14-25 mm)	17.0 %

Two types of cement were used: high initial strength Portland cement (I/45 A)and blended Portland cement (II-Z/35 A).

A fixed addition of microsilica has always been employed, fixed at 10% of the cement weight, plus five types of admixtures with different chemical natures:

Admixture A -   Carboxylic derivative modified lignosulphonate based flow agent

Admixture B -   Carboxylic polymer-based super flow agent.

Admixture C -   Melamine-formic aldehyde-based super flow agent.

Admixture D -   Vinsol resin-based plasticizer-aerator

Admixture E -   Naphtalene sulphate-based super flow agent.

The microsilica used corresponds to Elkem Materials' type 920-D.

The admixtures used are products formulated by Texsa, S.A.

We have also worked with two admixture doses (one high and the other low), and two different concrete plasticities (17 and 8 cm).

The following properties are measured for each concrete test: water/cement ratio, entrained aire percentage, compressive strength at 28 days, carbonation resistance and waterproofness at different ages.

Once the fresh concrete is obtained, the percentage of entrained air in same is determined and test pieces with a diameter of 15 cm and height of 30 cm are filled, and released the following day. These are then placed in a curing chamber under conditions of 20.8 $\pm$ 0.5°C .temperature and 97 $\pm$ 2% relative humidity. After 28 days, six test pieces from each test are broken up in a hydraulic compression press, with a 150-Tn capacity. All these tests were carried out according to ASTM C-494 Standard. At the same age, and with the help of a electric saw, two 15 x 30-cm test pieces are cut into 15 x 5-cm pieces, discarding the ends. The carbonation and waterproofness tests are carried out with these pieces, in triplicate, expressing the results as an average.

2.1 Waterproofness

The waterproofness test is carried out with three 15 x 5-cm discs from each concrete test. The test pieces are submerged in water-tight containers filled with water. Progressive water absorption is measured according to the immersion time of the test pieces: initial after 6 hours, 24 hours, 3 days and 14 days. After 14 days the test is concluded, as water absoption is practically detained. The scales used had a precision of 0.1 g and a 10-kg range.

2.2 Carbonation resistance

The Rilem CPC-18 recommendation headed "Measurement of hardened concrete carbonation depth" was followed to measure the carbonation depth of each test. For this, three 15 x 5-cm test pieces form each concrete test were placed in a hermetically sealed reactor, submitted to a CO_2 pressure of 4 Kg/cm² during three months, slightly moistening the test pieces once a week, as in this way the carbonation process is favoured. Once the time allocated for this test has elapsed, the discs are broken into 4 equal parts and the carbonation depth is measured on the two internal sides, using 1% phenolpthalein in ethyl alcohol as indicator, which shows red in the carbonation-free areas, and remaining colourless in the carbonated areas (ph < 9). Measurements are taken after 24 hours of having applied the alcoholic phenolphtalein solution on the concrete test piece, which is when, according to the authors, the separating margin appears more clearly between the carbonated region and carbonation-free region of the concrete, which is measured with the help of a 0.1-mm precision micrometer.

Carbonation depth is expressed in the table of results as an average of each test (all four parts of each the three test pieces), according to the formula:

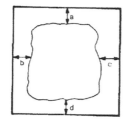

$$P = \frac{1}{24} \sum_{i=1}^{n} \frac{a_i + b_i + c_i + d_i}{4}$$

$i = 1$
$n = n^{\underline{o}}$ of sides measured

The transition area where the colour is slightly less intense than the rest after 24 hours, are considered carbonated.

2.3 Test tables
The tests carried out were the following:

Table 1. II-Z/35 A cement was used to produce the concrete (blended
Portland cement)

Mix n⁰	Microsilica (%)	Admixture (%)	W/C ratio	Air (%)	Plasticity (cm)	Compressive strength after 28 days (Kg/cm²)
101	–	–	0.469	2.2	9	228
102	10	–	0.516	1.8	8	296
103	10	A, 0.4	0.440	3.1	8.5	390
104	10	B, 0.7	0.438	2.8	8.5	373
105	10	C, 1.4	0.460	2.1	8	337
106	10	D, 0.03	0.491	4.5	8	286
107	10	E, 0.7	0.462	2.0	8	341
111	–	–	0.577	1.9	17.5	193
112	10	–	0.654	1.2	17	286
113	10	A, 0.8	0,526	3.6	17	397
114	10	B, 1.2	0,526	2.6	17.5	382
115	10	C, 2.4	0.554	1.6	18	326
116	10	D, 0.06	0.616	8.5	17	179
117	10	E, 1.2	0.572	1.7	17	289
121	–	–	0.460	2.2	8	231
122	10	–	0.507	1.9	8	302
123	10	A, 0.8	0.427	3.8	8.5	407
124	10	B, 1.2	0.422	3.3	8	393
125	10	C, 2.4	0.452	2.1	8	365
126	10	D, 0.06	0.490	6.4	8.5	274
127	10	E, 1.2	0,454	2.0	8	359

Table 2.   I/45 A cement, with high initial strength, was used to produced the concrete

Mix nº	Microsilica (%)	Admixture (%)	W/C ratio	Air (%)	Plasticity (cm)	Compressive strength after 28 days (Kg/cm²)
131	–	–	0.440	2.4	8	313
132	10	–	0.508	1.7	9	413
133	10	A, 0.4	0.451	2.8	10	478
134	10	B, 0.7	0.434	2.5	8	442
135	10	C, 1.4	0.470	1.8	8	421
136	10	D, 0.03	0.482	4.2	8	367
137	10	E, 0.7	0.483	1.9	8.5	404
141	–	–	0.567	2.3	17	259
142	10	–	0.651	1.4	18	353
143	10	A, 0.8	0.552	3.0	17	449
144	10	B, 1.2	0.553	2.5	17	434
145	10	C, 2.4	0.578	1.7	17	386
146	10	D, 0.06	0.619	7.5	17	251
147	10	E, 1.2	0.567	2.0	17	369
151	–	–	0.459	2.3	9	292
152	10	–	0.513	1.9	9	403
153	10	A, 0.8	0.427	3.0	9	524
154	10	B, 1.2	0.412	2.8	9	502
155	10	C, 2.4	0.432	2.2	9	451
156	10	D, 0.06	0.495	6.0	9	331
157	10	E, 1.2	0.437	2.2	8.5	418

The mix number is expressed in three figures, the first two refer to the fact that the concrete test has been carried out on the same day and where the percentage of humidity of the aggregates is the same for all the tests. For this reason, each time a batch of concrete is produced, we always start with a target or reference that does not contain either microsilica or water-reducing admixture. The third number varies depending on the addition, or otherwise, of microsilica and on the chemical nature of the water-reducing admixture used. The addition percentage depends on the cement weight.

3 Results

Waterproofness is measured in the form of percentage, using the following expression for the calculation:

$$\% \text{ permeability} = \frac{W_t - W_o}{W_o} \times 100$$

where: W_t = weight of the test piece at age t after being submerged in water.

W_o = initial weight of the test piece.

Carbonation depth is expressed in mm.
The result obtained are shown in the following tables:

Table 3. II-Z/35 A Cement

Mix nº	% permeability				carbonation depth (mm)
	6 hours	24 hours	3 days	14 days	
101	2.16	2.36	2.46	2.51	9.4
102	1.22	1.57	1.72	1.77	2.7
103	0.65	0.70	0.95	1.05	0.7
104	0.74	0.88	0.93	0.98	0.4
105	1.02	1.17	1.26	1.31	1.5
106	1.75	2.01	2.11	2.16	2.3
107	1.25	1.44	1.54	1.59	1.7
111	1.74	1.88	1.98	2.07	13.6
112	1.94	2.00	2.05	2.10	8.7
113	1.13	1.34	1.34	1.54	1.2
114	0.97	1.07	1.12	1.21	1.0
115	1.48	1.67	1.67	1.77	2.8
116	2.11	2.11	2.16	2.21	3.6
117	1.66	1.66	1.71	1.81	3.3
121	2.11	2.31	2.39	2.46	9.0
122	1.20	1.53	1.68	1.72	2.6
123	0.63	0.68	0.90	1.01	0.6
124	0.71	0.84	0.90	0.97	0.4
125	0.99	1.14	1.23	1.27	1.6
126	1.74	2.01	2.12	2.15	2.4
127	1.22	1.43	1.51	1.57	2.3

Table 4. I/45 A Cement

Mix nº	% permeability				carbonation depth (mm)
	6 hours	24 hours	3 days	14 days	
131	1.42	1.96	2.00	2.15	9.2
132	0.91	1.20	1.20	1.25	2.1
133	0.67	0.87	0.87	0.91	0.6
134	0.62	0.77	0.79	0.83	0.3
135	1.03	1.32	1.35	1.41	1.2
136	1.22	1.52	1.56	1.68	2.0
137	0.72	0.91	0.91	0.96	1.9
141	1.51	1.80	1.85	1.93	11.8
142	1.45	1.60	1.71	1.75	6.4
143	1.41	1.58	1.69	1.72	3.4
144	1.28	1.51	1.60	1.65	3.1
145	1.40	1.61	1.68	1.74	4.2
146	1.99	2.24	2.34	2.41	7.9
147	1.38	1.58	1.68	1.71	6.9
151	1.49	1.99	2.07	2.19	9.4
152	0.94	1.19	1.24	1.27	2.3
153	0.64	0.82	0.84	0.90	0.5
154	0.58	0.72	0.74	0.81	0.3
155	0.89	1.10	1.16	1.19	1.1
156	1.23	1.50	1.55	1.63	2.2
157	0.70	0.89	0.90	0.94	1.7

4 Conclusions

This study has shown that there is a very direct relation between the
water/cement ratio and the carbonation resistance and waterproofness
(see graphs 1 or 2). In the tables of results given in chapter 3 of
this paper one can see how the lesser the water/cement ratio, the
better the concrete protection, i.e. when the carbonation depth is
less, so is the percentage of water absorbed less, providing water-
reducing admixtures are used to favour the performance of the
microsilica in the heart of the concrete. Moreover, a fairly
intimate relationship can be established strength at the age of 28
days and the carbonation resistance and waterproofness (see graph 3).
    The admixtures showing greatest effectiveness are Admixture A
(carboxylic derivative-modified lignosulphonate based)and Admixture B
(carboxylic polymer-based super flow agent), more especially this
last.
    The melamine-based admixtures (Admixture C) and the naphtalene
sulphonate-based admixtures (Admixture E), each show similar
behaviour, poorer than the previous ones.

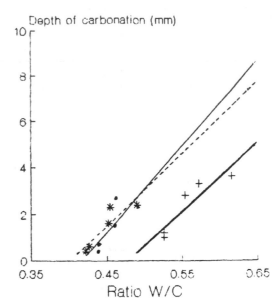

Depth of carbonation (mm)

- --*-- Slump 8.High dose.
- —+— Slump 17.High dose.
- —•— Slump 8.Low dose.

Ratio W/C

Graph 1. Influence of carbonation with
the ratio W/C. II-Z/35 A Cement.

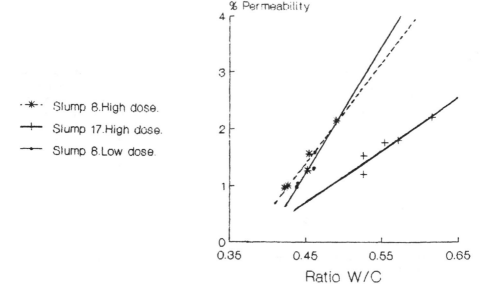

% Permeability

- --*-- Slump 8.High dose.
- —+— Slump 17.High dose.
- —•— Slump 8.Low dose.

Ratio W/C

Graph 2. Influence of permeability with
the ratio W/C. I/45 A Cement.

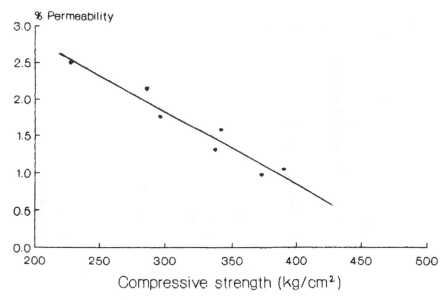

Graph 3. Influence of permeability with
the compressive strength. II-Z/35 A
cement. Slump 8. Low dose

Admixture D is the one that shows the worst behaviour, due to the
excess of entrained air it introduces into the heart of the concrete.
This is a Vinsol resin-based admixture with optimum aerating
properties, which hurt both the waterproofness and the carbonation
resistance.

From Tables 3 and 4 the following is deduced:

(a) The addition of microsilica in a percentage of 10% of the
cement weight, increases the water/cement ratio, but the change in
the concrete's microporous structure confers greater carbonation
resistance and a diminishing of permeability with respect to a
concrete without microsilica, as well as an increase in the
compressive strength.

(b) The combined addition of microsilica and water-reducing
admixtures, provokes a diminishing of the water/cement ratio, which
is translated into an evident improvement in the carbonation
resistance and waterproofness.

With a larger dose of admixture, for the same plasticity and
within certain limits, an obvious reduction in the w/c ratio is
observed, thus an improvement in the carbonation resistance and
waterproofness.

(c) When working with higher plasticities, the influence of the
water-reducing admixture also favours the durability of a concrete,
but in a less spectacular way than when making concrete with lower
plasticities, as the w/c ratio is already quite high (see graphs 4
and 5).

237

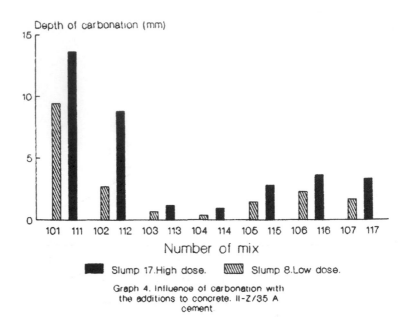

Graph 4. Influence of carbonation with
the additions to concrete. II-Z/35 A
cement.

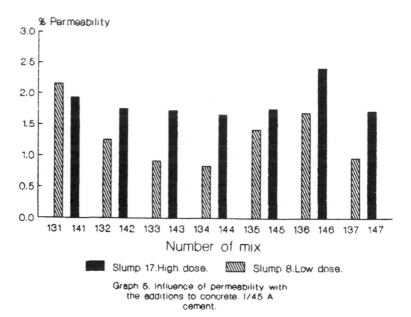

Graph 5. Influence of permeability with
the additions to concrete. I/45 A
cement.

(d) When comparing the cements used, type I/45 A cement confers greater carbonation resistance and better waterproofness to the concrete (see graphs 6 and 7).

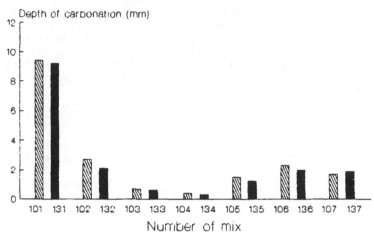

Carbonation vs additions to concrete

Depth of carbonation (mm)

Number of mix

■ I/45 A cement.   ▨ II-Z/35 A cement.

Graph 6. Comparison between cements
(II-Z/35 A and I/45 A). Slump 8.
Low dose.

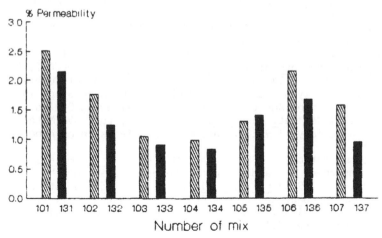

Permeability vs additions to concrete

% Permeability

Number of mix

■ I/45 A cement.   ▨ II-Z/35 A cement.

Graph 7. Comparison between cements
(II-Z/35 A and I/45 A).Slump 8.
Low dose.

References

Berry, E.E. and Malhotra, V.M. (1980) Fly ash for use in concrete. A critical review, ACI Journal, Proceedings, Vol. 77, 2, 59-73.

Carette, G.G. and Malhotra (1983) Mechanical Properties, Durability and Drying Shrinkage of Portland Cement Concrete Incorporating Silica Fume, American Society for Testing Materials, Vol. 5, 1.

Diamond, S., (1983) Effect of Microsilida (Silica Fume) on Pore Solution Chemistry of Cement Pastes, J. American Ceramic Society, Vol. 66, No. 5.

Dutruel, F. and Estoup, J.M. (1986) Effects of the Addition of Silica Fumes or Pozzolanic Fines on the Durability of Concrete Products, Madrid.

Gjorv, O.E. (1983) Durability of concrete containing condensed Silica Fume, American Concrete Institute, Special Publication, SP-79, Vol. II.

Hjorth, L. (1982) Microsilica in Concrete, Publication nr. 1, Nordic Concrete Federation, Nordic Concrete Research.

Hustad, T. and Loland, K.E. (1981) Report-4: Permeability, Report STF 65 A 31031, Cement and Concrete Research Institute, The Norwegian Institute of Technology, Norway.

Kroone, B. and Blakey, F.A. (1959) J. Am. Conc. Inst., 31, 497.

Maage, M., (1986) Mat. Res. Soc. Symp. Proc., Vol. 65, 193-198, Carbonation in concrete made of blended cements.

Mehta, P.K. and Gjorv, O.E. (1982) Properties of Portland Cement Concrete Containing Fly Ash and Condensed Silica-Fume, Cement and Concrete Research, V. 12, No. 5, 587-595.

Nagataki, S.; Ohga, H. and Kyum Kim E., (1986) Fly ash, silica fume, slag and natural pozzolans in concrete, Canmet/ACI, Ed. Malhotra, Vol. 1, 521-540, SP-91-24 Effect of Curing Condictions on the Carbonation of Concrete with Fly Ash and the Corrosion of Reinforcement in Long-Term Tests.

Paillere, A.M.; Raverdi, M. and Grimaldi G., (1986), Fly ash, silica fume, slag and natural pozzolans in concrete, Canmet/ACI, Ed. Malhotra, Vol. 1, 541-562, SP-91-25, Carbonation of Concrete with Low-Calcium. Fly ash and Granulated Blast Furnace Slag: Influence of Air-Entraining Agents and Freezing-and-Thawing Cycles.

Meland, I., (1986) Mat. Res. Soc. Symp. Proc., Vol. 65, Materials Research Society, 199-206, Carbonation effects in hardened fly ash cements.

Rilem Recommendations, (1988) TC-56, MHM Hydrocarbon Materials, CPC-18- Measurement of hardened concrete carbonation depth, 453-455.

Skjolsvold, O., (1986) Fly ash, silica fume, slag and natural pozzolans in concrete, Canmet/ACI, Ed. Malhotra, Vol. 2, 1031-1048, SP-91-51, Carbonation Depths of Concrete with and without Condensed Silica Fume.

MECHANICAL PROPERTIES AND DURABILITY
OF SUPERPLASTICIZED SILICA FUME MORTARS

G. BARONIO
DISET, Politechnico of Milan, Italy
G. MANTEGAZZA and G. CARMINATI
Ruredil SpA, San Donato, Milan, Italy

Abstract
Aim of this paper is to outline the resistance and durability characteristics of mortars manufactured with silica fume (microsilica)
and superplasticizing admixtures.
The silica fume, as well known, is a by-product of the silicon production and of the iron silicon alloys. This material is extremely
fine grained (of spheric shape), composed mostly by silica. This
silica is added to the mortar in variable percentage. The use of this
product causes an increase of water demand.
This inconvenient can be avoided by using superplasticizing admixtures
which, besides compensating the higher water demand due to silica
fume, permit to reduce the W/C ratio by increasing the mechanical
resistances.
The experiments we are considering intend to show that the introduction of the microsilica into the mortars with superplasticizer largely
increases the durability of the cement paste in different aggressive
conditions. Only carbonation action has been considered for this
preliminary part of the research.
In fact, being this material a highly reactive pozzolana, it reacts
with calcium hydroxide and water to form the calcium silicates
hydrates that whilst filling the vacuums among the cement particles,
sand and aggregates, lower the mortars porosity by increasing
durability.
The coupling of silica fume and superplasticizer allows to manufacture
mortars with low permeability and high resistance to chemical attacks.
Key words: Silica Fume, Superplasticizers, Workability, Porosity,
Pozzolanic Activity, Carbonation, Durability.

1 Introduction

When a mortar is to be used in strongly aggressive environmental
conditions it must possess definite characteristics of compactness and
durability.
The purpose of the following work is to prepare a mortar with the following characteristics: good workability, low permeability, sufficient
mechanical resistance at early ages of curing, and good durability.
It is known that condensed silica fume is highly reactive Pozzolana
(Aud Traette 1978) and because of its fineness (average particle
diameter = 0.15 μm) it reacts in a relatively short time with the

241

hydrated Ca(OH)2 present in the cement paste yielding by transforming lime into calcium silicate hydrate [Mehta and Gjørv 1982].
This work studied the influence of increasing additions of silica fume on the characteristics of the mortar. The use of this product causes an increase in the water requirement and a reduction in workability. This problem can be solved by using superplasticizers. The influence of the combined action of the silica fume and the superplasticizers was observed on the mechanical characteristics, the impermeability and above all on the behaviour of the accelerated carbonation of these mortars.
In particular the work investigated whether the calcium hydrate present in the cement paste was fixed prevalently by pozzolanic action of the silica fume or by the carbon dioxide.

2 Materials

In all the tests Portland commercial cement (C) type 425 was used with physiochemical and mechanical characteristics according to Italian Code as in Table 1.

Table 1. Physiochemical and mechanical characteristics of cement and silica fume.

	PTL 425	SILICA FUME
PHYSICAL TESTS		
Specific gravity g/cm^3	3.08	2.21
Fineness cm^2/g	3430 (Blaine)	250000 (B.E.T.)
Settings time min.		
Initial	165	
Final	215	
Compressive strength of		
40 mm cubes, MPa		
1 day	15.0	
3 days	28.5	
7 days	39.8	
28 days	56.0	
CHEMICAL ANALYSIS %		
Loss on ignition	3.40	1.50
Silicon dioxide	20.75	92.50
Ferric dioxide	3.12	0.65
Alluminium oxide	4.35	0.75
Calcium oxide	62.00	0.32
Magnesium oxide	1.93	0.85
Sulfur oxide	2.65	
Sodium oxide	0.45	0.55
Potassium oxide	0.88	0.35

Silica fume (SF): The silica fume "Durasil" was used with physio-chemical characteristics as shown in Table 1, and with high content of amorphous silica Fig. 1.

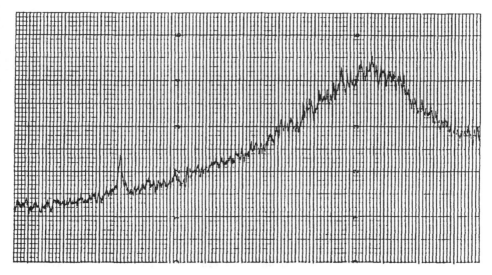

Figure 1. X-ray diffractogram of the silica fume with high content of amorphous silica

Superplasticizer (Adm): Fluiment 40, a sodium naphthalene sulfonate sodium melaminesulfonate based product, was employed.

Sand (A): The sand used was from Torre del Lago. It is prevalently silica by nature and with size grain distribution in the range 0.075 ÷ 2.0 mm as required by the Italian Code.

3 Mix proportion and experimental part

3.1 Preparation of the mixtures
Four series of mixture were made: (A, B, C, D) according to the Italian Code (Law n. 595 on 26/5/65) using a binder/aggregate ratio of 1:3. Mixture A (Table 2) was made with Portland 425 cement and sand (1:3) using a ratio W/C = 0.67. This ratio was selected deliberately high in order to reproduce a real situation of use, that is should one require a high workability without superplasticizers.
In the other 3 mixtures B, C, D, the cement was replaced by condensed silice fume in the proportions 5, 10, 15, 20% respectively (Table 2). These last 3 types of mixture were made with a ratio W/(C+SF) of 0.67, 0.48 and 0.40 respectively and superplasticizer was added in the percentages shown in (Table 2) to mixutres C and D in order to increase the workability (Flow = 110 about).
The superplasticizer additive was added to the mixtures before the last high velocity mixing cycle.

Table 2. Mix proportion.

Samples	SF * %	Adm. %	W/(C+SF)	Flow	Spec. grav ** (g/cm^3)	Impermeab. (cm)
A			0.67	115	2.14	10.0
B/5	5		0.67	86	2.12	6.3
B/10	10		0.67	57	2.12	5.5
B/15	15		0.67	37	2.13	5.0
B/20	20		0.67	23	2.12	4.0
C/5	5	0.95	0.48	112	2.14	5.0
C/10	10	1.17	0.48	113	2.16	4.5
C/15	15	1.52	0.48	116	2.19	3.2
C/20	20	1.73	0.48	117	2.2	2.6
D/5	5	4.69	0.4	112	1.83	3.2
D/10	10	5.73	0.4	112	1.86	3.0
D/15	15	7.13	0.4	114	1.91	2.9
D/20	20	9.56	0.4	112	1.94	2.5

* Replacing the cement   ** Of the fresh mortar

3.2 Preparation and curing of the samples

The samples for the mechanical and carbonation tests were made with
dimensions of 40x40x160 mm settling the material with 60 blows in two
equal bands.
The samples for the im permeability test were made with dimensions of
200x200x120 mm.
All the samples were stored for 24 hours in a conditioned environment
of 20°C and 90% relative humidity, then they were removed from the
molds and cured in water at 20°C for 28 days.

3.3 Impermeability test

This test was carried out according to DIN 1048 starting the test
immediately at 7 atm. The results are expressed as depth of
penetration.

3.4 Compressive and flexural strength tests

After 1, 3, 7, 28 days the samples were tested for compressive and flexural strength tests according to Italian Code. The results of these tests are shown in Table 3.

Table 3. Compressive and flexural strength

	MPa							
Samples	1 day		2 days		7 days		28 days	
	Flex.	Compr.	Flex.	Compr.	Flex.	Compr.	Flex.	Compr.
A	2.3	9.6	4.4	20.1	4.8	27.8	6.3	36.2
B/5	3	11	4.7	21.4	4.9	27.4	6.7	43.3
B/10	2.9	12.3	4.7	22.1	5.4	28.2	6.4	45.3
B/15	3.2	12.8	4.1	21.7	5	29.8	6.4	48.2
B/20	3.20	11.7	4	20.5	5.4	31.8	6.5	47.8
C/5	4.7	25.2	5.5	40	6.7	47.1	7.6	62.5
C/10	4.4	22.7	5.8	40	6.9	48.9	8.6	65.7
C/15	5	23.8	5.5	38.4	7.1	51.9	8.4	75.8
C/20	5.3	22.2	5.3	38.8	6.9	53.7	8.4	80.1
D/5	3.5	14.5	4.5	27.4	5	31.3	5.4	40.7
D/10	3.5	13.9	4.5	24.6	4.8	32.8	6.2	45.8
D/15	3.5	14.9	4.4	27.1	4.8	33.3	5.9	45.8
D/20	3.7	15.3	4.5	27.3	5.4	35.7	5.9	45.2

3.5 Accelerated carbonation test

After 28 days each sample was cut in 4 parts to obtain 4 cubes with sides of about 40 mm. A sample for each mixture type was tested to accelerated carbonation.

3.6 Reference samples

Another series of cubes (analogous to the ones to be carbonatised) were enveloped singularly in hermetically sealed plastic bags in order to avoid CO_2 penetration and therefore carbonation.

After 1, 2, 3 months a 5 mm thick slice was cut from each cube and a representative, homogeneous sample of the section was prepared for chemical analyses as quickly as possible.

3.7 Conditions of environmental exposure

The sample series to be subjected to accelerated carbonation were introduced to an appropriate environment with a continuous flow of CO_2. The velocity was 4 bubbles per second and the internal diameter of the pipe was 10 mm. The CO_2 was initially bubbled through dibutyl-phthalate and then was made to flow up the inside of the carbonation chamber. In order to accelerate the carbonation of the mortar cubes the relative humidity inside the chamber was as far as possible maintained within the range of 60 to 80%.

At the end of 1, 2, 3 months both the accelerated carbonation samples and the reference samples were titrated with HCl 0.1N using an alcoholic solution of phenolphtalein (1%) as the indicator.

From the volume of HCl used it is possible to calculate the meq/g of free calcium hydrate.

4 Discussion

As it can be seen from the data in Table 2, the addition of 5, 10, 15, 20% silica fume lowers the workability and increases the impermeability of the mortar in proportion to the value added (Fig. 2).

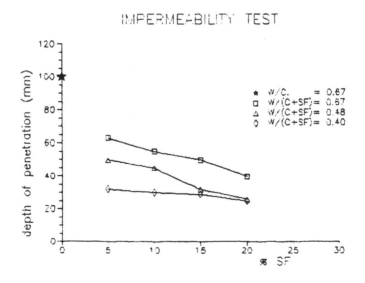

Fig. 2. Depth of penetration versus silica fume content.

The increase of both impermeability and mechanical resistance are due mainly to 2 phenomena, one physical and one chemical (Bache 1981) (Detwiler and Mehta 1989).
The physical phenomenon acts immediately, at early ages when the chemical phenomenon is still latent. According to some researchers (Detwiler and Mehta 1989) this physical phenomenon is due to the silica's fineness, its large specific surfaces and to the fact that its particles fill the existing spaces between the various granules of the cement, and those between the cement paste and the aggregate which are rich in exuded water and calcium hydrate (Monteiro 1985).
Later, the chemical reaction between the silica fume and the calcium hydrate also comes into action, producing hydrated calcium silicate which occupies a larger space to that of the two reagents, so reducing the porosity of the mortar.
The increase in mechanical resistance of mixture B, with respect to mortar sample A, is therefore due to physical phenomena at early ages and chemical phenomena after 28 days (Detwiler and Mehta 1989).
However, the addition of silica fume to the cement paste reduces the workability (Table 2), for which further addition of superplasticizer are necessary in order to achieve the workability of mix A. Such additions allow the same workability to be obtained with a lower W/C ratio.
The impermeability of mortar C is increased to 25% with respect to mortar A (Fig. 2), while the mechanical resistance roughly doubles.
The coupled action of the silica fume and superplasticizer gives a mortar with the following characteristics: high impermeability, good workability and high mechanical resistance even at early ages.
Solely for experimental purposes, the influence of higher addition of superplasticizer (1-10%) on the mortar was investigated. Such doses can further increase the workability, using a still lower W/C ratio (Table 2).
But such high doses reduce the mechanical resistance and the density while the impermeability remains almost equal to that of mortar made with smaller additions of superplasticizer. These results could suggest that the superplasticizer when used in such high doses behaves like an air entraining agent producing a closed-type porosity inside the mortar.
As for as carbonation tests, the data in Table 4 show that reference sample A is already carbonated by 83% after one month of accelerated carbonation. On the contrary, sample B shows a decrease in carbonation (from 82-54%) proportional to the silica added.
This decrease occurs for two reasons: firstly, the increased percentage of silica increases the quantity of calcium hydrate fixed by the pozzolan action of the silica, secondly, the increased quantity of reaction products reduces the porosity of the cement paste preventing the CO_2 from penetrating the paste. The carbonation is still prevalent over the pozzolan action even after 3 months of carbonation (Fig. 3c).

Table 4. Accelerated carbonation

% of Ca(OH)$_2$ reacted with pozzolana or CO$_2$								
% of Ca(OH)$_2$ Residual								
Samples	**1 month**			**2 months**			**3 months**		
	with Pozz.	with CO$_2$	Resid.	with Pozz.	with CO$_2$	Resid.	with Pozz.	with CO$_2$	Resid.
A	0.00	83.49	16.51	0.00	94.02	5.98	0.00	96.73	3.27
B/5	10.82	82.66	6.52	14.65	80.35	5.00	17.60	78.76	3.64
B/10	14.05	78.24	7.71	18.56	76.35	5.09	18.49	77.05	4.46
B/15	20.07	71.61	8.32	23.24	71.45	5.31	27,91	66,18	5,91
B/20	35.14	54.14	10.72	37.34	56.08	6.58	41.34	50.87	7.79
C/5	16.97	51.94	31.09	23.51	52.03	24.46	32.22	54.91	12.87
C/10	26.40	41.88	31.52	28.05	43.93	28.02	46.15	42.68	11.17
C/15	37.79	30.56	31.65	41.73	39.65	18.62	43.26	40.62	16.12
C/20	41.76	25.03	33.21	51.37	29.93	18.70	55.47	33.08	11.45
D/5	29.59	36.21	34.20	29.73	35.10	35.17	36.00	41.04	22.96
D/10	33.29	23.84	42.87	39.03	23.45	37.52	47.93	28.43	23.64
D/15	35.16	22.80	42.04	39.31	24.31	36.38	46.43	29.54	24.03
D/20	41.62	14.91	43.47	56.11	18.64	22.25	57.44	18.60	23.96

In the mortar type C, where the synergism of the silica and superplasticizer was studied, a further decrease in the carbonation is noted and, after only one month of exposure, the carbonation is still prevalent over the pozzolan action if the silica present is less than 15% and the superplasticizer is less than 1.52%. On increasing the percentage of silica the pozzolan action becomes prevalent over the carbonation (Figs. 3a, 3b).
This tendency also holds for mortar D.
It can be noted that as the impermeability increases, the calcium hydrate residue also increases, i.e. it is not fixed by the CO$_2$, nor

ACCELERATED CARBONATION
(1 month)

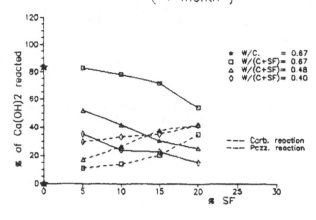

Fig. 3a - Accelerated carbonation test: distribution of reacted calcium hydrate between carbonation reaction and pozzolanic reaction, versus silica fume content.

ACCELERATED CARBONATION
(2 months)

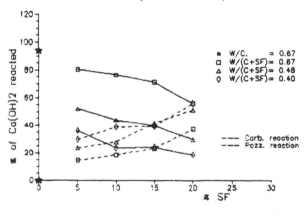

Fig. 3b - Accelerated carbonation test: distribution of reacted calcium hydrate between carbonation reaction and pozzolanic reaction, versus silica fume content.

ACCELERATED CARBONATION
(3 months)

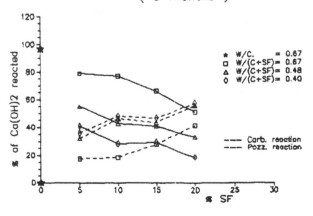

Fig. 3c - Accelerated carbonation test: distribution of reacted calcium hydrate between carbonation reaction and pozzolanic reaction, versus silica fume content.

by the pozzolan. In fact, the increase in impermeability reduces the amount of CO_2 which can penetrate inside the mortar. This phenomenon is particularly visible in mortar D.

5. Conclusions

The addition of 5, 10, 15, 20% of silica fume increases the mechanical resistance of the mortar even at early ages while reducing the workability and permeability. It reduces the carbonation to about 35% of that of the reference mortar.
The contemporary addition of silica fume and superplasticizer improves the workability and allows the use of a lower W/C ratio. If the addition of silica fume is over 15% and the superplasticizer over 1.5% the pozzolan action becomes prevalent over the carbonation reducing it by 70% after only 1 month.
The further addition of silica fume seems to stimulate this phenomenon.
The addition of superplasticizer in high percentages (1/10%) decreases the open porosity but also decreases the mechanical resistance.

Acknowledgements

This research was supported by CNR-PFE.
The Authors wish to thank Arch. Frigerio, Mr. Perolari and Mr. Sereni for their collaboration and assistance in the experimental work.

6. References

Aud Traette (1978) Silica fume as pozzolanic material. Il Cemento 3, 369-376.
Mehta, P.K. and Gjørv, O.E. (1982). Properties of portland cement containing fly ash and condensed silica-fume. Cement and Concrete Research, Vol. 12, 587-595.
Bache, H.H. (1981) Densified cement/ultra-fine particle-based materials. Second International Conference on Superplasticizers in Concrete, Ottawa, Ontario, Canada.
Detwiler, R.I. and Mehta, P.K. (1989) Chemical and physical effects of condensed silica fume in concrete, Third CANMET/ACI International Conference on Fly Asch, Silica Fume, Slag and Natural Pozzolans in Concrete. Supplementary Papers, Trandheim, Norway.
Monteiro, P.I.M. (1985) Microstructure of concrete and its influence on the mechanical properties. Ph. D. Thesis, University of California at Berkeley.

CALCIUM NITRITE CORROSION INHIBITOR IN CONCRETE

N. S. BERKE
Construction Products Division, W. R. Grace & Company, Connecticut, USA
A. ROSENBERG
Washington Research Center, W. R. Grace & Company, Connecticut, USA

Abstract

Extensive testing shows that calcium nitrite is an effective corrosion inhibitor for steel in concrete. Furthermore, in most cases, calcium nitrite improves the compressive strength of the concrete mix, and with proper air entrainment is freeze-thaw durable. In this paper data is presented documenting the effectiveness of calcium nitrite, and it is shows how corrosion performance can be further improved with the use of the proper concrete mix designs and concrete cover.

Key words: Concrete, Corrosion, Calcium nitrite, Corrosion inhibitor, Durability, Protection, Steel, Reinforcing

Introduction

Steel reinforced concrete is a widely used construction material. The concrete provides a protective environment for the steel and the steel provides tensile strength to the concrete. However, when reinforced concrete is subjected to chloride containing salt, corrosion of the steel reinforcement can occur. The corrosion of the steel eventually leads to significant deterioration of the reinforced concrete.

The effects of corrosion are staggering. Approximately half of the 500,000 plus bridges in the U.S. highway system are in need of repair[1]. The Strategic Highway Research Program pointed out that $450 to $500 million per year can be saved by correcting corrosion problems in current bridges[2].

Most bridge failures are due to deicing salts. However, bridges in marine environments are also susceptible to severe corrosion due to chloride ingress. The severity of the marine environment is such that damage may be evident in as little as 5 years[3].

In this paper the development of calcium nitrite as a corrosion inhibitor to protect steel in concrete from chloride induced corrosion is reviewed. Additionally, the beneficial effects of calcium nitrite on concrete such as improved mechanical and durability properties, will be discussed.

1   The history of the development

In 1961 it became apparent that there was a need for a non-corrosive accelerator for concrete. The main impetus was the definitive paper by Monfore and Verbeck[4] where they showed that admixed calcium chloride caused the Regina, Saskatchewan waterpipe failure. This followed closely a similar study by Evans[5] where he found chloride caused corrosion of steel in prestressed concrete.

The most obvious approach seemed to be a salt that would accelerate the setting time of concrete, but also be a corrosion inhibitor.

Treadaway and Russell[6] studied sodium nitrite and sodium benzoate. Craig and Wood [7] studied potassium chromate, sodium benzoate and sodium nitrite. These papers and others agreed that sodium nitrite offered the best protection in concrete from chloride induced corrosion but that the strength of the concrete was reduced when any of these materials were used. Figure 1 shows the reduction of strength with various inhibitors[7]. Further, the risk of the alkali-aggregate reaction was made worse by the addition of a sodium salt.

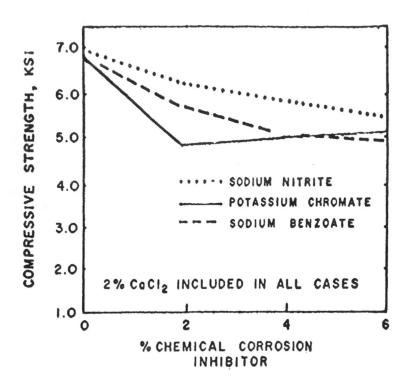

Figure 1.   Effects of some corrosion inhibitors on compressive strength[7].

Further studies revealed that in order to maintain or improve the strength of the concrete, the cation system of concrete, namely calcium, could not be changed. Thus, in 1964 a non-corrosive accelerator based on calcium formate and a small amount of sodium nitrite was introduced[8], which did not result in loss of strength. The calcium formate/10% sodium nitrite accelerator gave good setting time acceleration with no loss in compressive strength [9]. However, the product had to be handled as a dry powder. Ten years later the corrosion of steel in concrete became a serious problem in the U.S. and a process for the manufacture of calcium nitrite became available in Japan. The new corrosion threats to concrete placed in marine environments and the use of marine sand, along with the fact that calcium nitrite could be made up as a 40% stable solution, encouraged further study of calcium nitrite for use as an admixture in concrete.

First the mechanism of the protection of steel in concrete by calcium nitrite was determined. Several experts were consulted and laboratory experiments were conducted to confirm the mechanism. If corrosion was the dissolution of iron or

$$Fe^O \longrightarrow Fe^{++} + 2e^-, \tag{1}$$

a solid ferrous hydroxide (white) corrosion product would form in concrete (pH=12.5) preventing further dissolution of iron, in the absence of chlorides. In the absence of chlorides, the iron will convert to the ferric state. The resultant protective ferric oxide is less than 100 Å thick.

When not protected by concrete, iron continues to corrode because of available oxygen, carbon dioxide, moisture, and changing temperatures. The initial ferrous ions change to the more stable $Fe_2O_3 \times H_2O$ going through different intermediate stages such as Fe_3O_4. In well-rusted iron, layers of the various oxides of red and black rust are observed. The changing layers of oxides prevent a strong continuous passive film from protecting the iron from further corrosion.

Foley[10] has shown that iron forms soluble, light green complexes with chloride ions. These have been observed in concrete. These soluble forms of ferrous ions are able to migrate away from the reinforcing bar encouraging more iron to dissolve. This prevents the passive layer from forming. Thus, the corrosion process in concrete depends on chloride ion, water and oxygen content as borne out by Hart[11] where he found the worst corrosion of marine exposed concrete at the inter-tidal zone.

It was determined that calcium nitrite does not react with Fe^O or Fe^{+++}. It does react with Fe^{++} according to the following fast reaction:

$$2Fe^{++} + 2OH^- + 2NO_2^- \longrightarrow 2NO\uparrow + Fe_2O_3 + H_2O \tag{2}$$

Thus, if ferrous ions are produced in concrete, calcium nitrite changes them to a stable passive layer avoiding all the metastable intermediate forms. Chloride ions and nitrite ions compete for ferrous ions

produced in concrete. The relative concentrations of chloride and
nitrite determine the type of reaction that takes place. If the
nitrite ion concentration is sufficient, then nitrite reacts with
ferrous ions to form a passive layer which closes off the iron surface
stopping further reaction and, as such, the amount of nitrite reacting
in very small[12-17].

There was concern about the production of NO gas in concrete, nitrite
ions changing the size of the anode, and dangerous cathodic reactions.
NO gas, may change to NO_3^- eventually in the presence of O_2, but
a significant amount would only be produced after serious corrosion
would have taken place and the concrete destroyed. Nitrite does not
enter into the reactions involved in producing the anode, but reacts
with the resulting products of the anode. Thus, it cannot affect the
size of the anode. As will be discussed later, nitrite does not
increase the cathodic reaction rate.

The dangerous cathodic[18] reaction is:

$$NO_2^- + 6H^O \longrightarrow NH_3 + OH^- + H_2O \qquad\qquad (3)$$

which is a reaction that can take place only in an acid environment.
Therefore, the above concerns were unwarranted. However, it should be
noted that if hydrogen embrittlement were a concern nitrite would help
to control it.[19]

More important than the theory is the actual performance of calcium
nitrite in concrete. Calcium nitrite meets the requirements in ASTM
C494 and Table 1 shows the effect on setting time and compressive
strengths.

TABLE 1. ACCELERATION WITH CALCIUM NITRITE*

CEMENT BRAND	ADMIXTURE % BY WEIGHT OF CEMENT	3-DAY COMPRESSIVE STRENGTH PSI (kPa)	SETTING TIME		
			INITIAL HRS:MIN	FINAL HRS:MIN	% CHANGE RELATIVE TO CONTROL
A	none	1525 (10515)	8:45	12:21	–
A	1% calcium nitrite	1568 (10811)	6:00	10:20	31
A	1% calcium chloride	2151 (14831)	4:20	7:30	51
A	2% calcium nitrite	1924 (13266)	3:05	6:55	65
A	2% calcium chloride	2624 (18092)	2:10	4:55	75
B	none	1467 (10115)	8:38	--	–
B	1% calcium nitrite	1576 (10867)	5:24	9:05	37
B	1% calcium chloride	2220 (15307)	3:16	5:00	62
B	2% calcium nitrite	2075 (14307)	3:12	5:42	63
B	2% calcium chloride	2562 (17665)	2:15	3:40	74

*Water/Cement ratio was 0.56 to 0.57; slump loss was 4.0 + 0.5 in (100
+ 12mm); air content was 1.95 + 0.25 percent.

In the early 70's the use of the Partial Immersion Corrosion Test to study corrosion in concrete was popular[12]. Open circuit potentials of steel in mortars exposed to chloride using this method showed that calcium nitrite was a corrosion inhibitor[12].

In the late 70's, the Federal Highway Administration developed their own test for steel reinforced concrete. Small bridge decks were constructed with steel closely spaced to accelerate the test. Salting was done daily and potentials were measured over a long period. Several publications discussing the results from these tests can be found in the literature [13, 17, 20].

At the end of two years of accelerated testing, the most dramatic results with the corrosion inhibitor were found in the series where a water reducing admixture was used. Here the control showed serious corrosion at the end of six months, whereas the protected deck was just beginning to show a slight amount of corrosion at the end of two years of continuous salting. Thus, with calcium nitrite and a water reducing agent, the service life of a bridge deck under severe conditions will be significantly extended. However, a finite amount of calcium nitrite cannot protect against unlimited addition of chloride. It was, therefore, important to determine the chloride to nitrite effectiveness ratio.

Concrete cylinders were cast with embedded rebar using 6-inch I.D. plastic pipe, 12 inches long, using ASTM C-185 mortar with admixed sodium chloride and admixed calcium nitrite. All the specimens with Cl^-/NO_2^- ratio less than 2.0 performed satisfactory in the test without passing the -300 mV vs SCE level, while more than half the specimens with Cl^-/NO_2^- between 2.5 and 3.0 failed. It is apparent that if the chloride to nitrite ratio is below 2.0, or below 1.5 to be on the safe side, corrosion will be controlled.

However, if salt is continually put on bridge decks, the concentration of the salt will increase. Thus, it is important to know how the salt concentration builds up with time.

From the above it can be shown that 2% solids on cement by mass solids on solids (s/s) calcium nitrite will protect steel in concrete where 391 kg/m^3 of cement is used, corrosion will not initiate until the chloride ion concentration reaches 7.6 kg/m^3. It is generally believed that 0.6 to 1.2 kg of chloride ion per cubic meter of concrete will initiate corrosion of steel in concrete.

Thus, in concrete unprotected with calcium nitrite, corrosion will begin when the level of chloride is 0.6 to 1.2 kg/m^3, although high-quality concrete will take longer to get to that level[21]. Whereas in the concrete protected with calcium nitrite, the concentration must reach 7.6 kg/m^3.

The time it takes concrete to reach that level depends on the diffusion of chloride into the deck. Diffusion depends on several factors, for example, Ost and Monfore[22] found it to depend on water/cement ratio as shown in Table 2.

TABLE 2. [22] PERMEABILITY OF CHLORIDE ION IN CONCRETE

W/C	% $CaCl_2$ AT 2 IN. DEPTH AFTER 12 MONTHS SOAKING
0.61	5.3
0.45	1.4
0.37	0.1

Stratfull et al[23] examined chloride content in 16 bridges that had failed because of corrosion. They found the average chloride content at the steel to be 1.5 kg/m^3 and the average age of the decks was 13 years. If calcium nitrite had been used at 2% s/s the corrosion would not have started until the chloride level reached 7.6 kg/m^3 Since the chloride levels at which corrosion occurs with calcium nitrite are substantially higher then those in unprotected concrete there has been a recent switch to quality of inhibitor added independent of cement content. One liter of 30% $Ca(NO_2)_2$ solution provides 0.027 kg of nitrite. Chloride to nitrite ratios are now based upon the l/m^3 of 30% calcium nitrite solution added.

2   Recent corrosion testing in concrete

In recent years improved electrical measuring techniques have been used to examine the effectiveness of calcium nitrite. New studies examining the effectiveness of calcium nitrite as an inhibitor in concrete are reported[24-31] in this section.

One of the most extensive studies of calcium nitrite involving over 1200 samples, 15 mix designs and 0, 15, 30 liters per cubic meter of 30% calcium nitrite solution is nearing completion after four years of accelerated testing[24, 25]. Samples were partially submerged in a ·3% sodium chloride solution with 33 mm of concrete cover. This environment provided strong chloride wicking into the concrete with good access to oxygen, i.e., an extremely severe exposure.

Corrosion rates were determined using the polarization resistance technique, electrochemical impedance, and periodic removal of specimens so as to visually examine the reinforcing steel appearance. Corrosion currents expressed as $1/R_p$, inverse polarization resistance (μmho/cm^2), were integrated over time to determine the total corrosion in μmoh/cm^2. months (R_p is the polarization resistance and the corrosion current is equal to B/R_p where B is a constant.)

Figure 2 shows the latest data which are 1.5 years beyond that published in reference 25. A total corrosion measurement of 1000 μmho-month-cm² is roughly equivalent to 25 um of average corrosion on the steel, considered enough to cause rust staining or cracking of the concrete. This was observed for several specimens that reached this degree of corrosion.

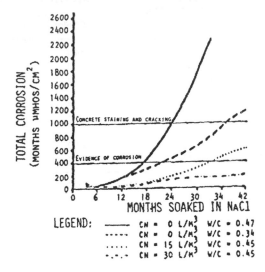

Figure 2. Total corrosion vs time in 3% NaCl as a function of w/c ratio and calcium nitrite content.

This study showed that calcium nitrite delayed the onset of corrosion, and in those cases where corrosion initiated, the rates remained lower than in unprotected cases. Furthermore, the benefits of calcium nitrite improved with lower water-to-cement ratios.

The FHWA examined the effectiveness of calcium nitrite in macrocell measurements made on minibridge decks, some of which had admixed chlorides [26]. Even though they used a water-to-cement ratio that was high (0.5), and admixed chlorides in, they concluded that ". . . calcium nitrite can provide more than an order of magnitude reduction in the corrosion rate".

Another major study in which calcium nitrite was tested was completed in 1987 [27]. In this study the macrocell corrosion rates of minislabs were investigated. The minislabs were subjected to chloride ponding and severe drying for 48 weeks. Additional slabs with calcium nitrite in concrete with different covers and water-to-cement ratios were added to the study and tests were extended from 48 to 116 weeks. A synopsis of reported results and the additional results of the extended test program are presented in Table 3.

TABLE 3. CORROSION INHIBITOR TESTS RESULTS [27]
TIME TO CORROSION AND CORROSION CURRENT

WATER/CEMENT RATIO	SAMPLE TYPE	TIME TO CORROSION (weeks)	CORROSION CURRENT, μA	
			CONCRETE COVER 25mm	
			48 Weeks	116 WEeks
0.50	Control	6	245	*
	20 1/m^3 CN	21	90	*
	30 1/m^3 CN	11	73	*
0.40	Control	5	128	*
	20 1/m^3 CN	21	10	19
0.32	Control	2.5	110	*
	20 1/m^3 CN	**	0	0

WATER/CEMENT RATIO	SAMPLE TYPE	TIME TO CORROSION (weeks)	CORROSION CURRENT, μa	
			CONCRETE COVER 51mm	
			48 Weeks	116 Weeks
0.50	Control	64	0	*
	20 1/m^3 CN	**	0	0
	30 1/m^3 CN	**	0	0
0.40	Control	***		
	20 1/m^3 CN	**	0	0
0.32	Control	***		
	20 1/m^3 CN	**	0	0

* Sample failed
** No corrosion activity after 116 weeks
*** Sample not made
CN-Calcium nitrite sample

Information Sources:
1. Protection Systems for New Prestressed and Substructure Concrete
   Report No. FHWA/RD-86/193, April 1987,
   US Department of Transportation

2. Corrosion Protection Tests on Reinforced Concrete
   September 1987
   Wiss, Janney, Elstner Associates, Inc.

The above study showed that as concrete quality improved (w/c ratio decreased and cover increased), the benefits of adding calcium nitrite corrosion inhibitor are greater.

The South Dakota DOT conducted a study in which rebars were embedded in concrete cylinders with/without admixed chloride[28]. Steel in cylinders with admixed chlorides without calcium nitrite went into corrosion almost immediately, whereas calcium nitrite containing samples remained passive for months. For samples without admixed chlorides, the benefits of calcium nitrite were significant.

Some of the above samples were further analyzed. Results of polarization resistance and electrochemical impedance tests using methods described in Reference 23 are given in Table 4. The corrosion rate is proportional to $1/R_p$ (R_p is the polarization resistance), and rates under 20 μmho-cm^2 are considered passive[25]. The corrosion rates were corrected for ohmic resistance[31], and clearly show that calcium nitrite significantly reduced corrosion.

TABLE 4.   SOUTH DAKOTA LOLLIPOPS CORROSION RATES AT 3.5 YEARS [28]

MIX	CONCRETE MIX (cf - 367Kg/m^3)	Ecorr (mV,SCE)	CONCRETE RESISTANCE (kohm-cm^2)	MEASURED POLARIZATION RESISTANCE (R_p), (kohm-cm^2)	CORRECTED (R_p), (kohm-cm^2)	$1/R_p$, (umhos/cm^2)
4	20.5 l/m^3 Ca(NO$_2$)$_2$	-210	35	230	195	5
6	20.5 l/m^3 Ca(NO$_2$)$_2$ + 16 kg/m^3 NaCl	-491	13	27.4	14.4	69
19	16 kg/m^3 NaCl	-609	6.9	8.5	1.6	625

*Concrete cylinder was cracked open by South Dakota DOT, which caused much higher corrosion rate. Note severe corrosion at $1/R_p$ = > 20 umhos/cm^2.

After the corrosion rates were measured, the samples in Table 4 were broken open. The samples without calcium nitrite were severely corroded. Chloride and nitrite analyses were performed and results showed that chloride levels ranged from 10.7 to 23.7 kg/m^3 at the rebar level and that at Cl^-/NO_2^- ratio of 1.6 to 2.2 calcium nitrite prevented corrosion.

This study showed that calcium nitrite significantly reduced corrosion of steel even when high chloride concentrations were present. Furthermore, it showed that until corrosion initiated there was no evidence of nitrite depletion at the reinforcement level.

Studies are now in progress examining the use of calcium nitrite with silica fume[29, 30]. Early results (one in two years) show that only the control samples without calcium nitrite and/or silica fume are corroding. Of interest is the observation that even though calcium nitrite increases the 12SHTO T277 Rapid Chloride Permeability coulomb value, Table 5, it either lowers or has no effect on the effective diffusion coefficient for chloride. Thus, calcium nitrite is compatible with silica fume, and should provide protection to the reinforcement when chlorides reach it. This study also showed the calcium nitrite increased compressive strength.

TABLE 5. EFFECTIVE CHLORIDE DIFFUSION COEFFICIENTS (D_{eff}) VERSUS AASHTO T227 COULOMBS [29]

MIX No.	CEMENT FACTOR kg/m^3 (pcy)		MICROSILICA %	CALCIUM NITRITE 1/m^3 (gpy)	W/C	AASHTO T227 COULOMBS	D_{eff} 10^{-8} cm^2/s
1	347	(587)	0	0	0.48	3663	11.0
2	343	(580)	0	20(4)	0.48	4220	6.0
5	350	(591)	15	0	0.48	198	0.7
6	359	(607)	15	20(4)	0.48	253	0.5
11	363	(614)	7.5	10(2)	0.43	380	0.8
13	338	(571)	0	0	0.38	3485	2.0
14	352	(595)	0	20(4)	0.38	1838	2.0
17	354	(599)	15	0	0.38	75	0.3
18	313	(580)	15	20(4)	0.38	119	0.3

NOTE: D_{eff} at 22°C (72°F).

Recent testing showed that calcium also protects galvanized steel and aluminum embedded in concrete[31]. This study involved alternate ponding of minibeams with 3% NaCl. Corrosion rates were measured by both polarization resistance and macrocell corrosion. Visual observations were in good agreement with the electrochemical test results shown in Table 6. Indeed analyses of the rebars or metals for which data are shown in Tables 3, 4 and 6 always showed significantly reduced corrosion (pitting and area corroded) when calcium nitrite was used.

Also, the benefits of calcium nitrite in cracked concrete have been demonstrated. Figure 3 shows the long-term total corrosion, to alternate ponding with 3% NaCl, for reinforced notched minibeams 406 mm X 102 mm X 76 mm with third point loaded cracks that extend down to the reinforcing bar, half-way to the bar, and not at all. Cracked samples with calcium nitrite took longer to go into corrosion or are corroding at substantially reduced rates. The uncracked samples are not yet corroding due to the good quality concrete employed as can be seen in Table 7. Cracks were less than 0.25 mm in width. From this study it was determined that calcium nitrite is effective in the presence of cracks.

TABLE 6. CORROSION CURRENTS OF METALS IN MINIBEAMS AFTER 24 MONTHS OF CYCLIC PONDING IN 3% NaCl

| | | | CORROSION CURRENT | |
| | | | POLARIZATION RESISTANCE ($\mu A/cm^2$) | MACROCELL TESTING ($\mu A/cm^2$) |
TYPE OF STEEL	COVER (mm)	CALCIUM NITRITE		
Black Steel	19	0	1.575	0.831
		2	0.026	0.005
		4	**	0.001
Galvanized Steel	19	0	0.722	0.041
		2	0.006	0.007-0.006*
		4	0.072	0
Chromate Treated	19	0	1.091	0.035-0.25*
Galvanized Steel		2	0.325	0.005
		4	0.253	0.001
Aluminum Conduit	35	0	2.200	0.370
(14 months)		2	0.013	0.008
		4	0.011	0.011

* Negative values are for corroding bottom bars.
** Not available.

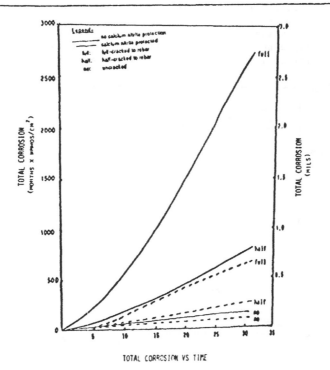

Figure 3.   Total corrosion vs time for precracked beams with/without calcium nitrite.

261

A recent study in Germany[33] showed that sodium nitrite was also
effective in reducing corrosion in cracked specimens, even at much
higher than recommended water-to-cement ratios and a low dosage rate
for the exposure conditions. This is not surprising since two other
studies showed that thin cracking was not a significant factor in
corrosion[34, 35].

TABLE 7. CONCRETE MIXTURE PROPORTIONS AND PROPERTIES OF SPECIMENS
          IN PRECRACKED CONCRETE. NOTE: Average for 3 mixes each design.

MIX	CF (kg/m³)	W/C	SLUMP (mm)	% PLASTIC AIR	f'c (28 days) MPa
Control	426	0.39	66	6.3	46.4
Calcium Nitrite (2% solids by mass of cement)	414	0.40	91	7.4	43.4

A small study was also conducted looking at the benefits of adding
calcium nitrite to concrete containing fly ash. Figure 4 shows the
long-term corrosion behavior of a control concrete compared to concrete
with fly ash or fly ash and calcium nitrite. Calcium nitrite
significantly improved corrosion resistance, and fly ash by itself
offered only minimal improvement. Mix designs are given in Table 8.

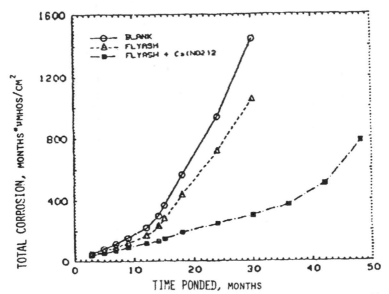

Figure 4.   Effect of fly ash and fly ash + calcium nitrite on the
             total corrosion of concrete rebars ponded in 3% NaCl.

TABLE 8. BASIC MIXTURE PROPORTIONS FOR FLY ASH-CALCIUM NITRITE STUDY

MIX No.	CF kg/m^3	W/C	ADMIXTURE
1	295	0.62	Control
2	240	0.74	CF/Fly Ash = CF/Fly Ash=240/72
3	240	0.73	$Ca(NO_2)_2$ = $Ca(NO_2)_2$=9.5 $1/m^3$ CF/Fly Ash = CF/Fly Ash = 240/72

3 Importance of mix design

ACI 201 states that low water-to-cement ratios and good concrete cover should be employed when chloride exposures are moderate to severe[36]. Several of the mentioned studies confirmed the need for low water-to-cement studies and adequate cover [24, 25, 28, 29, 30]. Furthermore, the almost complete and completed studies[24, 25, 27] clearly show that calcium nitrite becomes more effective as concrete quality improves. This is significant because the higher quality concretes outperform those of lesser quality, and thus, the improvements with calcium nitrite in good quality concrete are substantial.

ACI 201 also states the importance of air entrainment to protect the concrete against freeze-thaw damage. Concretes that are properly air intrained with calcium nitrite and high-range water reducers are resistant to freeze-thaw cycles, as shown in reference 29.

4 Compatibility with other corrosion protection systems

As shown above, calcium nitrite is very compatible with concrete containing microsilica or fly ash to provide reduced chloride permeability. Sealers or membranes would be beneficial in reducing chloride ingress.

5 Studies to confirm action of calcium nitrite

A major study examining the effects of calcium nitrite in saturated calcium hydroxide solutions, with and without sodium chloride was conducted[37]. This study showed that calcium nitrite significantly increased (more noble) the potential at which pitting initiates, and also increased the protection potential below which pitting and crevicing does not occur. This is indicative of the behavior of an anodic inhibitor and is beneficial in that the range in which steel cannot severely corrode in concrete is significantly broadened. Subsequent, unpublished work by one of the authors in higher pH, sodium and potassium hydroxide solutions showed no significant effect other than the transpassive potential becoming more active as would be expected for the decomposition of water[38].

Furthermore, the study showed that even extremely low concentrations of nitrite were beneficial. A recent study by Briesemann[33] also showed that low concentrations of nitrites are not detrimental in the high pH environment of concrete.

The studies in reference 36 were carried out in solutions purged with oxygen-free nitrogen. In the absence of oxygen the nitrite did not participate in a detrimental cathodic reaction as found at lower pH's by Cohen[18]. This point is important in that it shows that accelerated corrosion due to an increased cathodic raction does not occur.

In another unpublished laboratory study by one of the authors, there was no loss of nitrite in concrete samples exposed to fog room conditions for five years followed by two years of drying. In the absence of corrosion, nitrite is not consumed as shown by Cohen [18], by analysis of concrete in the South Dakota study [28], and a seven year old Illinois bridge deck. Thus, nitrite is stable in concrete.

6   How to use calcium nitrite

Corrosion data from several studies [13, 26, 28, 37] and chloride analyses performed on the concretes tested (or known from the solutions employed) were combined to develop a graph of chloride-to-nitrite ratio at which corrosion occurs versus admixed 30% calcium nitrite solution. The graph is given in Figure 5. Based upon Figure 5, Table 9 was developed as a guide as to how much calcium nitrite is needed to protect to a given chloride ingress.

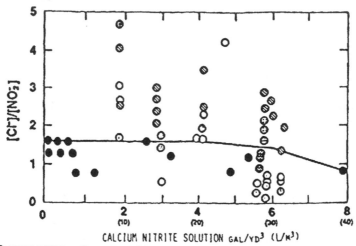

O WRC DECK at Corrosion Initiation   O Lollipops Test, Steel Not Corroded
● Solution Test, Steel Not Corroded   ⊖ FHWA Corrosion Initiation
⊘ Lollipops Test, Steel Corroded   ⊕ FHWA Corroded
                                       ⊖ FHWA No Corrosion

Figure 5.   Chloride to nitrite ratios at which corrosion occurs based on testing in solutions and concrete.

TABLE 9.   CALCIUM NITRITE DOSAGE RATES VS. CHLORIDE

CALCIUM NITRITE $(1/m^3)$	CHLORIDES (kg/m^3)
10	3.6
15	5.9
20	7.7
25	8.9
30	9.5

If one can estimate the amount of chloride that will reach the steel at the design life of a structure, an estimate of calcium nitrite needed to protect the structure can be given. Chloride levels can be estimated from experience and concrete properties, and several papers give methods and examples of predicting chloride levels (3, 29, 30, 31 39).

7   Conclusions

Nitrites have been used in concrete for over 30 years in Europe. Calcium nitrite has been available in the U.S. as a concrete admixture for over eight years. Extensive testing shows that it:

- protects steel against chloride induced corrosion;
- improves in efficiency as concrete quality improves;
- improves compressive strength;
- is beneficial to corrosion protection in good quality concrete and lowers corrosion rates once corrosion initiates;
- and works in the presence of cracks.

REFERENCES

1.   Strategic Highway Research Program, Participant Workbook--National Workshop, Strategic Highway Research Program: Washington, DC, Sept. 1985, p. TRA 4-1.

2.   Focus, Special Edition, Strategic Highway Research Program, Washington, DC, Sept. 18-20, 1985.

3. R.D. Browne, "Design Prediction of the Life for Reinforced Concrete in Marine and Other Chloride Environments," Durability of Building Materials, V. 1, pp. 113-125 (1982).

4. Monfore, G.E., and Verbeck, G.J., "Corrosion of Prestressed Wire in Concrete," ACI, 57, 5 (1960) p. 491.

5. Evans, R.H., "Uses of Calcium Chloride in Prestressed Concrete", Proceedings, World Conf. on Prestressed Concrete, San Fransisco, (1957) p. A31-1.

6. Treadaway, K.W.J., and Russel, A.D., Highways and Public Works, 36, 19-20 (1968), p. 40.

7. Craig, R.J., and Wood, L.E., Highway Research Record, No. 328 (1970), p. 77.

8. Dodson, V.H., Farkas, E., and Rosenberg, A.M., U.S. Patent No. 3, 210,207 (1965).

9. Technical Bulletin, W. R. Grace & Company, Cambridge, MA ADM-1, October 15, 1964.

10. Foley, R.T., J. Electrochem. Soc., 11, 122, (1975), p. 1493.

11. Hartt, W.H., and Rosenberg, A.M., "Influence of $Ca(NO_2)_2$ on Sea Water Corrosion of Reinforcing Steel in Concrete", ACI publication SP-65, (1980), p. 609.

12. Gaidis, J.M., Kossivas, T.G., Previte, R.W., and Rosenberg, A.M., ASTM Spec. Tech. Publ., No. 629 (1977), p. 89

13. Gaidis, J.M., and Rosenberg, A.M., TRB Record, No. 692 (1978), p. 28.

14. Gaidis, J.M., and Rosenberg. A.M., CEFRACOR, Paris, Fr. paper 154 (1979), p. 29.

15. Gaidis, J.M., and Lundquist, J.T., and Rosenberg, A.M., Mater. Perf., 18, 3 (1979), p. 36.

16. Gaidis, J.M., and Rosenberg. A.M., Mater Perf., 18, 11 (1979), p. 45.

17. Gaidis, J.M., Rosenberg. A.M., and Salek, J., ASTM, STP 713 (1980), p. 64.

18. Pyke, R., and Cohen, M., J. Electrochemical Soc., 93 (1948) p. 63.

19. Saito, A., Tokahino, Y., Voshizawa, S., and Yamokawa, K., Boshotygijutsu, Vol. 24, No. 8, (1980) pp. 385-391.

20. Gaidis, J.M., and Rosenberg, A.M., ASTM, Cement, Concrete & Aggregates, 9,1, (Summer 1987), p. 30.

21. Corrosion of Metals in Concrete. ACI, 222 R-85, Special Report by Comm. 222, (August 1985).

22. Ost, B., and Monfore, G.E., Mater. Perf., 12, 6 (1974), p. 21

23. Stratfull, R.F., Jurkovich, W.J., and Spellman, D.L., California Transportation Laboratory, Sacramento, Research Report, CA-DOT-TL-5116-12-75- 03, (1975).

24. Berke, N.S., "The Effects of Calcium Nitrite and Mix Design on the Corrosion Resistance of Steel in Concrete (Part 1)", NACE Corrosion-85, Paper No. 273, NACE, Houston, 1985.

25. Berke, N.S., "The Effects of Calcium Nitrite and Mix Design on the Corrosion Resistance of Steel in Concrete (Part 2, Long-Term Results)". Corrosion of Metals in Concrete, Proceedings of the Corrosion-87 Symposium on Corrosion of Metals in Concrete, NACE: Houston (1987), p. 134-144.

26. Virmani, Y. P., Clear, K.C., and Pesko, T.J., "Time-to-Corrosion of Reinforcing Steel in Concrete Slabs, Vol. 5-- Calcium Nitrite Admixture and Epoxy-coated Reinforcing Bars as Corrosion Protection Systems" Report No. FHWA-RD-83-012, Federal Highway Administration, Washington DC, September 1983, p. 71.

27. Pfeifer, D.W., and Lundgren, J.R., "Protective Systems for New Prestressed and Substructure Concrete", Report No. FHWA-86-193, Federal Highway Administration, Washington DC, April 1987, p. 113.

28. "Laboratory Performance on Corrosion Control Methods", Interim Report, Research Program Division of Planning, South Dakota Department of Transportation, Pierre, South Dakota, March 1984.

29. Berke, N.S., and Weil, T.C., "Corrosion Protection Through the Use of Concrete Admixture", Second International Conference on Performance of Concrete in Marine Environment, August 21-26, 1988, St. Andrews by the Sea, New Brunswick, Canada.

30. Berke, N.S., "Resistance of Microsilica Concrete to Steel Corrosion, Erosion and Chemical Attack", To be presented at the Third International Conference on the Use of Fly Ash, Silica Fume, Slag and Natural Pozzolans in Concrete", Trondheim, Norway, June 19-24, 1989.

31. Berke, N.S., Shen, D.F., and Sundberg, K.M., "Comparison of the Polarization Resistance Technique to Macrocell Corrosion Technique", Presented at the ASTM Symposium on Corrosion Rates of Steel in Concrete, Baltimore, MD, June 29, 1988.

32. Berke, N.S., Shen, D.F., and Sundberg, K.M., "Comparison of Current Interruption and Electrochemical Impedance Techniques in the Determination of the Corrosion Rates of Steel in Concrete", Presented at the ASTM Sympsoium on Ohmic Electrolyte Resistance Measurement and Compensation, Baltimore, MD, May 17, 1988.

33. Briesemann, D., "Corrosion Inhibitors for Steel in Concrete", Civil Engineering Department, Munich Technical University, Munich, West Germany.

34. Beeby, A.W., Concrete International, Vol. 5, No. 2, February, p. 35-40 (1983).

35. Margat, P.S., and Gurusamy, K., Cement & Concrete Research, Vol. 17, p. 385-396 (1987).

36. ACI 201 "Guide to Durable Concrete".

37. Berke, N.S., "The Use of Anodic Polarization to Determine the Effectiveness of Calcium Nitrite as an Anodic Inhibitor", Corrosion Effect of Stray Currents and the Techniques for Evaluating Corrosion of Rebars in Concrete, ASTM STP 906, V. Chaker, Ed., American Society for Testing and Materials, Philadelphia, 1986, p. 78-91.

38. Pourbaix, M. Atlas of Electrochemical Equilibria in Aqueous Solutions, National Association of Corrosion Engineers, Houston TX 1974.

39. Pereira, C.J., and Hegedus, L.L., Diffusion and Reaction of Chloride Ions in Porous Concrete", presented at the 8th International Symposium on Chemical Reaction Engineering, Edinburgh, Scotland. Sept. 10-13, 1984.

THE USE OF SUPERPLASTICIZERS AS STEEL CORROSION REDUCERS IN REINFORCED CONCRETE

M. COLLEPARDI, R. FRATESI and G. MORICONI
Department of Sciences of Materials and Earth, University of Ancona, Italy
S. BIAGINI
MAC-MBT, R&D Laboratories, Treviso, Italy

Abstract
The data of the present paper indicate that carbonation is a necessary process but not a sufficient one for the steel corrosion in reinforced concrete. The porous structure of concrete and the environmental relative humidity appear, instead, to be the determining factors in the corrosion process, since their increase causes capillary water to come into contact with steel reinforcement. So, superplasticizers can reduce steel corrosion as they allow a reduction of the water/cement ratio at the same concrete workability, and, at last, a reduction in the water content and the permeability to water of the concrete.
Key-words: Carbonation, Steel corrosion, Superplasticizers.

1 Introduction

Concrete is considered to have a particular protective action on the reinforcing steel bars in consequence of the formation of an adherent and passive oxide film in the high basicity conditions created by the hydration of the cement.

However, failures of structures are often attributed to reinforcement corrosion processes in aggressive environments. Aggressive phenomena are generally ascribed to the intrusion of either chloride ions, able to destroy the passive film, or carbon dioxide, which neutralises the alkalinity of the aqueous solution present in the concrete pores and thus removes the favourable passivation conditions.

The aim of the present work is, after firstly evaluating the real influence of the carbonation process in reinforcement corrosion phenomena, to verify if the use of superplasticizers as water reducers can reduce steel corrosion in reinforced concrete.

2. Experimental part

To acquire experimental data in a reasonably short time, the carbonation process was carried out in an accelerated way by artificially enriching in carbon dioxide the test environment.

Furthermore, experiments were performed on concretes, rather than mortars, as previously used by other Authors (1), in order to obtain results corresponding to real situation.

Finally, to avoid any contamination by factors, other than carbonation, particular attention was taken to prevent the contact of the reinforcement with chlorides.

Concretes with different water/cement ratio (w/c = 0.35 - 0.50 - 0.65 - 0.80) were casted. The binder was Portland cement type I, used separately or in combination with fly ash, an artificial pozzolan. The fly ash substituted partially (20%) the Portland cement or was added with no diminution of the Portland cement. In the case of partial substitution, a pozzolanic cement is, in fact, used.

Cubic concrete specimens (10 cm side) without rebars, were manufactured for the determination of the electrical resistivity and of the penetration depth of the carbon dioxide. Moreover, prismatic reinforced specimens (40x15x10 cm) were prepared for the determination of the corrosion electrochemical potential and the polarization resistance of the rebars. The rebars were of common manufacturing steel (8 mm diameter) and they were symmetrically positioned with concrete covers varying from 2 to 5 cm. During the casting, graphite bars as counterelectrodes were inserted at equal distances from the steel bars.

The specimens were cured for seven days in a saturated vapour atmosphere. Before their exposure to the artificially carbon dioxide enriched environment (carbon dioxide equal to 30%), kept at atmospheric pressure, the sides of the specimen perpendicular to the bars and the surfaces of the emerging bars were coated with epoxy resin. The relative humidity of the environment enriched in carbon dioxide was kept constant at a value of 75% in order to obtain the maximum carbonation velocity (2).

The penetration depth of carbon dioxide was measured after different times of exposure by means of the phenolphthalein test: an hydroalcoholic solution of 2% phenolphthalein was sprayed on the transversely split section of the cubic specimens.

The corrosion electrochemical potential of the reinforcement was measured with respect to mercurous sulphate (SSE) reference electrode by means of a differential electrometer with high input impedance, according to ASTM (3).

The polarization resistance measurements of the rebars (4-6) were carried out galvanodinamically, by polarizing the working electrode by means of the graphite counterelectrode, while the reference electrode was fixed on the external surface of the specimen; a wet sponge of sodium nitrate solution was placed between the reference electrode and the external surface of the specimen to improve surface contact; at last, the ohmic drop due to the concrete thickness

between the reference and the working electrode was
compensated (7).

3   Results and discussion

3.1 Carbonation depth

The results of the carbonation depth with exposure time
in carbon dioxide atmosphere, for concretes consisting
of Portland cement, are shown in Fig. 1. The data are
in agreement with those reported in literature (8,9),

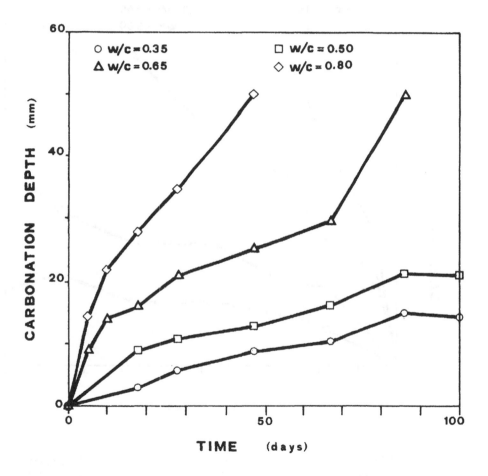

Fig. 1.   *The influence of exposure time and water/cement
ratio on the carbonation depth in Portland cement
concretes in carbon dioxide enriched atmosphere
(R.H. = 75%).*

demonstrating an increase in the permeability to carbon dioxide with the increase of water/cement ratio, that is with the porosity of the concrete.

Using pozzolanic cement with the same water/binder ratio (Fig. 2), the evolution appear similar to that shown in Fig. 1, but a higher penetration depth with the same exposure time is observed, due to a greater initial porosity derived from a higher effective water/cement ratio (10) and a lower amount of hydrolisis lime available (6).

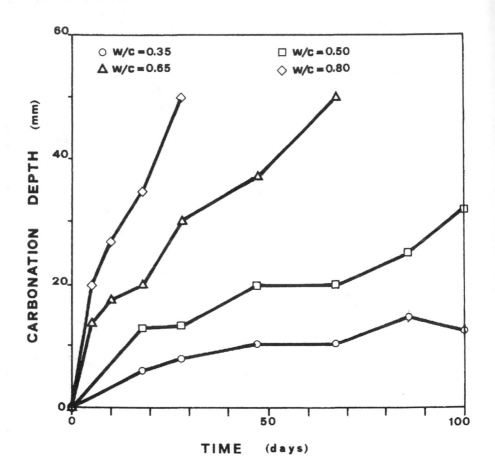

Fig. 2. *The influence of exposure time and water/cement ratio on the carbonation depth in pozzolanic cement concretes in carbon dioxide enriched atmosphere (R.H. = 75%).*

Instead, the use of fly ash in substitution to the fine inert, rather than cement, systematically decreases the depth of penetration of the carbon dioxide (Fig. 3).

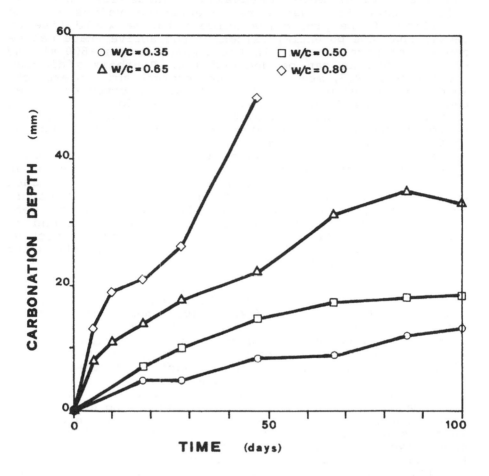

Fig. 3. *The influence of exposure time and water/cement ratio on the carbonation depth in Portland cement concretes with addition of fly ash in carbon dioxide enriched atmosphere (R.H. = 75%).*

3.2 Electrochemical potential

The potential of reinforcements subjected to accelerated carbonation shows significant changes with exposure time, as indicated in Fig. 4. However, since the evolution of the potentials was not modified by the carbonation front (point C on curves in Fig. 4), these potential variations could not be attributed to the carbonation process.

The potential trend, in carbonated concretes, depends on

273

the water/cement ratio and for its lower values (w/c = 0.35 and w/c = 0.50) moves gradually towards less negative values, that is towards representative values of the stability of the protective oxide film. We would like to recall that the potential values of the active state can be considered significative in the range between -600 mV and -1250 mV SSE as can be deduced from Pourbaix's diagram (10-13) at pH 8.3, measured on the concrete aqueous extract. No change in the potential tendency was observed even when the carbonation front had reached the reinforcements. This fact is probably justified by the low water/cement ratio and therefore by the lack of water necessary for the formation of a liquid layer on the rebars.

Instead, rebars embedded in concretes of higher water/cement ratio (w/c = 0.65 and w/c = 0.80) show an initial tendency towards more negative potentials. Such tendency is inverted afterwards and once again it occurs independently of the fact that the reinforcements were reached by the carbonation front. The initial trend is probably due to the growing instability of the protective oxide film, for a simultaneous availability of both water and oxygen in more porous concretes. The successive evolution can be explained by the progressive disappearance of the water layer on the rebars surface, caused by the drying of the porous concretes in an humidity unsaturated environment (R.H. = 75%). Effectively, a net decrease of the potentials is observed when the relative humidity is varied from 75% to 90%, due, probably, to the formation of a new liquid layer on the rebars surface. Such effect increases with water/cement ratio, the capacity of more porous concretes to absorb water being greater.

From the above-mentioned data, it would seem, therefore, that the determining factor for the progressing of corrosion in reinforced concrete is not much the carbonation process, which is of course essential for the dissolution of the protective oxide film, as the formation of a liquid layer on the rebars and obviously the presence of oxygen.

Such conclusions have been confirmed from results, not reported here, concerning reinforcements with concrete covers greater than 2 cm.

3.3 **Polarization resistance**

The polarization resistance values for reinforcements with 2 cm concrete cover, shown in Fig. 5, increase with time in carbon dioxide enriched environment at 75% relative humidity. An increase in relative humidity from 75% to 90% (dotted curves) causes a sudden drop of the polarization resistance values. The carbonation front (point C on Fig. 5) does not seem to have a direct influence on such evolution. On the contrary, the polarization resistance values for higher concrete covers remain constantly low with time. Such behaviour is actually very difficult to interprete, since

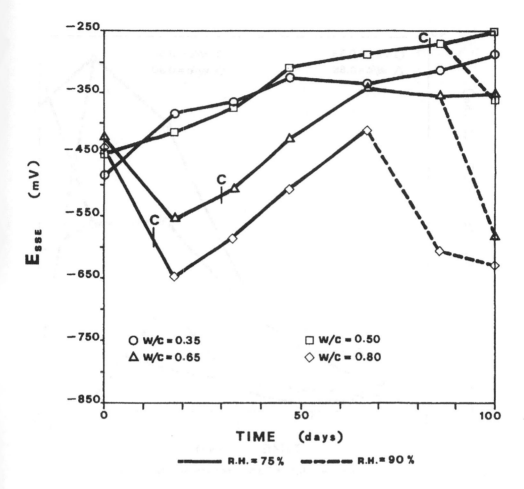

C = rebar reached by the carbonation front

Fig. 4. *The influence of exposure time, water/cement ratio and relative humidity on the electrochemical potential of the rebars in Portland cement concretes with 2 cm concrete cover in carbon dioxide enriched atmosphere.*

the low polarization resistance values, which should indicate corrosion active state, are not justified by the corresponding electrochemical potential values which, instead, indicate passive state of the rebars.

The evolution of the polarization resistance reported in Fig. 5 appears to be in agreement with the model proposed for the interpretation of the electrochemical potential

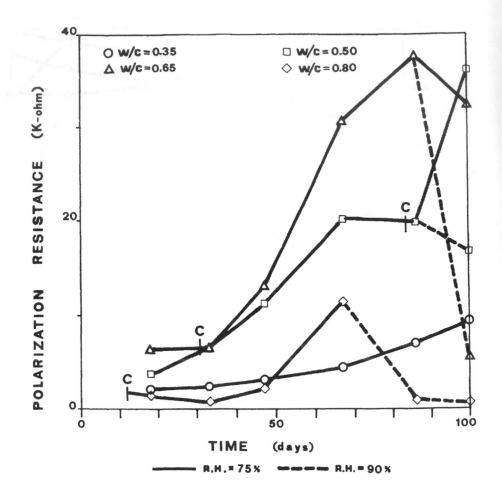

C = rebar reached by the carbonation front

Fig. 5. *The influence of exposure time, water/cement ratio and relative humidity on the polarization resistance of the rebars in Portland cement concretes with 2 cm concrete cover in carbon dioxide enriched atmosphere.*

trends, based on the possibility of formation of a liquid layer on the rebars surface.

Effectively, the polarization resistance increases with time, particularly for rebars embedded in concretes with high water/cement ratio, that is in concretes which are more porous and which dry easily. The anomaly shown by the polarization resistance values (whose trend is, however, consistent with the others) in concrete with 0.80

water/cement ratio is the object of further investigation. Such anomaly might probably be ascribed to the simultaneous complex influence of the various parameters which determine the equilibrium with the environmental relative humidity for the formation of a water liquid layer on the rebars surface. The changes in the potential and polarization resistance values of rebars placed at 2 cm from the concrete surface, with 0.65 water/cement ratio, and exposed to different relative humidities (75% and 90%) do suppose a different surface state of the rebars and have therefore suggested a direct observation of such presumed different surface conditions by splitting the concrete cover. Rebars exposed to 90% relative humidity have put into evidence corrosion traces, while those exposed to 75% relative humidity have shown surfaces covered with the protective oxide film. One can then deduce that low electrochemical potential values coupled with low polarization resistance values should indicate an effective corrosion state.

The visual observation of the corresponding rebars embedded in concrete with 0.35 water/cement ratio and 2 cm concrete cover, with low polarization resistance and high potential values, did not show any corrosion traces.

One can, therefore, conclude that low polarization resistance values indicate an effective corrosion activity if only they are coupled by low electrochemical potential values.

Briefly, the addition of fly ash (not reported in this paper) modifies the kinetics of carbonation but does not change at all the described results for concretes with Portland cement only.

4 Conclusions

The results of the present work show that the carbonation depth in concrete strongly depends on the water/cement ratio, increasing with it. Moreover, the addition of fly ash without diminution of the cement reduces the penetration depth of carbon dioxide.

In other words, the carbonation depth is anyway reduced by lowering the concrete porosity.

On the other hand, in the total absence of chlorides, electrochemical measurements proved the carbonation process to be a necessary condition but not a sufficient one to promote corrosion process in reinforced concrete, which seems, instead, to be strongly influenced by the concrete permeability to water and, obviously, to oxygen.

As a result, every admixture able to allow a reduction in concrete porosity and permeability can be considered a corrosion reducer admixture. By this point of view, superplasticizers are real steel corrosion reducers, when used as water reducers.

References

(1) Andrade, C., Castelo, V., Alonso, C., Gonzales, J.A. (1986), The determination of the corrosion rate of steel embedded in concrete by the polarization resistance and AC impedance methods, in _Corrosion effect of stray currents and the techniques for evaluating corrosion of rebars in concrete_, ASTM STP 906, 43.

(2) Pedeferri, P. (1987), Corrosione e protezione di strutture metalliche e in cemento armato negli ambienti naturali, Ed. Clup, Milano, 432.

(3) ASTM C876-80, (1980), Standard test method for half cell potentials of reinforcing steel in concrete.

(4) Stern, M., Geary, A.L. (1957), _J. Electrochem. Soc._, 104 (1), 56.

(5) Gonzales, J.A., Molina, A., Escudero, M.L., Andrade, C. (1985), _Corr. Sci._, 25 (10), 917.

(6) Gonzales, J.A., Alonso, C., Andrade, C. (1983), Corrosion rate of reinforcements during accelerated carbonation of mortars made with different types of cement, in _Corrosion of reinforcement in concrete construction_ (Ed. A.P. Crane, Ellis Horwood, Chichester), 159.

(7) Lauer, G., Osteryoung, R.A. (1966), _Anal. Chem._, 38, 1106.

(8) Rio, A. (1987), Influenza della carbonatazione e della penetrazione dei cloruri sulla durabilità dei conglomerati cementizi armati, Giornate A.I.C.A.P. sulla _Durabilità delle opere in c.a._, Padova, Italy.

(9) Treadaway, K.W.J., Macmillan, G., Hawkins, P., Fontenay, C. (1983), The influence of concrete quality on carbonation in Middle Eastern conditions – A preliminary study, in _Corrosion of reinforcement in concrete construction_ (Ed. A.P. Crane, Ellis Horwood, Chichester), 101.

(10) Pourbaix, M. (1973), Atlas of Electrochemical Equilibria in Aqueous Solutions, Pergamon Press, Oxford.

(11) Pourbaix, M. (1974), _Corr. Sci._, 14, 25.

(12) Hansson, C.M. (1984), _Cem. Concr. Res._, 14, 574.

(13) ACI Committee 222 (1985), _ACI J._, 82, 3.

USE OF NITRITE SALT AS CORROSION INHIBITOR ADMIXTURE IN REINFORCED CONCRETE STRUCTURES IMMERSED IN SEA-WATER

M. COLLEPARDI, R. FRATESI and G. MORICONI
Department of Sciences of Materials and Earth, University of Ancona, Italy
L. COPPOLA
Enco, Engineering Concrete, Spresiano, Italy
C. CORRADETTI
Snamprogetti S.p.A., Corrosion Protection Department, Fano, Italy

Abstract
Sodium nitrite is supposed to be a corrosion inhibitor of reinforced concrete structures in contact with chlorides. In the present work the effect of 4% sodium nitrite by weight of cement on the corrosion process of cracked reinforced concrete specimens (w/c=0.50) immersed in sea-water has been studied. It has been found that in cracked concrete specimens corrosion becomes more severe in the presence of sodium nitrite. The effect is quicker the larger the crack width of the concrete. Sodium nitrite does not substantially modify the chloride diffusion through uncracked concrete specimens, so that the above negative effect of sodium nitrite should be expected even in uncracked reinforced concrete areas but after longer immersion time in sea-water.
Key words: Corrosion inhibitor, Nitrite, Cracked reinforced concrete, Sea-water chloride diffusion.

1 Introduction

Nitrite salt is considered to be a corrosion inhibitor admixture for reinforced concrete structures containing chlorides (1). The available commercial nitrite salts are alkali nitrites such as sodium nitrite or calcium nitrite. The preferred form of nitrite should be the calcium salt when reactive aggregates are suspected to be present in a concrete, since sodium nitrite could aggravate the alkali-aggregate reaction effects. However if reactive aggregates are excluded, no substantial difference between sodium nitrite and calcium nitrite, at least for the inhibitive action, should be expected. The amount of nitrite required to counteract adequately the corrosive action increases when the amount of chlorides in the concrete is augmented. If the amount of nitrite is less than the required value, corrosion could act more severely than in its absence (2-4).

Therefore, the utilization of nitrite as a corrosion inhibitor could be to a certain extent successful only when the amount of chlorides in concrete is definitely known. This occurs when chloride is put into the mix as an impurity of one or more concrete ingredients (sand, mixing water, etc.). On the other hand, when chloride is diffusing into the concrete from the environment (sea-water or areas exposed to deicing salts), it is very difficult to evaluate the exact amount of chloride and then the required amount of nitrite. Some papers have been published on the effect of nitrite on the steel corrosion promoted by chloride. However, these works have been carried out by examining the behaviour of a steel bar in a calcium hydroxide saturated solution simulating the aqueous phase into the concrete, and by studying a steel bar immersed in a cement paste or mortar (5-9).

The main purpose of the present paper was to study the effect of nitrite on the chloride steel corrosion of reinforced concrete specimens immersed in sea-water. This approach appears to be more realistic, since a steel bar immersed in a water solution, cement paste, or mortar, does not reproduce the reinforced concrete structure, because the contact between coarse aggregate and steel reinforcement could change significantly the local electrochemical behaviour of the system. Moreover, the passivity current density of steel has been found to be much lower in concrete than in calcium hydroxide saturated aqueous solution (10,11).

The other important aim of the present paper was to examine the effect of cracks on the steel corrosion of reinforced concrete specimens immersed in sea-water. Indeed, in cracked reinforced concrete structures chloride should penetrate easily and quickly through cracks. Therefore chloride could attain to high concentration near the reinforcement, so that a very high and unrealistic amount of nitrite should be required to counteract the corrosion action.

2 Experimental part

Two concrete mixes with the same composition (*) were manufactured, the only difference being the absence or presence of sodium nitrite (4% by weight of cement). Since no reactive limestone aggregates were used, sodium nitrite was chosen as a potential corrosion inhibitor admixture.

Cubic specimens (100 mm) were produced by using the two above concrete mixes for compressive strength measurements at 1 to 28 days. Cubic specimens were also immersed in sea-water after a 45 days curing time, to determine the electrical resistivity and chloride penetration depth as a function of immersion time. The electrical resistivity was determined by measuring the current after inducing known voltages between two opposite faces covered by aluminium

(*) water = 150 Kg/m^3; type I Portland cement = 300 Kg/m^3; sand = 735 Kg/m^3; coarse aggregate = 1200 Kg/m^3; max size of coarse aggregate = 19.1 mm.

plates. The chloride penetration depth was determined through a colorimetric test by spraying fluorescein and silver nitrate after splitting the cubic specimen in two pieces. Pink and black coloured surfaces correspond to concrete areas penetrated by chloride or not respectively (12).

Prismatic reinforced specimens (400x150x100 mm) were also produced. These specimens were reinforced by a steel plate (320x130x1 mm) placed at 30 mm from the side of the concrete specimen containing a preformed notch 10 mm deep (Fig. 1). By loading the specimens on the surface opposite the notch, a flexural stress was induced and consequently, specimens were middle-cracked. The presence of the steel plate and different values in the load allowed the crack width to be controlled in the following ranges: 0.03-0.04, 0.25-0.40 and 0.95-1.10 mm.

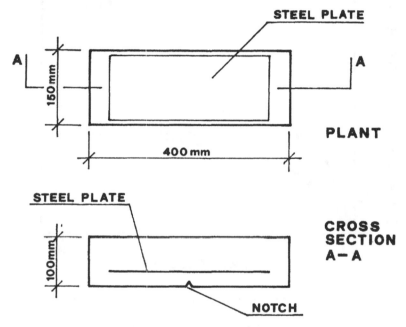

Fig. 1. Prismatic reinforced specimen.

Finally cracked reinforced specimens have been immersed in sea-water to evaluate the formation, the nature and the morphology of corrosion products by visual observations as well as to determine the corrosion extent and depth by microscopic measurements.

3 Results and discussion

The results of both plain concrete specimens and reinforced ones will be examined.

3.1 Plain concretes

Figure 2 indicates that 4% sodium nitrite addition reduces the concrete compressive strength by about 25%. The effect is similar to that of other sodium salts, such as sodium carbonate or sodium silicate, which accelerate set but retard cement hardening. Since the w/c ratio is the same for the plain mix and that containing sodium nitrite, the reduction in the compressive strength could be ascribed to a decrease in the degree of hydration and therefore, to a higher capillary porosity or a change in the cement paste microstructure.

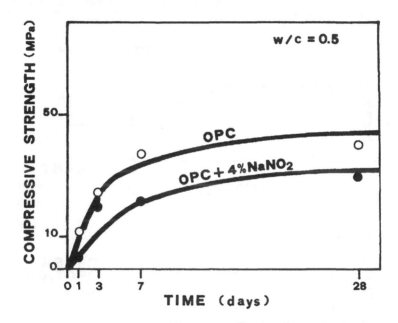

Fig. 2. Compressive strength versus time.

Figure 3 shows chloride penetration into the concrete as a function of immersion time in sea-water and indicates that chloride diffusion is not modified by sodium nitrite addition. These results would demonstrate that the degree of cement hydration and the capillary porosity of the cement paste is not changed by sodium nitrite. Alternatively, if one supposes that the degree of hydration is reduced and the capillary porosity is increased, then it should be admitted that the presence of Na^+ and NO_2^- ions in the capillary pores counterbalances the higher porosity of the cement paste, so that chloride diffusion becomes equal even if the capillary porosity is different. Figure 4 shows the change in concrete electrical resistivity as a function of immersion time in sea-water. For both concrete mixes, after immediate decrease, due to penetration of ions coming from sea-water, the electrical resistivity slightly increases. The precipitation of brucite crystals (13) into capillary pores could explain the slight resistivity increase. At a given immersion time in sea-water, sodium nitrite concrete always shows a lower electrical resistivity because of the higher ionic strength probably due to the presence of Na^+ and NO_2^- ions in the aqueous phase filling the capillary pores.

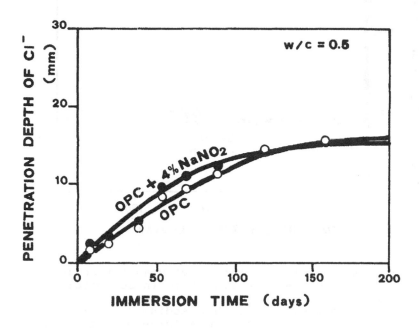

Fig. 3.   Chloride penetration as a function
of the immersion time in sea-water.

3.2 Reinforced concrete specimens

Table 1 summarizes all the tests carried out to evaluate the effect of sodium nitrite on the corrosion promoted by chloride penetration into the cracked concrete specimens. After 90 days of immersion in sea-water no corrosion product was observed in concrete specimens with cracks of 0.30 mm in width or less. Only in concrete specimens having cracks of 1 mm corrosion occurred. However, in concrete containing sodium nitrite, the corrosion of the embedded steel plate appeared to be much more severe: the corrosion products were incoherent and brown coloured, like FeOOH, with low extention but with deep attack (0.2 mm) for the corroded area (Fig. 5). On the other hand in the sodium nitrite free concrete, the corrosion products were dense and black coloured, like Fe_3O_4 with a higher extention and less deep attack (0.05 mm) for the corroded area (Fig. 6).

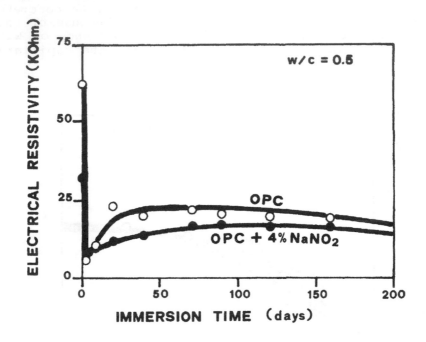

Fig. 4. Electrical resistivity as a function of the immersion time in sea-water.

These results demonstrate that sodium nitrite aggravates the local corrosion process when a large amount of chloride ions reaches the steel reinforcement.

By prolonging the immersion time in sea-water up to 150 days even the crack with the lowest width (0.03 mm) was to a certain extent involved in the corrosion process providing that sodium nitrite was present. Whereas, no corrosion was recorded in the absence of sodium nitrite (Table 1).

Figures 7 and 8 show the pitting corrosion and the absence of corrosion in the steel plate belonging to the concrete with and without sodium nitrite respectively. Again, the more severe steel corrosion in reinforced concrete specimens caused by sodium nitrite is confirmed even in concrete with low width cracks. To explain the negative effect of sodium nitrite on the corrosion process in the experimental conditions of the present work, it should be admitted that:

(a) a large amount of chloride ions penetrates through the cracks in a period of time which depends on the crack

Table 1. Visual observations and microscopic measurements on reinforced concrete specimens.

Crack width (mm)	Immersion time (days)	With NaNO$_2$		Without NaNO$_2$	
		Microscopic measurement	Visual observation	Microscopic measurement	Visual observation
1	90	0.2 mm corrosion depth 1.5 cm² corroded area	incoherent brown corrosion products (FeOOH)	0.05 mm corrosion depth 0.6 cm² corroded area	extensive compact black corrosion products (Fe$_3$O$_4$)
0.3	90	No corrosion		No corrosion	
0.03	90	No corrosion		No corrosion	
0.03	150	Pitting corrosion		No corrosion	

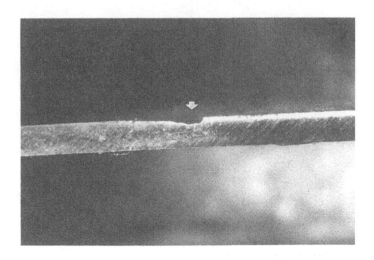

Fig. 5. Micrograph of the corrosion depth in the plate (cross section) embedded in the concrete with NaNO$_2$ after three months of immersion in sea-water.

Fig. 6. Micrograph of the corrosion depth in the plate
(cross section) embedded in the plain concrete
after three months of immersion in sea-water.

Fig. 7. Corrosion traces on the concrete (left) and on the
steel plate (right) embedded in the concrete
containing $NaNO_2$ after five months of immersion in
sea-water.

Fig. 8.   Steel plate embedded in the plain concrete
          after five months of immersion in sea-water
          (no corrosion trace).

width;
    (b) the chloride/nitrite concentration ratio near the
steel reinforcement becomes so high that the corrosion
inhibitor effect of nitrite is therefore vanished;
    (c) the different exposure conditions of steel
reinforcement in cracked areas and those near sound
uncracked concrete cause an electrical potential difference
which is greater in the presence of nitrite.

4   Conclusions

In cracked reinforced concrete specimens immersed in
sea-water the presence of sodium nitrite, supposed to be a
corrosion inhibitor, really makes chloride corrosion more
severe.
    The narrower the crack width, the longer the time
required to record such a negative effect. Since sodium
nitrite does not affect chloride diffusion in uncracked
concrete specimens, one can expect that the above mentioned
negative effects of sodium nitrite could be found even in
uncracked areas but at longer times of immersion.

References

(1) Maievaganam, U.P. (1984), Miscellaneous Admixtures, in Concrete Admixtures Handbook (Ed. V.S. Ramachandran, Noyes Publications, Park Ridge), Chapter 9, pp. 540-547.

(2) Gouda, V.K., Monfore, G.E. (1965), Journal of the Portland Cement Association Research and Development Laboratories, 7, 3, pp. 24-31.

(3) Treadaway, K.W.J., Russell, A.D. (1968), Highways and Public Works, 36, pp. 19-21 and pp. 40-41.

(4) Craig, R.J., Wood, L.E. (1970), Highway Research Record, 328, pp. 77-88.

(5) Rosenberg, A.M., Gaidis, S.M., Kossivas, T.G., Previte, R.W. (1977), A corrosion inhibitor formulated with calcium nitrite for use in reinforced concrete, in Chloride Corrosion of Steel in Concrete (Ed. D.E. Tonini and S.W. Dean, American Society for Testing and Materials), pp. 89-99.

(6) Gouda, K.V. (1966), British Corrosion Journal, 1, pp. 138-142.

(7) Berke, N.S. (1986), ASTM Standardization News, pp. 57-61.

(8) Berke, N.S. (1985), Corrosion, 85, p. 273.

(9) Berke, N.S. (1987), Corrosion, 87, p. 132.

(10) Grønvolt, F.O., Preece, C.M., Arup, H.(1981), Corrosion protection of steel by concrete, in particular by low porosity cement mortars, Proceedings of the International Congress on Metallic Corrosion, Frankfurt/Main, p. 1800.

(11) Verbeck, G.J. (1975), Mechanism of corrosion of steel in concrete, Am. Concr. Inst., Corrosion of Metal in Concrete, ACI Publication SP-49, Detroit.

(12) Collepardi, M., Marcialis, A., Turriziani, R. (1972), Il Cemento, 3, pp. 143-149.

(13) Buenfeld, N.R., Newmann, J.B. (1986), Cement and Concrete Research, 16, pp. 511-524.

INFLUENCE OF PLASTICIZERS ON CORROSION OF REINFORCING BARS IN CONCRETE

F. GOMÀ, J., VIVAR and J. MAURI
Laboratory of the Department of Construction, Polytechnic University of Barcelona, Spain
J. M. COSTA and M. VILARRASA
Department of Physical Chemistry, University of Barcelona, Spain

Abstract
Electrochemical corrosion rates of steel bars embedded in different concretes with and without plasticizers or superplasticizers were measured.

The influence of other properties of concretes such as: cement content, water/cement ratio, initial curing period, concentration of chloride ions in the concrete and exposure conditions, were also considered.

Relative increases in corrosion in different situations are presented and the requirements to maintain low corrosion rates of embedded steel in concretes are given. This work reports the results at early age.
Keywords: concretes; corrosion tests; reinforcing steels.

1 Introduction

In the literature on corrosion of embedded reinforcement in concrete, we have not found studies which consider the effects of organic plasticizers on the electrochemical behaviour of these steel bars including parameters such as the corrosion potential or the corrosion current etc.

The oxidation effects of organic products are particular to the type of molecule involved. It also depends on the type of hydrothermic conditions they undergo which itself is dependant on the type of cure to which the concrete is subjected : omega oxidation of the aliphatic chains, the breaking of molecules by oxidation, etc.

For these reasons it was considered worthwhile to try a thermic cure at temperatures around 50°C .

It is clear that the cure conditions in itself affects the behaviour of passivity of the reinforcements. However, it is also necessary to look into what happens in the presence of different types of plasticizers or superplasticizers most commonly used in this country.

This work only refers to results obtained at early ages after four months.Hence, from these results it is still not possible to calculate the corrosion rate, by the traditional method, from the weight loss of the reinforcements,with enough precision. This will be determined at least a year to give long term results.

This paper presents the results obtained from samples containing the following products : calcium salt of lignosulphonic acids (SLCa), sulphonated naphthalene formaldehyde condensates (SNF), and sulphonated melamine formaldehyde condensates (SMF).

This study takes into account the chloride content and its levels are near recently established standards.

The cement dosage chosen was 300 kg/m^3, but as 200 kg/m^3 is frequently used in reinforced concrete this has also been examined so the results can be compared.

Electrical measurements were carried out on the samples after four months preparation. However, the short — term results already show perceptible differences which justify this report.

We believe these results to be particularly helpful in the pathology of corrosion diagnosis of reinforcements in affected structures as they relate the already known variables to the presence of the type of additive.

2 Experimental

2.1 Materials
The chemical analysis of the cement and steel bars employed,which are normally used, are given in Tables 1, and 2.

Table 1.Chemical composition of cement (I-45-A),(%),

SiO_2	20.2	L.O.I	1.7
Al_2O_3	5.5	Insoluble residue	0.7
Fe_2O_3	3.2		
CaO	63.0	C_3S	52.8
MgO	1.6		
SO_3	3.0	C_2S	16.1
Na_2O	0.10	C_3A	9.2
K_2O	0.80		
free lime	1.3	C_4AF	9.7

Table 2. Chemical composition of steel bars, (%)..

AE 400	C	Mn	Cr	Si	P
	0.18	0.5	0.03	0.20	0.01

The rust of the bars was cleaned in 5% hydrochloric acid, the surface finishing was then dried in a heater and polished with a steel brush. They were painted with epoxy resin so as to leave an exposed surface of 11 cm^2.

The aggregates sand and gravel had a content a soluble chloride content of 0.015 % .

The composition of the cement used is given in Table 1. The chemical composition of the steel bars was the same as the normally employed in corrugated materials. The analysis is given in Table 2.

The composition of the concrete studied is shown in Table 3. The dosage used (an usual mix proportion employed in building structures) was calculated according to FAURY's method (1), and the so called "wall effect" was taken into account so that its porosity was not increased due to this effect. The concrete was compactated by hitting with a bar,according to the standards. A vibration machine was not used.

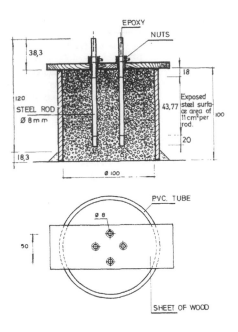

Fig. 1. CONCRETE SAMPLES FOR TESTING

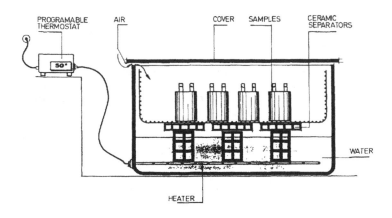

Fig. 2. HEAT-CURING CHAMBER ASSEMBLY

Table 3. Average characteristics values of concretes.

Cement content (kg/m³)	W/C	slump (mm)	porosity (Vol.%)	strength (N/mm²)	Ca(OH)$_2$ (%) carbonated	Ca(OH)$_2$ (%) not carbonated
200	0.65	30	17	18-19	0.8	1.7
300	0.40	30	12	26-27	1.3	2.4

The content of additives were 0.04 % of pure substances referred to weight of concrete.

The minimum contents of chloride ions in concrete were 0.02 % and the samples with chloride addition were 0.10 %, both referred to the weight of concrete.

The samples were prepared without admixtures. 60 samples test distributed in 10 series of six samples each were examined.

They were 100 mm in diameter and 100 mm high cylinders. In the Fig.1 is shown the characteristics of the samples. All were stored in moulds for 24 hours in a chamber with near 100 % R.H. at 20°C with a variation less than 2°C.

After taking them out of their moulds the different series were kept under different conditions: in immersion (at room temperature);in accelerated carbonation chamber, with a constant flow of carbonic gas, at the same temperature at 25 ~ 30 °C, and at 90% of R.H.; steam chamber at 50°C.

The steam chamber in Fig. 2 consisted of a tank of water with a heater controlled by a temperature programmed thermostat. There was no contact between the water and the samples.

2.2 Procedure

Electrochemical measurements were carried out using a Tacussel System assembled with a H.P. computer. A steel rod was used as the auxiliary electrode and a saturated calomel electrode (SCE) resting on a thin wet sponge and placed on the surface of the sample was used as a reference electrode. (2),(3).

The free corrosion potential was measured, and when its change was less than 5 mV / 10 min,the desired polarization curve was measured, applying a scan speed of 10 mV / min.,within the range ± 20 mV with respect to the free

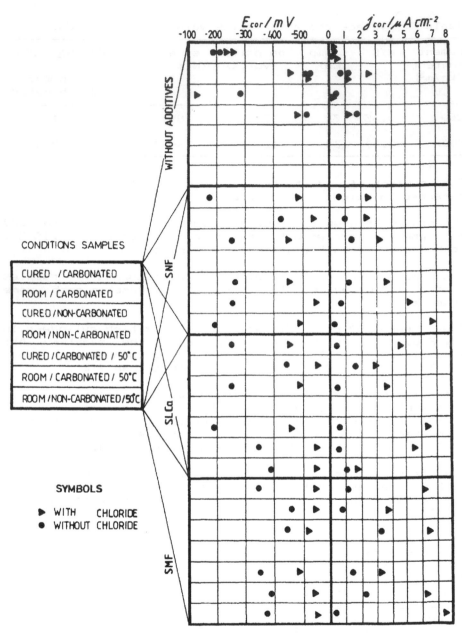

Fig. 3. RESULTS OF CONCRETE CORROSION TESTS.

corrosion potential. From the polarization resistance
obtained, the corrosion current density was calculated
using B = 0.026 V in STERN—GUEARY's equation **(4)**.

3 Results and discussion

The results of the electrochemical behaviour of the samples
is summarised and corrosion potentials and corrosion
current density are given in Fig. 3.

In all cases after 90 days a well stabilized reading was
obtained.

As far as the samples **without additives** were concerned,
they behaved as expected, with corrosion potentials in the
passive zone and very small corrosion intensities.

The results of the quantitative measurements relative to
the **carbonation** are similarly reproduced.

However,there exists a great increase in the corrosion
potential and intensity in the samples which were not
cured, being even more marked in the potentials than in
intensities. In the carbonated samples the intensities were
somewhat greater than the potentials.

In the tests with only **200 kg/m^3** of cement and the
additives **SNF** and **SLCa** the same or similar results were
obtained to those in the samples without additives. That is
corrosion intensities of less than 1.4 μ A. The samples
with **SMF** displayed high disperse potentials between −330,
−460 mV, thus producing a non-passive state.
The corrosion intensities were also appreciably higher than
those of the other additives.

Samples with **additives and chloride ions** had great
corrosive activity reflected by the corrosion potentials
and intensities.

With equal concentration with additives and having
undergone the same conditions of treatment the samples with
SNF and **SLCa** gave similar potential results, although **SLCa**
gave higher corrosion intensities.

Under the conditions studied, the **SMF** gave large
electronegative potentials and higher corrosion intensities
than the other additives.

In respect to **curing conditions** the behaviour of the
different additives was as follows: the potentials of the
samples with **SLCa** and **SNF** had a clear tendency to remain
below 250 mV, and the intensities were less than
1 μ A, when there are a previous cured by immersion.

In the early ages the results of the steam chamber curing effect without previous immersion coincided with those immersion curing at room temperature. This is important as the steam chamber causes a complementary curing effect.

The samples with **SNF** or **SLCa** displayed a similar behaviour to that the other additives. There was an appreciable dispersion in the potential in the uncured and uncarbonated potentials,but without proportionally increasing the corrosion intensity.

The favourable effect of immersion curing was counteracted in the samples which contained chloride due to the attack of this ion on the reinforcements in humid environments. This was demonstrated more in the corrosion intensities than in the potentials.

The previous curing by immersion of the samples gave the most favourable effect for the electrochemical behaviour.

As happens with other properties, immersion curing produces the best electrochemical behaviour.We would interpret this as an increase the SCH phases in cement paste thus the steel is covered better and as CH, also increases, a true cathodic protection is given.

Ageing at the temperature of 50 °C

The samples with **LSCa** and **SNF** ageing at 50 °C did not display a disperse negative effect. On the contrary, a post-curing effect was seen in those which had not been previously immersion cured. However, **SMF** caused a greater electronegative potential and slightly increased intensities. The most unfavourable cases occurred in those which had not been cured,carbonated or aged.

Ageing with chloride ions

Due to the hot, humid ageing process, samples with chloride concentrations of 0.10 %, referred to the weight of concrete, gave particularly unfavourable results in all cases,especially marked in the corrosion intensities.

Carbonation with chloride ions

In the previous cured tests,no large effect of carbonation can be observed, either in potentials or in intensities, but there is a notable influence in the samples which are not cured, especially on the intensities of corrosion.

The favourable behaviour of the concretes which are in accordance with the new EH-88 Standard with respect to corrosion can be confirmed. Therefore by following the standards as to the quantity of portland cement, W/C relation, cover curing,and chloride ion limit, the samples with cement only gave potential values in the passive zone and the corrosion current is very small.

Samples with 0.04 % (referred to weight of concrete) of **LSCa** and **SNF**, and without chlorides are showing good electrochemical behaviour at early age.

4.Conclusions

-Immersion curing is the main factor in obtaining a state of passivity in the reinforcements.

-Results at early ages show that samples without initial immersion curing had a considerably greater degree of corrosion.The potentials were higher than the corresponding intensities.

-At early ages,the **SLCa** and **SNF** additives do not modify the above results although **SMF** appears to produce an increase in corrosion and gives more disperse values under the conditions used.

-The steam chamber produces a noticeable curing effect but it is inferior to that of immersion.

-The worst results are given by the effect of carbonation, aged in a steam chamber without previous immersion.

-We propose that the determination of the calcium hydroxide remaining in the concrete should be done quantitatively by selective attack with ethylene-glycol at 70 °C, rather than a qualitative determination of colouring with phenolphthalein as in commonly used. **(5)**.

References

1. Faury, J. (1958) Le béton Dunod, Paris.
2. Andrade,C., Gonzales,J.A.: (1978)Quantitative measurements of corrosion rate of reinforcing steels embedded in concrete using polarization resistence measurements. Werkstoffe und Korrosion,Vol. 29 ,pp 515-519.

3. ASTM C-876- (1982) Standard test method for half cell potentials of reinforcing steel in concrete,14,concrete and minerals aggregates.

4. Stern,M., Geary,Al L. (1957) Electrochemical Polarization,I.A theoretical analysis of the shape of polarization curves..Journal of the electrochemical society
   Vol.104,Jan. pp.56-63.
5. Gomà,F.(1989) The chemical analysis of hardened concrete
   containing fly ashes,slags,natural pozzolans ,etc.
   Third CANMET/ACI International Conference on Fly
   Ash,Silica Fume,Slag,and Natural Pozzolans in
   concrete. Supplementary Papers pp 828 - 845.
   Trondheim NORWAY.

EFFECT OF CALCIUM NITRITE AND SODIUM MOLYBDATE ON CORROSION INHIBITION OF STEEL IN SIMULATED CONCRETE ENVIRONMENT

B. B. HOPE
Department of Civil Engineering, Queen's University at Kingston, Ontario, Canada
A. K. C. IP
Trow Consulting Engineers, Brampton, Ontario, Canada

Abstract
The synergistic effect of calcium nitrite and sodium molybdate on corrosion of steel specimens in oxygenated lime water containing sodium chloride has been investigated. The combined inhibitor consisted of four and one half parts of calcium nitrite and one part of sodium molybdate. Visual inspection and electrochemical methods were used to evaluate the effectiveness of the inhibitor.

The preliminary test results showed that the combination of calcium nitrite and sodium molybdate effectively protected the steel specimens against corrosion when the ratio of calcium nitrite and sodium molybdate to chloride ions was about 1:11. The combination of calcium nitrite and sodium molybdate appeared to be more effective in corrosion protection than calcium nitrite alone.
Key words: Corrosion, Corrosion inhibitor, Calcium nitrite, Sodium molybdate, Chloride.

1 Introduction

Corrosion of steel in concrete structures, mainly highway bridges and parking structures, is a major problem in many Canadian and U.S. cities where deicing salts are used during the winter months. In Canada, it is currently estimated that the cost of restoration of highway bridges and parking structures damaged by the corrosion is more than ten billion dollars.

The pH of concrete is approximately 12.5. Under this high alkaline environment, an oxide film is developed on the surface of reinforcing steel embedded in concrete and this inhibits corrosion of the steel. However, when a sufficient amount of chloride ions from the deicing salts penetrate through the concrete to the steel surface, the protective film is destroyed, and the steel corrodes when oxygen and moisture are present. The corrosion products occupy several times the volume of the steel and this expansion in volume creates substantial internal stress which eventually cause the concrete to crack, spall or delaminate.

Corrosion inhibitors have been studied for protecting reinforcing steel against corrosion by Arber and Vivian (1961), Berke (1986), Berke and Stark (1985), Chen et al (1983), Craig and Wood (1970), Gonzalez et al (1980), Gouda (1966 and 1970), Gouda and Monfore (1965), Hartt and Rosenberg (1980), Hope and Ip (1987), Lewis et al (1956), Lundquist et al (1977 and 1979), Rosenberg et al (1977 and 1979),

Treadaway and Russell (1968), and Virmani et al (1983). A calcium nitrite based corrosion inhibitor is available commercially; it is a fine white powder and is added to the mixing water of fresh concrete. Hope and Ip (1987) examined this calcium nitrite inhibitor for steel samples submerged in oxygenated lime water; they found that the calcium nitrite inhibitor was easily dissolved in water and, under the test conditions, it effectively protected the steel against corrosion when the ratio of calcium nitrite to chloride ions was about 1:8.

Molybdate compounds have been widely used as corrosion inhibitors for many applications, such as engine coolants, paints and coatings, metalworking and hydraulic fluids, and cooling waters. A review of molybdate in corrosion inhibition is described by Vukasovich and Farr (1986). Although molybdates have been successfully applied in the above areas, these compounds have not been used for corrosion protection of reinforcing steel in concrete structures to date.

The solubility of different molybdates varies significantly. This was reported by Vukasovich and Farr (1986) and their data is reproduced in Table 1. For example, the solubility of potassium molybdate and sodium molybdate are 65 and 39 grams/100 grams of water, respectively, whereas calcium molybdate is almost insoluble. To protect reinforcing steel in concrete effectively, a corrosion inhibitor must be soluble in concrete mixing water so that the inhibitor can be distributed uniformly throughout the concrete.

Table 1. Properties of some simple molybdates

	MoO_3	Na_2MoO_4	K_2MoO_4	$(NH_4)_2MoO_4$	$CaMoO_4$	$ZnMoO_4$
Formula Weight	143.94	205.94	237.97	195.94	200.02	225.33
Colour	white	white	white	white	white	white
Density, g/cm^3	4.69	3.28	2.34	2.28	4.28	
Crystal Form	orthorhombic	spinel	monoclinic	monoclinic	scheelite	tetragonal
Melting Point °C	795	686	919	40(dec)*	965(dec)*	~1000
Solubility,g/100g H_2O	0.5	39.38	64.57	~74	0,005	0.4

* Decomposes rather than melting

A laboratory investigation was carried out at Queen's University to examine the synergistic effect of calcium nitrite and sodium molybdate on corrosion of steel specimens in oxygenated lime water containing sodium chloride. This paper summarizes the findings of the study.

2 Test program

2.1 Test specimens
The corrosion inhibitor under study consisted of five parts of a calcium nitrite admixture and one part of sodium molybdate. The sodium molybdate was laboratory grade, fine, white powder. The calcium nitrite admixture was the same as that used perviously by

Hope and Ip (1987); it was a fine white powder and contained 90% calcium nitrite. Hence, the combined inhibitor comprised 4 1/2 parts of calcium nitrite and 1 part of sodium molybdate.

Sodium chloride was used as a corrosion initiator in the experiment. It was a laboratory grade, fine granulated, white powder.

A reaction kettle was used to contain the test solution. Four ports were available and were used for a working electrode (steel sample), a platinum counter electrode, a calomel reference electrode, and a gas inlet through which oxygen was introduced to the test solution. The arrangement of the test cell was similar to that described by ASTM (1981). Fig (1) shows a schematic diagram of the test cell.

The steel sample was machined from 19 mm diameter Grade 400 reinforcing steel. Each specimen was 12.7 mm in length and 9.5 mm in diameter and was drilled, tapped, and mounted to an electrode holder. Prior to testing, each steel sample was cleaned with paper towel, polished with 600-grit SiC paper, and rinsed with distilled water. The active surface area of the steel sample was approximately 4.5 cm^2.

The counter electrode was a platinum circular mesh type with a platinum wire extension at one end. It was 30 mm in length and 15 mm in diameter and was mounted to an electrode holder at the wire extension.

The reference electrode was a saturated calomel electrode and was connected with a salt bridge filled with saturated lime water. The salt bridge extended to about 1 cm from the surface of the steel sample.

The test solution was a saturated lime water containing the inhibitor and the sodium chloride. The concentration of the inhibitor remained constant (0.12% by mass of lime water),

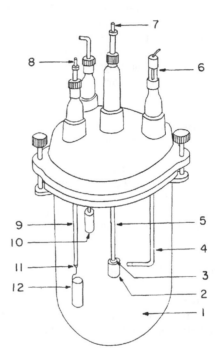

Fig 1. Schematic diagram of test cell
1. Reaction kettle; 2. Steel specimen; 3. Epoxy and rubber o-ring connection; 4. Salt bridge; 5. Glass electrode holder; 6. Reference electrode; 7. Steel rod connected to specimen; 8. Steel rod connected to counter electrode; 9. Glass electrode holder; 10. Gas inlet; 11. Epoxy seal; 12. Counter electrode.

while the amounts of sodium chloride was increased gradually from zero to 2.0% by mass of lime water.

The lime water was prepared from calcium hydroxide powder and distilled water. Precautions were taken to prevent formation of calcium carbonate, which results from the reaction between lime and carbon dioxide in the air. The lime water was filtered into the reaction kettle, and the inhibitor and sodium chloride were added to the filtered lime water. The solution was stirred continuously and saturated with oxygen during testing.

2.2 Corrosion Measurements

Visual inspection, AC Impedance, and half cell potential measurements were used to determine the corrosion conditions of the steel samples. The electrochemical measurement equipment consisted of an EG&G PARC Model 273 potentiostat, and Model 5208 lock-in analyzer, and a Zenith microcomputer. Both the potentiostat and lock-in analyzer were remotely controlled by the computer via a National Instrument PC-2A IEEE interface. A software program was used to control the test procedures and parameters and to analyses the experimental data.

The AC Impedance measurements were performed over frequencies ranging from 10 mHz to 100 kHz. The magnitude of applied AC signal was 10 mV. The total time for a complete set of measurements was approximately 40 minutes.

2.3 Test Procedure

Initially, 0.1% (0.9 g) calcium nitrite admixture and 0.02% (0.18 g) sodium molybdate were added to 900 g oxygenated lime water. After a 2-day conditioning period, 1% (9 g) sodium chloride was added to the solution. After two weeks an additional 0.5% sodium chloride was added to the solution. The concentration of sodium chloride was finally increased to 2.0% after a further 9-week conditioning period.

Visual inspection, half-cell potential, and AC impedance were performed at each concentration level of the inhibitor and sodium chloride.

3 Results and discussion

3.1 Solubility

The calcium nitrite admixture and sodium molybdate were soluble in the oxygenated lime water after stirring, except that a few small white residue particles were observed in the solution.

3.2 Corrosion Inhibition

After the addition of the combined inhibitor, a thin silvery-white film was formed on the steel surface and no rust spots were observed. Steel potentials varied from -50 mV SCE to 90 mV SCE, showing low probability of corrosion.

The coating formed on the steel surface was likely to be a passive oxide layer formed as a result of the chemical reactions between the ferrous ions, nitrite ions and molybdate ions.

After the addition of 1.0% sodium chloride, no rust spots were seen on the steel surface. Steel potentials varied from -210 mV SCE initially to -120 mV SCE at the end of the two-week period, indicating low probability of corrosion.

AC impedance diagram showed a passive film formation, as indicated by a large incomplete semi-circle in Fig. 2. The passive film resistance, given by the chord length of the semi-circle, was estimated to be 600 kohm-cm^2.

When a further 0.5% sodium chloride was added, there was little change on the steel surface. Steel potentials gradually became less negative and varied from -130 mV SCE to -105 mV SCE, showing low probability of corrosion.

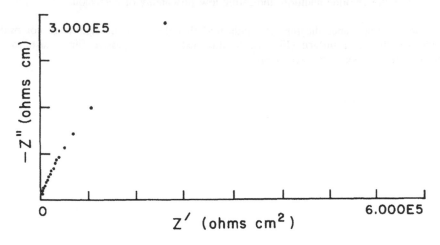

Fig. 2
AC Impedance Spectrum for steel in inhibitor plus 1.0% sodium chloride

The AC Impedance diagram (Fig. 3) was very similar to that shown in Fig. 2, confirming the formation of a passive film on the steel surface.

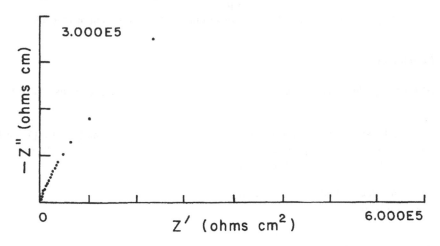

Fig. 3
AC Impedance Spectrum for steel in inhibitor plus 1.5% sodium chloride

When the concentration of sodium chloride was increased to 2% (the ratio of the combined inhibitor to chloride ions was about 1:11), a few tiny rust spots were seen. These rust spots were formed shortly after the sodium chloride concentration was increased to 2.0%. However, the rust spots did not grow and remained the same for the remainder of the time the steel was in the solution.

Steel potentials continuously became less negative and was about -70 mV SCE nine weeks after the chloride addition, indicating low probability of corrosion.

The AC Impedance diagram also indicated that there was no significant corrosion activity on the steel surface (Fig. 4). In this instance the passive film resistance is estimated to be about 2400 kohm-cm^2.

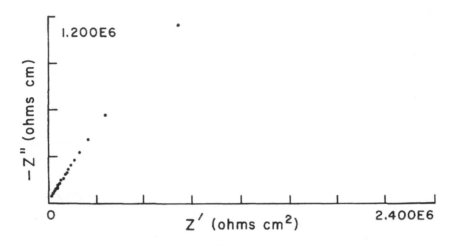

Fig. 4
AC Impedance Spectrum for steel in inhibitor plus 2.0% sodium chloride

4 Conclusions

Based on the preliminary test results from this study, the following conclusions can be drawn:

1. The combination of calcium nitrite (4.5 parts) and sodium molybdate (1 part) effectively protected steel specimens in oxygenated lime water when the ratio of the combined inhibitor to chloride ions was about 1:11.

2. The combined inhibitor appeared to be more effective than calcium nitrite in corrosion protection.

5 Further research

Investigations have been undertaken at Queen's University to verify the test results of this study. Work is currently being performed on reinforcing steel embedded in concrete specimens.

6 Acknowledgements

The authors wish to acknowledge the support of the Natural Sciences and Engineering Research Council which provided funds in the form of an equipment grant and an operating grant for the research described in this paper.

References

Arber, M.G., and Vivian, H.E. (1961) Inhibition of the Corrosion of Steel Embedded in Mortar, Australian Journal of Applied Science, V.12, 1961, pp. 339-347.

ASTM (1981) Standard Reference Method for Making Potentiostatic and Potentiodynamic Anodic Polarization Measurements, ASTM G5-78, Annual Book of ASTM Standards, Part 10, 1981, pp. 864-874.

Berke, N.S., and Stark, P. (1985) Calcium Nitrate as an Inhibitor: Evaluating and Testing for Corrosion Resistance, Concrete International, September 1985. pp. 42-47.

Berke, N.S. (1986) The Use of Anodic Polarization to Determine the Effectiveness of Calcium Nitrite as an Anodic Inhibitor, ASTM STP 906, 1986, pp. 78-91.

Chen, B., Hong, D., Guo, H., and Zhuang, Y. (1983) Ten-Year Field Exposure Tests on the Endurance of Reinforced Concrete in Harbor Works, Cement and Concrete Research V.13, 1983, pp. 603-610.

Craig, R.J., and Wood, L.E. (1970) Effectiveness of Corrosion Inhibitors and Their Influence on the Physical Properties of Portland Cement Mortars. Highway Research Record, No. 328, 1970, pp. 77-88.

Gonzalez, J.A., Algaba, S., and Andrade, C. (1980) Corrosion of Reinforcing Bars in Carbonated Concrete, British Corrosion Journal, V.15, 1980, pp. 135- 139.

Gouda, V.K., and Monfore, G.E. (1965) A Rapid Method for Studying Corrosion Inhibition of Steel in Concrete, PCA Journal, September 1965, pp. 24-31.

Gouda, V.K. (1966) Anodic Polarization Measurements of Corrosion and Corrosion Inhibition of Steel in Concrete, British Corrosion Journal, V.1, 1966. pp. 138-142.

Gouda, V.K. (1970) Corrosion and Corrosion Inhibition of Reinforcing Steel; I. Immersed in Alkaline Solutions; II. Embedded in Concrete, British Corrosion Journal, V.5, 1970, pp. 98-208.

Hartt, W.H., and Rosenberg, A.M. (1980) Influence of $Ca(NO_2)_2$ on Sea Water Corrosion of Reinforcing Steel in Concrete, ACI STP 65-33, 1980, pp. 609-615.

Hope, B.B., and Ip, A.K.C. (1987) Corrosion Inhibitors for Use in New Concrete Construction, Report ME-87-09, The Research and Development Branch, Ontario Ministry of Transportation, December 1987, 23 pp.

Lewis, J.M., Mason, C.E., and Brereton, D. (1956) Sodium Benzoate in Concrete, Civil Engineering and Public Works, V.51, 1956, pp. 881-882.

Lundquist, J.T. Jr., Rosenberg, A.M., and Gaidis, J.M. (1977) Corrosion Inhibitor Formulated with Calcium Nitrite for Chloride Containing Concrete - 2. Improved Electrochemical Test Procedure. National Association of Corrosion Engineers, Corrosion '77 - International Corrosion Forum Devoted Exclusively to the Protection and Performance of Materials, March 1977, Paper 126, pp. 126.1-126.13.

Lundquist, J.T. Jr., Rosenberg, A.M., and Gaidis, J.M. (1979) Calcium Nitrite as an Inhibitor of Rebar Corrosion in Chloride Containing Concrete, Material Performance, V. 18, 1979, pp. 36-40.

Rosenberg, A.M., Gaidis, J.M., Kossivas, T.G., and Previte, R.W. (1977) Corrosion Inhibitor Formulated with Calcium Nitrite for Use in Reinforced Concrete, ASTM STP No. 629, 1977, pp. 89-99.

Rosenberg, A.M., and Gaidis, J.M. (1979) The Mechanism of Nitrite Inhibition of Chloride Attack on Reinforcing Steel in Alkaline Aqueous Environments, Material Performance, V. 18, 1979, pp. 45-48.

Treadaway, K.W.J., and Russell, A.D. (1968) Inhibition of the Corrosion of Steel in Concrete, Highway and Public Works, V.63, 1968, pp. 19-21, 40-41.

Virmani,Y.P., Clear, K.C., and Pasko, T.J. (1983) Time-to-Corrosion of Reinforcing Steel in Concrete Slabs, V.5 - Calcium Nitrite Admixture and Epoxy- Coated Reinforcing Bars as Corrosion Protection Systems, Report No. FHWA/RD- 83/012, Federal Highway Administration, Washington, D.C., September 1983, 71 pp.

Vukasovich, M.S., and Farr, J.P.G. (1986) Molybdate in Corrosion Inhibition - A Review, AMAX Inc. Publication C-140, Michigan, March 1986, 15pp.

INFLUENCE DES ADJUVANTS SUR LA DURABILITE DU BETON

(Influence of admixtures on the durability of concrete)

M. MAMILLAN
CEBTP, St-Rémy-Lès-Chevreuses, France

Résumé

Un important programme de recherches effectuées au cours de ces 3 dernières années a permis de mettre en évidence l'influence bénéfique de l'incorporation d'adjuvants sur les propriétés du béton : en particulier sur sa durabilité. Les études expérimentales ont été réalisées sur des bétons de différentes compositions (nature et dosage du ciment... etc.) avec plusieurs catégories d'adjuvants : fluidifiants, fumées de silice, retardateurs. Tous ces bétons avaient une maniabilité très élevée pendant la première heure.

En plus des propriétés mécaniques (résistance en compression et en traction) l'étude a été concentrée sur la durabilité, en approfondissant l'interprétation des résultats : Le rôle de la microstructure du béton : porosité accessible à l'eau, absorption par capillarité, pouvoir de rétention ... ont été mis en évidence.

La durabilité a été déterminée par l'étude du comportement des éprouvettes à des cycles de gel dans l'air et de dégel dans l'eau (selon les Recommandations de la RILEM 4.C.D.C.). L'intérêt de cette technique expérimentale réside dans la précision, la reproductibilité de la mesure, mais surtout dans la possibilité d'une interprétation nuancée des résultats. Une longue expérience nous a permis d'établir une corrélation digne de confiance entre le classement des résultats d'essais obtenus en laboratoire et les références réelles de tenue dans le temps des matériaux utilisés dans des édifices français.

Mots clefs : Durabilité, Porosité, absorption capillaire, teneur en eau critique.

1 Introduction

Un important programme de recherches expérimentales, sur le béton à haute performance a été réalisé au cours de ces trois dernières années au C.E.B.T.P. Le but était de mettre en évidence l'influence de l'incorporation d'adjuvants pour améliorer les propriétés du béton. Le rôle des fluidifiants, fumées de silice et retardateur a été déterminé sur plusieurs compositions de béton : aussi bien à l'état frais (mesure de la rhéologie du mélange pendant la première heure) et à l'état durci évolution dans le temps des résistances mécaniques. Ce rapport est spécialement destiné à montrer l'amélioration de la durabilité, en étudiant, pour interpréter les résultats, les propriétés liées à la micro-

structure (porosité globale) et la détection des vides accessibles à
l'eau par pénétration capillaire. L'intérêt économique du développe-
ment de ces bétons à très haute compacité est justifié non seulement
par la possibilité de réduction des sections, la résistance mécanique
est trois fois supérieure à celle des bétons réalisés il y a une dou-
zaine d'années, mais surtout par la très importante amélioration de la
durabilité. La mauvaise qualité du béton de recouvrement des armatures,
confectionné depuis ces cinquantes dernières, a montré que l'enrobage
des aciers constituait jusqu'à ce jour, le point faible des ouvrages
en béton armé exposés à l'air extérieur. Ce béton de peau trop poreux
et trop perméable constitue une enveloppe facilementpénétrable par
l'humidité, la vapeur d'eau, le gaz carbonique... etc..etc., ce qui
engendre rapidement la corrosion des armatures. De mêem, de nombreux
exemples de béton détruit par l'effet du gel confirment que des pro-
grès dans cette voix étaient nécessaires, pour réduire le coût de la
maintenance si on souhaite que le béton de ciemnt hydraulique continue
d'être le matériau porteur le plus compétitif pour des constructions
de l'an 2.000.

2   Conditions d'exécution de l'étude expérimentale

2.1 Les matériaux étudiés
L'étude expérimentale a été réalisée avec une composition granulométrique continue, constituée de granulats silico-calcaire de Seine. Les
liants utilisés étaient des Ciments CPA à Haute Résistance, dosé à
350 kg, 400 kg et 425 kg/m³. La composition granulométrique était la
suivante :
Gravillon   5/8   mm   5 %   Sable   0 – 1,6 mm   31 %
            8/12,5 mm   22 %   1,6/5   7 %
           12,5/20 mm   31 %

2.2 Adjuvants étudiés
Les fluidifiants étaient des mélanges à base de résine de mélamine
et de naphtalène sulfoné présenté sous forme liquide contenant 30 %
d'extrait sec. La fumée de silice, avait une teneur en silice de
l'ordre de 84 %. Ce condensat extrêmement fin (de 0,01 microns à
qualques microns) a pour rôle de remplir les vides existant entre les
grains de ciment.
L'activité pouzzolamique de la fumée de silice permet aussi la créa-
tion de C.S.H. supplémentaire qui favorise la fixation de la chaux
par la silice. L'ajout d'un retardateur à base d'hydroxy-carboxyde à
faible dosage (0,4 % du poids de ciment), a été nécessaire pour con-
server à 20°C la maniabilité du béton pendant 1 heure à 20°C. Le slump
était fixé à 20 cm $=$ 3 (norme NF-P.18.451).
L'ensemble des dosages étudiés est donné sur le tableau 1

2.3 Méthodes expérimentales
La confection des éprouvettes et des essais de résistances ont été
réalisés selon les prescriptions normalisées (norme NF-P.18.403 -
P.18.421). Les essais de durabilité par mesure directe de résistance

Tableau 1.

Nature du ciment	Dosage kg/m³	Référence	Dosage en fluidifiant % du poids de ciment	Dosage en fumée de silice % du ciment	Rapport $\dfrac{E}{C + \text{fumée de silice}}$
CPA HP		T	0	0	0,5
		A	2	0	0,339
prise mer	425	B	2,5	6	0,290
		C	3	10	0,283
Usine SV		D	5	15	0,252
CPA HP	400		2	0	0,364
			2,5	6	0,317
			3,5	10	0,306
			5	15	0,275
Usine O			2	0	0,414
			2,5	6	0,369
	350		3,5	10	0,353
			5	15	0,316

aux cycles de gel-dégel ont été réalisés selon les modalités suivantes : Les éprouvettes prismatiques de 7 x 7 x 28 cm ont été conservées
en salle humide à 20°C jusqu'à 28 jours. Après cette durée de conservation, les 5 éprouvettes (par composition de béton) ont été soumises
à des cycles de 6 heures. Dans l'air à - 15°C et immergées dans l'eau
à + 5°C. Périodiquement un contrôle de l'évolution de la qualité est
effectué par des essais non destructifs : variation de masse, variation
de volume (par pesées hydrostatiques) et mesure de la fréquence fondamentale de résonance (norme NF-B.10.513).

2.4 Résultats des essais de durabilité
L'évolution de la qualité (par la mesure de la fréquence fondamentale
de résonance) a été étudiée pendant les cycles de gel-dégel. Le pourcentage de variation de fréquence est porté en ordonnées en fonction
du nombre de cycles portés en abscisse.
Les résultats de l'étude réalisée avec un ciment CPA HP dosé à 425 kg/m³
sont portés sur la figure 1. On voit que le béton témoin T (sans adjuvant présente après 300 cycles une réduction de qualité de 10 %. De
même, la composition A (contenant 2 % de fluidifiant) présente un comportement analogue. La perte de cohésion correspondant à 5 % de réduction de la fréquence de résonance est atteinte après 200 cycles, elle
atteint 8 à 10 % après 300 cycles. Par contre, les 3 compositions contenant de la fumée de silice et du fluidifiant présentent un excellent
comportement pendant 400 cycles. Au-delà de cette limite les bétons
D (10 % de silice) et C (15 % de silice) subissent une légère réduction de qualité (2 %).
A ce stade, l'examen attentif de la surface permet de déceler à quelques emplacements une microfissuration de pâte qui englobe certains
granulats, plus poreux, de composition minéralogique plus riche en
calcaire qu'en silice. Cette observation confirme que la durabilité
du béton, à long terme, même avec une pâte de liaison très compacte
risque d'être vulnérable, car elle est limitée par les propriétés

d'absorption d'eau, due à la microporosité des granulats qui sont en-
globés par cette pâte.

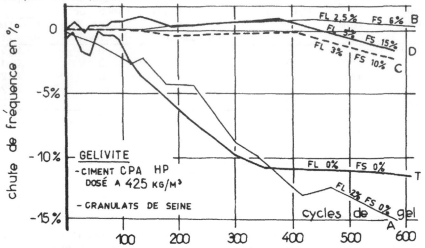

Fig. 1. Evolution de la fréquence de résonance en fonction du nombre
de cycles.

Les résultats de l'étude effectuée avec un béton dosé à 350 kg de ciment CPA HP (Usine O) sont présentés sur la figure 2. Le béton témoin
sans adjuvant atteint après 240 cycles une diminution de qualité de
16 %, mais avec 6 % de fumée de silice la réduction après 400 cycles
est très faible (3 %).

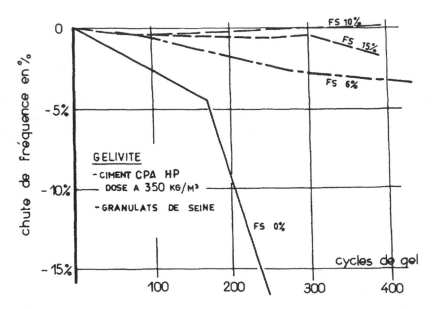

Fig. 2. Evolution de la fréquence en fonction des cycles de gel-dégel.

Sur la figure 3, les résultats concernent la composition contenant 400 kg/m³ de ciment CPA HP (Usine 0). L'influence de l'incorporation de fumée de silice (6, 10 et 15 %) est toujours confirmée. Aucune réduction de qualité après 400 cycles. Le béton sans fumée de silice mais avec 2 % de fluidifiant s'altère après 200 cycles.
D'une façon générale, il est possible de conclure que la durabilité des bétons avec fumée de silice et fluidifiant est nettement améliorée. Les résultats obtenus avec du ciment CPA HP provenant de 2 usines différentes sont identiques. Même le béton qui n'est dosé qu'à 350 kg de ciment par mètre cube est aussi durable. Au point de vue économique, pour la composition étudiée,un dosage en fumée de silice de 6 % paraît suffisant

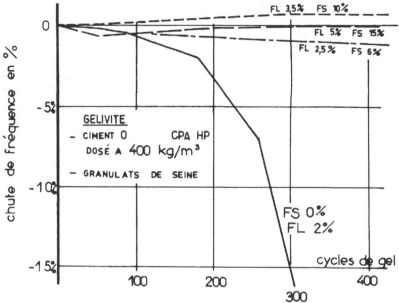

Fig. 3. Evolution de la fréquence en fonction des cycles de gel-dégel.

2.5 Correspondance des essais de gel par cycles accélérés et tenue réelle in situ.
A titre comparatif nous présentons figure 4 le comportement, à ces cycles de gel-dégel, de quelques qualités de pierre calcaire utilisée en France pour la construction de monuments. Comme nous connaissons le comportement réel depuis plus d'un siècle d'exposition aux intempéries dans des positions bien définies, nous avons pu établir la corrélation entre la tenue réelle et la résistance à ces cycles accélérés de gel-dégel en laboratoire. Sur cette base, nous avons proposé une interprétation des résultats avec 4 degrés de sévérité différents d'exposition : l'élévation, le rajaillissement, le couronnement et le dallage extérieur.
Par exemple, la pierre de Souppes qui résiste à plus de 240 cycles est exposée à l'Arc de Triomphe de Paris depuis plus de 150 ans.

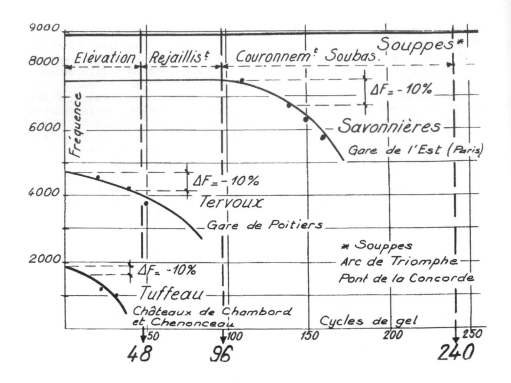

Fig.4. Evolution de la fréquence de résonance en fonction du nombre de
cycles de gel pour des pierres calcaires de référence d'emploi
connue.

3   Interprétation des résultats

3.1 Porosité.
L'explication physique de l'excellente durabilité des bétons avec
ajout de fumée de silice réside essentiellement dans la réduction de
leur porosité (la porosité mesurée selon la norme NF-B.10.500, exprime
le rapport du volume des vides accessibles à l'eau sous vide au volume
apparent). La diminution des vides accessibles à l'eau par absorption
capillaire, a été mise en évidence par cette étude ; cela justifie
l'amélioration de la durabilité.
Les valeurs de porosité des bétons contenant 425 kg de CPA HP par
mètre cube sont présentées sur la figure 5 pour les 5 compositions. On
voit que la porosité varie de 13 % (béton témoin) à 5,3 % pour le béton contenant 15 % de fumée de silice et 5 % de fluidifiant. Il faut
souligner que tous ces bétons ont été réalisés à maniabilité constante
(affaissement 20 cm).

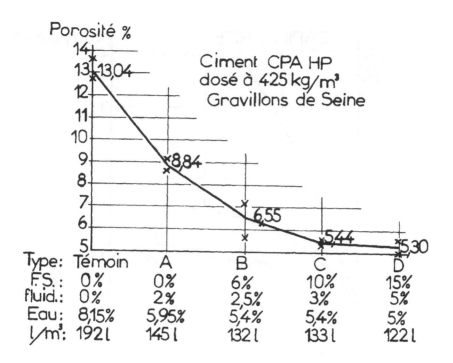

Fig. 5. Evolution de la porosité en fonction des dosages en fumée de
silice.

3.2 Absorption d'eau par capillarité
La diminution de l'absorption d'eau par capillarité qui a été mise en
évidence par cette étude, explique l'amélioration de la durabilité.
Les résultats sont présentés sur la figure 6, la quantité d'eau absor-
bée par remontée capillaire après plus de 13 jours se stabilise. Le
béton témoin sans adjuvant absorbe 4 % d'eau par rapport à son poids
sec. Cette quantité représente un remplissage de 82 % des vides accessibles sous vide. Le béton A, sans fumée de silice (mais avec 2,5 % de
fluidifiant) absorbe 1,6 % d'eau (soit 45 % du volume des vides).
Toutes les compositions avec de la fumée de silice (B, C, D)n'absorbent
que 1 % d'eau, ce qui correspond à 40 % du volume des vides.

3.3 Absorption pendant les cycles de dégel dans l'eau
Pendant la durée des cycles de dégel dans l'eau, le suivi des variations de masse nous a permis d'interpréter ce phénomène ; les résultats sont portés sur la figure 7.

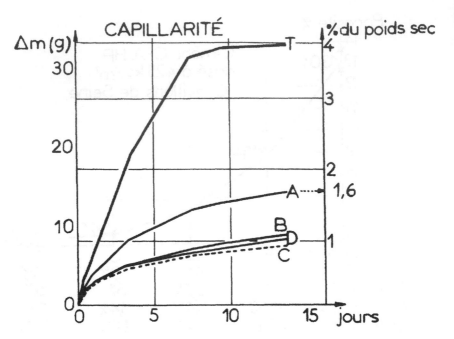

Fig. 6. Absorption d'eau par capillarité en fonction du temps.

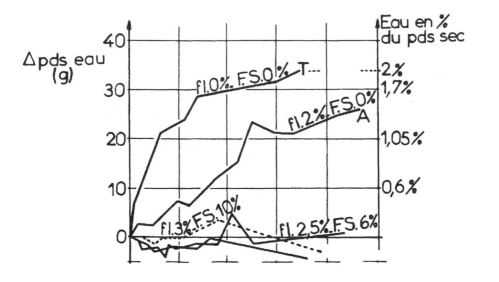

Fig. 7. Absorption pendant les cycles de dégel en eau.

On observe que le béton témoin et le béton A (sans fumée mais avec fluidifiant) accroissent leur teneur en eau, durant le dégel par immersion. Par contre, les 3 compositions avec fumée de silice se stabilisent à poids constant. Ce qui confirme que la diminution de la qualité est due à l'accroissement de la quantité d'eau dans les vides. Ce mécanisme est connu. Il est attribuable au phénomène de cryosuccion durant le dégel en présence d'eau. Les cristaux de glace ont l'aptitude pendant cet état transitoire, d'accroître la teneur en eau locale par leur propre pouvoir de succion. La lente variation de température que nous avons adoptée pour les modalités expérimentales pour le réchauffement (dégel à + 5°C) est plus néfaste. L'augmentation de la teneur en eau permet d'atteindre plus facilement la teneur en eau critique dans les pores. Ainsi la cinétique de la température durant la phase de dégel est plus représentative de la réalité. Cette technique expérimentale permet de réduire l'effet du choc thermique, qui conduirait plutôt à provoquer des sollicitations dues aux brusques écarts (entre la paroi et le coeur), et aux différences de coefficients de contractions thermiques des corps en contact (béton, glace, eau, air...).
Notre soucis était de reproduire des conditions de transferts dans les pores des matériaux en contact avec de la glace ou de la neige en cours de dégel, prochent des conditions réelles. Les périodes d'alternances gel et dégel sollicitent beaucoup plus les matériaux directement en contact avec de l'eau. Nous pensons que ces conditions expérimentales plus proches de la réalité sont à la base de l'excellente corrélation que nous observons entre la tenue réelle aux intempéries et les résultats obtenus sur éprouvettes en laboratoire.

3.4 La teneur en eau critique dans les pores des matériaux
De nombreux chercheurs [FAGUERLUND 1977, HELMUTH 1960, LITVAN 1981, POWERS 1945] ont proposé différentes théories pour expliquer l'action dangereuse due à l'existence de la teneur en eau dite "critique", qui provoque l'expansion des matériaux. Mais dans la structure poreuse des matériaux, le problème est complexe : la transformation en glace n'est pas uniforme. Pour une température négative donnée, l'eau ne gèle pas immédiatement dans tous les capillaires. Dans les vides des plus gros l'eau se solidifie, d'abord, en s'accroissant de volume, ce qui engendre des pressions hydrauliques dans les autres capillaires. Si les pores sont saturés ou si leurs dimensions sont très faibles (< 0,01 micron) la perméabilité est réduite. La pression exercée par la glace entraîne des tensions considérables, dépassant les limites d'expansion des matériaux minéraux actuels utilisés dans la construction.
Une décohésion de leur structure se produit : des fissures, des éclatements, des ruptures entraînent la ruine du béton.
L'importance de cette teneur en eau critique, qui provoque l'expansion et la destruction des matériaux a été signalée depuis plusieurs années [MAMILLAN 1967, FAGUERLUND 1975]. Des essais en commun réalisés par plusieurs chercheurs dans le cadre de la RILEM ont montré l'intérêt de l'air entraîné. Il a été possible de déduire de l'expérimentation que la différence entre la teneur en eau critique et la teneur en eau accessible par capillarité passe de 2 % [sans air entraîné] à 16 % avec des bulles obtenues par l'ajout d'un entraîneur d'air. Les études et

l'examen de plusieurs cas pathologiques de destruction par le gel ont mis en évidence que les matériaux exposés dans une partie horizontale ou en rejaillissement, sont durables si la morphologie de leur structure contient des bulles d'air de 50 microns sans espacement excessif (de 300 à 500 microns) ou si leur porosité est faible (inférieure à 7 %) comme cette recherche l'a montré.

4 Conclusions

L'amélioration de la durabilité par l'incorporation d'adjuvants dans le béton a nettement été mise en évidence. La combinaison de l'utilisation d'un ciment C.P.A. HP, avec un fluidifiant, de la fumée de silice, un retardateur permet d'obtenir un béton très durable et facile à mettre en oeuvre. Avec cette composition, on confectionne un béton de recouvrement des armatures très compact. Cette performance confère une très faible perméabilité au béton ainsi la protection des armatures contre la corrosion est assurée, si l'épaisseur est de 2 cm pour les parements exposés aux intempéries, aux condensations ou en contact d'un liquide. Pour les ouvrages exposés aux embruns marins, l'enrobage des aciers doit être au moins de 4 cm.
Le respect de ces mesures préventives est une garantie de la pérennité des ouvrages en béton armé.

Références
FAGUERLUND G. Studies of the destruction mechanism et freezing of porous materials - Congress FFEN. Le Havre 1975.

FAGUERLUND G. The critical degree of saturation method of assessing the freeze/thaw resistance of concrete. Materiaux et construction n° 58 - Juillet/Août 1977.

HELMUTH R.A. Cement pastes Proceeding 4th Symp. Chemistry Cement Washington DC - 1960.

LITVAN Gh. Frost Action in Porous Systems. College International des Sciences de la construction. Novembre 1981.

MAMILLAN M. Influence de la microstructure des roches sur leur durabilité - Colloque "de la Science des Matériaux au Génie des Matériaux de Construction" Septembre 1987.

POWERS A working hypotheses for further studies of frost resistance concrete. J. A.C.I. Volume 16 n° 2 Février 1945.

PROPERTIES OF SUPER-DURABLE MORTARS WITH ADMIXTURES

Y. OHAMA, K. DEMURA, Y. SATOH, K. TACHIBANA and T. ENDOH
College of Engineering, Nihon University, Japan
Y. MIYAZAKI
Mitsui Petrochemical Industries Ltd, Japan

Abstract
The present paper deals with a basic study which is conducted to
develop effective admixtures for improving the durability of
reinforced concrete structures by using a good combination of an
amino alcohol derivative, an alkyl alkoxy silane and a calcium
nitrite. Experimental admixtures are prepared with various
formulations of these chemical compounds. Mortars are prepared with
the experimental admixtures, and tested for strength, water
absorption, chloride ion penetration and carbonation. It is
concluded from the test results that the simultaneous use of the
amino alcohol derivative, alkyl alkoxy silane and calcium nitrite is
effective for producing super-durable mortar with excellent
resistance to water absorption, chloride ion penetration and
carbonation.
Key words: Super-durable mortar, Amino alcohol derivative, Alkyl
alkoxy silane, Calcium nitrite, Strength, Water absorption, Chloride
ion penetration, Carbonation

1 Introduction

In recent years, the rapid deterioration of various reinforced
concrete structures has become a serious social problem in the world.
For the purpose of preventing the rapid deterioration, the
penetration process of water, chloride ions, carbon dioxide and
oxygen into the concrete structures should be significantly retarded
or inhibited. In order to achieve this purpose, Sakuta et al.(1987)
and Weil and Berke(1987) reported the use of amino alcohol derivative
and calcium nitrite as concrete admixtures. On the other hand,
Ohama et al.(1988 and 1989) have reported that the addition of alkyl
alkoxy silane to mortar and concrete results in excellent
waterproofness and resistance to chloride ion penetration.

Mortars are prepared with various contents of an amino alcohol
derivative, an alkyl alkoxy silane and a calcium nitrite, and tested
for strength, water absorption, chloride ion penetration and
carbonation. From the test results, the effectiveness of the
simultaneous addition of the amino alcohol derivative, alkyl alkoxy
silane and calcium nitrite for producing super-durable mortars
is discussed.

2 Materials

2.1 Cement and aggregate

Ordinary portland cement and Toyoura standard sand as specified in JIS (Japanese Industrial Standards) were used in all the mixes. The properties of the ordinary portland cement are shown in Table 1.

Table 1. Properties of cement.

Specific gravity	Blaine's specific surface area (cm^2/g)	Setting time (h-min)		Soundness	Chemical compositions (%)		
		Initial set	Final set		MgO	SO_2	ig. loss
3.16	3310	2-27	3-25	Good	1.7	1.9	1.1

Compressive strength of mortar (kgf/cm^2)		
3d	7d	28d
155	251	412

2.2 Chemical admixtures

An amino alcohol derivative (AM), an alkyl alkoxy silane (AAS) and a 35% aqueous solution of calcium nitrite (CN) were used as chemical admixtures. Their basic functions are listed in Table 2.

Table 2. Basic functions of chemical admixtures.

Type of admixture	Basic function of admixture
Amino alcohol derivative (AM)	Prevention of the penetration of CO_2 and Cl^- due to the adsorption of CO_2 and Cl^- inside mortar and concrete
Alkyl alkoxy silane (AAS)	Inhibition of the penetration of H_2O and Cl^- due to the water-repellent capillaries formed inside mortar and concrete
Calcium nitrite (CN)	Acceleration of cement hydration and the inhibition of corrosion of reinforcing steel bars inside mortar and concrete

3 Testing procedures

3.1 Preparation of specimens

According to JIS R 5201(Physical Testing Methods for Cement), mortars were mixed with the mix proportions as shown in Table 3. Beam specimens 40x40x160mm and 40x40x80mm were molded, and then given a 2-day-20 °C-80%R.H.-moist, 5-day-20 °C-water and 21-day-20 °C-50%R.H.-dry cure.

Table 3. Mix proportions of mortars with AM, AAS and CN.

Mix no.	Cement:Sand (by weight)	Admixture content (wt% of cement)			Water-cement ratio (%)	Air content (%)	Flow
		AM	AAS	CN			
1		0	0	0	78.0	5.7	170
2		1.0	0	0	73.2	5.7	168
3		2.0	0	0	73.2	5.6	171
4		3.0	0	0	73.2	5.2	168
5		0	0.5	0	65.5	14.3	166
6		1.0	0.5	0	67.5	10.2	173
7		2.0	0.5	0	66.2	10.0	169
8		3.0	0.5	0	66.2	10.1	170
9		0	1.0	0	70.5	9.4	170
10	1 : 3	1.0	1.0	0	70.5	12.7	172
11		2.0	1.0	0	70.5	9.8	171
12		3.0	1.0	0	72.5	7.6	173
13		0	0	4.0	73.9	5.2	170
14		1.0	0	4.0	71.2	6.5	169
15		2.0	0	4.0	71.2	6.6	170
16		3.0	0	4.0	71.2	6.6	174
17		0	1.0	4.0	66.2	10.2	169
18		1.0	1.0	4.0	68.7	9.3	170
19		2.0	1.0	4.0	68.7	8.9	174
20		3.0	1.0	4.0	68.7	8.4	172

3.2 Strength tests
The cured beam specimens 40x40x160mm were tested for flexural and compressive strengths in accordance with JIS R 5201.

3.3 Water absorption test
According to JIS A 6203 (Polymer Dispersions for Cement Modifiers), the cured beam specimens 40x40x160mm were dried at 80 °C for 24 hours, immersed in water at 20 °C for 48 hours, and their water absorption was determined.
3.4 Chloride ion penetration test
The cured beam specimens 40x40x80mm, whose both ends, top and bottom surfaces were coated with an epoxy resin paint, were immersed in 2.5% NaCl solution at 20°C for 7 days for chloride ion penetration. After immersion, the beam specimens were split, and the split crosssections were sprayed with 0.1% sodium fluorescein and 0.1N silver nitrate solutions as prescribed in UNI 7928 (Concrete-Determination of the Ion Chloride Penetration). The depth of the rim of each crosssection

319

changed to white color was measured with slide calipers as a chloride ion penetration depth as shown in Fig. 1.

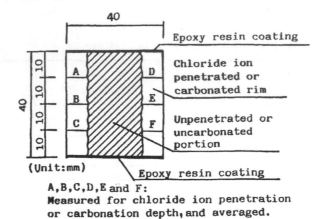

(Unit:mm)

A,B,C,D,E and F:
Measured for chloride ion penetration
or carbonation depth, and averaged.

Fig. 1. Crosssection of specimen after chloride ion penetration or carbonation test.

3.5 Accelerated carbonation test

The cured beam specimens 40x40x80mm, which were coated with an epoxy resin paint like the beam specimens for chloride ion penetration test, were placed in a nonpressurizing carbonation test chamber for 14 days, in which temperature, humidity and CO_2 gas concentration were controlled to be 30 °C, 60%R.H. and 5% respectively. After accelerated carbonation, the beam specimens were split, and the split crosssections were sprayed with 1% phenolphthalein alcoholic solution. The depth of the rim of each crosssection without color change was measured with slide calipers as a carbonation depth as shown in Fig. 1.

4 Test results and discussion

Figs. 2 and 3 represent the flexural and compressive strengths of mortars with AM, AAS and CN. In general, the flexural and compressive strengths of mortars with AM and with AM and CN are higher than those of plain mortar regardless of AM content. In particular, the flexural and compressive strengths of the mortars with an AM content of 2.0% and a CN content of 4.0% are 84kgf/cm^2 and 360kgf/cm^2 respectively, and about 1.3 and 1.5 times those of the

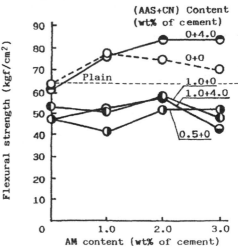

Fig. 2. Flexural strength of mortars with AM, AAS and CN.

320

plain mortar respectively. On the other hand, the flexural and compressive strengths of the mortars with AM and AAS and with AM, AAS and CN are lower than those of the plain mortar regardless of the AM content. This may be due to the fact that the addition of AAS causes an increase in entrained air in the mortars with and without AM as seen in Table 3. Irrespective of the AM content, the 4.0% CN addition causes an increase in the compressive strength of the mortars with AM and AAS, whereas the effect of the CN addition on the flexural strength of the mortars with AM and AAS is hardly recognized. Fig. 4 shows the 48-h water absorption of mortars with AM, AAS and CN. In comparison with plain mortar, the 48-h water absorption of the mortars with AM tends to somewhat decrease with raising AM content. The 48-h water absorption of the mortars with AM and CN is almost the same as that of the mortars with AM regardless of the AM content. Regardless of the AM and AAS contents, the 48-h water absorption of the mortars with AM and AAS and with AM, AAS and CN is considerably lower than that of the plain mortar, and is about 1/4 of that of the plain mortar. This is attributed to the improved resistance to water absorption of the mortars with and without AM because of the water repellency of AAS. However, the effect of the CN addition on the 48-h water absorption of the mortars with AM and AAS is hardly recognized.

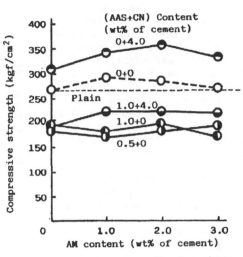

Fig. 3. Compressive strength of mortars with AM, AAS and CN.

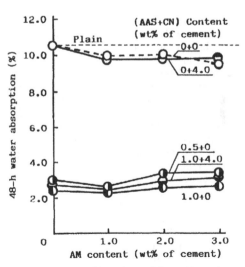

Fig. 4. 48-h water absorption of mortars with AM, AAS and CN.

Figs. 5 and 6 exhibit the weight change and chloride ion (Cl^-) penetration of mortars with AM, AAS and CN, immersed in 2.5% NaCl solution for 7 days. Like their water absorption test results, the

weight change of the mortars
with AM and with AM and CN is
somewhat lower than that of
plain mortar. The weight
change of the mortars with AM
and with AM and CN is markedly
reduced by AAS addition
irrespective of AM content. On
the other hand, the chloride
ion penetration depth of the
mortars with AM and with AM and
CN decreases with an increase
in the AM content. This may
be explained to be due to the
adsorption of chloride ions by
AM in the mortars. Regardless
of the AM content, the chloride
ion penetration depth of the
mortars with AM and CN is
somewhat smaller than that of
the mortars with AM. In
general, the chloride ion
penetration depth of the plain
mortar is considerably reduced
by the addition of AM and AAS
and of AM, AAS and CN. This
barrier effect of the mortars
with AM and AAS on the chloride
ion penetration depth is found
to be based on the chloride ion
adsorption due to AM and the
water repellency due to AAS in
the mortars. However, the
chloride ion penetration depth
of the mortars with AM and AAS
and with AM, AAS and CN increases
with raising AM content. The
reason for this may be due to
a difference in the
microstructures between the
mortars with and without AAS.
Fig. 7 illustrates the
carbonation depth of mortars
with AM, AAS and CN. The
carbonation depth of the
mortars with AM tends to
decrease with raising AM

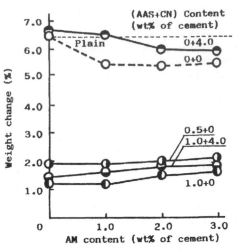

Fig. 5. Weight change of mortars with
AM, AAS and CN, immersed in
2.5% NaCl solution for 7 days.

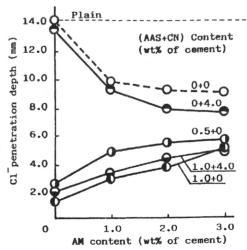

Fig. 6. Cl⁻penetration depth of mortars
with AM, AAS and CN, immersed in
2.5% NaCl solution for 7 days.

content because AM absorbs carbon dioxide in the mortars, and the
carbonation depth at an AM content of 3.0% is about 1/2 of that of
plain mortar. The carbonation depth of the mortars with AM and CN is
almost the same as that of the mortars with AM and with AM, AAS and
CN irrespective of the AM content. The carbonation depth of the
mortars with AM and AAS is much larger than that of the mortars with

AM except for an AM content of
3.0%. However, the carbonation
depth of the mortars with AM
and AAS is markedly decreased
by the 4.0% CN addition.

5 Conclusions

The conclusions obtained from
the above test results are
summarized as follows:

(1) The flexural and
compressive strengths of
mortars with AM and with
AM and CN are higher than
those of plain mortar, but
tend to somewhat decrease
with raising AAS content.

(2) The resistance to water
absorption and chloride ion
penetration of mortars with
AM and with AM and CN and of
plain mortar is remarkably
improved by AAS addition.

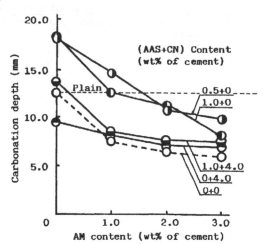

Fig. 7. Carbonation depth of mortars with
AM, AAS and CN, exposed to air
with CO_2 gas concentration of 5%
at 30°C and 60% R.H. for 14 days.

(3) The carbonation depth of mortars with AAS and with AAS and CN is
considerably decreased with increasing AM content.

(4) Accordingly, the simultaneous use of AM, AAS and CN is effective
for producing super-durable mortars with excellent resistance to
water absorption, chloride ion penetration and carbonation. The
corrosion-inhibiting property of the mortars with AM, AAS and CN
is tested at present.

References

Sakuta, M., Urano, T., Izumi, I., Sugiyama, M. and Tanaka, K.(1987)
    Measures to Restrain Rate of Carbonation in Concrete. Concrete
    Durability, Katharine and Bryant Mather International Conference,
    SP-100, V.2(ed. J.M.Scanlon), American Concrete Institute,
    Detroit, pp.1963-1977.
Weil, T.G. and Berke, N.S.(1987) Calcium Nitrite: Effective Corrosion
    Protection for Prestressed Concrete. Concrete Products, (9),
    34-35,42.
Ohama, Y., Demura, K., Satoh, Y., Tachibana, K. and Miyazaki, Y.
    (1988) Properties of Polymer-Modified Mortars Containing Alkyl
    Alkoxy Silane (in Japanese). 15th Semento Konkurito Kenkyu
    Toronkai Kenkyu Hokokushu, Semento Konkurito Kenkyukai, Hiroshima,
    Japan, pp.29-34.
Ohama, Y., Demura, K., Satoh, Y., Tachibana, K. and Miyazaki, Y.

(1989) Development of Admixtures for Highly-Durable Concrete.
<u>Proceedings</u> <u>of</u> <u>the</u> <u>Third</u> <u>International</u> <u>Conference,</u> <u>SP-119</u>
(ed.V.M. Malhotra), American Concrete Institute, Detroit,
pp.321-342.

EFFETS DE SUPERFLUIDIFIANTS SUR DES BETONS RESISTANT AUX SULFATES
(Effects of superplasticizers on sulphate-resistant concretes)

Ph. SIMONIN and F. ALOU
Laboratoire des Matériaux de Construction, EPF, Lausanne, Switzerland
M. ENDERLI
Bonnard & Gardel, Lausanne, Switzerland

Résumé
Le présent article décrit les résultats d'une étude sur des bétons à haute performance (BHP) résistant aux sulfates. Trois adjuvants superfluidifiants utilisés en tant que réducteurs d'eau ont permis d'améliorer la plupart des caractéristiques du béton. Trois types de ciment ont été pris en considération dans cette étude : un ciment Portland à faible teneur en C_3A (CPHS) correspondant au ciment type V ASTM, un ciment au laitier (CLK) ainsi qu'un ciment de mélange obtenu avec 75 % de clinker CPHS et 25 % de cendres volantes (CV). La résistance à la compression sur cylindres 160/320 mm à 90 jours sur les bétons "adjuvantés" est supérieure à 50 N/mm² et ceci pour des bétons pouvant être mis en place à la pompe. Les bétons "adjuvantés" avec les ciments CPHS et CPHS+CV présentent un retrait de dessiccation jusqu'à 300 jours légèrement plus élevé que celui des bétons témoins; avec le ciment CLK, les adjuvants n'ont que peu d'influence. En ce qui concerne la pénétration d'eau sous pression et la capillarité, l'effet des adjuvants est très variable selon le type de ciment utilisé : amélioration avec le ciment CPHS, pratiquement pas de modification avec le ciment CPHS+CV, comportement très inégal avec le ciment CLK.
Mots clés : Sulfates, Ciment résistant aux sulfates, Pouzzolane, Laitier, Cendres volantes, Superfluidifiants, Résistance à la compression, Retrait de dessiccation, Capillarité, Pénétration d'eau, Porosité.

1 Introduction

Dans le cadre de la construction du réseau des routes nationales suisses, un tronçon de celles-ci traverse la chaîne montagneuse du Jura. Le tracé choisi entraînera la construction de plusieurs tunnels qui traverseront des couches géologiques à teneur élevée en sulfates. D'autre part, la présence de marnes gonflantes pourrait provoquer des contraintes importantes sur la paroi de l'ouvrage.

Des dispositions constructives particulières ont été prévues par le projeteur et les exigences en ce qui concerne le béton sont sévères. En particulier, le cahier des charges prévoit :

- choix d'un ciment présentant la meilleure stabilité vis-à-vis de l'agressivité des eaux sulfatées
- pompabilité du béton
- rapport E/C aussi faible que possible
- résistances à la compression supérieures à 12 N/mm^2 à 24 heures et à 60 N/mm^2 à un an
- faible perméabilité à l'eau (étanchéité)

La résistance aux sulfates est principalement conditionnée par deux facteurs : le choix du type de ciment et la perméabilité du béton.

Le ciment à faible teneur en C_3A habituellement recommandé pour des ouvrages en contact avec des eaux séléniteuses s'est avéré non satisfaisant dans un certain nombre de cas en Suisse (Houst Y. 1979). Il a donc été envisagé de recourir à d'autres types de ciment, essentiellement des ciments de mélange et de procéder à des essais sur éprouvettes de mortier et de béton en vue de caractériser au mieux les propriétés de ceux-ci.

Au total, onze ciments ont fait l'objet d'essais; tous ont été préparés à base de ciment portland à faible teneur en C_3A et contiennent dans des proportions variables, soit du laitier de haut fourneau, des cendres volantes ou des pouzzolanes. Ces essais ont fait l'objet d'un rapport interne LMC.

L'emploi d'adjuvants du type superfluidifiant a été retenu dans la mesure où ils permettent de réduire de façon significative la teneur en eau du béton, améliorant ainsi sa compacité et par conséquent sa perméabilité. Les paramètres suivants ont été déterminés sur les bétons avec et sans adjuvant :

- teneur en eau et consistance du béton frais
- résistance à la compression à 2, 28 et 90 jours
- retrait de dessiccation jusqu'à 1 an
- absorption capillaire et pénétration d'eau sous pression
- porosité

Les essais se sont déroulés en deux phases :

a) Phase préliminaire dont le but était d'évaluer la réduction d'eau en fonction du dosage en adjuvant en maintenant dans la mesure du possible, la consistance du béton constante.
b) Phase définitive afin de déterminer certaines caractéristiques du béton frais et/ou durci comprenant :
- essai de gâchage : masse volumique apparente, teneur en air, consistance
- résistances à la compression
- mesure du retrait
- comportement en rapport avec l'eau (étanchéité) :
          - capillarité
          - pénétration d'eau sous pression
          - porosité

2 Matériaux utilisés

2.1 Liants

Les concentrations en sulfates des eaux prélevées dans les forages effectués dans les couches géologiques traversées sont de l'ordre de 600 à 1300 mg/l. D'après DIN, de telles concentrations présentent un degré d'agressivité qualifié de fort à très fort, vis-à-vis du béton, ce qui nécessite l'emploi de ciments spéciaux résistant aux sulfates en plus des précautions particulières que l'on doit attacher à la confection des bétons (Delisle J.P. et Alou F. 1978).

L'emploi de ciment du type CPHS à faible teneur en C_3A est généralement recommandé et considéré comme suffisant pour prévenir toute dégradation dans les cas courants. Toutefois, l'expérience a montré que des dégâts ne sont pas à exclure lorsque de tels ciments sont utilisés; c'est pourquoi d'autres types de ciments ont été envisagés.

L'on a donc choisi d'étudier trois types de ciment dont on pouvait estimer sur la base des connaissances acquises, qu'ils présentaient les meilleures caractéristiques potentielles en vue de satisfaire les exigences posées. Ces choix ont été fixés en tenant compte des possibilités réelles de production ou d'approvisionnement. Les ciments suivants ont été retenus :

Ciment C_1 : ciment portland à résistance élevée aux eaux sulfatées de dénomination CPHS selon la norme SIA 215. Ce ciment est comparable au ciment type V des normes ASTM. Avec les limites indiquées précédemment, ce ciment est recommandé pour des ouvrages en contact avec des eaux sulfatées.

Ciment C_2 : ciment de laitier au clinker (CLK) dont la teneur en laitier est supérieure à 80 % en masse.
Selon Adam M. "Ces ciments utilisés traditionnellement dans les travaux en terrains gypseux ou exposés aux eaux séléniteuses, on fait la preuve de leur bon comportement".

Ciment C_3 : ciment de mélange constitué de 75 % de ciment CHPS et de 25% de pouzzolane naturelle ou artificielle.
Selon la norme UNI 9156 ce type de liant présente une très haute résistance à l'action des sulfates.
Dans un premier temps, on a choisi une pouzzolane naturelle d'origine volcanique, activée par un traitement thermique.
Au terme des essais préliminaires, il s'est avéré que les performances des bétons confectionnés avec ce liant étaient inférieures à celles obtenues sur les bétons avec ciment CPHS.

Ciment C_4 : le ciment C_3 décrit ci-dessus n'ayant pas donné satisfaction, on a remplacé cette pouzzolane naturelle par des cendres volantes tout en conservant le même rapport ciment CPHS/pouzzolane (75 % et 25 % respectivement).

Remarque : Pour les ciments C_3 et C_4 on a choisi les proportions de 75 % clinker et de 25 % pouzzolane afin que les bétons, dosés à 400 kg/m³ de liant, contiennent au moins 300 kg/m³ de ciment. Cette valeur de 300 kg/m³ de ciment portland est le minimum exigé par le norme SIA 162 pour les bétons 0/32 exposés à des agents aggressifs (norme SIA 162).

Des essais de résistance aux sulfates ont été effectués sur des mortiers normalisés selon ASTM confectionnés avec les 4 liants. Les résultats ont fait l'objet d'un rapport interne.

2.2 Granulats

Pour la confection de bétons nous avions à disposition des granulats silico-calcaires séparés en 4 composantes : 0/4, 4/8, 8/16 et 16/32. Les analyses granulométriques de ces composantes sont représentées sur la figure 1. Il s'agit d'un mélange de granulats roulés (50 %) et concassés (50 %).

La composition granulométrique (figure 2) a été choisie de manière à obtenir des bétons pouvant être mis en place par pompage.

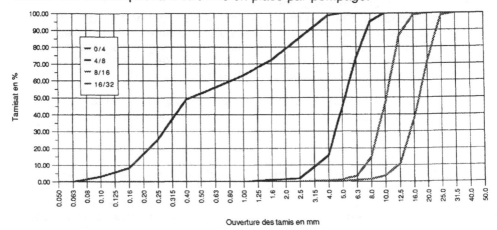

Fig. 1. Analyses granulométriques

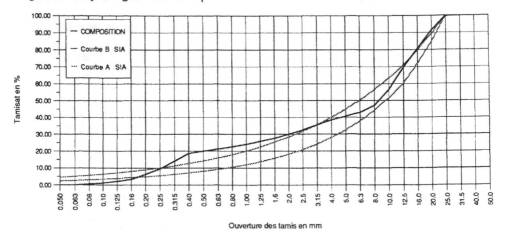

Fig. 2. Composition granulométrique des bétons.

2.3 Adjuvants

Trois adjuvants de type superfluidifiant, provenant de fournisseurs différents, ont été utilisés. Les constituants de base sont :

Adjuvant A_1 : condensat de naphtalène-formaldéhyde sulfonée
Adjuvant A_2 : condensat de naphtalène-formaldéhyde sulfonée
Adjuvant A_3 : condensat de mélamine-formaldéhyde sulfonée et d'un sel d'acide hydroxycarboxylique

La consistance de tous les bétons a été maintenue constante dans la mesure du possible : les adjuvants ont donc été utilisés en tant que réducteurs d'eau.

3 Essais préliminaires

3.1 But des essais

Le but de ces essais était d'évaluer le taux de réduction de l'eau de gâchage en fonction du dosage en adjuvant lorsque la consistance du béton est maintenue constante. Il a donc uniquement été procédé à des essais de gâchage pour mesurer quelques caractéristiques du béton frais et à des essais de résistance à la compression à 2, 7 et 28 jours.

3.2 Résultats des essais

Les résultats figurent sur les tableaux 1 à 4.

Tableau 1. Résultats obtenus sur les bétons sans l'adjuvant (bétons témoins)

| Liant | Dosage de l'adjuvant (%) | Eau (l/m³) | Slump (cm) | Résistance à la compression | | |
				2 jours (N/mm²)	7 jours (N/mm²)	28 jours (N/mm²)
C_1	néant	186	10	20.3	28.9	38.9
C_2	néant	185	12.5	19.4	26.4	36.7
C_3	néant	184	10	15.9	23.8	31.1

Tableau 2. Résultats obtenus sur les bétons avec l'adjuvant A_1.

Liant	Dosage de l'adjuvant (%)	Eau (l/m^3)	Slump (cm)	Résistance à la compression		
				2 jours (N/mm^2)	7 jours (N/mm^2)	28 jours (N/mm^2)
C_1	0.8	177	8.5	29.3	39.5	41.5
	1.0	173	9.0	30.8	40.6	48.3
	1.2	163	8.5	33.8	45.4	53.0
C_2	0.8	158	18.0	30.8	41.3	46.5
	1.0	148	18.0	33.5	46.8	52.5
	1.2	145	>20.0	36.8	47.8	58.2
C_3	0.8	172	9.0	22.3	33.7	38.5
	1.0	166	9.5	24.0	36.1	43.5
	1.2	158	8.0	27.3	37.8	47.0

Tableau 3. Résultats obtenus sur les bétons avec l'adjuvant A_2.

Liant	Dosage de l'adjuvant (%)	Eau (l/m^3)	Slump (cm)	Résistance à la compression		
				2 jours (N/mm^2)	7 jours (N/mm^2)	28 jours (N/mm^2)
C_1	0.8	162	13.5	33.5	44.6	55.0
	1.2	152	10.5	40.5	53.6	56.0
	1.5	148	9.0	41.0	50.9	59.5
C_2	0.8	157	18.0	33.0	46.2	56.0
	1.2	148	>18.0	11.5	51.9	65.0
	1.5	151	>20.0	7.8*	48.2	62.0
C_3	0.8	166	9.0	25.8	33.0	43.8
	1.2	155	9.0	28.0	39.7	51.5
	1.5	153	10.5	25.0	41.7	53.5

* essai à 3 jours; à 2 jours la prise n'était pas terminée

Tableau 4. Résultats obtenus sur les bétons avec l'adjuvant A_3.

| Liant | Dosage de l'adjuvant (%) | Eau (l/m^3) | Slump (cm) | Résistance à la compression | | |
				2 jours (N/mm^2)	7 jours (N/mm^2)	28 jours (N/mm^2)
C_1	0.8	174	8.5	31.6	43.1	49.5
	1.2	174	8.5	32.5	42.5	53.8
	1.5	146	10.5	40.1	58.4	66.0
C_2	0.8	158	12.0	30.2	44.2	51.0
	1.2	145	>20.0	34.1	52.3	61.5
	1.5	139	>20.0	32.5	49.2	60.0
C_3	0.8	176	8.0	23.0	32.9	41.0
	1.2	159	13.0	23.5	38.7	45.5
	1.5	150	>20.0	15.6	40.7	51.8

3.3 Remarques et commentaires

a) Lors du malaxage, on a d'abord introduit la quantité d'eau calculée sur la base de 185 l/m^3 moins la réduction estimée en fonction du dosage de l'adjuvant selon le tableau suivant :

Dosage adjuvant	Réduction	Eau l/m^3
~ 0.8	12 %	162
~ 1.0 à 1.2	18 %	152
> 1.2	23 %	142

L'introduction de l'eau dans le malaxeur s'est faite en deux temps; on a d'abord introduit environ 80 % de l'eau, puis l'adjuvant, et ensuite le 20 % d'eau restant.

Après deux minutes de malaxage, on a effectué la mesure de la consistance au moyen du cône d'Abrams. La valeur de l'affaissement était fixée à 10 cm. Lorsque cette valeur n'a pas été atteinte, le béton a été réintroduit dans le malaxeur puis remélangé en y ajoutant une quantité d'eau supplémentaire de façon à obtenir la consistance souhaitée (voir tableau 5).

En procédant de la sorte, on constate que l'amélioration de la maniabilité n'est pas très importante. Par exemple, si l'on examine les résultats obtenus avec le ciment C_1 et les adjuvants A_1 et A_3, on remarque que l'affaissement au cône n'augmente que de 1,5 à 3-4 cm pour un supplément d'eau de l'ordre de 20 l/m^3. Lors de gâchées de répétition, nous avons pu vérifier que l'efficacité de l'adjuvant, dans le sens où il permet une réduction d'eau par rapport à une consistance donnée, dépend essentiellement de la quantité d'eau de départ. Toute adjonction d'eau supplémentaire après malaxage ne se traduit que par un gain négligeable de maniabilité.

b) L'influence des adjuvants sur la consistance du béton frais varie selon le type de ciment, l'effet le plus prononcé étant obtenu avec le ciment CLK, qui a la particularité, entre autres, d'être très fin (Blaine ~ 4000 cm^2/g). Les deux autres liants ont un comportement sensiblement égal.

c) En ce qui concerne les résistances à la compression, on se bornera à relever qu'à 2 et 28 jours, elles sont suffisantes pour satisfaire aux exigences du cahier des charges à une exception près : les bétons avec ciment CLK et adjuvant A_2 ne durcissent que très lentement. En effet, avec cet adjuvant, dosé à raison de 1,2 %, la résistance à la compression à 2 jours est de 11,5 N/mm^2; avec un dosage de 1,5 %, la résistance à 3 jours n'atteint que 7,8 N/mm^2. Ceci est certainement dû à l'effet retardateur de cet adjuvant.

4 Essais définitifs

4.1 Béton frais
Sur la base des résultats des essais préliminaires, les dosages en adjuvants figurant dans le tableau 6 ont été retenus.

Les caractéristiques des différents bétons frais figurent dans le tableau 7.

4.2 Remarques sur les bétons frais
L'effet fluidifiant le plus prononcé est obtenu avec le produit A_2, ceci indépendemment du type de ciment; les deux autres produits ont un comportement à peu près semblable.

La diminution de l'affaissement mesuré au cône d'Abrams au cours des 20 premières minutes suivant le malaxage est d'autant plus prononcée que la mesure de départ est élevée.

Le liant C_2 à base de laitier combiné avec un superfluidifiant conduit à des bétons nettement plus fluides que les deux autres liants testés. Toutefois, il y a lieu de relever que même avec des valeurs de l'affaissement de l'ordre de 20 cm, le béton peut être difficile à ouvrer, à la main par exemple, car il se produit un auto-serrage par suite de sa fluidité élevée.

Les masses volumiques étant plus élevées que celles obtenues lors des essais préliminaires, le dosage réel en ciment se trouve légèrement augmenté et avoisine 420 kg/m^3. Rappelons que le dosage visé était fixé à 400 kg/m^3.

Tableau 5. Essais préliminaires de gâchage.
Dosage en eau et consistance des bétons.

Liant	Adjuvant			Eau de départ		1ère adjonction		2ème adjonction	
	Marque	Dosage (%)	l/m^3	Slump (cm)	l/m^3	Slump (cm)	l/m^3	Slump (cm)	
C_1	A_1	0.8	161	5.0	169	6.0	177	8.5
	A_2	0.8	162	13.5	--	--	--	--
	A_3	0.8	155	5.0	162	6.0	174	8.5
	A_1	1.0	149	4.5	165	6.5	173	9.0
	A_2	1.2	152	10.5	--	--	--	--
	A_3	1.2	150	7.0	158	5.0	174	8.5
	A_1	1.2	155	6.5	163	8.5	--	--
	A_2	1.5	148	9.0	--	--	--	--
	A_3	1.8	146	10.0	--	--	--	--
C_2	A_1	0.8	158	18.0	--	--	--	--
	A_2	0.8	157	18.0	--	--	--	--
	A_3	0.8	158	12.0	--	--	--	--
	A_1	1.0	148	18.0	--	--	--	--
	A_2	1.2	148	>18.0	--	--	--	--
	A_3	1.2	145	>20.0	--	--	--	--
	A_1	1.2	145	>20.0	--	--	--	--
	A_2	1.5	151	>20.0	--	--	--	--
	A_3	1.6	139	>20.0	--	--	--	--
C_3	A_1	0.8	164	8.0	172	9.0	--	--
	A_2	0.8	166	9.0	--	--	--	--
	A_3	0.8	168	6.5	176	8.0	--	--
	A_1	1.0	166	9.5	--	--	--	--
	A_2	1.2	155	9.0	--	--	--	--
	A_3	1.2	159	13.0	--	--	--	--
	A_1	1.2	158	8.0	--	--	--	--
	A_2	1.5	153	10.5	--	--	--	--
	A_3	1.8	150	>20.0	--	--	--	--

Tableau 6. Dosage des divers adjuvants.

Liant	C_1	C_2	C_4
A_1	1.0 %	0.8 %	0.8 %
A_2	1.2 %	1.0 %	1.2 %
A_3	1.0 %	1.0 %	1.0 %

Tableau 7. Caractéristiques des bétons frais

Adjuvant	Caractéristiques du béton		Liant C_1 CPHS	Liant C_2 CLK	Liant C_4 CPHS+CV
Néant	Dosage réel	kg/m³	411	403	414
	Dosage en eau	l/m³	179	182	165
	E/C		0.44	0.45	0.40
	Consistance : slump	cm	10.5	14	10
	slump après (t)min	cm	--	--	--
	étalement	cm	40	42	39
	Masse volumique	kg/dm³	2.42	2.38	2.42
A_1	Dosage réel	kg/m³	420	417	417
1,0 % C_1	Dosage en eau	l/m³	150	152	145
0,8 % C_2	E/C		0.357	0.365	0.348
0,8 % C_4	Consistance : slump	cm	15.5	>20	8.5
	slump après (t)min	cm	11 (10)	--	--
	étalement	cm	45	51	37
	Masse volumique	kg/dm³	2.44	2.43	2.42
A_2	Dosage réel	kg/m³	414	418	416
1,2 % C_1	Dosage en eau	l/m³	145	150	142
1,0 % C_2	E/C		0.350	0.359	0.342
1,2 % C_4	Consistance : slump	cm	>20	>20	>20
	slump après (t)min	cm	13 (15)	>20 (10)	14 (10)
	étalement	cm	58	56	44
	Masse volumique	kg/dm³	2.41	2.435	2.415
A_3	Dosage réel	kg/m³	419	415	417
1,0 % C_1	Dosage en eau	l/m³	151	149	148
1,0 % C_2	E/C		0.360	0.359	0.356
1,0 % C_4	Consistance : slump	cm	9	>20	10
	slump après (t)min	cm	8 (15)	>20 (15)	6.5 (23)
	étalement	cm	35	54	41
	Masse volumique	kg/dm³	2.44	2.42	2.42

4.3 Bétons durcis
4.3.1 Résistance à la compression

les résistances à la compression ont été déterminées à 2, 28 et 90 jours sur des éprouvettes cylindriques de 16 cm de diamètre et 32 cm de hauteur. Afin de pouvoir établir des comparaisons, nous avons calculé d'une part, pour chaque type de béton et adjuvant, le rapport de la résistance à 2 et 90 jours en pourcentage de celle obtenue à 28 jours, et d'autre part les rapports entre les résistances des bétons "adjuvantés" et celles des bétons témoins, à 2, 28 et 90 jours. Tous les résultats sont regroupés sur le tableau 8 et illustrés graphiquement aux figures 3, 4 et 5.

Tableau 8. Résistances à la compression : résultats.

Adjuvant	Age Jours	C_1 CPHS N/mm^2	% de R_{28j}	% de $R_{témoin}$	C_2 CLK N/mm^2	% de R_{28j}	% de $R_{témoin}$	C_4 CPHS + CV N/mm^2	% de R_{28j}	% de $R_{témoin}$
Néant (témoin)	2	21.8	53	100	20.3	52	100	22.1	61	100
	28	41.3	100	100	39.0	100	100	36.3	100	100
	90	49.3	119	100	44.3	114	100	49.7	137	100
A_1	2	32.3	71	148	27.9	62	137	23.6	57	107
	28	45.3	100	110	45.3	100	116	41.3	100	114
	90	57.2	126	116	52.8	117	119	53.5	130	108
A_2	2	32.1	65	147	18.1	32	89	20.6	46	93
	28	49.7	100	120	56.3	100	144	45.1	100	124
	90	54.7	100	111	49.1	87	111	56.3	125	113
A_3	2	33.5	73	154	29.5	70	145	21.9	57	99
	28	45.8	100	111	42.3	100	108	38.6	100	106
	90	54.5	119	111	53.3	126	120	56.2	146	113

L'examen des résultats permet de faire les commentaires suivants :

- Bétons avec liant C_1 (CPHS)

Les bétons avec adjuvants se distinguent par une forte augmentation de la résistance à 2 jours, de 47 à 54 %, par rapport au béton témoin. L'accroissement de résistance de 2 à 28 jours étant nettement plus prononcé pour le béton témoin, le gain de résistance à 90 jours des bétons "adjuvantés", toujours par rapport au béton témoin, n'est plus que de 11 à 16 %.

- Bétons avec liant C_2 (CLK)

a) Bétons avec adjuvants A_1 et A_3

Ces deux adjuvants permettent d'améliorer la résistance à 2 jours de 37 et 45 % par rapport à celle du béton témoin. A 90 jours, le gain de résistance est encore de 20 % environ.

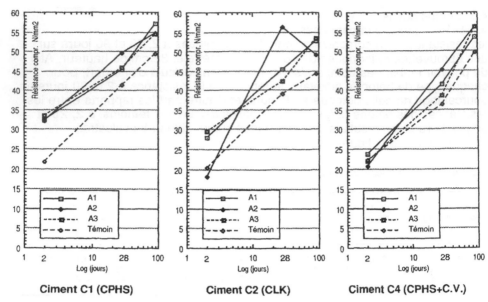

Ciment C1 (CPHS) **Ciment C2 (CLK)** **Ciment C4 (CPHS+C.V.)**

Figure 3. Résistance à la compression : résultats par type de ciment.

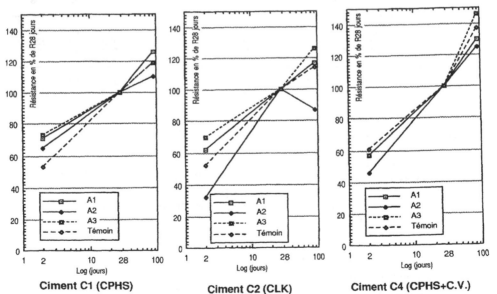

Ciment C1 (CPHS) **Ciment C2 (CLK)** **Ciment C4 (CPHS+C.V.)**

Figure 4. Evolution de la résistance à la compression par rapport à la
résistance à 28 jours (base 100).

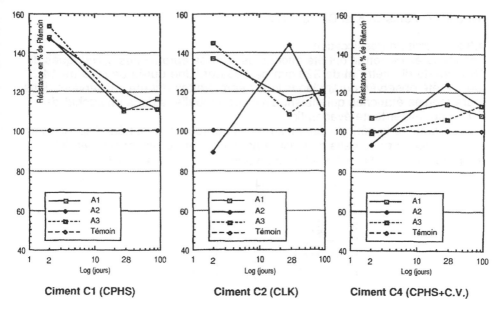

Ciment C1 (CPHS) Ciment C2 (CLK) Ciment C4 (CPHS+C.V.)

Figure 5. Résistance à la compression des bétons adjuvantés par rapport aux bétons témoins (base 100).

b) Béton avec adjuvant A_2
Les résistances obtenues sur le béton gâché avec cet adjuvant présentent par contre deux singularités qu'il convient de relever :
• nous avons observé un retard de prise et de durcissement des bétons. Cet effet est amplifié en raison des spécificités de ce ciment.
• l'accroissement des résistances dans le temps est très élevé entre 2 et 28 jours puis la résistance diminue d'environ 10 % entre 28 et 90 jours.
S'agit-il d'anomalies liées à l'incompatibilité entre le ciment CLK et l'adjuvant A_2 ?

- Bétons avec ciment C_4 (ciment CHPS avec cendres volantes)
L'influence de l'adjonction d'adjuvants sur les résistances est moins marquée avec ce ciment qu'avec les deux autres.
A 2 jours, l'écart entre le résistance du témoin et celle des bétons "adjuvantés" est compris entre -7 et +7 %; à 28 et 90 jours, les écarts sont compris entre +6 et +24 et entre 8 et 13 % respectivement.

- Remarques générales
Si l'on examine indépendemment du type de ciment, l'effet des adjuvants sur les résistances et leur évolution, les remarques suivantes s'imposent :
- Les produits A_1 et A_3 conduisent à des résultats très voisins.
- Le produit A_2 se distingue par son effet retardateur; les résistances à court terme sont donc inférieures à celles des bétons témoins pour les gâchées

avec les ciments CLK et CPHS + CV.

4.3.2 Retrait de dessiccation

Les mesures de retrait ont été effectuées sur des éprouvettes cylindriques de 160 mm de diamètre et de 320 mm de hauteur. Une durée de cure de 14 jours a été observée durant laquelle les éprouvettes ont été conservées dans leurs moules étanches qui étaient d'ailleurs emballées dans un sachet de PVC pour éviter toute évaporation.

Durant toute la période des mesures, les éprouvettes ont été disposées en permanence sur des bâtis de mesure ad hoc (figure 6) et conservées à la température de 19°C ± 1°C et à une humidité relative de 55 ± 5 %.

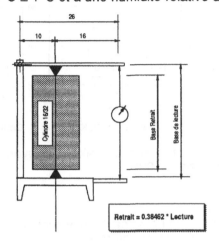

Figure 6. Schéma des bâtis de mesure de retrait.

On relèvera que les déformations mesurées correspondent au retrait axial de l'éprouvette.

Les résultats des mesures effectuées jusqu'à 300 jours sont donnés au tableau 9 et représentés à la figure 7.

On notera que :

- Pour un liant donné, quel que soit l'adjuvant, tous les bétons conduisent à des retraits pratiquement égaux, la concordance des mesures étant même remarquable.

- Les bétons "adjuvantés" avec liants C_1 et C_4 donnent des retraits plus élevés (+ 13 % et + 25 % respectivement) que ceux du béton témoin. Avec le liant C_2, par contre, le béton témoin présente un retrait légèrement supérieur aux bétons "adjuvantés".

On constate donc que, sauf pour le liant C_2 (CLK), l'utilisation des adjuvants, malgré une réduction sensible du rapport E/C, provoque une augmentation du retrait de dessiccation. Ceci peut paraître surprenant à première vue, mais ce fait avait déjà été signalé précédemment (Massazza F. et Testolin M. 1980, Alou F., Ferraris C.F. et Wittmann F.H. 1987).

338

Tableau 9. Retrait en o/oo à 300 jours.

| Liant | Avec adjuvant | | | Témoin |
	A_1	A_2	A_3	
C_1	0.470	0.478	0.470	0.418*
C_2	0.320	0.352	0.317	0.372*
C_4	0.496	0.472	0.494	0.390*

* valeurs obtenues par interpolation

4.3.3 Absorption d'eau par capillarité

L'absorption d'eau par capillarité a été effectuée en se basant sur la norme DIN 56617. Le principe de l'essai est illustré à la figure 8 et les résultats sont représentés sous forme d'histogramme à la figure 9.

On remarque qu'à une exception près (ciment C_2 avec adjuvant A_1), tous les bétons "adjuvantés" présentent, en pourcentage, une absorption capillaire plus faible que celle des bétons témoins. Ce phénomène est plus marqué pour les bétons avec les ciments C_1 (CPHS) et C_4 (CPHS+CV) que pour les bétons avec le ciment C_2 (CLK).

4.3.4 Pénétration d'eau sous pression

Les essais ont été réalisés sur des cylindres 16/32. La pression de 1 MP (10 bars) a été appliquée pendant 48 heures.

Pour mettre en évidence la différence de structure du béton entre le "bas" et le "haut" des éprouvettes provoquée entre autre par le ressuage, l'essai a été effectué sur les deux faces des cylindres après sciage de ceux-ci. Un schéma du dispositif d'essai est indiqué à la figure 10.

Sitôt l'essai terminé, les échantillons sont fendus selon un plan diamétral et le front de pénétration d'eau est relevé. Le résultat est exprimé par la moyenne des profondeurs maximales de pénétration mesurées lors des 4 essais. Ils sont illustrés sous forme d'histogramme à la figure 11.

En comparant les résultats avec ceux obtenus sur les bétons témoins, les points suivants sont à souligner :

Bétons avec ciment C_1 (CPHS)

La profondeur de pénétration est nettement plus faible lorsque les bétons sont "adjuvantés". Les trois adjuvants utilisés produisent des effets semblables.

Bétons avec ciment C_2 (CLK)

Les bétons gâchés avec les adjuvants A_1 et A_2 présentent des défauts d'étanchéité très marqués. L'examen des échantillons après fendage a révélé la présence de petits canaux probablement formés par suite du ressuage de l'eau avant la prise du béton. Ce phénomène pourrait être lié à l'effet retardateur de prise des adjuvants.

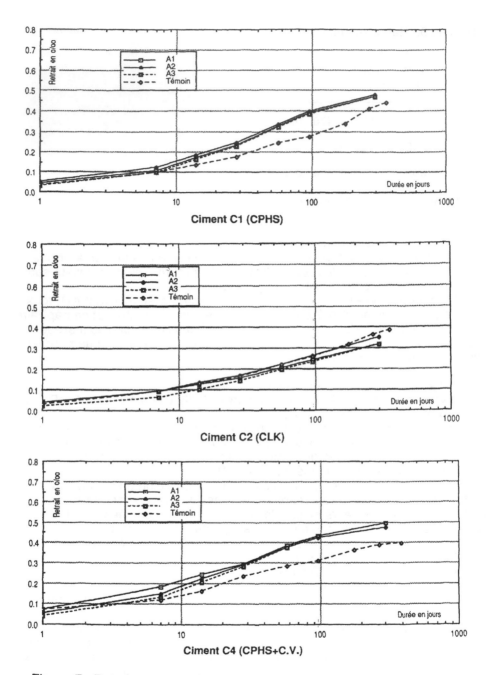

Figure 7. Retraits en fonction du temps.

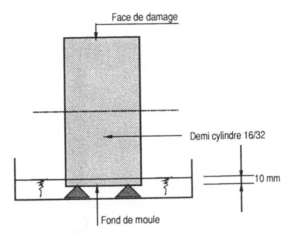

Figure 8. Schéma du dispositif de mesure de l'absorption capillaire.

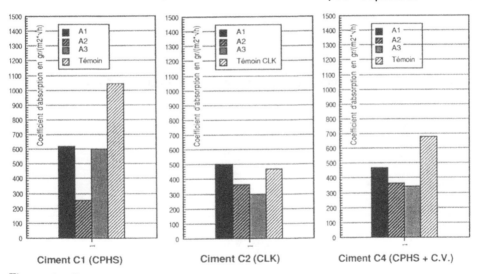

Figure 9. Coefficient d'absorption capillaire.

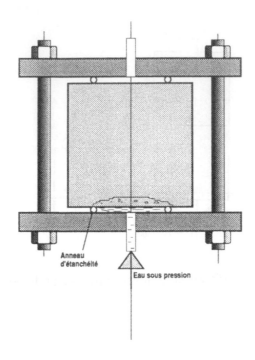

Anneau
d'étanchéité

Eau sous pression

Figure 10. Pénétration d'eau sous pression. Principe de l'essai.

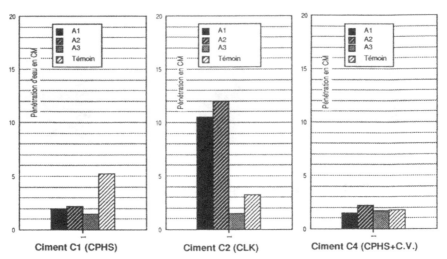

Figure 11. Profondeur moyenne du front de pénétration d'eau.

Bétons avec ciment C_4 (CPHS+CV)
Pour ces bétons, l'emploi des adjuvants n'a pas permis d'améliorer
l'étanchéité de façon sensible car l'adjonction de cendres volantes est déjà en
soi une méthode pour diminuer la perméabilité des bétons. En effet, si l'on
compare les trois bétons témoins, on observe que le béton avec ce ciment est
nettement moins perméable que les deux autres.

4.3.5 Porosité

La porosité accessible à l'eau a été déterminée sur tous les bétons. Les
résultats figurent dans le tableau 10 et sont représentés sous forme
d'histogramme à la figure 12.

Tableau 10. Résultats des mesures de porosité.

| Liant | Porosité en % du volume apparent | | | |
	A_1	A_2	A_3	Témoin
C_1	11.4	13.6	12.3	12.7
C_2	12.4	11.9	12.3	14.5
C_4	12.4	11.3	12.3	12.8

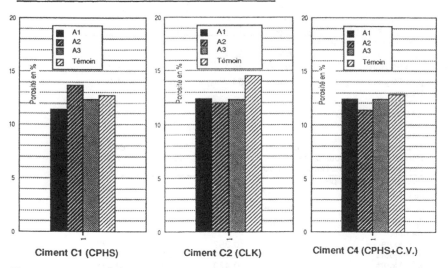

Figure 12. Résultats des mesures de porosité.

On se bornera à relever que d'une façon générale, les porosités relevées sur
les bétons adjuvantés sont légèrement plus faibles que celles des bétons
témoins. Toutefois, il n'est pas possible d'établir des corrélations entre
porosité et absorption capillaire, ou porosité et pénétration d'eau.

4.3.6 Tableau récapitulatif
Tous les résultats des essais sur les bétons durcis figurent dans le tableau 11.

Tableau 11. Résumé de tous les résultats d'essais.

Adjuvant	Caractéristiques du béton		Liant C_1 CPHS	Liant C_2 CLK	Liant C_4 CPHS+CV
Néant	Résistance à la compression :				
	β_p à 48 heures	N/mm²	21.8	20.3	22.1
	β_p à 28 jours	N/mm²	41.3	39.0	36.3
	β_p à 90 jours	N/mm²	49.3	44.3	49.7
	Retrait :				
	à 96 jours		0.273	0.259	0.310
	à 180 jours		0.337	0.319	0.362
	à 300 jours		0.418	0.372	0.390
	Absorption d'eau :				
	Capillarité	g/cm²•√t	1041	469	680
	Pénétration d'eau sous pression	cm	5.2	3.2	1.8
	Porosité accessible à l'eau	% vol	12.7	14.5	12.8
A_1	Résistance à la compression :				
1,0 % C_1	β_p à 48 heures	N/mm²	32.3	27.9	23.6
0,8 % C_2	β_p à 28 jours	N/mm²	45.3	45.3	41.3
0,8 % C_4	β_p à 90 jours	N/mm²	57.2	52.8	53.5
	Retrait :				
	à 96 jours		0.386	0.245	0.432
	à 180 jours		0.448	0.284	0.473
	à 300 jours		0.470	0.320	0.496
	Absorption d'eau :				
	Capillarité	g/cm²•√t	618	501	469
	Pénétration d'eau sous pression	cm	2.0	10.5à4	1.5
	Porosité accessible à l'eau	% vol	11.4	12.4	12.4
A_2	Résistance à la compression :				
1,2 % C_1	β_p à 48 heures	N/mm²	32.1	18.1	20.6
1,0 % C_2	β_p à 28 jours	N/mm²	49.7	56.3	45.1
1,2 % C_4	β_p à 90 jours	N/mm²	54.7	49.1	56.3
	Retrait :				
	à 96 jours		0.396	0.262	0.423
	à 180 jours		0.456	0.310	0.447
	à 300 jours		0.478	0.352	0.472
	Absorption d'eau :				
	Capillarité	g/cm²•√t	253	343	364
	Pénétration d'eau sous pression	cm	2.2	12à2.5	2.2
	Porosité accessible à l'eau	% vol	13.6	11.9	11.3
A_3	Résistance à la compression :				
1,0 % C_1	β_p à 48 heures	N/mm²	33.5	29.5	21.9
1,0 % C_2	β_p à 28 jours	N/mm²	45.8	42.3	38.6
1,0 % C_4	β_p à 90 jours	N/mm²	54.5	53.3	56.2
	Retrait :				
	à 96 jours		0.382	0.234	0.429
	à 180 jours		0.430	0.280	0.455
	à 300 jours		0.470	0.317	0.494
	Absorption d'eau :				
	Capillarité	g/cm²•√t	599	501	343
	Pénétration d'eau sous pression	cm	1.5	1.5	1.7
	Porosité accessible à l'eau	% vol	12.3	12.3	12.1

5 Conclusions

5.1 Concernant la réduction d'eau
Les adjuvants A_1 et A_3, malgré leur base chimique différente, ont pratiquement le même pouvoir réducteur d'eau.

L'adjuvant A2, semblable à A1 du point de vue chimique, permet une réduction d'eau supérieure.

5.2 Concernant la résistance à la compression
- Quel que soit l'adjuvant utilisé, les bétons "adjuvantés" présentent des résistances à la compression supérieure à celles obtenues sur les bétons témoins.
- L'augmentation de résistance est très forte à 2 jours, sauf pour les bétons avec le liant C_4 (CPHS+CV); à 28 jours et 90 jours l'écart en % diminue.
- Les résistances obtenues à 28 et 90 jours sur les bétons "adjuvantés" permettent d'espérer que pratiquement tous ces bétons présenteront à 1 an des résistances égales ou supérieures à 60 N/mm^2; ces bétons peuvent donc être considérés comme des BHP.

5.3 Concernant le retrait de dessiccation
- Les trois adjuvants utilisés ont un comportement à peu près semblable du point de vue retrait.
- Pour les ciments C_1 et C_4 le retrait des bétons "adjuvantés" est supérieur à celui des bétons témoins et la différence atteint même 25 % (C_4).
- Le retrait des bétons témoins, pour une conservation de 55 % HR, est relativement faible, inférieur dans tous les cas à 0,45 o/oo après 1 an, et ceci malgré un dosage en liant élevé de 400 kg/m^3.

5.4 Concernant l'étanchéité
- La capillarité du béton, sauf pour le ciment C_2, diminue fortement lors de l'emploi des adjuvants. Du point de vue de la capillarité, l'adjuvant A_2 est le plus efficace.
- La porosité accessible à l'eau n'est que très faiblement modifiée par l'utilisation des adjuvants.
- Pour le béton avec le liant C_1, la pénétration d'eau est nettement diminuée lors de l'utilisation des adjuvants.

Références

Alou, F., Ferraris, C.F. et Wittmann, F.H. (1987) Etude expérimentale du retrait du béton, Matériaux et Constructions, 20, 323-333.
Adam, M., Guide pratique pour l'emploi des ciments, Ed. Eyrolles.
Delisle, J.P. et Alou, F. (1978) Technologie des bétons et mortiers, cours EPFL.
Houst, Y. (1979) Deux cas d'attaque de bétons par les sulfates. Revue Chantiers, 2, 1979.
Massazza, F. et Testolin, M. (1980) Recenti sviluppi nell'impiego degli additivi per cemento e calcestruzzo, Il Cemento, 77, 73-146.
Norme DIN 4030.
Norme italienne, Unicemento UNI 9156.
Norme SIA 162, édition 1968.

MICROSILICA BASED ADMIXTURES FOR CONCRETE

P. J. SVENKERUD and P. FIDJESTØL
Elkem Materials a/s, Kristiansand, Norway
J. C. ARTIGUES
TEXSA, Barcelona, Spain

ABSTRACT
A new class of admixtures has been developed and commercialized based on the combination of microsilica and high range water reducers.

Microsilica is amorphous silicon dioxide with a mean particle size of 0.15 micrometers. This makes the material suited both as a pozzolan and as a filler in concrete. Water reducing admixtures give concrete improved properties due to low water requirements. These two materials are combined to give a new class of admixtures which has special properties in the fresh and hardened concrete.

This paper discusses the combined effects of microsilica and water reducing admixtures on the durability of concrete. Specifically, the sulphate resistance is discussed through an exposure test of concretes exposed to moderate sulphate attack. Even moderate dosages of the new class of admixtures give sulphate resistance to concretes made with ordinary portland cement. This concrete compare favourably with concrete made with sulphate resistant cement.

Keywords: Durability, Microsilica, Sulphate attack, Corrosion, Water reducing admixtures, Chlorides.

1 Introduction to microsilica

Elkem started the research and development of silica fume 40 years ago. Besides refractory concrete, ordinary concrete was of interest. Silica fume is produced when quartz reacts with coal, coke and wood chips in an electric arc smelting furnace, giving SiO-gas, further reacting to form the amorphous SiO_2-spheres, the pure microsilica particle. Average particle diameter is 0.15 micrometers, and the reactive surface is approximately 20 m^2/g. The process mixture of heated air, gasses, airborne disintegrated raw materials and micro-silica particles are referred to as silica fume.

Harmful and unusable particles are removed in a classification process to purify the silica fume. Microsilica and condensed silica fume (CSF) are often used as equal terms. Lately microsilica are used as the generic term for quality assured and refined silica fume. Chemical composition and physical data for microsilica are given in Table 1.

CHEMICAL COMPOSITION, in percent   PHYSICAL DATA

	Elkem Microsilica® Grade 920	Cement OPC		Elkem Microsilica® Grade 920	Cement OPC
SiO_2	86—96	20	Particle density (kg/m³)	2200	3150
AlO_3	0.4—1.0	4	Bulk density (kg/m³)		1200—1400
Fe_2O_3	0.1—1.5	3	• undesified 920-U	200—300	
CaO	0.1—0.5	63	• densified 920-D	500—700	
Na_2O	0.4—0.5	0.4	• slurry EMSAC® 500 S	1300—1400	
K_2O	0.3—3.0	1	• slurry (powder content)	650—700	
MgO	0.3—2.0	2	Specific surface (m²/g)	18—28	0.2—0.5
S	0.1—0.4		Coarse particles > 44 µm	<1%	
C	0.5—2.5				

Table 1. Chemical composition and technical data for
        Elkem Microsilica grade 920.

2 How microsilica works in concrete

Microsilica are both a filler and a pozzolan, changing both fresh and hardened properties of concrete.

As a **filler** - the ultra fine microsilica spheres act like ball-bearings in the concrete, and form a part of the pore water solutions. Use of microsilica requires water reducing admixtures to maintain a given slump at the same water/cement ratio. In a typical microsilica concrete mix containing 10 % microsilica, there will be at least 50.000 microsilica particles for each grain of cement. This improvement in the pore particle distribution results in a more dense, stronger and less permeable concrete.

As a **pozzolan** - the highly reactive microsilica with a large specific area and high content of SiO_2 alters the cement paste significantly. The microsilica reacts with the free lime in the hydration process. This effect is used to increase the concrete strength, lower cement content (heat of hydration), control alkali silica reactions or reduce the content of $Ca(OH)_2$.

The most important effect of the pozzolanic reaction is without any doubt the improvement of the microstructure of the concrete. This reaction creates a dense and almost impermeable matrix between the aggregates.

3   Water reducing admixtures

The earliest known published reference to the use of small amounts of organic materials to increase the fluidity of cement containing compositions dates back to 1932 [1]. Today water reducing admixtures are an important part of a good concrete mix design.

The effect of water reducing admixtures can be utilized in three ways:

1)   To reduce the water/cement ratio, obtaining a higher strength for the same workability.

2)   To improve the workability, measured as a higher slump, without having to add water so the strength is maintained.

3)   To reduce the cement content, without reducing the concrete strength. This is referred to as the more economical concrete.

The high range water reducing admixtures are formulated products. They can be more effective or added at a higher dosage than ordinary water reducers. The results can be both high slump concrete at the same w/c ratio, or considerable reductions in the w/c ratio to obtain high strength concrete.

4   Combination of microsilica and water reducing admixtures

When adding microsilica to a concrete mix the water demand is increasing [2]. The very high surface area of microsilica needs to be wetted. However, the increased water demand can easily be eliminated by using a water reducing admixture.

These admixtures are normally very effective used in combination with microsilica. How adding a wet water reducing admixture influence the water demand for a certain concrete mix is shown in figure 1.

Certain chemicals are very suitable for microsilica, and selected after testing to make a formulated product, EMSAC (Elkem MicroSilica Admixture Compositions).
EMSAC, which is a registered trademark owned by Elkem, is admixtures widely used in the United States. There are more than 100 large project where EMSAC admixtures has been used successfully.

These admixtures are highly specialized products. Different types are developed for different applications, and new microsilica based additives will be introduced to the market in the coming years. The products can be delivered both as wet and dry admixtures and these products are now available world wide.

5   Effect on fresh concrete

The use of microsilica based admixtures will result in a more cohesive concrete with no bleeding or segregation. The most interesting here is the effect on the pumpability of concrete.

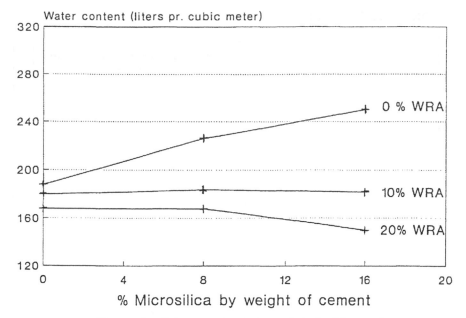

Figure 1. Water content to maintain 12 cm slump
as a function of microsilica addition and use
of a liquid water reducing admixture.
The admixture total dosages are percentages
of the microsilica addition.
(From reference 2)

Norwegian Contractors [3], a specialist in construction of offshore platforms, used microsilica in the construction of "Gullfaks C". The mix design was:

Material	kg/m³
Cement OPC	380-400
Microsilica	8
Sand (0-5 mm)	810-905
Aggregate (5-20 mm)	980-1075
High range water red.	8
Water	160-170

The addition of 2% (by weight of cement) microsilica revealed a distinct stabilizing effect on the flowing mix. The slump was more than 250 mm. Use of microsilica reduced the pump pressure by 15 to 30%, which gave rise to the production rate. The monitored concrete pressure for a mix with and without microsilica is presented in figure 2.For slipforming the 4 tall shafts to a maximum height of 180 meters, 40.000 m³ microsilica concrete was pumped continuously over a 50 day period.

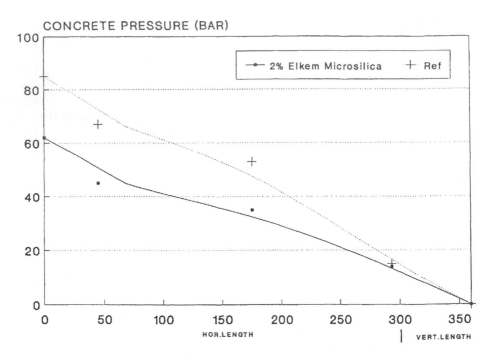

CONCRETE PRESSURE (BAR)

Figure 2. Monitored concrete pressure, concrete
with and without microsilica (from ref. 3).

6 Hardened concrete

6.1 Concrete strength

Microsilica contributes to the concrete strength. This effect can be
used in different ways, but most interesting is the possibility to
make a high strength concrete. Adding microsilica will increase
concrete strength significant [4], as shown in figure 3.

Contractors and architects can use the increase in compressive
strength to make concrete a more competitive material. In the
construction of the world's tallest reinforced concrete building, 311
South Wacker Drive in Chicago, EMSAC was used to make a concrete of
12.000 psi (83 MPa) [5]. The mix design was:

Material	Kg/m³
Cement OPC	391
EMSAC F-100	62 (contains 45% microsilica)
w/c ratio	0.28

The contractor could slipform the major columns by using a high
strength concrete in the first 14 floors. Concrete strength was the
reduced in the columns up to the total height of 962 ft (293 m).

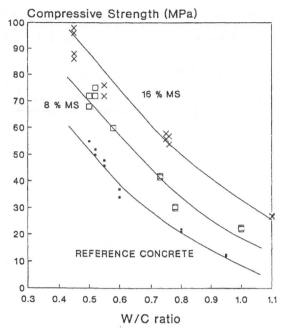

Figure 3. 28 days compressive strength versus
w/c ratio for concrete with different microsilica
contents. Data from ACI, SP-79 Vol. II.

Another way of using the additional strength is to reduce the cement
content to have a lower heat of hydration. This is a problem in
massive constructions like concrete dams. At the Alta Dam in Norway,
microsilica was used [6]. The dam had a total of 135.000 m³ concrete,
and is 145 meters high and 15 meters tick. Data for the concrete:

		Core	Surface
Cement	(kg/m³)	150	200
Microsilica	"	10	10
Concrete comp. strength (Mpa)		29.7	31.5

Cracks caused by temperature tensions from the heat of hydration are
the most common problem for concrete dams. To prevent shrinkage and
thermal cracks special cement, microsilica ice and low water content
was used in the concrete mix. The dam was completed in 1987, and no
leakage was found.

6.2 Permeability
The water permeability of concrete is often used as a reference to the
durability. However, for concretes with a lower w/c ratio than 0.5
permeability to water is not a problem [7].

The resistance to chloride ion penetration are more interesting. The
chloride ions will attack the steel in the reinforced concrete, and
give a rapid deterioration. Permeability to concrete ions is tested
using the rapid chloride permeability test (AASHTO T 277). Tests done

in reference [7] shows that microsilica concrete at a water/cement
ratio of 0.4 preformed equivalent to a latex -modified concrete. Test
results form the concretes with w/c ratio 0.5 and 0.4 is shown in
figure 4.

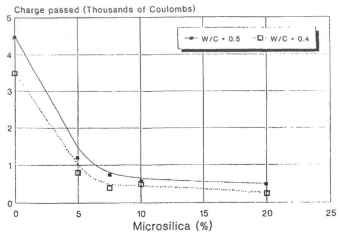

Figure 4. Chloride ion permeability as a
function of microsilica dosage.
(From ref. 7)

For a w/c ratio of 0.24 the microsilica concrete had a total amount of
electric charges of 150 Coulombs in six hours, at the same level as
polymer-impregnated concretes.

Whiting reports [8] a good correlation between the AASHTO T 277 test
results and long term exposure.

6.3 Resistance to sulphate attack.
The first test results dates back to 1952 [9]. Concrete specimens made
by ordinary portland cement (OPC), OPC + microsilica or sulphate
resistant cement SRC was made, and placed in the Alum-shale part of
the Oslo-metro. All concretes had a w/c ratio of 0.5, except the
microsilica, which had a W/C ratio of 0.62, because the lack of water
reducing admixtures. A large number of 100 x 100 x 400 mm concrete
prism was made of each concrete. The groundwater contains up to 4 g/l
SO₃ and the pH varies from 2.5 to neutral. The last published report
of 20 years exposure shows that the OPC concrete is very deteriorated,
but the combination of OPC + microsilica preformed as good as the SRC
concretes. The volume reduction of the concretes are shown in Figure
5.

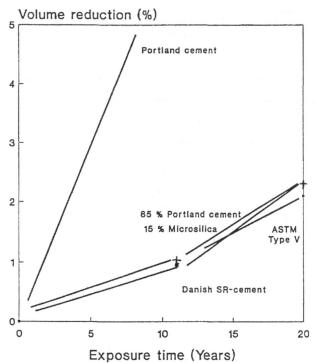

Figure 5. Volume reduction of 100 x 100 x 400 mm
concrete prism stored in sulphate rich water
for more than 20 years.

7   New laboratory experiments

7.1 Introduction
There has been done a lot of testing in laboratories to confirm the
strong resistance to sulphate attack provided by the combination of
microsilica and OPC-concretes. A laboratory test, a part of a major
Norwegian research program, "Serviceability of Concrete" have now
three years results of testing different concretes in a moderate and
realistic sulphate environment.

7.2 Realistic exposure conditions
Instead of having another test on accelerated exposure these concretes
were tested under moderately aggressive environments. The CEB Model
Code describes moderate to be between 600 to 3000 milligram SO_4 pr.
litre. It was decided to use a concentration of 1500 mg/l, which
should be reasonable for groundwater or sewage systems.

Exposure	Material	Weight/600 l	Fresh pH
Magnesium sulphate	$MgSO_4$	1.9 kg	6.3
Sodium sulphate	Na_2SO_4	2.3 kg	8.6

7.3 Concrete mix designs
The whole projects includes 26 different concrete mixes. Table 2 shows
the 14 mixes discussed in this paper.

Table 2. Mix design for concretes used in the test program.

No	Cement type	Cement kg/m^3	Microsilica kg/m^3	Water kg/m^3	Compr.str. Mpa
1	OPC	200	0	160	22.0
2	OPC	450	0	225	50.5
7	OPC	200	10	160	28.3
8	OPC	450	22.5	225	62.1
9	OPC	200	20	160	34.3
10	OPC	450	45	225	65.9
15	OPC	200	40	160	40.0
16	OPC	450	90	225	70.9
17	MPC	200	20	160	32.5
18	MPC	450	45	225	64.9
23	MPC	200	0	160	21.0
24	MPC	450	0	225	52.2
25	SRC	200	0	160	21.4
26	SRC	450	0	225	56.5

Following selected materials were used:

Cement: All produced of Norcem A/S, Norway. OPC is an ordinary portland cement, (P-30). MPC is an OPC with a content of 20 % pulverised fuel ash, (MP-30). SRC is a sulphate resistant cement, SR.

Microsilica: The microsilica was grade 920 undensified from Elkem Materials.

Chemicals: Water reducing admixture was Lomar D, 40 % solution 1-2.5 % by weight of cement. Air entraining agent was Beraid 741, 20% solution, 0.15 % by weight of cement.

Aggregates: Quality aggregates with a D-max of 16 mm.

7.4 Preparation of specimens and reading of results.
The specimens were cylinders, 102 mm in diameter and 400 mm long, casted in acrylic tubes. For this exposure, an inspection after 1 year revealed no sign of deterioration, and changes came first after 30 months. The most useful procedure for inspection has been visual examination. The specimens are evaluated and ranked in a scale from 1 to 10, where 10 is severe attack and 1 is no visible attack. Scanning Electron Microscopy, SEM verifies the visual inspection. These results are not finalized, and will be reported later.

The results from the testing program are shown in table 3.

7.5 Analysis of results
Sodium sulphate

The effect of the different cement type are shown in figure 7.

The attack was high in some series, and the worst was remarkably the specimens with SRC-cement. Using MPC-cement with 20 % fly ash has no significant influence on the resistance to moderate sulphate attack. Low cement content seems to be more beneficial than use of low w/c ratio.

These unexpected observations is to be more examined using SEM. The results shows that sulphate attack is more complex than to specify SRC, which is the common practice.

Rasheeduzzafar et. al. [9] recommends, based on their investigations from the Middle East area, use of microsilica or fly ash to improve the concrete durability. Recommended dosages are 20-30 percent of fly ash or 10-15 percent of microsilica. W/C ratio should generally be less than 0.45, and preferably 0.40.This article discuss the resistance to steel corrosion as well.

The laboratory tests shows that adequate protection is to use 10 % or more microsilica. This addition is effective both for OPC and MPC. Figure 8 shows the effect of microsilica.

Table 3. Results from test program.

No	Type	Na_2SO_4 rating	$MgSO_4$ colour
1	OPC, 200,0	2	6
2	OPC, 450,0	7.5	5
7	OPC, 200, 10	1.5	3.5
8	OPC, 450,22.5	5.5	1
9	OPC, 200, 20	1	3
10	OPC, 450, 45	1.5	1
15	OPC, 200, 40	1	2
16	OPC, 450, 90	1	1
17	MPC, 200, 20	1	1
18	MPC, 450, 45	1	1
23	MPC, 200, 0	1.5	4
24	MPC, 450, 0	8	4
25	SRC, 200, 0	7	5
26	SRC, 450, 0	9	3

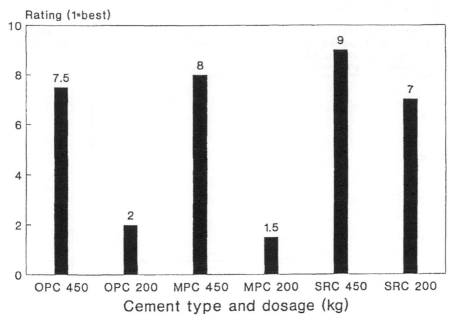

Figure 7. The effect of cement type and cement content.

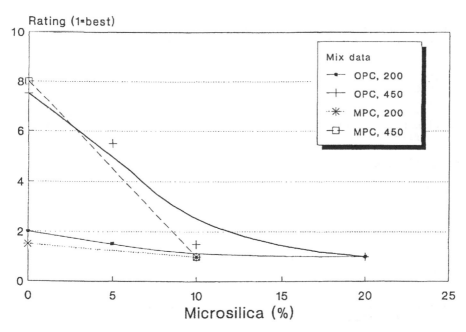

Figure 8. The effect of microsilica on resistance
to moderate sulphate attack.
Results from laboratory tests.

357

Magnesium sulphate

There is difficult to draw firm conclusions from the weak attack. However, addition of microsilica is beneficial, and the effect of microsilica is shown in figure 9.

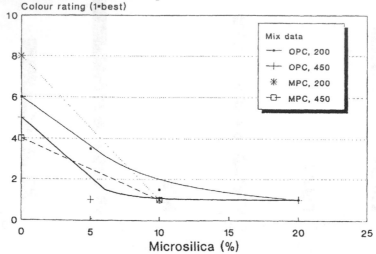

Figure 9. The effect of microsilica.

7.6 Conclusions for the test program
The effect of cement type for moderate exposure is of minor importance.

Adding microsilica will improve the sulphate resistance, both for OPC and MPC.

Further testing at moderate and realistic exposures are needed to confirm the results. Use of SEM or other special equipment is helpful to confirm the visual inspections.

8   Conclusion

A new product series EMSAC, formulated by microsilica and water reducing agents has been developed.

Adding the EMSAC admixture will have a stabilising effect on the fresh properties of the concrete. The pumpability is improved.

High strength concrete can be utilized and the concrete durability is improved. Combinations of the EMSAC admixture and ordinary portland cement makes a sulphate resistant concrete.

The EMSAC product is available world wide, through Elkem, and make the use of microsilica much easier.

References

1. US Patent (1932). 643 740.

2. Hjorth, L., "Microsilica in concrete," Nordic Concrete Federation, Oslo 1982, Publication No. 1.

3. M. Sandvik et. al., "Condensed Silica Fume in High Strength Concrete for Offshore Structures - A Case Record," Paper presented at the Third International Conference on Fly Ash, Silica Fume, Slag and Natural Pozzolans in Concrete, ACI, SP-114, 1989.

4. Sellevold E.J. and Radjy F.F., "Condensed Silica Fume (Microsilica) in Concrete: Water Demand and Strength Development," ACI, SP-79, 1983.

5. Orth, W.A. Jr., "311 S. Wacker reaches SKYWARD on climb towards "tallest" title," Construction Digest, September 5, 1988.

6. Børseth, Ivar, "Store betongdammer i Norge", NORWEGIAN, Journal of the Nordic Concrete Federation, vol 6-1988.

7. Perraton, D. et. al., "Permeabilities of Silica Fume Concrete," ACI, SP-108, 1988.

8. Whiting. D., "Permeability of selected Concretes," ACI, SP-108, 1988.

9. Rasheeduzzafar et. al., "Proposal for a Code Practice to Ensure Durability of Concrete Construction in the Arabian Gulf Environment."

ADMIXTURES FOR CONCRETE UNDER SEA-WATER ACTION

F. TAFLAN and I. FACAOARU
Building Research Institute, Bucharest, Romania

Abstract
The paper presents the influence of different admixtures
upon the concrete strength under the sea water action.
The following admixtures have been utilised: a mixed-plas-
tifier and air entrainer-admixture, on base of natrium lignosulphonate mixed with natrium alchil-sulphonate, called
DISAN, a retarde admixture on base of calcium gluconate,
called RETARGOL and another retarder admixture on base of
natrium hexameta-phosphate called REPLAST. Laboratory
tests have followed the capillary ascent, the degree of impermeability, the freezing-thawing resistance, the dimen-
sional variation, the mechanical strength of specimens,
prepared with different cements. The mechanical strength
evolution has been checked both by destructive and non-
destructive methods. When using non-destructive methods
the most reliable results have been obtained by longitudinal and flexural resonance methods. On the base of obtained results a correlation between the mechanical
strength, given by destructive and non-destructive methods,
has been established, which facilitates in-situ quality control of concrete structures. Valuable informations has
been obtained also concerning the best cement and the best
admixture to be used under the sea water action.
Key words: admixture, concrete, cement, sea water, resonance, strength, aggressivity.

1 General

Deterioration of concrete in sea water is a complex process
in which beside the chemical attack, due to the salts contained in sea water, an important role is played by wetting-
drying phenomena, accompanied by salts concentration and
precipitation, by mechanical phenomena of loading and erro-
sion, due to waves action and by freezing-thawing phenomena.
The chemical aggressivity of sea water is the result of
its complex ionic composition from which we can select, according their aggressivity on concrete, the ions: sulphate
(SO_4^{--}), chloride(Cl^-), dicarbonate (HCO_3^-) and magnesium

(Mg^{++}). Their action on concrete cannot be deducted by summing up their individual action. The sulphate corrosion, which is probably the dominant chemical phenomenon is attenuated by the presence of other ions and combined with an attack specific for magnesium. The attenuation of sulphate corrosion is due according some authors, to the chloride ions which intensify the processes of gelly formation formation impermeable to sulphate ions difusion. According to other authors the dicarbonate ions are responsable for the tightening of concrete surface and they slow down the sulphate ions attack by the formation of calcite crystals.

The magnesium ions, in large concentrations and in the presence of sulphates, are extremely dangerous for concrete due to the fact that gypsum crystalisation reduces the tightening properties of concrete: chloride ions, existing in sea water nevertheless favorise the solubility of gypsum formed by sulphate attack. In small concentrations the magnesium ions form, with the calcium hydroxide existing in concrete, magnesium hydroxide, an insoluble product, which closes up the existing pores on concrete surface.

The permanent movement, existing always in sea water, due to streams and waves, favorises a deterioration process initiated on concrete surface, which once started, is continued in a sustained rhythm, by removing the damaged parts and by exposing new surfaces of concrete, to the sea water attack.

In concrete units, subjected to sea water attack, we can distinguish, according the physical and chemical conditions of aggressivity to which concrete is exposed, the following zones:
- a concrete zone situated in the area of variable level of sea water
- a concrete zone situated permanently under sea water level
- a concrete zone situated permanently above maximum sea water level, which, under normal conditions, doesn't come in direct contact with water. Only incidentally this zone might be wetted by splashing from waves or rains but for a short time and rarely.

The most difficult conditions are met by concrete which is situated in the zone of variable level of water. In this zone a progressive concentration of the salts, deposited in concrete pores occurs, leading to an accelerated degradation of concrete structure. It must be added the effect or repeated wetting-drying, bringing additional alternative stresses of shrinkage-swelling and the effect of freezing-thawing phenomena.

For the concrete situated permanently under sea water level such phenomena does not exist and its main degradation source is the chemical attack, usually cement matrix.

The concrete situated permanently above maximum sea water level is also subjected to volume variations, due to changes

in the atmosfere moisture content and temperature, but in
less extent then in the zone of variable sea level.

Phenomena of alternative shrinkage-swelling of concrete
are accentuated by exposure to sun. The continous evaporation of sea water, which is mounting in concrete by capillarity, produces the crystalisation of the salts concentrated in pores and exerts a splitting force upon concrete.
When studying the influence of sea water aggressivity upon
concrete or cement mortar, it is necessary to take into
consideration all the particularities of the type of aggressivity.

The factors which can delay the destruction of hydro-
technical marine concrete are: high compaction, high impermeability and low gelivity, all interconected.

In order to obtain such conditions different means can
be used. An efficient way is the use of admixtures.

In certain nordic countries with frequent freezing-
thawing cycles, an increase of freezing-thawing resistance
has been obtained by preparing concrete with 5% entrained
air using admixtures.

2 Description of research programme

Building Research Institute - from Bucharest has promoted
a research programme aiming to obtain concrete with higher
resistance in sea water, by using admixtures manufactured
in Romania. In this scope the influence of the following
admixtures upon concrete behaviour in sea water and marine
climate was studied:
  - admixture DISAN on base of natrium lignosulphonate
    mixed with natrium alchil-sulphonate acting as plasti-
    ciser and air entrainer
  - admixture RETARGOL on base of calcium gluconate acting
    as retarder
  - admixture replast on base of natriumhexameta-phosphate
    acting as retarder
Retarder admixtures are used only when such requirements
are needed on technological grounds.

The influence of admixtures upon concrete subjected to
the sea water action has been studied on concrete specimens
prepared with two kinds of cement:
  - sulphate resistant cement, aditionated with slag, called SMA. This cement has a reduced content of tricalcium
aluminate (under 6%) and is frequently used for hydrotechni-
cal marine works in Romania
  - cement with high initial strengths called RIM. This
cement has a relatively high tricalcium aluminate content
allowing an early appearence of degradation phenomena.

The concrete composition was characterised by a medium
dosage (275 kg/m^3) in order to obtain a closed structure
of concrete but an early appearence of degradations.

The sea water was prepared in laboratory conditions according the following formula, which corresponds to the concentration of the water in Mediteranean Sea:

NaCl	30 mg
MgCl$_2$.6H$_2$O	6 mg
MgSO$_4$.7H$_2$O	5 mg
CaSO$_4$.2H$_2$O	1,5 mg
KHCO$_3$	0,2 mg
Water	.1000 mg

The following properties governing the behaviour of concrete under the sea water action has been measured:
- capillary ascension
- water permeability
- length variation
- freezing-thawing resistance
- mechanical properties in sea water conditions
- longitudinal and flexural natural frequency of specimens, in order to estimate the evolution of structural degradations in time

3 Capillary ascension

The sea water penetrates into concrete above sea level by capillary ascension and it is concentrated forming salts by evaporation. The capillary ascension depends on the compactness and structure of concrete. Due to the influence of admixtures upon these two properties of concrete it was necessary to verify how the capillary ascension changes if admixtures are used.

Fig. 1 shows the variation of capillary ascension on prisms 100 x 100 x 550 mm in the first three days. It can be seen that all admixtures reduce the capillary ascension the difference increasing in time.

4 Water permeability

Water permeability is a main property of concrete governing its behaviour in the variable level and permanently immersed zones.

The tests were performed on cubes 20 cm side at a pressure of 4 Bars and the results are presented in fig. 2 for specimens prepared with SMA cement and in fig. 3 for specimens prepared with RIM cement. It can be seen that, in the case of SMA cement, the depth of water penetration shows an important reduction if admixtures are used, the most efficient admixture being DISAN. In the case of RIM cement this reduction is not so important due to the high finess-grinding characterising this cement, which leads to a higher compactness of concrete.

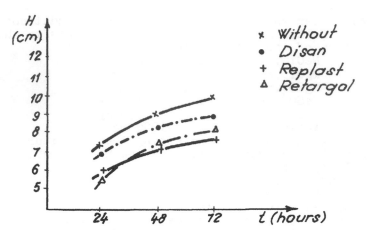

Fig. 1.   Capillary ascension of concrete specimens
          in sea water

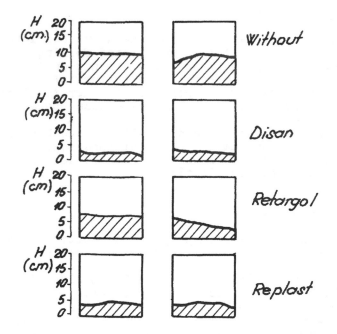

Fig. 2.   Height of water penetration in concrete
          specimens with SMA cement at 4 Bars pressure

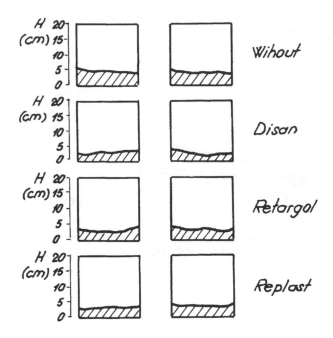

Fig. 3. Height of water penetration in concrete
specimens with RIM cement at 4 Bars pressure

5 Length variation

This measurement is aiming to ilustrate the process of concrete degradation by the formation of calcium sulpho-aluminate under the chemical attack of sea water. This attack
is developed by a large number of chemical reactions between the components of cement matrix and salts contained in
sea water. As a function of the amount of calcium aluminate, contained in cement and of sea water content in the
ions: sulphate, chloride, magnesium, bicarbonate etc., the
formation of sulpho-aluminate might be slowed or accelerated.
   In fig. 4 are represented the length variation in time of
concrete specimens prepared with SMA cement and in fig. 5 of
concrete specimens, prepared with RIM cement. It is obvious
that destructive phenomena are accelerated when using RIM
cement with a high tricalcium aluminate cement and that, regardless of the cement, all the admixtures improve the behaviour of concrete in sea water. For this kind of test
the best results have been obtained, in order, by the following admixtures: Retargol; Disan; Replast.

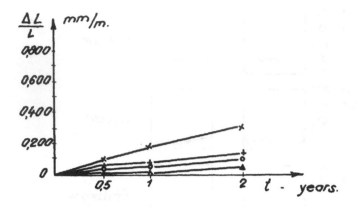

Fig. 4.   Length variation of specimens prepared with
SMA cement immersed in sea water.

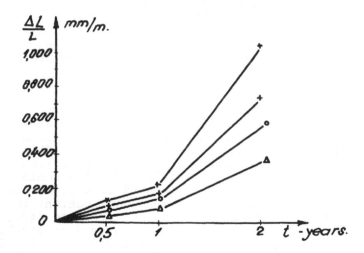

Fig. 5.   Length variation of specimens prepared with
RIM cement immersed in sea water.

6 Freezing-thawing resistance

In order to ilustrate the combined action of sea water chemical aggressivity and of repeated freezing-thawing the specimens were subjected to freezing-thawing cycles, the thawing being performed in sea water. A cycle is consisting from: curing of specimens at a temperature $-15^{\circ}C$ for 6 hours followed by 18 hours immersion in sea water, at a temperature $+20^{\circ}C$.

Non-destructive tests were performed after each 50 cycles, by pulse velocity measurements. Finally after 200 cycles also destructive compressive tests were performed. The results of destructive tests are presented in table 1.

Table 1.   Compressive strength after 200 cycles.

Nr.	Cement	Admixture	R_c^w MPa	R_c^{fr-th} MPa	R_c MPa	$\dfrac{R_c}{R_c^w}$ %
1	RIM	–	39.5	Destroyed (150 cycl)	-39.5	100.0
2	RIM	DISAN	38.5	34.2	-4.3	11.2
3	RIM	RETARGOL	55.0	Destroyed (150 cycl)	-55.0	100.0
4	RIM	REPLAST	37.5	23.0	-14.5	38.7
5	SMA	–	50.5	39.2	-11.3	22.4
6	SMA	DISAN	47.7	45.2	-2.5	5.2
7	SMA	RETARGOL	59.2	49.2	-10.0	16.7
8	SMA	REPLAST	55.5	Destroyed (150 cycl)	-55.5	100.0

The results obtained by non-destructive tests are reproduced in fig. 6 for concrete prepared with SMA cement, with and without admixtures and in fig. 7 for concrete prepared with RIM cement with and without admixtures. They confirmed in general lines the results obtained by destructive tests giving a more detailed picture of the evolution of the structural degradation of concrete, in time. Summarising these results the following conclusions can be derived:
   (a)   Concrete prepared with SMA cement has a better behaviour than concrete prepared with RIM cement to freezing-thawing cycles, in the conditions of sea water aggressivity.
   (b)   The use of DISAN admixture improves in a spectacular way the behaviour of concrete at freezing-thawing cycles, in the presence of sea water aggressivity, regardeless of the type of cement used in concrete.
   (c)   The admixture RETARGOL acts also in the direction of

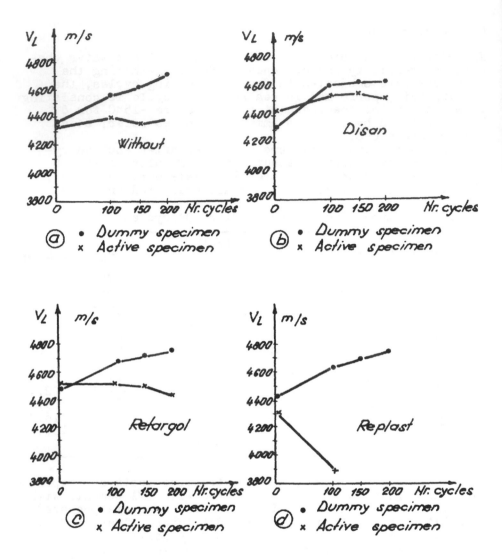

Fig. 6. Ultrasonic pulse velocity variation as a
function of the number of freezing-thawing
cycles for specimens prepared with SMA cement

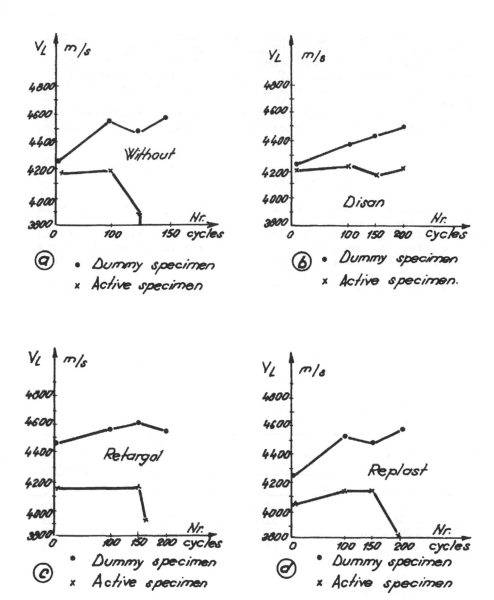

Fig. 7. Ultrasonic pulse velocity variation as a
function of the number of freezing-thawing
cycles for specimens prepared with RIM cement

improving the freezing-thawing resistance of concrete but
it is less efficient than DISAN (about 3 times).
(d) The admixture REPLAST does not improve or it may
even damage the behaviour of concrete at freezing-thawing
action, as it is shown in fig. 6, for concrete prepared with
SMA cement.
Visual examination of the specimens have confirmed the
results obtained by destructive and non-destructive tests.

5 Physico-mechanical properties of concrete under long
   term action of sea water.

The influence of the chemical aggressivity of sea water has
been studied by measuring the variation of compressive and
tensile strengths and of longitudinal and flexural natural
frequencies, by resonance method. The results obtained when
measuring mechanical strengths are given in table 2 under
the form of relative values.
When examining the results presented in table 2 the following conclusions might be deduced:
(a) The chemical aggressivity is a much slower attack
than freezing-thawing action.
(b) The main role for a given chemical aggressivity is
olayed by the mineralogical composition of cement. In this
order of idea SMA cement has a much better behaviour than
RIM cement, under sea water chemical aggressivity.

Table 2. Relative compressive strength variation of
         concrete in sea water.

| Cement | Admixture | Relative strength | | | | | |
| | | Initial | | 1 year | | 2 years | |
		R_t	R_c	R_t	R_c	R_t	R_c
RIM	–	100	100	92	108	92	91
	DISAN	100	100	95	104	62	108
	RETARGOL	100	100	88	108	66	102
	REPLAST	100	100	85	109	37	78
SMA	–	100	100	105	139	107	141
	DISAN	100	100	117	117	126	121
	RETARGOL	100	100	118	128	107	136
	REPLAST	100	100	104	116	109	115

(c) If the judgement is based on tensile strength evolution the best behaviour is shown by specimens prepared
with DISAN and RETARGOL admixtures.
(d) If the judgement is based on compressive strength
evolution the best behaviour is also shown by specimens prepared also with DISAN and RETARGOL admixtures.

(e) All specimens prepared with SMA cement show no decrease of compressive and tensile strength after 2 years of immersion in sea water, in respect with initial values but show slight decreases of tensile strength values in respect to those measured after 1 year for specimens prepared with RETARGOL and REFLAST admixtures.

The non-destructive tests are presented in fig. 8 in terms of longitudinal frequencies variations for specimens prepared with RIM cement, in fig. 9 in terms of flexural

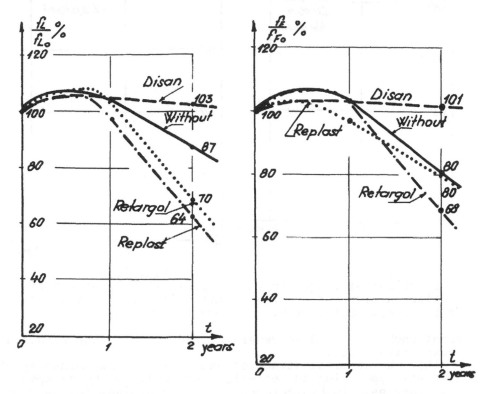

Fig. 8.
Longitudinal frequency variation in time for specimens prepared with RIM cement.

Fig. 9.
Flexural frequency variation in time for specimens prepared with RIM cement

frequencies variations of the same specimens, in fig. 10 for longitudinal frequencies of specimens prepared with SMA cement and in fig. 11 for flexural frequencies of the same specimens. When examining these results the following conclusions can be deduced:

(a) Non-destructive testing results by the two different resonance methods, longitudinal and flexural, are in agre-

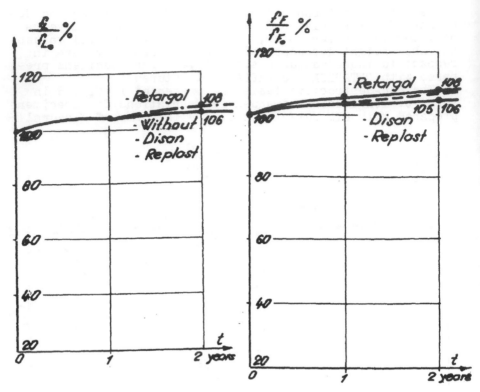

Fig. 10.
Longitudinal frequency variation in time for specimens
prepared with SMA cement.

Fig. 11.
Flexural frequency vari-
ation in time for specimens
prepared with SMA cement.

ement each other and in general agreement with the result of
destructive tests.

(b) The chemical attack of sea water is predominantly of
sulphate type being characterised at early age by an apparent increase of physico-mechanical properties, followed at
later ages by a marked decrease of these properties. This
tendency is obvious for specimens prepared with RIM cement.

(c) Non-destructive tests have confirmed that all specimens prepared with SMA cement have a good behaviour at the
chemical attack of sea water at least up to 2 years, regardeless of of type of admixture used and even regardless
if these is or not an admixture added. The utility of an
admixture is obvious only when RIM cement is used.

(d) Both longitudinal and flexural frequencies measurements have shown that, for specimens prepared with RIM cement, the best behaviour is given by the DISAN admixture
which shows no decrease of the initial values of frequencies.
Specimens prepared without admixtures or with the admixtures

RETARGOL or REPLAST showed marked decreases of the initial
values of both frequencies.
   The differences remarked between the behaviour of specimens prepared with RIM and SMA cements confirm the special
role played by the calcium aluminates in the mineralogical
composition of cement, for the cement behaviour in sea water as well as the character predominantly sulphate of sea
water chemical attack.
   Comparing the results obtained by non-destructive and
destructive test results, we come to the conclusion that
the following aproximation formulae can be adopted, in order to estimate the the variation of concrete strength,
in the condition of sea water attack:
   (a) For compressive strength variations:

$$\frac{\Delta R_c}{R_c} = \alpha \frac{\Delta f_L}{f_L}$$

where: R_c - is the compressive strength variation

           f_L - the longitudinal frequency variation

           α - a coefficient of correlation having
                the value $\alpha = 1.5$
   (b) For tensile strength variations:

$$\frac{\Delta R_t}{R_t} = \beta \frac{\Delta f_L}{f_L}$$

where: R_t - is the tensile strength variation

           β - a coefficient of correlation having
                the value $\beta = 2$
It can be seen that relative tensile strength variations
expected to be larger than relative compressive strength,
in the same conditions.

6. Conclusions

The researches performed on the behaviour of concrete prepared with different cement and admixtures have lead to the
following conclusions:
   (a) The use of the studied admixtures has a positive effect upon the mechanical properties, the permeability and
the capillary ascension for concrete prepared either with
RIM or with SMA cement.
   (b) The influence of the studied admixtures upon the
durability of concrete prepared with SMA (sulphate resis-
tent) cement can be summarised as follows:
   - admixture DISAN - increases the resistance to freezing-
thawing action and maintains a good chemical resistance to
the sea water action

- admixture REPLAST - decreases the resistance to freezing-thawing action but maintains an acceptable chemical resistance to the sea water action
- admixture RETARGOL - maintains the same resistance to freezing-thawing action and about the same good chemical resistance to the sea water.

(c) The influence of the studied admixtures upon durability of concrete prepared with RIM (rapid hardening) cement can be summarised as follows:
- admixture DISAN - increases the resistance to freezing-thawing action and also increases the chemical resistance to the sea water action
- admixture REPLAST - improves slightly the resistance to freezing-thawing action and damages the chemical resistance to the sea water action
- admixture RETARGOL - maintains the same bad behaviour to freezing-thawing action and increases the chemical resistance to the sea water action.

(d) It would be useful to extend the experimental programme presented in this paper towards higher ages because a two years duration seems to us too short for the type of aggressivity specific to sea water action.

(e) The non-destructive methods, especially those based on longitudinal and flexural resonance seemed to be very adequate for estimating the structural degradation produced by the sea water chemical attack but they seemed less sensitive to ilustrate the combined effects of freezing-thawing and chemical aggressivity. This situation might be the consequence of filling the cracks, produced by freezing-thawing action, with the products of chemical attack.

EFFECT OF ZINC OXIDE ADMIXTURE ON CORROSION INHIBITION OF FERROCEMENT

C. TASHIRO, K. UEOKA and K. KOZAI
Faculty of Engineering, Yamaguchi University, Ube, Japan
M. KONNO
Nippon Steel Corporation, Ooita, Japan

Abstract

Ungalvanized wire mesh were embedded in cement mortars mixed with
sweet water, sea water and 3% solution of calcium chloride,respec-
tively.   Corrosion state was monitored by the corrosion potential,
and degree of the wire mesh corrosion was observed visually.   Another
electrochemical method was used; the anodic polarization curves of
iron plate embedded in cement mortars were measured by the potentiodynamic method.   When the zinc oxide was added to the ferrocement in
small amounts,the corrosion potential moved to noble or ignoble value
and the current density increased.   And the corrosion of wire mesh
was inhibited.   Zinc oxide as an admixture was very effective for inhibition of ferrocement corrosion.
Key words: Ferrocement, Corrosion, Admixture, Zinc oxide.

1 Introduction

The authors have made various investigation aimed at the effect of
zinc oxides on hydration of cement, microstructures and properties of
hardened cement materials(1-3).   From these investigation,it was
found that the small amount of zinc oxide admixture effects the inhibition of corrosion of iron(1),but several points on the corrosion inhibition still remain unsoluble.

This study deals with the   inhibition of corrosion of iron in ferro-

375

cement.   The corrosion state was monitored by the corrosion potential and degree of the wire mesh corrosion was observed visiually.   And anodic polarization curves of iron plate embedded in cement mortars were measured by the potentiodynamic method.

2 Experimental method

Wire (Ø 0.5 mm)mesh having a net of 2.5 mm meshes was coated with cement mortar as shown in Fig.1.   The wire net was obtained from galvanized wire net which was washed by HCl solution. A commercially availably ordinary portland cement was used . The zinc oxide admixture was of commercial first grade quality. Standard sand from Toyoura in Yamaguchi Prefecture in Japan was employed for making the mortar.

The cement mortar was made with cement to sand ratio of 1:3,and water to cement ratio of 0.8.   Sweet water,sea water and 3% solution of cal- cium chloride were used as mixing water,were designated as I,S and C, respectively.   The amounts of the zinc oxide admixture used were 0.1 and 0.2 wt.% of cement,and were designated as Z1 and Z2,respectively.

Steam curing at 65°C for 90 days,after precuring at 20°C for 4 hours, was used for accelerating the corrosion process.

The spontaneous electromotive potential of specimens was measured by use of milli voltmeter, saturated potassium chloride solution and silver chloride electrode as a reference electrode, as shown in Fig.2. The measurement was carried out in a sealed case every three days after holding at 20°C for 0.5 hours.   The rust area was measured on the wire mesh after curing the specimens.   The measurements were made at 7 14, 28, 56 and 90 days.

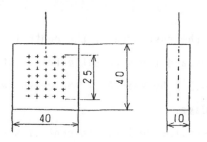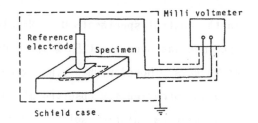

Fig.1. Schematic diagram of specimen.   Fig.2. Schematic representation
of apparatus for measurment of spontaneous potential.

3. Results and discussion

Table 1 shows the data on the amount of rust area of the wire mesh.
When sweet water was used, corrosion was not observed in all the specimens(I-0,I-Z1 and I-Z2). It seems that the amount of corrosion was appreciable under the conditions of water/cement ratio, thickness of mortar cover,and curing used in this study.

Table 1. Rust area of the iron mesh (%).

Specimens	Curing time (days)		
	30	60	90
I - 0	0	0	0
I - Z1	0	0	0
I - Z2	0	0	0
S - 0	3	55	cut
S - Z1	0	0	60
S - Z2	0	1	5
C - Q	5	60	80
C - Z1	1	20	60
C - Z2	0	0	0

On the other hand,when sea water was used,the corrosion was observed

in the control specimen(S-0). Namely,the rust area of the control
specimen was over about 55 % at 60 days,and the wire ultimately cut
at 90 days. However,when zinc oxide 0.2 % by weight of the cement was
used,no corrosion was observed at 60 days.

When calcium chloride solution was used,the rust of the control specimen was over about 60 % at 60 days as the case of sea water. But when
zinc oxide admixture was used, no corrosion was even observed at 90
days.

The data on spontaneous potential of the specimens were shown in Fig.3
-5. When the sweet water was used, the spontaneous potentials of I-0,
I-Z1 and I-Z2 were about -300mV,except the specimen 1-Z2 after 45days.
The specimen I-Z2 moved to ignoble after about 45 days.

On the other hand ,when the sea water was used spontaneous potential
of the all specimens shown about -200--400 mV. The control specimen
moved to ignoble with curing time, and the specimen contained zinc
oxide admixture moved to noble with curing time. And the calcium chloride was used,the spontaneous potential of C-0 was about -400 mV
until 28 days after mixing and it moved to noble with curing time,and
also it again moved to ignoble after 70 days. On the other hand, the

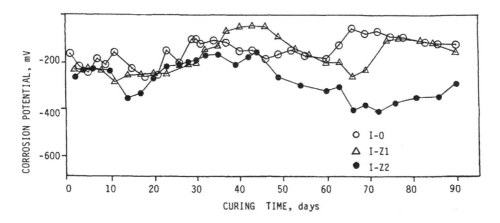

Fig.3.Corrosion potential of specimens mixed with sweet water.

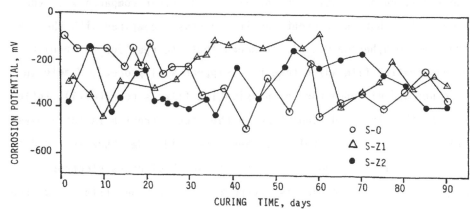

Fig.4. Corrosion potential of specimens mixed with sea water.

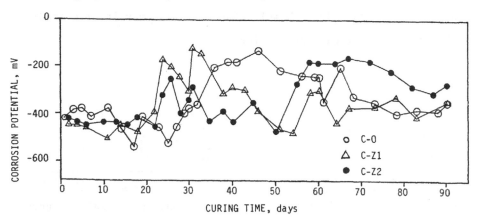

Fig.5. Corrosion potential of specimens mixed with CaCl$_2$ solution.

specimen(C-Z2) contained zinc oxide admixture moved to noble from 50 days. From above data, zinc oxide admixture effect a change of the spontaneous potential.

Fig.6-8 show the anodic polarization curves of the specimens at 7 and 28 days. The curves changed with the curing time. When the sweet water used passivating potentials were -200 --250 mV, and these current density were 0.1 μA. On the other hand, when the sea water was used, the passivating potential of S-0 was 350 mV and current density was 0.15 μ A. And also, when the specimen contained the zinc oxide admixture,the passivating potential was not clear compared with the sweet water. But the current density increased compared with the control specimen,became 0.3 μA. It seems that the reaction for repair of passivating film which take to disappear will proceed. Namely,appearance and disappearance of passivating film is repeated(4). Influence of the zinc oxide admixture on the iron corrosion is not always clear, but it seems as follows, when zinc oxide admixture is added to ferrocement,it is dissolved as zinc ion which migrates electrophoretically to anode, and anode is covered with inhibition film. On the other hand, the oxygen ion from the zinc oxide or zinc compound passivates the cathode. Namely, the mechanism may involve anodic protection and cathodic protection.

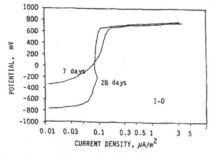

Fig.6. Anodic polarization curves of specimens mixed with sweet water.

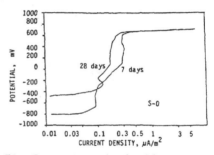

Fig.7. Anodic polarization curves of specimens mixed with sea water.

4 Conclusion

When the zinc oxide added to cement mortar which was made with sea water or calcium chloride solution, corrosion of wire mesh embedded in the mortar was not observed. And the

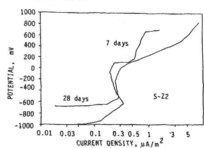

Fig.8. Anodic polarization curves of specimens mixed with sea water and zinc oxide admixture 0.2%.

corrosion potential moved to noble value.   Anodic polarization
curves show current density for passivating increased.   Zinc oxide
admixture may mainly serve as anodic protection.
The zinc oxide admixture is very effective for   inhibition of   ferrocement corrosion.

References
1)Tashiro C. and Fukushima Y.(1986) Effect of zinc oxide admixture on
corrosion inhibition of iron in cement mortar containing sea water.
Proceeding of 8th International Congress on the Chemistry of Cement,
5.pp 226-230.
2)Tashiro C. and Ueoka K.and Tachibana S.(1985)Effects of heavy metal
oxide on corrosion of iron in cement mortar.   Proceeding of 14th Conference on silicate Industry and Silicate Science.III,pp 281-192.
3)Tashiro C.and Ueoka K.(1981)Bond strength betweenC3A paste and iron
,copper or zinc wire and microstructure of interface.Cement and Concrete Research ,11,pp 619-624.
4) Wranglen G.(1972) An introduction to corrosion and protection of
metals(in japanease). pp 151-157.

EXPERIMENTAL STUDY ON THE EFFECTIVE-NESS OF CORROSION INHIBITOR IN REINFORCED CONCRETE

F. TOMOSAWA
Department of Architecture, University of Tokyo, Japan
Y. MASUDA
Building Research Institute, Ministry of Construction, Japan
I. FUKUSHI
Housing and Urban Development Corporation, Japan
M. TAKAKURA and T. HORI
Nissan Chemical Industries Ltd, Japan

Abstract
The purpose of this research is to evaluate the effectiveness of calcium nitrite as a corrosion inhibitor in concrete containing chloride.

We carried out two experiments to examine the relationship between the quantity of calcium nitrite and the inhibition effect on corrosion. In Experiment 1, test specimens of concrete to which chloride and calcium nitrite had been added, were subjected to accelerated corrosion by means of repeated drying and wetting.

In Experiment 2, reinforced concrete specimens containing calcium nitrite were repeatedly immersed in chloride solution and dried. This procedure accelerated the penetration of chlorides and the corrosion of embedded steels.

From the results of the experiments described above, it was confirmed that the occurrence of corrosion can be prevented and the corrosion rate can be suppressed, if more chlorides are penetrated, depending on the addition amount of nitrite ions.
Key words: Corrosion rate, Calcium nitrite, Chlorides, molar ratio of (Cl/NO_2), accelerated test.

Introduction

Early deterioration of structures due to corrosion of steel rebars has become a social problem recently in Japan. Corrosion of concrete structures by chloride affects directly their lifetime, and is therefore one of the most serious problems.

There are two ways in which chloride can invade concrete; one, via construction materials such as marine sand, mixing water and admixture, and the other, via adhesion of salt spray or scattering chloride to the concrete surface.

Marine sand is traditionally used in the construction industry in Japan due to the lack of river sand. In this regard, the use of $Ca(No_2)_2$ has been contemplated to prevent infiltration of chloride, and good results have in fact already been achieved over the last 10 years or so.

Secondly, chloride from scattering sea water, or salt spray and de-icing agents would cause far higher concentrations of chloride to the concrete, consequently more corrosion inhibitor is required in this case than in the case of marine sand.

Therefore, the authors carried out two experiment in order to examine the relationship between the quantity of calcium nitrite and the inhibition of corrosion.

Experiment 1 – Accelerated Corrosion Tests in the presence of a large amount of salt

1 Object

A series of accelerated corrosion tests were conducted on concrete containing large quantities of salt and a corrosion inhibitor to ascertain the effectiveness of the corrosion inhibitor.

2 Outline of the tests

Salt (NaCl) Level and Corrosion Inhibitor Content is shown in Table 1.

Table 1. NaCl level and corrosion inhibitor content

NaCl level	Corrosion inhibitor ℓ/m^3			
%/Concrete	0	11.8	23.5	35.3
0	O	—	—	—
0.1	O	O	—	—
0.3	O	O	O	—
0.5	O	O	O	O

Corrosion inhibitor : $Ca(NO_2)_2$ 30% water solution

Table 2 shows the Concrete Preparation and the Properties of Concrete.

Table 2. Concrete Preparation and Properties of Concrete

Inhibitor Content $(1/m^3)$	NaCl Content (kg/m^3)	W/C (%)	S/a (%)	Added Water (kg/m^3)	Slump (cm)	Air Content (%)	Compressive Strength 28day (kg/cm^2)
0	0			186	18.0	4.7	271
0	2.25			186	18.0	4.7	266
11.8	2.25			176	19.0	5.0	266
0	6.75			187	19.0	4.7	266
11.8	6.75	60	46	177	19.5	4.1	269
23.5	6.75			166	19.5	4.1	289
0	11.25			185	18.5	5.0	245
11.8	11.25			175	18.5	4.7	268
23.5	11.25			163	18.0	4.6	274
35.3	11.25			152	*14.0	4.0	331

* Stiffened after eight minutes.

Materials used are as follows:

    Corrosion inhibitor: A calcium nitrite-based corrosion inhibitor
                      having a solid content of 33%
    Steel: Deformed bar having a diameter of 13mm.
                                      steel Weight: abe.189g/20cm
    Cement: Portland cement
    Aggregate: Coarse - Crushed stone; Fine - River sand

A test specimen was prepared by curing a cement mixture in a moist
room for four weeks after curing, and sealing the end faces and the
upper and lower faces of the resulting body of concrete with an epoxy
resin.

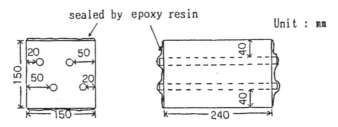

Fig. 1. Size of specimens

An alternately wet and dry test was conducted at a constant temperature of 80°C.

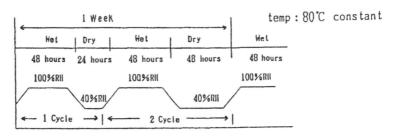

Fig. 2. Conditions of wet and dry cycles

The specimen was visually inspected for the presence of any crack. A
30mm long portion was cut off each end of the steel bar, that portion
of each exposed end surface of the bar which was obviously rusty was
copied onto a transparent piece of vinyl, and the ratio of the
corroded area was obtained by an image analyzer. The weight loss of
the bar was obtained after removing the rust by dipping it into a 10%
solution of di-ammonium citrate.

3 Test Results
(1) Number of cycles which had been repeated before cracking occurred:

Table 3. Number of cycles which had been repeated
before cracking occured

NaCl level	Corrosion inhibitor ℓ/m^2			
%/Concrete	0	11.8	23.5	35.3
0	○	—	—	—
0.1	○	○	—	—
0.3	10	15	○	—
0.5	10	15	15	○

○: No cracking occured even when 25 cycles had
been repeated.

(2) Amount of corrosion inhibitor, ratio of corroded area and weight
loss by corrosion: As the cracking of concrete apparently has a closer
correlation to the weight loss of the steel bars by corrosion than to
the ratio of the corroded area, the degree of corrosion is preferably
represented by the weight loss by corrosion. When the concrete
contained 0.1% of sodium chloride however, the steel bars showed so
small a weight loss by corrosion that no definite corrosion tendency
could be determined, and the ratio of the corroded area therefore, was
employed for representing the degree of corrosion. The degree of
corrosion is shown by the average amount of corrosion of the four
steel bars in each specimen.

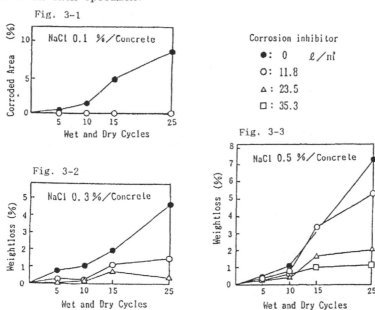

Fig. 3-1

Fig. 3-2

Fig. 3-3

Corrosion inhibitor
●: 0 ℓ/m^2
○: 11.8
△: 23.5
□: 35.3

Fig. 3. Relationship between the amount of $Ca(NO_2)_2$
and corrosion loss of rebar

Experiment 2 - Effect of a Corrosion Inhibitor under Accelerated Salt Penetration

1 Object

A series of tests were conducted to determine the effectiveness of a corrosion inhibitor against corrosion of reinforcing steel mainly by way of a variation in half-cell potential by repeating the dipping of concrete into a salt solution and its drying to accelerate the penetration of salt into the concrete.

2 Outline of the Tests

Materials used are as follows.
　　Corrosion inhibitor: A calcium nitrite-based corrosion inhibitor
　　　　　　　　　　　　having a solid content of 33%
　　Steel: Deformed bar having a diameter of 13mm.
　　Cement: Portland cement
　　Aggregate: Coarse - Crushed stone; Fine - River sand

Table 4. Concrete Preparation and Properties of Concrete

Inhibitor Content (1/ m³)	W／C (%)	S／a (%)	Added Water (kg/m³)	A・E Agent (C×%)	Slump (cm)	Air Content (%)	Compressive Strength 28day (kg/ cm³)
0			186	0.010	17.0	4.4	289
1 0	6 0	4 5	177	0.008	18.0	4.7	291
2 0			169	0.007	18.5	4.8	300
3 0			160	0.006	19.0	4.7	309
0			190	0.008	18.0	3.9	352
1 0	5 0	4 1	179	0.010	18.0	4.6	406
2 0			171	0.008	18.5	4.5	401
3 0			162	0.007	19.0	4.1	399

A test specimen was prepared by curing a cement mixture in a moist room for four weeks after pouring, connecting a lead wire to an exposed steel portion on one end surface of the resulting body of concrete with an electrically conductive adhesive and sealing the end, upper and lower faces with a silicone resin.
　　The specimen was repeatedly dipped into a 3%salt solution and dried, whereby salt was caused to penetrate the concrete. Each of the first 25 cycles consisted of four days of dipping and three days of drying, and each cycle thereafter which was intended for accelerating the penetration of salt consisted of one day of dipping and two days of drying.

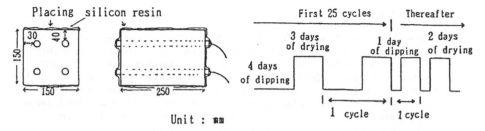

Fig. 4. Size of specimens   Fig. 5. Conditions of dipping and drying cycles

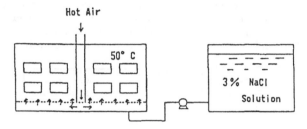

Fig. 6. Testing apparatus

The same results were obtained by bringing an electrode into direct contact with the surface of the concrete specimen which had been removed from the salt solution and by leaving both the specimen and the electrode in the salt solution as shown in figure 7.

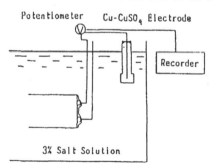

Fig. 7. Measurement of potential

The amount of chloride ions in the concrete was determined by taking a sample from the specimen at various depths below its surface having a distance of 10mm apart, and employing a potentiometric titration method.

    The ratio of a corroded area was determined by breaking the specimen upon a lapse of 126, 220 and 286 days after the test started, and employing the method outlined under Experiment 1.

3　Test Results

(1) Penetration of salt: The rate of salt penetration did not show any substantial change when different numbers of cycles were employed but was in proportion to the length of time elapsed When the test ended after 385 days, the concrete specimen had an NaCl content of 1.5 - 2.0% in the vicinity of its surface and an NaCl content of about 0.5% at a depth of 30 to 40mm below its surface where the steel bars were embedded. The specimens which had been prepared employing a water to cement ratio of 0.5 had a higher NaCl content in the vicinity of their surfaces than those which had been prepared employing a ratio of 0.6, but had a lower NaCl content on the interior. Therefore, it follows that concrete prepared by employing a lower water to cement ratio makes the penetration of salt more difficult. No great difference in salt penetration was found between concrete containing a corrosion inhibitor and one not containing a corrosion inhibitor.

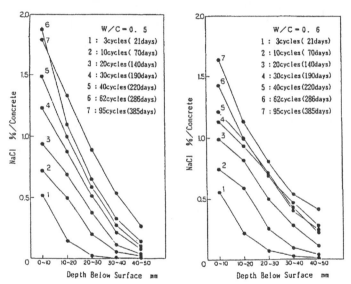

Fig. 8.　Chloride Penetration

(2) Relation between half-cell potential and ratio of corroded area: According to ASTM C876-80, there is a probability of at least 90% that no corrosion occurs when the potential is smaller than - 200mV, and that corrosion occurs when it is larger than -300mV, while nothing definite can be said when it is from -200mV to -350mV. The present experiment also revealed that corrosion started to occur when the potential was in the region of -350mV.

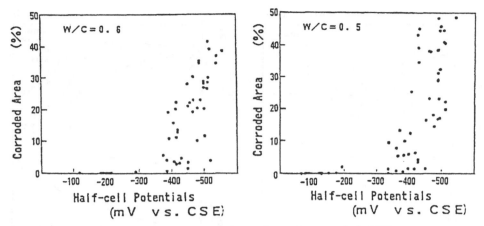

Fig. 9. Half-cell potentials and corroded area

(3) Amount of corrosion inhibitor and potential variation: While the potential changes from a small to a large level with the penetration of NaCl, the addition of a corrosion inhibitor makes it possible to delay the change of potential to a large level until a higher degree of salt penetration occurs.

The following relation was derived between the salt content of concrete and the amount of corrosion inhibitor used from the relation between the potential and the ratio of the corroded area indicating that the steel bar started to be corroded when it had a potential of -350mV.

Table 5. Amount of corrosion inhibitor and the salt content of concrete which caused corrosion

CANI-30(ℓ/m^3)	0		1 0		2 0		3 0	
W/C	0.5	0.6	0.5	0.6	0.5	0.6	0.5	0.6
NaCl content %	0.06	0.11	0.24	0.29	0.32	0.43	0.49	0.45
Age (days)	88	60	242	134	278	192	355	234
Mol.ratio NO_2/Cl	—	—	0.63	0.52	0.94	0.70	0.92	1.00

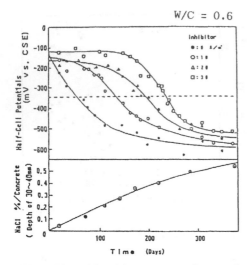

W/C = 0.6

W/C = 0.5

Fig. 10. Chloride penetration and
half-cell potentials

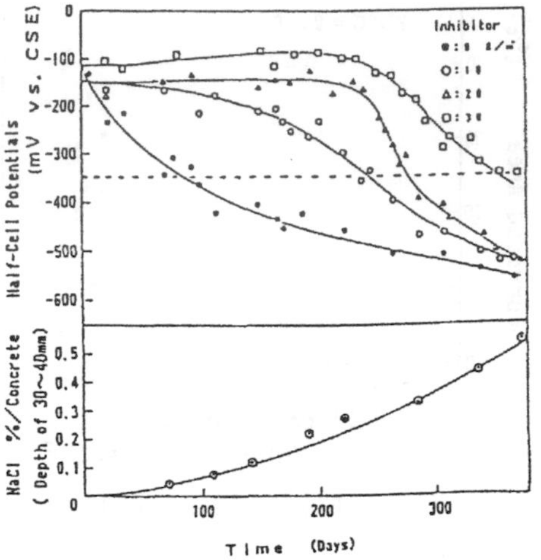

Fig. 11. Chloride penetration and
half-cell potentials

Summary

   1. Even when the accelerated penetration of salt gives rise to a
far larger amount of salt than when sea sand is used, it is possible
to restrict a rate of corrosion to a very low level if an increased
amount of inhibitor is used. The amount of corrosion inhibitor to be
employed is such that nitrous acid and chloride ions may have a molar
ratio of (NO_2/Cl) of 0.6 when 0.3% or less salt is assumed to be
accumulated where the steel bars are located, or of 1.0 when a larger
amount of salt is assumed to be accumulated.
   2. Even if an insufficiency of corrosion inhibitor results from an
unexpectedly heavy accumulation of salt, the inhibitor controls
corrosion to the extent that any corrosion may be just as if the
concrete did not contain   any corrosion inhibitor.
   3. The same amount of corrosion inhibitor can effectively prevent
corrosion for a longer period of time if the concrete has a lower
cement to water ratio, allowing salt to penetrate only at a lower
rate.

Others:

   1. The addition of a corrosion inhibitor hardly affects the consistency of concrete, but tends to accelerate its setting to some extent
and cause it to shrink to a slightly greater extent upon drying. The
corrosion inhibitor accelerates the hardening of concrete to a degree
depending on the amount of inhibitor which it contains, but there is

no substantial difference between the 28-day strength of concrete containing a corrosion inhibitor and that of concrete not containing any such inhibitor.

2. When concrete containing nitrous acid ions had been left outside for two years, about 70% of the ions could be detected from the interior, though a some what smaller amount of ions was found in the vicinity of its surface. If the amount of ions which had been absorbed or fixed in the concrete is taken into account, it can be said that it still retained nearly all of the ions which had been added two years before.

5 OTHER PROPERTIES

PERMEABILITY OF THE CEMENT-AGGREGATE INTERFACE: INFLUENCE OF THE TYPE OF CEMENT, WATER/CEMENT RATIO AND SUPERPLASTICIZER

U. COSTA, M. FACOETTI and F. MASSAZZA
Italcementi S.p.A., Bergamo, Italy

Abstract
Air permeability and porosity of portland and pozzolanic cement pastes, prepared with w/c ratios of 0.32 and 0.40 and with or without superplasticizer, were determined. Permeability measurements have also been taken in parallel on disks consisting of a ring made up of the same paste and cast around a limestone core. The comparison between the permeability of pastes as received and the permeability of composites emphasized the interface area's contribution to this characteristic.
The interface permeability is reduced as the w/c ratio decreases. The results have shown that, the w/c ratio being equal, the permeability of the interface is markedly higher than that of the pure paste in the first 7 days of curing but then the differences diminish.
The addition of superplasticizer, the w/c ratio being equal, has little effect on the permeability of the portland and pozzolanic cement pastes even though it slightly increases porosity. However, the superplasticizer highly reduces the interface permeability not only when it is used as water reducer but also as fluidifier owing to the increased workability of the cement paste, which favours its adhesiveness to the aggregate surfaces.

1 INTRODUCTION

Mechanical strength, porosity, permeability, ion diffusion and chemical resistance of concrete depend on the physical and mechanical characteristics of the cement matrix and of the aggregate. However, since the role of the aggregate is a subordinate one, concrete properties are generally correlated to those of cement matrix (cement paste plus fine aggregate) and particularly to the type of cement, to the w/c ratio, to the presence of admixtures, etc. (Oberholster, 1986; Feldman, 1986).

In fairly recent times researchers have turned their attention to the cement matrix aggregate interface.

It was demonstrated that the structure of the cement paste which surrounds the aggregate (transition halo) differs considerably from that of the bulk of the paste (Grandet et Ollivier, 1980).

The transition halo includes two layers. The former, very thin and with a fine texture, is intimately fixed to the aggregate through bonds which are sometimes not only physical but also chemical (Barnes et al., 1978; Barnes et al., 1979). The latter, chiefly made up of comparatively large-sized and fairly flaky portlandite crystals, is highly porous and exhibits weak cohesion (Farran et al., 1972; Massazza and Costa, 1986).

For these reasons the layer at the interface constitutes a zone of higher porosity and therefore of preferential permeation of aggressive ions as chloride (Tognon and Cangiano, 1980).

Studies conducted on factors which modify the structure of the halo and on the nature of the components may improve significantly both mechanical strength and durability of concrete.

For example, it has been demonstrated that a reduction in the interface porosity is obtained with calcareous aggregates in normal curing and with quartz aggregates in autoclaved concrete (Tognon and Cangiano, 1980).

This work has investigated the influence of a superplasticizer on the porosity and permeability to gas of the paste-aggregate interface. The admixture has been used both to increase the workability of the paste, while keeping the w/c ratio constant, and to reduce water, while keeping the workability constant. The action of the admixture has been examined on Portland and pozzolanic cement pastes.

2   EXPERIMENTAL

2.1 Materials

Portland and pozzolanic cements were used for the test. The pozzolanic cement has been obtained by mixing portland cement with Bacoli (Naples) pozzolana at a 70:30 ratio. The chemical composition and fineness (Blaine) of the two cements are shown on table 1.

The superplasticizer was an aqueous solution of sodium naphthalenesulphonate condensated with formaldehyde having a low sodium sulphate content. The active matter contained in the solution was about 40%.

2.2 Cement pastes

Cements were mixed with pure water or water containing 1% of admixture referred to the weight of the cement. The w/c ratios applied were 0.32 and 0.40.

Cement pastes made up with the lower w/c ratio and the superplasticizer, had the same consistency, determined by Vicat's needle, as the pastes made up with the higher w/c ratio and without admixture. The pastes were mixed by hand for 3 minutes and then quickly poured into cylindrical moulds measuring 28 mm in diameter and 100 mm in height. The samples were settled by striking lightly until no more large bubbles surfaced.

TABLE 1. Characteristics of cements

DETERMINATIONS	Portland	Pozzolanic
L. o. I. (%)	1.25	1.76
SiO2	22.41	32.49
Al2O3	5.11	9.39
Fe2O3	2.26	3.19
CaO	62.32	44.55
MgO	2.56	2.14
SO3	2.48	1.75
Na2O	0.12	1.09
K2O	0.91	3.14
SrO	0.15	0.12
Mn2O3	0.08	0.09
P2O5	0.07	0.11
TiO2	0.27	0.33
Insoluble residue	0.10	9.95
Pozzolanicity test		positive
Vol. mass (g/cm3)	3.16	2.99
Blaine's S.S.(cm2/g)	3490	4910

Other samples were prepared, as shown on fig. 1. A limestone cylinder with an 18 mm diameter, obtained via rock coring, was placed at the centre of the moulds. In this case the cement paste has been quickly inserted in the space between the mould's inner surface and the limestone cylinder. In any case the moulds were then closed and kept turning slowly for 24 hours around the axis parallel to the cylinder's base, to prevent settling. The hardened pastes were then drawn out of the moulds and cured in water at 20°C for 28 days.

2.3 Preparation of the specimens

After the scheduled cure was completed, the cylinders were cut off with a diamond saw into disks measuring 28 mm in diameter and 10 mm in thickness.

The disks were made up of pure cement paste or composites including the limestone core surrounded by a 5 mm thick cement paste ring. In this way the cement paste-limestone interface had the same geometry for all specimens.

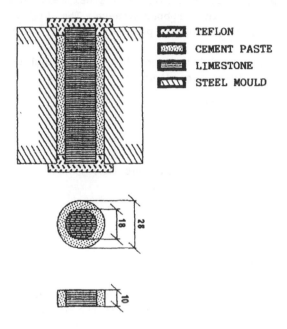

TEFLON
CEMENT PASTE
LIMESTONE
STEEL MOULD

FIG. 1. Knock-down mould for the preparation of cement paste-limestone core composites.   Scale 1:2

Before being subjected to pore and permeability measurements, the
specimens were dried by displacing the water by means of isopropanol
and n-penthane (Marsh et al., 1983). After drying, the disks were
conditioned in a chamber containing CO_2-free air with 50% R.H.

Drying of the disks with limestone core turned out to be critical
for the samples'integrity. It has not always been possible to obtain
disks free of microcrackings and fit to undergo permeability testing.

2.4 Permeability measurements

Measurements were carried out by air permeation, since a gas flows
through a paste more easily than a liquid does, and therefore both
pressure gradient and time required to achieve stationary flow
conditions are less.

Nevertheless this method involves drying the specimens before testing. If the operation is not carried out taking some particular precautions, the structure of the paste may be modified and cracking induced (Day and Marsh, 1988).

The measuring apparatus consists of a steel cell in which the disk to be tested is placed in a rubber sleeve. The tightness round the disk is assured by a pneumatic pressure higher than that of the permeation air.

The permeability coefficients of both plain and composite paste specimens were calculated according to Darcy's law

$$Vm = \frac{K}{\mu} \frac{\Delta p}{L} \qquad\qquad 1)$$

where

K = specific permeability coefficient (m^2)
μ = air viscosity = $1.83 \cdot 10^{-10}$ bar.s
ΔP = pressure differential (bar)
L = thickness of the specimen in the direction of flow (m)
Vm = average velocity of the fluid (m^3/s)

$$Vm = \frac{V2 \; P2}{(P1+P2)/2} \qquad\qquad 2)$$

Vm has been obtained from equation (2) measuring the velocity of the gas flow issuing from the cell (V2), once the pressures of air within the cell (P1) and issuing from it (P2) were known.

The measurements have been carried out at different $\Delta P/L$ pressure gradients ranging from 100 to 1000 bar/m and the value of K has been obtained from the angular coefficient of the regression line $Vm = f\left(\frac{\Delta p}{L}\right)$

3   RESULTS

3.1 Permeability of the pastes
Table 2 shows the permeability coefficients of the plain paste specimens and of the paste-aggregate composites made up of Portland cement. Table 3 shows the corresponding values relating to pozzolanic cement pastes.

The permeability coefficients of plain pastes depend on cement type, w/c ratio, curing time and on the presence of superplasticizer.

Pozzolanic cement pastes exhibit permeability coefficients which are up to 7 days higher than those of portland cement pastes with equal w/c ratio. Differences between values, however, drop considerably as curing time increases. After 28 days, while portland cement pastes still exhibit an albeit very low but still measurable degree of permeability, pozzolanic cement pastes no longer show an appreciable permeability, at least by the measurement technique followed.

396

TABLE 2. Coefficient of specific permeability ($K = m^2 \times 10^{17}$) measured on samples based on portland cement

MATERIAL		PLAIN PASTE DISKS				DISKS WITH LIMESTONE CORE			
CURING (days)		1	3	7	28	1	3	7	28
without admix.	R w/c 0.32	0.154	0.073	0.036	0.011	3.48	1.15	0.41	0.135
+ 1% NSF		0.312	0.080	0.050	0.013	0.23	0.08	0.06	0.038
without admix.	R w/c 0.40	1.717	0.333	0.133	0.070	117.5	33.4	10.29	2.528
+ 1% NSF		1.970	0.252	0.096	0.061	5.28	0.78	0.21	0.118

TABLE 3. Coefficient of specific permeability ($K = m^2 \times 10^{17}$) measured on samples based on pozzolanic cement

MATERIAL		PLAIN PASTE DISKS				DISKS WITH LIMESTONE CORE			
CURING (days)		1	3	7	28	1	3	7	28
without admxi.	R w/c 0.32	2.078	0.366	0.113	0	3.243	0.506	0.222	0.21
+ 1% NSF		1.634	0.262	0.125	0	2.2	0.2	0.08	0.07
without admix.	R w/c 0.40	10.110	1.095	0.390	0	15.7	12.8	3.6	0.58
+ 1% NSF		11.150	1.638	0.470	0	14.6	1.58	0.41	0.01

The influence of the w/c ratio on the permeability of the pastes is stronger after short ageing, especially with regard to portland cement pastes. It is observed that while the permeability of Portland cement pastes grows by a factor from 4 to 11 when the w/c ratio increases from 0.32 to 0.40, that of pozzolanic cement pastes rises by a factor ranging from 0 to 5, depending on age.

In the majority of cases the addition of a superplasticizer with the same w/c ratio increases slightly the permeability after short curing. This effect is less strong in pozzolanic cement pastes.

3.2 Permeability of the paste-aggregate composites

The permeability of the disks with limestone core is generally higher than that of the corresponding plain-paste specimens. It depends to a marked degree on cement type, w/c ratio, curing and admixture.

By way of example, the portland cement composite made up with a w/c ratio = 0.32 and no admixtures exhibits a permeability coefficient which is — depending on curing time — from 11 to 22 times as high as that of the corresponding plain paste. The composite permeability is from 36 to 100 times as high as that of the plain paste when the w/c ratio is 0.40.

The permeability of the composite diminishes progressively in time, but it still remains comparatively high even after 28 days.

With regard to the influence of the cement type, it is remarked that the increase in permeability of the composite against the plain paste is much lesser when pozzolanic cement is used. Even with these cements, however, interface permeability after 28 days still retains relatively high values.

The superfluidizing admixture is highly effective in reducing interface permeability since in many cases it causes the permeability of the composites to revert to values closely approaching those of the plain paste. The effect of the admixture, however, is less stressed after long curing.

The improvement of short-term impermeability, also when admixture is added at same w/c ratio, might be the consequence of the improvement in workability, which allows a better adhesiveness of the paste to the surface of the aggregate.

3.3 Porosity of the pastes
The plain paste disks used for the permeability tests have been ground to a few millimetres size and subjected to total porosity (P) and pore distribution determination by mercury-intrusion porosimetry.

Tables 4 and 5 show the values of P, of the maximum continuous diameter (M.C.D.) and of the threshold diameter (T.D.).

The data referred to admixture-free paste confirm the findings of earlier works (Costa, Massazza, 1987) according to which pozzolanic cement pastes, w/c ratio being equal, exhibit higher total porosity but smaller final pore sizes than portland cement do.

The addition of the superplasticizer causes an increase in porosity, particularly in the case of portland cement pastes made up with lower w/c ratio but does not affect M.C.D. and T.D. of the porous system to any significant degree, especially after long curing.

It may therefore be assumed that the increase in porosity induced by the superplasticizer is chiefly ascribable to its aerating effect. This supplementary porosity due to entrained air has at any case a limited effect on the permeability of the paste curing and does not cause any changes in the average pore size.

TABLE 4. Total porosity, threshold diameter and maximum continuous diameter on portland cement paste disks

		TOTAL POROSITY (%)				M.C.D. (nm)				THRESHOLD D (nm)			
CURING (days)		1	3	7	28	1	3	7	28	1	3	7	28
without admix.	R w/c 0.32	19.0	18.4	13.6	8.3	50	36	30	19	380	70	60	40
+ 1% NSF		33.7	28.8	25.2	15.4	50	40	36	18	380	120	70	40
without admix.	R w/c 0.40	33.7	28.8	25.2	13.9	84	64	50	25	660	440	120	75
+ 1% NSF		33.7	30.2	26.5	15.3	65	48	36	15	700	200	120	70

TABLE 5. Total porosity, threshold diameter and maximum continuous diameter on pozzolanic cement paste disks

		TOTAL POROSITY (%)				M.C.D. (nm)				THRESHOLD D (nm)			
CURING (days)		1	3	7	28	1	3	7	28	1	3	7	28
without admix.	R w/c 0.32	32.6	25.4	20.6	10.6	170	72	40	10	700	200	75	20
+ 1% NSF		32.2	27.2	23.2	12.5	130	60	20	10	400	120	70	40
without admix.	R w/c 0.40	38.3	32.5	30.3	20.4	500	100	75	15	1200	380	120	70
+ 1% NSF		40.5	35.1	29.6	22.6	360	170	65	11	1200	400	124	68

4   CONCLUSIONS

The interface area between cement paste and aggregate contributes to a considerable extent to the gas permeability of the paste-aggregate composite, and the closing process of open porosity, through which gas and liquid flow is more slowly in this area with respect to the bulk of the paste.

As a matter of fact, while pozzolanic cement pastes made up with a w/c ratio of 0.32 or 0.40 become practically impermeable to air after 28 days, the interface area exhibits appreciable permeability even later.

The use of a superplasticizer has proved to be a highly effective means to reduce the permeability of the interface not only as might be expected when used to lower the w/c ratio, but also when the amount of mixing water is kept unchanged.

The higher fluidity and plasticity conveyed to the paste by the admixture would improve the interface adhesion, but it cannot be excluded that the presence of the admixture causes changes in the interface structure because of the morphology changes in the products of hydration.

The reduction of permeability caused by the superplasticizer is specifically performed in the interface area, since the addition of admixture to the plain paste, w/c ratio being constant, slightly increases the permeability.

As porosity measurements have pointed out, this effect may be ascribed to an increase in the porosity of the paste due to an aerating by-effect exerted by the admixture.

This investigation has therefore stressed the prominent role of interface areas on concrete permeability while pointing out the important contribution of superplasticizers to its reduction.

References

Oberholster, R.E. (1986), Pore structure, permeability and diffusivity of hardened cement paste and concrete in relation to durability: status and prospects. 8th Int. Congr. on the Chem. of Cement (Rio de Janeiro, September), vol. I, 323

Feldman, R.F. (1986), Pore structure, permeability and diffusivity as related to durability. 8th Int. Congr. on the Chem. of Cement (Rio de Janeiro, September), vol. I, 336

Grandet, J. and Ollivier, J.P. (1980), Orientation of hydration products near aggregate surfaces. 7th Int. Congr. on the Chem. of Cement (Paris, 30 June-5 July), vol. III, VII-63

Grandet, J. and Ollivier, J.P. (1980), New method for the study of cement-aggregate interfaces. 7th Int. Congr. on the Chem. of Cement (Paris, 30 June-5 July), vol. III, VII-85

Barnes, B.D.; Diamond, S. and Dolch W.L. (1978), The contact zone between portland cement paste and glass "aggregate" surfaces. Cem. Concr. Res. 8 (2), 233

Barnes, B.D.; Diamond, S. and Dolch, W.L. (1979), Micromorphology of the interfacial zone around aggregates in portland cement mortar. J. Am. Cer. Soc. 62 (1-2), 21

Farran, J.; Javelas, R.; Maso, J.C. and Perrin, B. (1972), Etude de l'aureole de transition existant entre les granulats d'un mortier et la masse de la pâte de ciment hydraté. Coll. Int. RILEM/INSA: Liaison de Contact dans les Matériaux Composites utilisés en Génie Civil (Toulouse, November), vol. I, 60

Massazza, F. and Costa, U. (1986), Bond: paste-aggregate, paste-reinforcement and paste-fibres. 8th Int. Congr. on the Chem. of Cement (Rio de Janeiro, September), vol. I, 158

Tognon, G.P. and Cangiano, S. (1980), Interface phenomena and durability of concrete. 7th Int. Congr. on the Chem. of Cement (Paris, 30 June-5 July), vol. III, VII-133

Marsh, B.K.; Day, R.L.; Bonner, D.G. and Illston, J.M. (1983), Principles and applications of pore structural characterization. Proc. of the RILEM/CNR Int. Symp. (Milan, April), 365

Day, R.L. and March, B.K. (1988), Measurement of porosity in blended cement pastes. Cem. Concr. Res. 18, (1), 63

Costa, U. and Massazza, F. (1987), Remarks on the determination of the pore distribution of portland and pozzolanic cement pastes. Proc. of the First Int. RILEM Congress: Pore Structure and Materials Properties (Versailles, September), vol. 1 (Chapman and Hall Ed.), 159

REDUCTION OF DEFORMATIONS WITH THE USE OF CONCRETE ADMIXTURES

H. CHARIF, J.-P. JACCOUD and F. ALOU
Department of Civil Engineering, Reinforced and Prestressed Concrete Institute,
École Polytechnique Fédérale de Lausanne, Switzerland

Abstract
As part of a vast research programme concerning the calculation and the reduction of concrete slabs deformations, an experimental study on high strength concrete, H.S.C., and its use in the case of slabs has been undertaken. Rather than looking for concretes with an exceptional compressive strength, we investigated types of concrete characterized by a good tensile strength, and by a weak deformability, i.e. where the elastic modulus is as high as possible and the time-dependent deformations are as small as possible. We also tried to make concretes with good workability. In order to obtain the "optimum" high performances of these concretes, we varied the following parameters : the cement dosage, the dosage of a superplasticizer used as a water reducing agent, the eventual addition of silica fume, the maximum diameter and the shape of the aggregates. The experimental study, which lasted for two years (April 87 - April 89), was divided into three phases. The first two phases consisted of different time duration (short or long) tests on cylinders. These tests allowed us to choose the different types of concrete for the third phase which consisted of long-term experiments with simply supported slabs and with cylinders made at the same time as the slabs.

The objective of this paper is to present only the principal results of the tests carried out on cylinders during the three phases of the experimental study. The final results permit us to draw the following conclusions about the improved performances offered by the H.S. concretes in comparison to ordinary concretes :

    a) The compressive strength can be increased by 100 to 150 %.
    b) The elastic modulus can be increased by 20 to 30 %.
    c) The tensile strength can be increased by 40 to 60 %.
    d) The shrinkage deformations are more important at an
        early age, but attain the same values after one year.
    e) The creep deformations can be reduced by 30 to 60 %.

Key words :Elastic modulus, Tensile strength, Compressive strength, Creep, Shrinkage, Superplasticizer, Silica fume.

1   Introduction

The problem of excessive deformations in concrete slabs has become increasingly important in the last few years; abroad [1] as well as in Switzerland, as the expert appraisals can testify (to which we have to refer to occasionally). Today, the fact that the long-term real deflection of a concrete slab could be six to twelve times greater than the deflection resulting from a linear elastic calculation, is now internationally accepted [1],[2]. This amplification factor which increases the elastic deflection calculated by formulas, by numerical tables or resulting from a linear finite element program, is due to cracking as well as the time dependent effects (creep and shrinkage).
How can we reduce the deflections ? Several well known technical solutions have been applied in certain cases by engineers. We can mention for example, the resource to the prestressing or the trussing of the reinforcement. Today, a new approach to concrete material technology is rapidly developing. This approach seemed to us an interesting proposal for the reduction of deformations. In addition to high strengths, the H.S. concrete offers advantages hardly accessible to the ordinary concrete, we can notably mention :

   a) An excellent workability which permits a greater facility of concreting even in the case of elements with a high density of reinforcement.
   b) A binder paste extremely compact which results in a reduction of the porosity and the permeability.
   c) A very high resistance against chemical agents.
   d) A weak elastic and time dependent deformations.

To reduce the slab deformations by the use of H.S. concretes, we tried to exploit four performances offered by these concretes, three with a mechanical nature and one with a rheological nature :

   a) A better steel-concrete bond, which increases the tension stiffening effects and favourably influences the deflection and the crack widths.
   b) A better tensile strength, from which an increase of the cracking moment and a limited and delayed appearance of the cracking can be seen.
   c) A better elastic modulus, which decreases the elastic deformations.
   d) Less creep, which decreases the time dependent deformations.

Note that the compressive strength is not directly included in these parameters.

2   Experimental study objective

Many scientific publications concerning H.S. concretes show in general interesting results, but are sometimes contradictory. These interpretation problems come principally from the different test conditions and from the use of different products. For this raison, it seemed to us reasonable to make H.S. concretes with local materials and products. This allowed us to verify the feasibility of these kinds of concretes and to make the necessary corrections in the case of problems.

It is not a question in this paper of comparing different equivalent products (cement, superplasticizer, silica fume, etc...) and to state whether or not the products are better or worse than one another. The principal objective is to show that with certain types of good quality materials and products (without them being exceptional) and a certain concrete composition mixed according to the state of the art technology, that we can obtain a H.S. concrete which when used will allow us to reduce notably the long-term deformations of R.C. slabs.

3   Experimental   study   programme

To obtain the high performances of the concretes and achieve our aim, we tried various technical combinations of the following :

   a)   An increase of cement dosage without exceeding 450 kg/m^3.
   b)   An increase of the maximum diameter of the aggregates.
   c)   The use of angular coarse aggregates.
   d)   The use of a superplasticizer as a water reducing agent.
   e)   The addition of silica fume.

Note that other parameters concerning the concrete material (aggregate nature, grading curve, etc...), or curing rules may play an important role.

The experimental study, which lasted two years (April 87 - April 89), can be divided into three phases [3] :

   a) Phase 1 : short-term preliminary tests with cylinders in order to determine the mechanical properties of different H.S. or ordinary concretes and to optimise the design mix. These tests allowed us to select a limited number of concretes for the second phase.
   b) Phase 2 : long-term preliminary tests with cylinders. These tests allowed us to determine the long-term mechanical and rheological properties of several H.S. or ordinary concretes and to make the final choice of the concretes for the third phase.
   c) Phase 3 : long-term tests with simply supported slabs and cylinders (made at the same time as the slabs). The cylindrical tests allowed us to control the mechanical and rheological properties, while the slab measurements enabled us to determine the deflections, the curvatures and the cracking for different loading levels and for different H.S. or ordinary concretes.

In this paper, we only present and analyse the results obtained from the cylinders during the three phases of this study.

4 Characteristics of the materials used

4.1 Cement

One type of cement was used for all the tests : ordinary portland cement = OPC. This cement was prefered to the RHPC (rapid-hardening portland cement) because of its lower drying shrinkage. In addition, OPC represents about 93 % of the total consumption of all cements in Switzerland. The principal characteristics of the OPC used are as follows :

a) Specific surface (Blaine) $\cong 0,3$ (m^2/g)

b) Chemical composition (%) :

SiO_2	:	19,8	MgO	:	2,5
Al_2O_3	:	5,0	K_2O	:	0,3
Fe_2O_3	:	2,3	Na_2O	:	0,4
CaO	:	63,0	SO_3	:	2,5

c) Mineralogical composition (anhydrous cement) (%) :

C_3S	:	56,4	C_4AF	:	9,8
C_2S	:	14,3	$CaSO_4$	:	4,3
C_3A	:	7,8			

d) Indexes (-) :

hydraulic modulus : 2,3
siliceous modulus : 2,4
aluminous modulus : 1,6
lime saturation : 33,3

e) Compressive strength of mortar :
SIA mortar [4] (Swiss code) : sand 0/5 mm, mortar 1/3 (one part cement, 3 parts sand), W/C = 0,44.
f_C on half-prism 40/40/160 mm :

age (days)	:	2	7	28
f_C (average value on 30 cyl.) (N/mm^2)	:	32	48	58
coefficient of variation (%)	:	2,5	2,2	1,9

ISO (RILEM) mortar : sand 0/1,6 mm, mortar 1/3, W/C = 0,5
fc28 on half-prism 40/40/160 mm $\cong 37$ (N/mm^2)

The difference of obtained strengths at 28 days of SIA and ISO (RILEM) mortars comes from the differences of the mortar compositions and from the differences in the making and the conservation of the prisms.

4.2 Aggregates

The sand and the coarse aggregates used in this study are siliceous limestone moraine materials. We used four ranges :

a) rounded sand : 0/3 and 3/8 mm

b) rounded or angular coarse aggregates : 8/16 and 16/32 (or 16/40) mm.

The angular coarse aggregates were obtained by block crushing. Figure 1 shows the grading curves of the four rounded grading ranges : 0/3, 3/8, 8/16 and 16/32 mm.

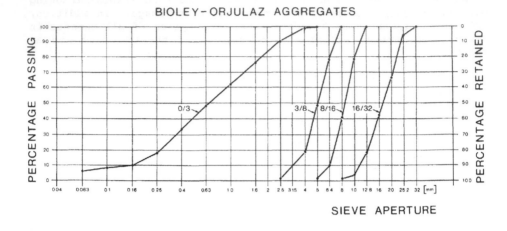

Fig. 1. Sieve grading curves

4.3 Superplasticizer

We used only one type of superplasticizer principally as a water reducing agent for all the tests : The Sikament 320 supplied by the Swiss company SIKA. It has the following composition :

a) Sulfonated melamine formaldehyde condensate
b) Hydroxylated carboxylic acid salt (sodium salt).
c) The dry extract of this superplasticizer is 40 %.

It was added to the mixing water used for all the mixes made with this superplasticizer.

4.4 Silica fume

The silica fume which was used was also supplied by SIKA; it was imported from the Germany.F.R. (trademark : V.A.W. RW-Füller)in a grey powder slightly densified form. The principal characteristics of this silica fume are as follows :

a) Chemical composition (%) :
SiO_2 : 96,5
K_2O : 0,4
Si (metal) : 0,4
MgO : 0,3
Al_2O_3 : 0,3

```
            SO₄⁻⁻                    :  0,25
            Na₂O                     :  0,1
            CaO                      :  0,1
            Fe₂O₃                    :  0,05
            loss of weight at 1100 °C  :  1,2
            C                        :  0,9
            H₂O                      :  0,2
```

b)   Bulk density (uncompacted)   :   170 - 190 (g/l)
c)   Bulk density (compacted)   :   260 - 290 (g/l)
d)   Absolute density   : 2,2 (g/cm³)
e)   PH   :   7-7,5 (-)
f)   X-ray diffraction sprectrum   :   amorphous structure
g)   Specific surface (BET)   :   20 - 22 (m²/g)
h)   Grading :
     secondary particles   (Ø > 1 µm)   : 30 (%)
     (agglomerate)   (Ø > 10 µm)   : 5 (%)
     Ø maximum   : 50 (µm)
     primary particles   : 0,1 - 0,3 (µm)
i)   Packaging   :   - bagged
                       - bulk supply in silos or lorries

This silica fume was added into the mixer just after the aggregates and before the cement for all the mixes made with silica fume.

5 Fixed parameters for concrete proportioning

5.1 Grading curve
As a reference curve, we chose the Fuller curve :

$$p\ (\%) = 100\sqrt{\frac{d}{D}}$$

This curve is situated inside the grading range defined in the new Swiss code SIA 162 [5].

5.2 Proportioning method
For the concretes with 300 kg/m³ of cement, we used the Fuller curve without any changes. For concretes with a higher dosage of cement or which contain silica fume, we used the empirical method of J. Bolomey (1879 - 1952). The reference curve of this method uses the Fuller curve, but takes into account the weight of cement. The equation for the Bolomey - Fuller curve is :

$$p\ (\%) = 100\ \sqrt{\frac{d}{D}} + K$$

$$K = A\ (1 - \sqrt{\frac{d}{D}})$$

where A = 5 to 16 according to the consistency of concrete and the aggregates shape. Bolomey's method is characterised by proportionings which contain small quantities of sand.

5.3 Consistency of fresh concrete and W/C ratio

We aimed for a slump of 70 mm (acceptable range : 50 to 90 mm) for all the concretes. This slump corresponds to a plastic to slighty fluid value (class S2 fresh concrete according to ISO/DIS 4103).

5.4 Mixing

Concerning the mixing of mixes, we used Eirich mixers with a vertical axis, from 180 to 400 litres capacity, according to the quantity of mix. The time of mixing was two minutes after all the components had been introduced.

6 Principal results

6.1 Phase 1 : short-term preliminary tests with cylinders

6.1.1 Composition of concretes

To fabricate the concretes of the first phase, we varied the five following parameters :

   a)   The cement dosage : 300,375 or 450 kg/m^3.
   b)   The maximum diameter of aggregates   : 32 or 40 mm.
   c)   The shape of coarse aggregate (Ø > 8mm) : rounded or angular.
   d)   The use of a superplasticizer : 0 or 2 % of cement weight.
   e)   The addition of silica fume   : 0 or 10 % of cement weight

Table 1 shows the composition of all the concretes.

6.1.2 Making and Conservation of cylinders

We used cylindrical moulds 160/320 mm made from cardboard, with a plastic film inside. All the specimens were conserved for three days in their moulds, then they were removed from their mould and their faces were rectified; after which they were put into water at 18 °C until the day of the test.

6.1.3 Mechanical Characteristics

Tables 2a, 2b and 2c show the principal results of the strengths and moduli in relation to the cement dosage and the age of concrete. The average values and the percentage increase with respect to the concrete A1 are also given.

6.1.4 Complementary tests

We carried out some complementary tests to investigate the influence of the silica fume dosage or the superplasticizer dosage. Table 3 presents the results of the strengths with the following dosages of silica fume : 0, 5, 10, 15 and 20 % of cement weight (c.w.); according to two cement dosages : 375 and 450 kg/m^3 with 2 % of c.w. of superplasticizer.

Table 1. Phase 1 : Composition of concretes

Name	Grading [%]				Cement [Kg/m3]	Sikamemt 320 [% p.c.]	Silica Fume [% p.c.]	Water [l/m3]	W/C [-]	Slump [cm]
	0/3	3/8	8/16	16/32(40)						
	0/32 rounded									
A1	32	22	20	26	300	0	0	179	0.60	5.5
A2	29	23	21	27	375	0	0	177	0.47	7
A3	26	24	22	28	450	0	0	184	0.41	7.5
	0/32 rounded									
BS1	32	22	20	26	300	2	0	148	0.49	***
BS2	29	23	21	27	375	2	0	144	0.38	***
BS3	26	24	22	28	450	2	0	156	0.35	***
BF1	31	22	22	26	300	2	10	160	0.53	5.5
BF2	27	24	21	28	375	2	10	156	0.41	5.5
BF3	23	25	25	29	450	2	10	150	0.33	5.5
	0/8 rounded 8/32 angular									
CS1	32	22	20	26	300	2	0	155	0.52	***
CS2	29	23	21	27	375	2	0	150	0.40	5.5
CS3	26	24	22	28	450	2	0	159	0.35	5
CF1	31	22	20	26	300	2	10	159	0.53	5
CF2	27	24	21	28	375	2	10	150	0.40	9
CF3	23	25	23	29	450	2	10	147	0.32	6
	0/8 rounded 8/40 angular									
ES1	29	19	18	34	300	2	0	147	0.49	***
ES2	26	20	19	35	375	2	0	140	0.37	***
ES3	22	21	20	37	450	2	0	140	0.31	***
EF1	27	19	19	34	300	2	10	146	0.48	4
EF2	27	20	19	36	375	2	10	146	0.38	4
EF3	20	22	20	38	450	2	10	141	0.31	7

*** = collapse or shear slump

Table 2a. Phase 1 : Compressive strength

* Average strengths of cylinders 160/320 mm [N/mm2]
** Relative increase with regard to concrete A1 [%]

Cement dosage [Kg/m3]	300		375		450		
Age of concrete [days]	3	28	3	28	3	28	
Concretes " A "	14.0	27.9	21.4	39.7	27.8	43.7	*
	0 %	0 %	+53 %	+42 %	+99 %	+57 %	**
Concretes " BS "	19.1	36.5	33.1	51.8	38.5	57.2	
	+36 %	+31 %	+136%	+86 %	+175%	+105%	
Concretes " BF "	25.4	55.3	34.7	63.5	43.8	70.8	
	+81 %	+98 %	+148%	+128%	+213%	+154%	
Concretes " CS "	22.4	39.3	37.1	55.6	41.1	59.0	
	+60 %	+41 %	+165%	+99 %	+194%	+111%	
Concretes " CF "	28.4	57.2	37.1	69.4	50.7	81.8	
	+103%	+105%	+165%	+149%	+262%	+193%	
Concretes " ES "	16.2	34.6	39.4	60.1	44.8	65.8	
	+16 %	+24 %	+181	+115	+220	+136%	
Concretes " EF "	32.0	62.9	40.5	74.0	51.2	82.5	
	+129%	+125%	+189%	+165%	+266%	+196%	

Table 2b . Phase 1 : Splitting strength

* Average strengths of cylinders 160/320 mm [N/mm2]
** Relative increase with regard to concrete A1 [%]

Cement dosage [Kg/m3]	300		375		450		
Age of concrete [days]	3	28	3	28	3	28	
Concretes " A "	1.87	3.44	2.74	4.02	3.11	4.14	*
	0 %	0 %	+47 %	+17 %	+66 %	+20 %	**
Concretes " BS "	2.36	3.90	3.75	4.19	3.61	4.77	
	+26 %	+13 %	+101%	+22 %	+93 %	+39 %	
Concretes " BF "	2.61	4.72	3.61	5.26	3.85	5.26	
	+40 %	+37 %	+93 %	+53 %	+106%	+53 %	
Concretes " CS "	2.74	4.06	3.61	5.22	3.36	5.51	
	+47 %	+18 %	+93 %	+52 %	+80 %	+60 %	
Concretes " CF "	3.11	4.89	3.48	5.10	4.60	5.43	
	+66 %	+42 %	+86 %	+48 %	+146%	+58 %	
Concretes " ES "	2.11	3.90	4.35	4.77	3.85	5.46	
	+13 %	+13 %	+133%	+39 %	+106%	+59 %	
Concretes " EF "	3.75	5.22	3.23	5.51	4.10	5.26	
	+101%	+52 %	+73 %	+60 %	+119%	+53 %	

Table 2c . Phase 1 : Elastic modulus

* Average elastic modulus of cylinders 160/320 mm [KN/mm2]
** Relative increase with regard to concrete A1 [%]

Cement dosage [Kg/m3]	300		375		450		
Age of concrete [days]	3	28	3	28	3	28	
Concretes " A "		24.6		30.6		31.4	*
		0 %		+24 %		+28 %	**
Concretes " BS "		28.4		32.2		34.6	
		+15 %		+31 %		+41 %	
Concretes " BF "		33.6		36.5		35.1	
		+37 %		+48 %		+43 %	
Concretes " CS "		34.1		41.0		39.7	
		+39 %		+67 %		+61 %	
Concretes " CF "		37.9		38.6		42.2	
		+54 %		+57 %		+72 %	
Concretes " ES "		33.2		40.2		41.2	
		+35 %		+63 %		+67 %	
Concretes " EF "		38.5		41.5		42.6	
		+57 %		+69 %		+73 %	

Table 3. Phase 1 - Complementary tests : Influence of silica fume

Age of concrete = 28 days
Aggregates 0/32 rounded
Dosage of superplasticizer = 2 % c.w.

* Average strengths of cylinders 160/320 mm [N/mm2]
** Relative increase with regard to concrete without silica fume [%]

			Cement dosage [Kg/m3]						
			375			450			
Silica F. [% c.w.]	Slump [cm]	W/C [-]	Splitting [N/mm2]	Comp. [N/mm2]	Slump [cm]	W/C [-]	Splitting [N/mm2]	Comp. [N/mm2]	
0	***	0.38	4.19	51.8	***	0.35	4.77	57.2	*
			+100 %	+100 %			+100 %	+100 %	**
5	9	0.35	5.16	64.7	7	0.28	5.10	70.4	
			+123 %	+125 %			+107 %	+123 %	
10	5.5	0.41	5.26	63.8	5.5	0.33	5.26	71.2	
			+126 %	+123 %			+110 %	+124 %	
15	5.5	0.37	5.04	72.6	6	0.32	4.97	75.3	
			+120 %	+140 %			+104 %	+132 %	
20	5	0.39	4.97	72.1	7	0.33	4.72	75.1	
			+119 %	+140 %			+99 %	+131 %	

*** = collapse or shear slump

411

Table 4 presents the results of the strengths with the following dosages of superplasticizer : 0, 1, 1,5, 2 and 3 % of c. w.; for an intermediary cement dosage of 425 kg/m³.

From the results obtained, we can say that the optimum dosage of the silica fume is about 8 % of c. w. for the splitting strength, while it is about 13 % of c.w. for the compressive strength. The optimum dosage of the superplasticizer seems to be about 1,8 % of c. w. for both the splitting and the compressive strengths.

Table 4. Phase 1 - Complementary tests: Influence of superplasticizer

Age of concrete = 28 days
Aggregates 0/32 rounded
Cement dosage = 425 [Kg/m3]
Silica fume dosage = 0 [% c.w.]

* Average strengths of cylinders 160/320 mm [N/mm2]
** Rel. increase with regard to concrete without superplasticizer[%]

Slump [cm]	W/C [-]	Super. [%]	Splitting [N/mm2]	Compression [N/mm2]	
7	0.45	0	4.10	42.4	*
			+100 %	+100 %	**
7	0.42	1	3.98	45.7	
			+97 %	+108 %	
9	0.38	1.5	4.60	54.4	
			+112 %	+128 %	
6.5	0.36	2	4.58	55.4	
			+112 %	+131 %	
20	0.35	3	4.23	49.4	
			+103 %	+117 %	

6.2 Phase 2 : long-term preliminary tests with cylinders

6.2.1 Composition of concretes
From the preliminary tests of phase 1, we chose three concretes : one ordinary concrete A1 and two H.S. concretes CS2 and CF2.   Table 5 shows the composition of these three concretes.

Table 5. Phase 2 : Composition of concretes

Name	Grading [%]				Cement [Kg/m3]	Sikamemt 320 [% p.c.]	Silica Fume [% p.c.]	Water [l/m3]	W/C [-]	Slump [cm]
	0/3	3/8	8/16	16/32						
A1	0/32 rounded				300	0	0	180	0.60	5
	32	22	20	26						
CS2	0/8 rounded 8/32 angular				375	2	0	150	0.42	18
	40	19	18	23						
CF2	0/8 rounded 8/32 angular				375	2	10	147	0.39	6
	32	22	20	26						

Because of some segregation problems, we corrected the grading of the concrete CS2, with superplasticizer, by increasing the fine sand 0/3 quantity from 29 to 40 % of the total mass of aggregates. We had some difficulties to obtain a slump of 70 mm for the concrete CS2. The consistency was too sensitive to the quantity of the added water : 2 or 3 liters/m^3 can fundamentaly change the consistency from stiff plastic to completely fluid.

6.2.2 Making and conservation of cylinders
The type and the making of cylinders are the same as what was described in paragraph 6.1.2. We only changed the conservation of the specimens : the cylinders were kept three days in their moulds, after this time their faces were rectified without removing the moulds and then they were placed into an air-conditioned hall with the following conditions : RH = 50 ± 5 % and T = 20 ± 1˚C.

6.2.3 Mechanical characteristics
We carried out four types of long-term tests : the compressive strength and the splitting strength at 3, 28, 182 and 364 days; as well as the direct tensile strength and the elastic modulus at 28, 182 and 364 days.
    Table 6 presents the obtained results; for each concrete we

Table 6. Phase 2 : Mechanical characteristics

* Average values of cylinders 160/320 mm
** Relative increase with regard to concrete A1 [%]

Age of concrete		3 days	28 days	6 months	1 year	
A1	Split.N/mm2	1.89	3.34	3.85	3.77	
	Comp. N/mm2	15.6	25.4	29.3	36.3	
	Tens. N/mm2		2.20	2.40	2.46	
	Mod. KN/mm2		25.1	31.2	29.7	
CS2	Splitting [N/mm2]	3.50	4.43	5.60	5.68	*
		+86 %	+33 %	+45 %	+51 %	**
	Comp. [N/mm2]	33.5	54.4	66.1	69.0	
		+116 %	+114 %	+125 %	+90 %	
	Tensile [N/mm2]		3.34	3.96	4.21	
			+52 %	+65 %	+71 %	
	Modulus [KN/mm2]		32.9	39.1	37.4	
			+31 %	+25 %	+26 %	
CF2	Splitting [N/mm2]	2.87	5.37	5.40	5.51	
		+52 %	+61 %	+40 %	+46 %	
	Comp. [N/mm2]	43.1	67.0	71.7	76.7	
		+177 %	+163 %	+144 %	+112 %	
	Tensile [N/mm2]		3.86	4.05	4.13	
			+76 %	+68 %	+68 %	
	Modulus [KN/mm2]		35.1	40.3	35.9	
			+40 %	+29 %	+21 %	

indicate the average values of the strengths and moduli as well as the relative percentage increase of the concretes CS2 and CF2 with regard to the concrete Al.

6.2.4 Shrinkage

We measured the axial shrinkage of cylindrical specimens from the age of 3 days. The deduced deformation can be named total shrinkage deformation, ε_{cs}.

For each of the three concretes, we measured the shrinkage on two types of cylindrical specimens :

a) Three cylinders demoulded (shrinkage "without carton")

b) Six cylinders partially conserved in their mould (shrinkage "with carton")

All the specimens were placed in an air-conditioned hall (RH = 50 ± 5 %; T = 20 ± 1°C). Figure 2 shows the evolution of shrinkage deformation of the three concrete Al, CS2 and CF2, "with" and "without" carton.

6.2.5 Creep

A compressive force of 150 KN was applied at the age of 28 days on three specimens of each concrete. This force represents a compressive stress of 7,5 N/mm² for all the specimens. The cylinders were partially conserved in their carton mould and were placed in the same air-conditioned hall as the shrinkage cylinders (RH = 50 ± 5 %, T = 20 ± 1 °C). Figure 3 shows the creep deformations, ε_{cc}, of the three concretes Al, CS2 and CF2.

We also determined also the instantaneous deformation, ε_{co}, and deduced the instantaneous modulus, E_{co}. Table 7 indicates for each of the three concretes the value of ε_{co} as well the value of E_{co} calculated from the force - displacement graphs at the time of loading. Also shown are the modulus values measured according to RILEM procedure.

From the instantaneous deformations, ε_{co}, and those of creep, ε_{cc}, we calculated the creep coefficient, φ, for each of the 3 concretes. Figure 4 shows the coefficient, $\varphi = \dfrac{\varepsilon_{cc}}{\varepsilon_{co}}$, of the three concretes Al, CS2 and CF2.

Table 7. Phase 2 : Instantaneous deformations, instantaneous and secant moduli

Concrete	A 1	CS2	CF2
Inst. measured deformation ε_{co} [‰]	0,333	0,211	0,191
E_{co} inst. modulus [KN/mm²]	26,6	36,0	39,5
E_c secant modulus RILEM [KN/mm²]	25,3	32,9	35,1

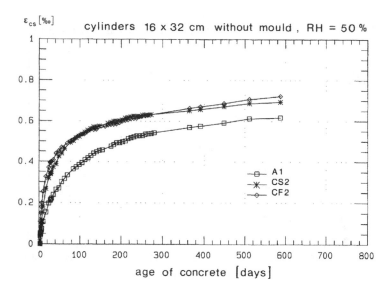

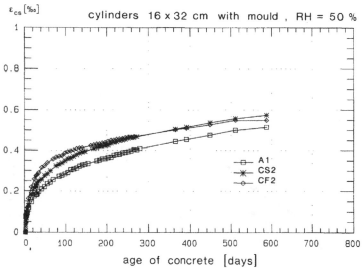

Fig. 2. Phase 2 : Shrinkage deformations

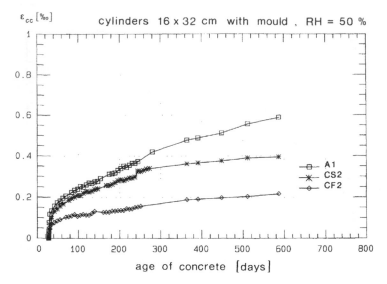

Fig. 3. Phase 2 : Creep deformations

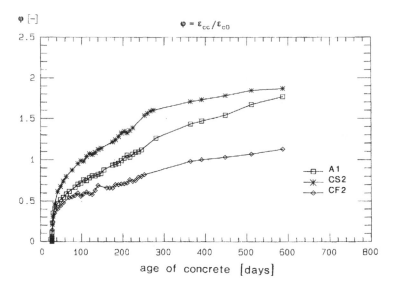

Fig. 4. Phase 2 : Creep coefficients

Table 8 summarizes all the deformations of the three concretes up to one year of measurments.

Figure 5 compares the total deformations, $\varepsilon_{tot} = \varepsilon_{co} + \varepsilon_{cc} + \varepsilon_{cs}$, of concrete A1 with those of concretes CS2 and CF2.

Table 8. Phase 2 : Average relative deformations from t-t_0 = 0 to 1 year [‰]

* Creep coefficient [-]. t_0 = 28 days (σ_c = 7.5 N/mm^2)
** Relative difference with regard to concrete A1 [%]
RH = 50 ± 5 %, T = 20 ± 1°C

Concrete	"A1"					"CS2"					"CF2"				
t-t_0	ε_{co}	ε_{cs}	ε_{cc}	ε_{ctot}	φ [-]*	ε_{co}	ε_{cs}	ε_{cc}	ε_{ctot}	φ [-]*	ε_{co}	ε_{cs}	ε_{cc}	ε_{ctot}	φ [-]*
0	0.333	0.182	0	0.515	0	0.211	0.222	0	0.433	0	0.191	0.275	0	0.466	0
						-37 %	+22 %	-	-16 %	-	-43 %	+51 %	-	-10 %	-
1 day	0.333	0.184	0.076	0.593	0.23	0.211	0.227	0.044	0.482	0.21	0.191	0.276	0.027	0.494	0.14
						-37 %	+23 %	-42 %	-19 %	-9 %	-43 %	+50 %	-64 %	-17 %	-39 %
7 days	0.333	0.197	0.133	0.663	0.40	0.211	0.249	0.097	0.557	0.46	0.191	0.291	0.067	0.549	0.35
						-37 %	+26 %	-27 %	-16 %	+15 %	-43 %	+48 %	-50 %	-17 %	-13 %
28 days	0.333	0.236	0.182	0.751	0.55	0.211	0.285	0.157	0.653	0.74	0.191	0.340	0.091	0.622	0.48
						-37 %	+21 %	-14 %	-13 %	+35 %	-43 %	+44 %	-50 %	-17 %	-13 %
3 months	0.333	0.308	0.260	0.901	0.78	0.211	0.370	0.227	0.858	1.08	0.191	0.403	0.113	0.707	0.59
						-37 %	+20 %	-13 %	-5 %	+38 %	-43 %	+31 %	-57 %	-22 %	-24 %
6 months	0.333	0.366	0.351	1.050	1.05	0.211	0.436	0.281	0.928	1.33	0.191	0.450	0.137	0.778	0.72
						-37 %	+19 %	-20 %	-12 %	+27 %	-43 %	+23 %	-61 %	-26 %	-31 %
1 year	0.333	0.455	0.488	1.276	1.47	0.211	0.515	0.366	1.092	1.74	0.191	0.510	0.191	0.892	1.0
						-37 %	+13 %	-25 %	-14 %	+18 %	-43 %	+12 %	-61 %	-30 %	-32 %

6.2.6 Complementary test

The creep of the concrete CS2 seemed quite high with respect to the ordinary concrete A1. For this reason we concluded that it was possible to reduce the creep of this concrete by reducing the quantity of sand 0/3 mm. In fact the quantity of sand in the concrete CS2 was high (40 % of the total mass of aggregates) and we knew that the fine aggregates favoured creep and shrinkage.

In this light, complementary tests were carried out while varying the dosage of the fine sand, 0/3 mm, from 29 to 40 %, with a cement dosage of 425 kg/m^3 and with 2 % c. w. of a superplasticizer without silica fume.

Table 9 indicates the results of the splitting strength and the compressive strength at 28 days.

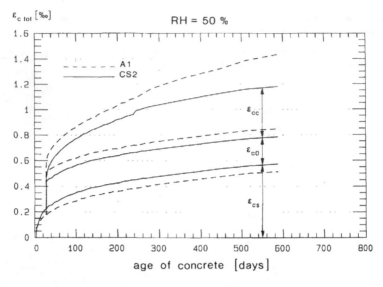

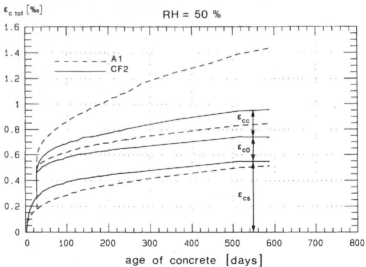

Fig. 5. Phase 2 : Total deformations

Table 9. Phase 2 - Complementary tests : Influence of fine sand 0/3

Age of concrete = 28 days
Aggregates 0/32 rounded
Cement dosage = 425 [Kg/m3]
Dosage of silica fume = 0 [% c.w.]
Dosage of superplasticizer = 2 [% c.w.]

* Average strengths of cylinders 160/320 mm [N/mm2]
** Relative increase with regard to concrete with 29% of sand 0/3 [%]

Slump [cm]	W/C [-]	fine sand [%]	Splitting [N/mm2]	Compression [N/mm2]	
***	0.36	29	4.58	55.4	*
			+100 %	+100 %	**
6.5	0.34	32	5.78	62.8	
			+126 %	+113 %	
6.5	0.36	35	4.60	60.1	
			+100 %	+107 %	
9.0	0.38	37	4.10	58.3	
			+90 %	+105 %	
11.0	0.39	40	3.98	57.0	
			+87 %	+103 %	

*** = collapse or shear slump

We notice that when we increased the sand content from 29 to 32 % the strengths (splitting and compressive) increased by an average of 20 %. This result can be explained by the fact that we passed from a concrete which presents some segregation signs to a concrete with a good cohesion. From 32 % just to 40 % of sand content, we noted a decrease of 40 % in the tensile strength and 10 % in the compressive strength. For the third phase, we chose a 0/3 mm sand content representing 35 % of the total mass of aggregates.

6.3 Phase 3 : long-term tests with cylinders (and slabs)

6.3.1 Composition of concretes
Table 10 shows the composition of the four concretes chosen for the slab experiments in this final phase.

Table 10. Phase 3 : Composition of concretes

Name	Grading [%]				Cement [Kg/m3]	Sikamemt 320 [% c.w.]	Silica Fume [% c.w.]	Water [l/m3]	W/C [-]	Slump [cm]
	0/3	3/8	8/16	16/32						
A1	0/32 rounded				300	0	0	180	0.60	6.5
	32	22	20	26						
A4	0/32 rounded				375	0	0	179	0.48	5.5
	32	22	20	26						
CS4	0/8 rounded 8/32 angular				425	2	0	159	0.38	11
	35	21	19	25						
CF4	0/8 rounded 8/32 angular				388	2	10	156	0.40	6
	35	21	19	25						

To pass from concrete CS2 in phase 2 to concrete CS4 in phase 3, we added 50 kg/m^3 of cement and reduced the content of the fine sand, 0/3 mm, from 40 to 35 % of the total mass of aggregates. The W/C ratio was reduced from 0,42 for CS2 to 0,38 for CS4.

The total quantity of binder of the concrete CF4 is 425 kg/m^3 (388 kg/m^3 of cement plus 37 kg/m^3 of silica fume). This concrete is almost identical to the concrete CF2 of phase 2 (difference of 13 kg/m^3 of cement). The W/C ratios for these two concretes were almost identical (0,39 for CF4 and 0,4 for CF2).

6.3.2 Making and conservation of cylinders

We made 249 cylinders for the four concretes. The type and the making of cylinders were described in paragraph 6.2.2. The cylinders were kept in their moulds for 6 days (same curing regime as for the slabs), then the faces were rectified without removing the mould. The cylinders were then stored in an air-conditioned hall (RH = 50 ± 5 %, T = 20 ± 1 ·C) until the day of testing.

6.3.3 Heat of hydration

Figure 6 shows the evolution of temperatures between 0 and 48 hours for each of the four concretes. We notice that the concrete CS4 presents an important retarding set. The induction period (or "sleeping" period) lasted until 17 hours. This can be explained by the use of a high dosage of superplasticizer and by the combined retarding effect of the superplasticizer itself and the added retarder (Sodium salt).

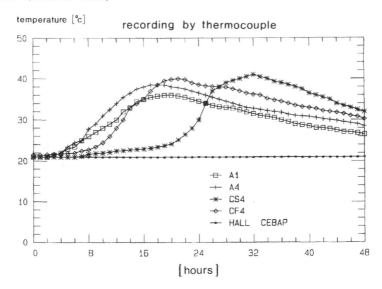

Fig. 6. Phase 3 : Temperature evolution

We notice also that the hydration rate of concretes CS4 and CF4 is greater than that of concretes A1 and A4, but globally speaking the total quantity of heat given off was kept more or less the same for the four concretes. Figure 7 shows the evolution curves of the compressive strength ("skin" strength), deduced from the schmidt hammer measurments. We notice that the retarding set of the concrete CS4 does not have a detrimental influence on its hardening.

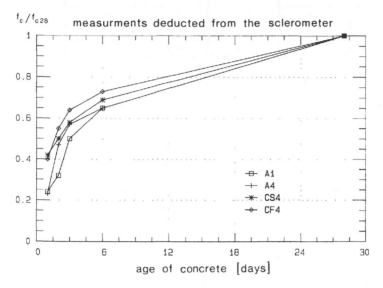

Fig. 7. Phase 3 : Cube compressive strength evolution

6.3.4 Mechanical characteristics

Table 11 shows the average values of the strengths and moduli results of the four concretes A1, A4, CS4 and CF4, as well as the percentage increases for concretes A4, CS4 and CF4 with regard to the concrete A1.

Table 11. Phase 3 : Mechanical characteristics.
* Average values of cylinders 160x320 mm
** Relative increase with respect to concrete A1 [%]

Age of concrete		3 days	28 days	6 months	1 year	
A1	Split.N/mm2	1.87	3.85	3.85	4.62	
	Comp. N/mm2	13.6	28.0	29.3	40.5	
	Tens. N/mm2		2.09	2.40	3.06	
	Mod. KN/mm2		29.0	31.2	33.1	
A4	Splitting [N/mm2]	2.74	4.04	4.48	4.97	*
		+47 %	+5 %	+16 %	+8 %	**
	Comp. [N/mm2]	23.0	36.4	44.1	49.5	
		+70 %	+30 %	+50 %	+22 %	
	Tensile [N/mm2]		2.47	2.89	3.61	
			+18 %	+20 %	+18 %	
	Modulus [KN/mm2]		31.0	33.6	34.5	
			+7 %	+14 %	+4 %	
CS4	Splitting [N/mm2]	3.48	4.56	5.17	5.35	
		+87 %	+18 %	+34 %	+16 %	
	Comp. [N/mm2]	32.6	56.7	67.1	70.1	
		+140 %	+103 %	+129 %	+73 %	
	Tensile [N/mm2]		3.19	3.99	3.97	
			+53 %	+66 %	+30 %	
	Modulus [KN/mm2]		33.8	35.6	37.7	
			+17 %	+14 %	+14 %	
CF4	Splitting [N/mm2]	3.63	4.56	5.18	5.78	
		+94 %	+18 %	+34 %	+25 %	
	Comp. [N/mm2]	39.0	71.4	71.7	78.2	
		+187 %	+155 %	+144 %	+93 %	
	Tensile [N/mm2]		3.60	4.05	4.52	
			+73 %	+68 %	+48 %	
	Modulus [KN/mm2]		34.6	40.3	37.9	
			+19 %	+29 %	+15 %	

6.3.5 Shrinkage

In this phase, we started to measure the shrinkage deformations of the four concretes A1, A4, CS4 and CF4 at the age of 6 days. This age corresponds to the age of the slabs at which their formwork was struck.

Figure 8 shows the evolution of shrinkage deformations, ε_{cs}, for the four concretes A1, A4, CS4 and CF4 with or without "carton".

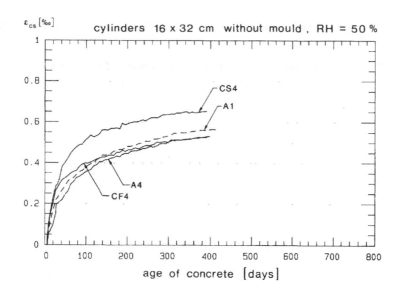

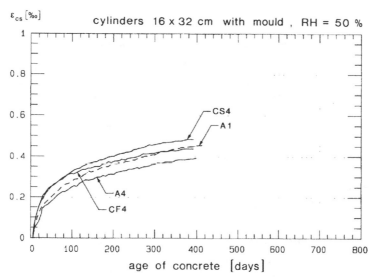

Fig. 8. Phase 3 : Shrinkage deformations

6.3.6 Creep

We carried out creep tests identical to those described in paragraph
6.2.5, on three concretes A4, CS4 and CF4. During this phase, we did
not load cylinders of concrete A1 and we took the results obtained in
phase 2.

Figure 9 shows the creep deformations of the four concretes A1,
A4, CS4 and CF4.

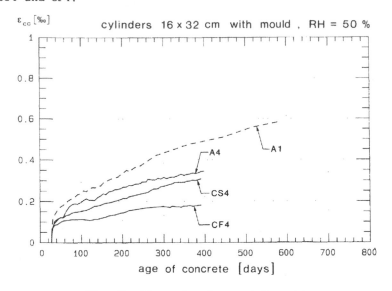

Fig. 9. Phase 3 : Creep deformations

Table 12 indicates for each of the four concretesthe values of the
measured instantaneous deformations, ε_{co}, and the calculated modulus,
E_{co}, when the load was applied (to = 28 days), as well as the measured
modulus according to RILEM.

Figure 10 shows the creep coefficients, $\varphi = \dfrac{\varepsilon_{cc}}{\varepsilon_{co}}$, of the four
concretes.

Table 13 summarizes the total deformations of the three concretes
A4, CS4 and CF4.

Figure 11 compares the total deformations of concrete A1 with
those of concretes A4, CS4 and CF4.

Table 12. Phase 3 : Instantaneous deformations, instantaneous and secant moduli

Concrete	A1 *	A4	CS4	CF4
Inst. measured deformation ε_{co} [‰]	0,333	0,298	0,220	0,205
E_{co} inst. modulus [KN/mm^2]	26,6	27,3	30,2	37,5
E_c secant modulus RILEM [KN/mm^2]	25,3	31,0	33,8	34,6

* Values taken from phase 2

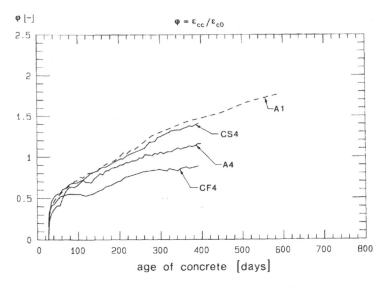

φ [-]   φ = ε_cc / ε_cO

age of concrete [days]

Fig 10. Phase 3 : Creep coefficients

Table 13. Phase 3 : Average relative deformations from $t-t_0 = 0$ to 1 year [%]

* Creep coefficient [-], t_0 = 28 days (σ_c = 7,5 N/mm^2)
** Relative difference with regard to concrete A1 [%]
RH = 50 ± 5 %, T = 20 ± 1°C

Concrete	"A4"					"CS4"					"CF4"				
$t-t_0$	ε_{co}	ε_{cs}	ε_{cc}	ε_{ctot}	φ [-]*	ε_{co}	ε_{cs}	ε_{cc}	ε_{ctot}	φ [-]*	ε_{co}	ε_{cs}	ε_{cc}	ε_{ctot}	φ [-]*
0	0,298	0,139	0	0,437	0	0,220	0,189	0	0,409	0	0,205	0,195	0	0,400	0
	- 11 %	- 24 %	-	- 15 %	-	- 34 %	+ 4 %	-	- 21 %	-	- 38 %	+ 7 %	-	- 22 %	-
1 day	0,298	0,147	0,046	0,491	0,15	0,220	0,190	0,053	0,463	0,24	0,205	0,198	0,067	0,470	0,33
	- 11 %	- 20 %	- 40 %	- 17 %	- 35 %	- 34 %	+ 3 %	- 30 %	- 22 %	+ 4 %	- 38 %	+ 9 %	- 12 %	- 21 %	+ 44 %
7 days	0,298	0,157	0,094	0,549	0,32	0,220	0,213	0,102	0,535	0,46	0,205	0,219	0,084	0,508	0,41
	- 11 %	- 20 %	- 29 %	- 17 %	- 20 %	- 34 %	+ 8 %	- 23 %	- 19 %	+ 15 %	- 38 %	+ 11 %	- 37 %	- 23 %	+ 3 %
28 days	0,298	0,191	0,123	0,612	0,41	0,220	0,264	0,122	0,606	0,56	0,205	0,260	0,106	0,571	0,52
	- 11 %	- 19 %	- 32 %	- 19 %	- 26 %	- 34 %	+ 12 %	- 33 %	- 19 %	- 4 %	- 38 %	+ 10 %	- 42 %	- 24 %	- 6 %
3 months	0,298	0,269	0,207	0,774	0,69	0,220	0,343	0,164	0,727	0,75	0,205	0,329	0,109	0,643	0,53
	- 11 %	- 13 %	- 20 %	- 14 %	- 12 %	- 34 %	+ 11 %	- 37 %	- 19 %	- 4 %	- 38 %	+ 7 %	- 58 %	- 29 %	- 32 %
6 months	0,298	0,324	0,271	0,893	0,91	0,220	0,409	0,215	0,844	0,98	0,205	0,379	0,152	0,736	0,74
	- 11 %	- 12 %	- 23 %	- 15 %	- 13 %	- 34 %	+ 12 %	- 39 %	- 20 %	- 7 %	- 38 %	+ 4 %	- 57 %	- 30 %	- 30 %
1 year	0,298	0,390	0,345	1,033	1,16	0,220	0,485	0,310	1,015	1,41	0,205	0,439	0,183	0,827	0,89
	- 11 %	- 14 %	- 29 %	- 19 %	- 21 %	- 34 %	+ 7 %	- 37 %	- 20 %	- 4 %	- 38 %	- 4 %	- 63 %	- 35 %	- 40 %

425

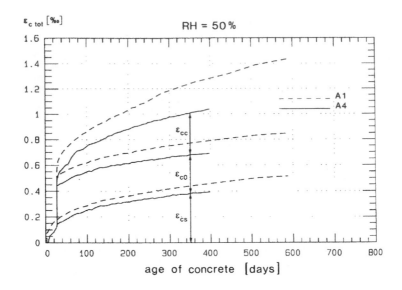

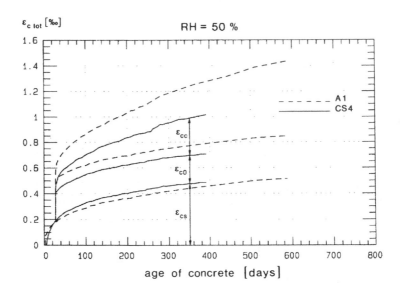

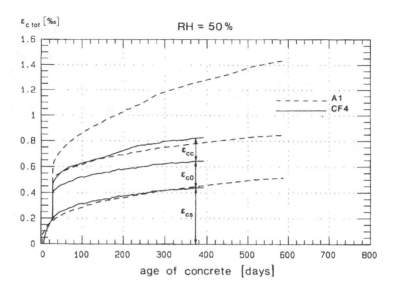

Fig. 11. Phase 3 : Total deformations

7 Conclusions

The conclusions which are given below are based on the comparison between an ordinary concrete which is currently used in the construction industry in Switzerland and the H.S. concretes with or without silica fume, but always with a superplasticizer used as a water reducing agent, made from good quality local products and materials without them being exceptional. All the concrete, ordinary or H.S., have also the same consistency.

Concerning the long-term mechanical properties of the H.S. concretes, the experiments with cylinders revealed the following conclusions :

a) The compressive strength of the H.S. concretes with or without silica fume increased by large proportions with regard to an ordinary concrete (100' - 150 % increase).

b) The tensile strength of the H.S. concretes increased by smaller proportions (40 - 60 % increase).

c) The elastic modulus increased by 20 - 30 %.

d) The cement dosage, the use of superplasticizer as a water reducing agent, and the addition of silica fume seemed to play an important role in these improvements (85 - 90 % of the increases).

e) The substitution of rounded coarse aggregates by angular coarse aggregates increases slightly the strengths (10 % of the increases). On the other hand it plays an important role concerning the increasing of the modulus (30 % of the increases).

f) The fact of increasing the maximum diameter from 32 to 40 mm does not gave any favorable effect in our tests on strengths or moduli.

g) These meechanical characteristics, especially the elastic modulus of the H.S. concretes with or without fume rapidly increased during the first month with relation to those of the ordinary concrete but the rate subsequently slowed down.

h) The effect of the silica fume is very benefical at an early age but seems to decrease with time.

Concerning the rheological properties of the H.S. concretes, the experiments with cylinders revealed the following conclusions :

a) At an early age, the shrinkage deformations, ε_{cs}, of the H.S. concretes with or without silica are greater than those of the ordinary concrete (25 - 50 % increase). This difference appears to decrease with time and after one year the shrinkage deformations were of the same order or even less for the H.S. concrete with silica fume.

b) The creep deformations, ε_{cc}, of the H.S. concrete without silica fume are notably reduced (30 % reduction) to those of the ordinary concrete but the creep coefficients have the same values. The creep deformations of the H.S. concrete with silica fume are greatly reduced (60 % reduction) and the creep coefficient also (40 % reduction).

c) The evolution with time (the kinetic) of the shrinkage of the H.S. concretes with or without silica fume is quicker than that of the ordinary concrete. This is also valid for the evolution of creep with time but at a slower rate.

8 Acknowledgments

The authors acknowledge all the institutions and personnel who have helped with this research programme. In particular we wish to thank the Swiss cement industry (VSZKGF) and the firm SIKA for their financial support.

References
[1] ACI Committee 435 : Observed deflections of reinforced concrete slabs systems, and causes of large deflections. Deflection of concrete structures, ACI, SP-86, 1985.
[2] CEB Manuel : Cracking and Deformations, EPFL, Lausanne, 1985.
[3] H. Charif, J.-P. Jaccoud : Réduction des déformations et amélioration de la qualité des dalles en béton. Rapport final sur les études expérimentales, Publication IBAP N° 133, juillet 1989.
[4] Norme SIA 215 : Liants minéraux, Société suisse des ingénieurs et des architectes, Zurich, 1989.
[5] Norme SIA 162 : Ouvrages en béton, Société suisse des ingénieurs et des architectes, Zurich, 1989.
[6] F. Alou, H. Charif et J.-P. Jaccoud : Bétons à hautes performances. Revue "Chantiers / Suisse", Vol. 19, pages 725-730, 9/1988.

INFLUENCE OF SUPERPLASTICIZERS ON THE BLEEDING CHARACTERISTICS OF FLOWING CONCRETE

E. G. F. CHORINSKY
Chemotechnik Abstatt GmbH, Abstatt, West Germany

Abstract
The great potential of super-plasticizers as water reducing agents has brought some disadvantages to job sites where concrete is placed at flowing conistency. The difficulties which might arise from bleeding and segregation in concrete mixes with high slump can be avoided by using combinations of stabilizing agents and melamine or naphthalene sulfonates. Air entrainment can be used for improving homogenity and stability of the concrete mix and thereby improve workablity and frost/thaw-resistance. The enhanced properties of the stabilized wet concrete will directly lead to a better performance and a longer service-time of concrete structures even under severe climate conditions.

1. Introduction

The remarkable progress which was made by the introduction of super-plasticizers to the concrete industry in the early Seventies has also brought some disadvantages specially in the field of application of flowing concrete. It was certainly not the fault of the chemical industry that specially the bleeding characteristics of super-plasticizers were neglected in the last years, even concrete experts sometimes forget how hazzardous bleeding can be.

It seems quite natural that the easy placeability of flowing concrete together with the increase compression strength and the related mechanical data could become so impressive that bleeding seemed to be of minor importance. That this is probably a very expensive error will be explained in this paper.

2 Reduction of the mixing water

The first synthetic super-plasticizers introduced to the markets in Japan, North America and the United States were Mighty and Melment, the first on naphthalene sulfonate basis, the later on melamine sulfonate basis. Both materials were favoured over the well known lignin sulfonates because of their great ability to reduce the mixing water. As the ultimate goal of a good concrete design is to use a minimum amount of water to obtain a given consistency, these synthetic materials were rather enthusiastically introduced to the market. They were designed to reduce mixing water up to 50 %, without any remarkable side effects. The use of minimum mixing water permits a reduction in cement and thereby a tremendous increase in strength together with the positive effects on shrinkage, creep and durability. The problem of access mixing water in the cement paste seemed to be solved and it became obvious that this benefits together with a use of a minimum paste content could lead to a great progress in concrete technology.

Many theories were brought up and discussed about the mechanisms of super-plasticizers and its chemical activities. But still not all phenomena are completely understood. To simplify our considerations we can say that super-plasticizers are acting as dispersing aids and do not entrain air to the paste. Their surface activity is highly increased when they can react on wetted cement particles and therefore it is strictly recommended to add super-plasticizers to a pre-mixed wet concrete and not to the mixing water.

The dispersion of cement offers a better grading of the fine particles to the mix and influences considerably the lubrication and reduction of mixing water.

Some authors have described the mechanisms of super-plasticizers as surface active agents which are selectively absorbed on cement grains and thereby influence the formation of certain initial hydration products. It was also said that release of entraped air and mixing water between aggregated and clumped cement particles could effect the workability and strength development. These and similar explanations of the water reducing effect of super-plasticizers are very interesting but do not help to improve understanding in practice. In this respect the theory of electrical charges on the cement particles which improve dispersion by surfactant absorption is more helpful as it explains the separation of the very fine particles in a very understandable way. Nevertheless we should keep in mind that all theories are scientifically not proved yet and that all super-plasticizers react more effectively on a water-shell than on the solid state material itself.

3 Effects of entrained air

As super-plasticizers improve the packing in a concrete mix by an improved grading in the range of the fines, specially the cement, a self-compacting effect results at high consistencies. As a flowable cement paste tends to release the entraped air unmodified super-plasticizers will take the air out of a concrete mix. As this effect is related to an increase in compression strength, it was considered as one of the benefits of super-plasticizers for a long time. Probably the high air entrainment by lignin sulfonates and other natural plasticizing agents, which were only available before the synthetic materials appeared on the market has influenced this approach.

The benefits of entrained air on the physical properties of concrete are commonly neglected in practice. Air entrainment usually is applied together with super-plasticizers only when improved frost/thaw resistance is required.

The most important benefits of controlled air entrainment, mainly to improve stability and homogenity of the concrete, are often forgotten. This is somewhat contradictory because it was already known since the late Thirties that controlled air entrainment could greatly improve the durability of concrete even though it decreased its unit weight. It is also known that entrained air can reduce the mixing water and gives a more homogeneous, less permeable concrete by distribution of microscopic fine spheroids in the cement paste. These small air-voids decrease porosity by reducing the capillarity and the water channel structure of the cement matrix. The workability of the concrete is improved although no self-compacting effect by flowability can be achieved if the air-content is high enough.

4 Effects of homogenity on frost/thaw-resistance

The U.S. Army Corps of Engineers has run field tests on frost/thaw-resistance by exposing several hundreds of concrete specimen to sea water splashing in a rather frosty climate. As these tests were done over a period of more than 10 years now, the results give an impressive view of the disability of super-plasticizers to reduce ice-scaling. They show that all specimen with controlled air entrainment show better frost/thaw resistance at comparable compression strength. But there is also a tendency that high slump concretes without air entrainment show poorer frost/thaw resistance than concretes with the same strength but prepared at a lower slump.

To understand this phenomena we should concentrate on the conditions in the fresh cement paste.

Without air entrainment we usually achieve bleeding concretes when we work at high consistencies even if high cement contents are provided. Certainly the greater fineness of the cement grain can reduce bleeding and also other fine particles as grinded limestone, fly-ash or micro-silica can help to avoid bleeding. Nevertheless on normal job sites where flowing consistency is practiced almost all concretes are bleeding. Of course this goes together with segregation and honey combing. The problem simply is that there are not enough fines to stabilize flowing concrete and if the fines are increased, the mixing water is usually increased too. When the placed concrete begins to settle the normal effect is increased bleeding.

It was even thought that this sort of bleeding could be helpful because it lowers the water/cement ratio in the inner parts of the concrete and could prevent drying shrinkage by keeping the concrete surface wet. Unfortunately with the bleeding water a lot of fines and specially these which are not or less reactive than the cement come to the surface. The increased water/cement ratio later would increase drying shrinkage and cracking. In countries where it is not prohibited to grind inreactive materials to the cement, concrete surfaces of that kind cause a lot of costs and troubles and restrict the use of flowing concrete. Such surfaces are not resistant to frost.

Even more important for the durability of a concrete can be the micro-cracking which occurs by the internal bleeding of concretes which are not homogeneous and stable enough and by the use of super-plasticizers tend to segregate. At the micro crack frost scaling usually starts and therefore a concrete which is exposed to severe weather conditions should never show any external bleeding because it will have without any doubt internal bleeding, too.

431

5 Minimizing bleeding effects

There are wonderful materials available to stabilize cement paste, such as high molecular polyethylenoxides, which are able to homogenize wet concrete to such an extend that almost no bleeding is possible. Bleed water formation is eliminated by binding of water to all components of the mix and thickening the cement paste. Unfortunately these materials cannot be combined with the now available super-plasticizers.

The common method of entraining air together with a high water reducing agent is the most suitable way in practice. The problem of compatibility still remains but it can be overcome by using combinations of air entraining materials which can give a grading of fine air bubbles. These air bubbles act in the concrete as fine aggregate, distributed throughout the cement paste in the concrete. As the air bubbles add to the workability of the plastic concrete they can help the super-plasticizer component to disperse the cement and other fine particles. This lubricating effect results in a highly workable concrete which cannot bleed or cause aggregate segregation and thereby provide homogenity throughout the total mix. The intentionally entrained air cannot contribute to the strength of the concrete and therefore the air entrainment has to be limited to the amount necessary for stabilizing and the corresponding frost protection. Super-plasticizers combined with air entraining agents can be formulated to keep the air content between 3 and 4 % depending on the cement, mixing time, consistency and temperature.

6 Conclusion

The important features of a highly workable concretes as plasticity, homogenity and strength development can be achieved by super-plasticizers with stabilizing properties through controlled air entrainment. External and internal bleeding can be eliminated even for high slump concrete. The enhanced properties of the plastic concrete will pay out in greater durability of the hardened concrete. Considerably reduced micro-cracking and permeability will be achieved to provide better frost/thaw resistance.

References

Czernin, W. Gf. (1977) Zementchemie für Bauingenieure, Bauverlag GmbH, Wiesbaden und Berlin
Woermann, H. H. (1977) Beton, Vorschrift und Praxis, Verlag Wilhelm Ernst & Sohn, Berlin, München, Düsseldorf
Sellevold, E. J. (1987) The Function of Condensed Silica Fume in High Strength Concrete, in Utilization of High Strength Concrete (ed. I. Holand, St. Helland, B. Jakobsen and R. Lenschow, Symposium in Stavanger, Norway, pp 39-49.
Malhotra, V. M, Carette, G. G. and Bremmer, T W. (1988) Current Status of CANMET's Studies on the Durability of Concrete Containing Supplementary Cementing Materials in Marine Environment, in Concrete in Marine Environment, Proceedings (ed. V. M. Malhotra), Second International Conference St. Andrews by-the-Sea, Canada, pp 31-72.

UNE ETUDE DE L'INFLUENCE DU POURCENTAGE DE GRANULAT SUR LA RESISTANCE A LA COMPRESSION DU BETON, LORQUE L'ON UTILISE DES EXPANSIFS NON-METALLIQUES

(A study of the influence of aggregates on the compressive strength of concrete containing non-metallic expansive agents)

S. B. DOS SANTOS and N. P. BARBOSA
Department Tecnologia da Construção Civil, Universidade Federal da Paraíba, Brèsil

Résumé
Dans ce travail, on presente les résultats d'une étude sur la variation de la resistance du béton preparé avec des expansifs non metalliques ('graut'), en fonction du pourcentage des granulats additionnés. On a utilisé des melanges ciment-expansif-materiau inerte produits par trois fabricants différents. Comme agregat, ont été employés des galets divisés selon deux granulometries. Sept melanges du produit industrialisé avec des galets ont été preparés, le taux d'agregat allant jusqu'a 70%. Les resistances à compression ont été obtenues a un, trois et sept jours. Les résultats sont comparés entre eux et montrent qu'on peut beaucoup reduire le taux de ciment-expansif dans chaque metre cube de béton, sans perte considerable de sa resistance.
Key words: Béton, Expansif, 'Graut', Résistance à compression

1 Introduction

Dans certains travaux de Genie Civil, y compris restauration et renforcement des structures, on a besoin d'un béton de haute resistance initiale et grande fluidité. Pour cela, l'industrie chimique a developpé des produits à être incorporés au béton. Parmi ces produits, se trouvent les expansifs non metalliques, lesquels, melangés au ciment, donnent un béton de haute resistance, haute adderence, sans retrait, impermeable et de grande fluidité.

Au Brésil, les expansifs non metalliques sont fournis par les fabricants dejá melangés au ciment et aux materiaux inertes. A ce melange, l'on ne doit ajouter que de l'eau. Quand on a affaire à des pieces un peu plus epaisses, les fabricants permetten d'additioner certaines quantités de granulat de diametre de tois a sept millimetres.

Dans les pays en developpement, l'utilisation directe de

ce produit industrialisé conduit a des coûts incompatibles avec la realité economique locale. On a donc besoin de diminuer le taux de melange ciment-expansif pour chaque metre cube de béton par l'addition d'agregats de plus gran diamètre. Dans ce travail, on présente les résultats de l'influence de cette addition de granulats sur la resistance du béton avec des expansifs.

2 Materiaux et méthodes

On a utilisé trois expansifs non metalliques, produits par trois differents fabricants, ici apellés A, B, C.

Comme agregat, ont été employeés des galets venus de Gramame, prés de João Pessoa, Brèsil. Deux granulometries ont été considerés: une apellé galet O, l'autre microgalet, dont on peut voir certaines caracteristiques au tableau 1.

Tableau 1. Caracteristiques des granulats

granulat	Dmax (mm)	Module finesse	Densité reele	Densité apparent	Absorp. (%)
Galet O	9.5	5.2	2.56	1.64	0.4
Microgalet	4.8	4.1	2.56	1.54	0.6

Les agregats ont été melangés au produit industrialisé dans les pourcentages suivantes, en poids: 0, 20, 30, 40, 50, 60, 70.

La quantité d'eau a été toujours prés de 13% du poids total du melange.

La resistance à compression du béton a été obtenue par essais sur eprouvettes cylindriques de 5 cm x 10 cm, aux temps de 24 h, 72 h et 168 h (7 jours). Chaque point obtenu represente la moyenne des résultats de cinq eprouvettes.

Le béton a été preparé comme suit: d'abbord on a melangé, a sec, l'expansif et les granulats jusqu'a obtenir une bonne homogénéité. En suite on a additionné le deux tiers d'eau, en laissant la betonniere travailler pendant deux minutes. En fin, on a rajouter le reste de l'eau.

3 Résultats obtenus

Les figures 1 et 2 montrent la variation de la resistance du béton avec le taux d'agregat additionné aux produits industrialisés. On peut noter la difference de comportement

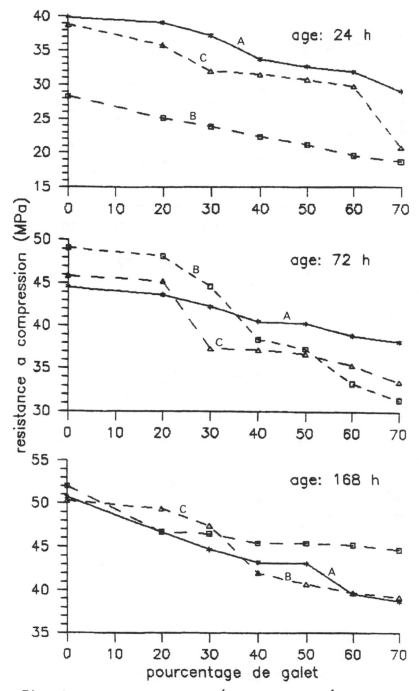

Fig. 1. Variation de la résistance du béton avec le taux de galet O additionné.

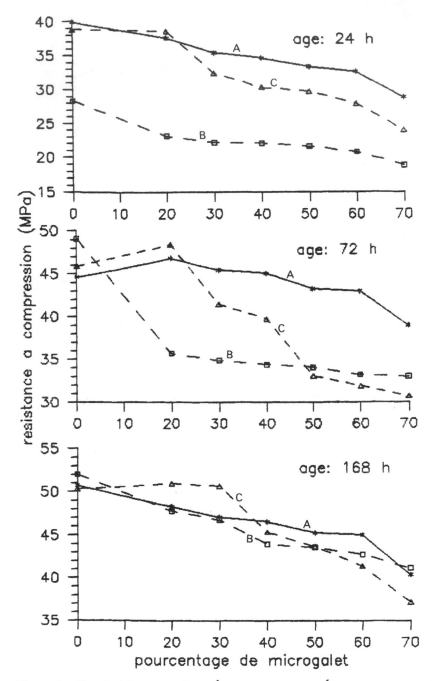

Fig. 2. Variation de la résistance du béton avec le taux de microgalet additionné.

des trois produits utilisés. Le produit B, qui a la plus petite resistance à compression à 24 heures, devient le plus resistante à l'âge de 7 jours, pour un taux de granulat de 70%. Le produit B a toujours eu un décroissement de resistance avec l'addition de granulats. Par contre, les produits A et C, à l'âge de 3 jours, ont eu une petite augmentation de resistance quand on a additionné le 20% de microgalet.

Le tableau 2 montre les valeurs numeriques du rapport entre la resistance à compression du béton avec l'addition de granulats et la resistance du produit pur. On note que, memme avec un haut taux d'agregats, la resistance à compression decroit seulement de 14 a 26%, à l'âge de 7 jours.

Tableau 2. Rapport entre résistances à compression

granulat	%	âge 24 h			âge 168 h		
		A	B	C	A	B	C
	0	1.00	1.00	1.00	1.00	1.00	1.00
	20	0.98	0.89	0.92	0.92	0.90	0.98
	30	0.93	0.84	0.82	0.88	0.89	0.94
galet 0	40	0.84	0.79	0.81	0.85	0.87	0.84
	50	0.82	0.75	0.79	0.85	0.87	0.81
	60	0.80	0.69	0.77	0.78	0.87	0.79
	70	0.73	0.66	0.53	0.76	0.86	0.78
	20	0.94	0.82	0.99	0.95	0.92	1.01
	30	0.88	0.78	0.83	0.93	0.90	1.01
micro-	40	0.86	0.77	0.78	0.92	0.85	0.90
galet	50	0.83	0.76	0.76	0.89	0.83	0.87
	60	0.81	0.73	0.72	0.89	0.82	0.82
	70	0.72	0.66	0.62	0.80	0.79	0.74

Pour un taux de granulats additionné de 50%, on a toujours une resistance plus grande que 75% de la resistance du produit pur.

La figure 3 montre l'evolution de la resistance à compression du béton avec l'âge, pour les trois produits étudiés et pour plusieurs taux de granulats additionnés.

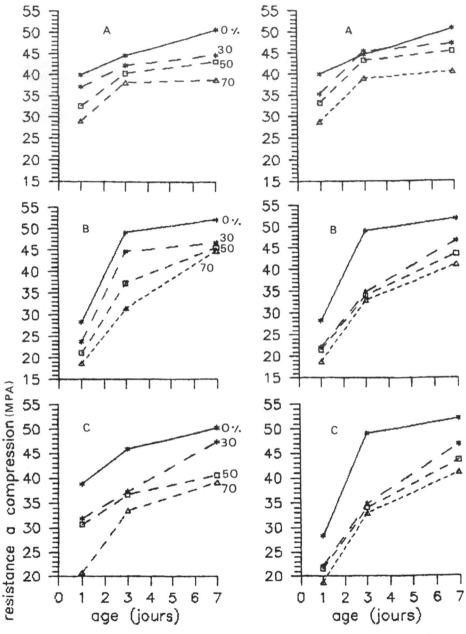

Fig. 3. Variation de la résistance du béton avec l'âge
pour les trois expansifs utilisés et differents
taux de granulats (a gauche galet 0, a droite
microgalet)

C'est le produit B qui a la plus grande augmentation de résistance entre 24 et 72 heures de cure.

4. Conclusions

Les résultats de l'étude developpée montrent que:

La résistance du béton préparé avec des expansifs non metalliques, en général, decroit avec le taux de granulats additionnés;

La loi de decroissement de la résistance à la compression n'est pas la même pour les trois produits employés et elle varie avec l'âge du béton;

À l'âge de 24 heures et de 72 heures, c'est le produit A qui donne la plus haute résistance, si le taux de granulats est 70 %; à l'âge de 7 jours c'est le produit B;

On ne peut pas établir une relation générale entre la résistance du béton et la granulometrie des agregats additionnés;

À l'âge de sept jours, pour un taux de granulats de 50 % les résistances obtenues avec les trois produits utilisés sont toujours plus hautes que 80 % de la résistance du produit pur;

En additionnant des granulats aux produits industrialisés avec les expansifs non meta. iques, on peut obtenir un béton beaucoup moins cher, sans pertes considerables de sa résistance.

ADMIXING EFFECT OF HIGH FINENESS SLAG ON THE PROPERTIES OF UNDERWATER CONCRETE

M. HARA and K. SATO
Steel Research Center, NKK Corporation, Japan
M. SAKAMOTO and T. HATSUZAKI
Technical Research Institute, Taisei Corporation, Japan

Abstract
Trial cements for underwater concrete were prepared by mixing various fineness (Blaine specific surface area of 4080-19800 cm^2/g) of water granulated blast furnace slag with ordinary Portland cement (OPC). Strength development of mortar and segregation resistance property under coexistence of segregation controlling agent of mortar were studied. As the results, cement mixed high fineness slag with OPC is found to be applicable to the cement for underwater concrete because of high hydraulic hardening capacity of the fine slag and sticky property of it by which outbreak of suspended solid during falling through water was suppressed. Then, applicability for underwater concrete of the trial cement was examined by concrete test using 13200 cm^2/g Blaine fineness slag. As the results, compressive strength of concrete specimen aged 28 days increased by 75 kgf/cm^2 by use of the slag and the outbreak of suspended solids was less than 150 mg/l. The results obtained show that the dosage of special admixing agent to underwater concrete as well as cement (OPC) content of the concrete can be reduced by the use of high fineness slag.
Key words : Underwater concrete, Blast furnace slag, Slag, high fineness, hydraulic hardening, Segregation resistance property, Segregation controlling agent, Compressive strength.

1 Introduction

The hydraulic hardening property as well as other properties of water granulated blast furnace slag (hereafter referred to as slag) has been widely used for cementing materials. Recently, there have been efforts to fully utilize these properties through fineness control. In the meanwhile, underwater concrete has been developed mainly for marine constructions. The mix proportions of underwater concrete is commonly designed by admixing segregation controlling agent, for the purpose of prevention against materials segregation. On the other hand, high fineness slag powder shows both sticky property when mixed with water and high hydraulic hardening property when it reacts with water under sufficient alkaline activator. The aim of this study is to examine the possibility of the use of high fineness slag powder to underwater concrete as a main admixture, in order to reduce both the segregation controlling agent and the cement content.

2 Experimental

2.1 Used materials
The chemical compositions of slag and ordinary Portland cement (hereafter reffered to as OPC) used are shown in Table1.

Table 1. Chemical compositions of slag and OPC. (wt%)

	SiO_2	Al_2O_3	Fe_2O_3	CaO	MgO	SO_3	Na_2O	K_2O	TiO_2	S
Slag	33.7	12.9	0.46	42.1	7.29	tr.	0.25	0.41	1.50	0.89
OPC	21.4	5.1	2.75	64.5	1.57	1.92	0.41	0.47	0.35	0.78

Slag was pulverized to a fineness of 4080 cm^2/g (Blaine specific surface area) by ball-milling (denoted as ①, in Fig.1). The high fineness slag powders were prepared from the slag powder ① under various classifying conditions with a forced voltex type centrifugal classifier (resulted powders are denoted as ② - ⑥). The particle size distribution, measured by laser diffraction method, and specific surface area of these slag powders are shown in Fig.1.

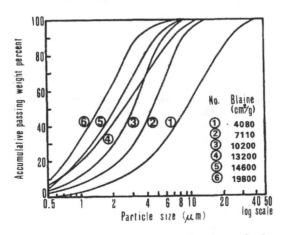

Fig.1. Particle size distribution of slags.

Properties of materials, together with main components, used for concrete test are shown in Table 2.

Table 2. Properties of materials for concrete test.

Materials	Properties and main components
Ordinary Portland cement	Fineness : 3200cm^2/g(Blaine)
Slag	Fineness :13200cm^2/g(Blaine)
Fine aggregate (River sand)	Max.size : 5mm , FM : 3.02 Specific gravity : 2.61 (under saturated surface-dry) Absorption : 1.38%
Coarse aggregate (Crushed stone)	Max.size : 20mm , FM : 6.48 Specific gravity : 2.65 (under saturated surface-dry) Absorption : 0.75%
Segregation controller	Cellulose ether
Superplasticizer A	Triazine
Superplasticizer B	Polyether carboxylic acid
Retarder	Salt of hydrocarboxylic acid

2.2 Compressive strength test of mortar with various fineness of slag powder

A trial cement was prepared by admixing various fineness of slag powder to OPC. The compressive strength test was made for trial cements and for OPC in accordance with the Japanese Industrial Standard (JIS R 5201).

2.3 Segregation resistance test for mortar and concrete in water

Segregation resistance in water were tested according to the procedure illustrated in Fig.2.

Fig.2. Testing procedure for segregation resistance property.

Here, suspended solid in sampled water was measured in accordance with JIS (JIS K 0102) to evaluate the outbreak of suspended solids. The segregation resistance was tested for mortar with some kinds of slag fineness and admixing ratio (slag content). The results were compared with it of OPC. The admixing agent mainly composed of cellulose ether was used for giving segregation resistance to mortar. The dosage of the agent was set as 0.5 wt% to every trial cements. The suitable fineness and admixing ratio of the slag for the concrete was determined by above mentioned mortar test. The segregation resistance was tested for concrete according to the same method as for mortar. Then, the admixing effect of trial cement for segregation resistance capacity of concrete was evaluated.

2.4 Slump flow and compressive strength test of underwater concrete mainly composed of high fineness slag
Mixing conditions of trial concrete were designed as follows : slump flow of 50 ± 2.5cm, air content of $4\pm1\%$, and cement (OPC+slag) content of 370 kg/m^3. Mix proportions of tested concrete, including admixture dosage, are presented in Table.3.

Table 3. Mix proportions of tested concrete.

Test No.	$\dfrac{W}{C+B}$ (%)	Mix composition(kg/m^3)					Admixture dosage(kg/m^3)		
		Water W	OPC C	Slag B	Fine aggr- egate	Coarse aggr- egate	Segre- gation contr- oller	Super- plasti- cizer A	Reta- rder B
1	55.1	204	370		650	1033	2.5	9.25	
2	55.9	207	222	148	627	1038	2.5	9.25	
3	51.6	191	222	148	642	1064	2.0	9.25	
4	45.9	170	222	148	663	1099	1.5	9.25	
5	51.1	189	222	148	661	1050	2.0	5.55	
6	51.1	189	222	148	661	1050	2.0	9.25	3.7

The concrete test pieces were prepared by the procedure illustrated as Fig.3, and provided to the following tests.

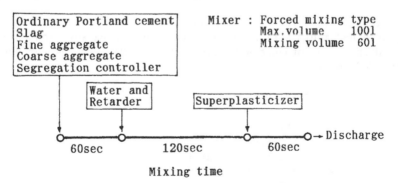

Fig.3. Mixing Procedure for concrete test.

A value of slump-flow was used as an index of consistency of concrete, because a slump value would not make a distinction for the slight difference of consistency due to the high viscosity of underwater concrete. The slump flow was measured in accordance with JIS (JIS A 1101 ; Method of test for slump of concrete). The slump flow is

defined as a spreading width of concrete at after 5 minutes, still-
holding period, from pulling up of slump cone. The slump flow were
measured at immediately after mixing, and at after 30 and 60 minute
from mixing. The compressive strength of concrete was measured in
accordance with JIS (JIS A 1108) on the specimen moulded both in-air
and under-water. The evaluation of strength developing was done in
terms of the strength ratio of under-water/in-air. The concrete
through water was moulded as shown in Fig.4.

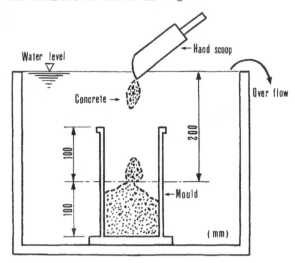

Fig.4.   Moulding of concrete through water.

A specimen moulded through water was drawn out from water tank with
the mould and mould-stripping, capping, and curing were sequentially
carried out in accordance with JIS (JIS A 1132), same as for specimen
of moulded in air.

3 Results and discussion

3.1 Compressive strength of mortar
The relation between fineness of slag and compressive strength of
mortar is shown in Fig.5. In that figure, compressive strength of OPC
mortar is shown horizontally as broken line. Both 3-and 7-day
strengths increase with the increase of slag fineness. It is shown
that the lower strength at early stage of the OPC-slag system can be
improved by use of high fineness slag. The 28-day strength have a
maximum of value at the Blaine fineness of 7110 cm^2/g, and decrease
with the increase of Blaine fineness within the range from 7110 to
19800 cm^2/g. The decrease of the strength is considered to be related
to the very rapid hydration rate of finer slag particle and to a way
of structural formation.

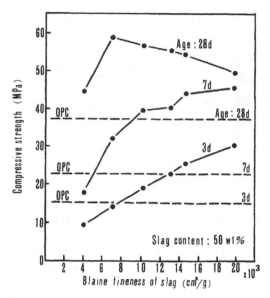

Fig.5.   Fineness of slag and compressive strength of mortar.

3.2 Segregation resistance property in water of mortar and concrete

3.2.1 Mortar test
The relation between slag fineness and suspended solids of segregation
resistance test is shown in Fig.6.

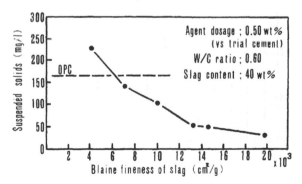

Fig.6.   Fineness of slag and suspended solids.
(segregation resistance test for mortar)

From the results, the increase of slag fineness cause to increase of
segregation resistance, then reduce the suspended solids. However,
the high slag fineness of over 13200 cm^2/g causes a small increase of
segregation resistance.

The relation between admixing amount of slag and suspended solids is shown in Fig.7.

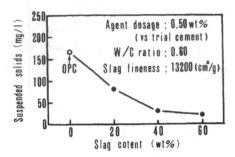

Fig.7.   Slag content and suspended solids.
         (segregation resistance test for mortar)

From the results shown in Fig.7, the suspended solids decrease with the increase of slag content ; however, over 40 wt% of admixing of slag shows a little effect.   Considering these results, the trial cement of which slag admixing conditions of 13200 cm^2/g (fineness) and 40 wt% (slag content) was used in the following concrete test.
3.2.2 Concrete test
The results of the segregation resistance test are presented in Fig.8. Here, the Test Nos. show the mix proportions of tested concrete prescribed in Table 3.

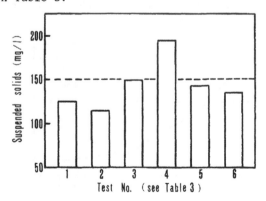

Fig.8.   Suspended solids of tested concrete.

The concrete composed of high fineness slag gives a low suspended solids value comparing to that of OPC concrete under the same dosage of segregation controlling agent (Test Nos.1,2).   In the case of a permissible suspended solids value is 150 mg/l, the use of the concrete composed of high fineness slag reduces the dosage of segregation controlling agent by about 0.5 kg/m^3 comparing to the case of the OPC concrete (Test Nos.3,5,6).

3.3 Slump flow and compressive strength

3.3.1 Slump flow
The change of slump flow for various mix proportions (shown as Test Nos. prescribed in Table 3) are shown in Fig.9.

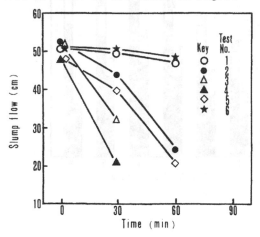

Fig.9. Change of slump flow.

According to the slump flow shown in Fig.9, the concrete composed of high fineness slag (Test Nos.2 - 5) gives a large loss of slump flow referring to that of OPC concrete (Test No.1). By the use of superplasticizer B (prescribed in Table 2), slump loss prevention type admixing agent, gives little effect for prevention of slump loss (Test No.5), while the use of retarder (prescribed in Table 2) gives a remarkable effect for that (Test No.6).

3.3.2 Compressive strength of concrete moulded in-air and under-water
The compressive strength of concrete with various mix proportions (shown as Test Nos. prescribed in Table 3) are shown in Fig.10. From the result shown in Fig.10, it is found that under the constant conditions of water-to-cement ratio and of admixing amount of segregation controlling agent, the concrete composed of high fineness slag gives a larger 28-day compressive strength, as large as 7.3 MPa than that of OPC concrete (Test Nos.1,2). Furthermore, the concrete composed of high fineness slag shows high compressive strength (Test Nos.2 - 6). From the fact that the strength ratio, moulded under-water to in-air, of each mix proportion are larger than 0.8, it is presumed that the concrete composed of high fineness slag is applicable to actual use.

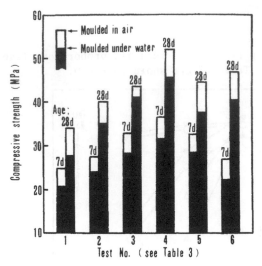

Fig.10. Compressive strength of concrete moulded
in air and under water.

4 Conclusions

The admixing effect of high fineness slag on the properties of
underwater concrete was studied. It is concluded that the underwater
concrete composed of high fineness slag has good segregation
resistance and strength developing properties. The results obtained
are summarized as follows:

(1)The cement composed of high fineness slag has excellent
segregation resistance, so it is possible to reduce the amount of
segregation controlling agent by use of that cement.

(2)The concrete using the above mentioned-cement shows high
strength development. The 28-day compressive strength of the concrete
increased by 75 kgf/cm^2 by use of the 13200 cm^2/g Blaine fineness
slag, comparing to that of OPC concrete. This result shows that the
cement (OPC) content of the concrete can be reduced by use of high
fineness slag.

(3)The large loss of slump flow of the underwater concrete admixed
high fineness slag can be improved by using the retarder containing
salt of hydroxycarboxylic acid.

References
Sato,K. et al. (1986) Properties of very fine blast furnace slag
    prepared by classification., 8th International Congress on the
    Chemistry of Cement., Rio de Janeiro, Vol.IV, pp.239-244.
Coastal Development Institute of Technology and Japanese Institute of
    Technology on Fishing Ports and Communities (1986) "Tokushu Suichū
    Konkurīto Manyuaru" (in Japanese), Sankaidō.

IMPROVED AIR ENTRAINING AGENTS FOR USE IN CONCRETES CONTAINING PULVERISED FUEL ASHES

J. T. HOARTY and L. HODGKINSON
Cormix Construction Chemicals Lt, England

Abstract
It is a well established fact that entraining air into concretes containing certain pulverised fuel ashes can present technical problems.
Published work has demonstrated that the most likely cause of these problems relates to the presence of small quantities of residual carbon resulting from the combustion of the pulverised fuel.
This paper has further investigated the nature of these problems and concluded that the problems derive from two related variables.
Variation of the level and nature of the carbon from source to source has been identified as a significant problem. This problem may be insurmountable where large variations of carbon content occur or where uncontrolled changes in source of ash are made.
A second problem has been identified as being related to the destablilisation of the air after it has been entrained. This variable has been investigated and probably relates to the adsorption of either surfactant or the air voids themselves onto the highly active surface of the carbon particles.
Novel air entraining agents have been developed which overcome this second, de-stabilising problem. These air entraining agents contain deliberate additions of polar species which are preferentially adsorbed onto the carbon surface, satisfying this activity, leaving the surfactant free to fulfil its prime role.
Data is presented relating to the performance of such air entraining agents with various commonly available United Kingdom pulverised fuel ashes.
Key words: Air entraining agent, Concrete, Pulverised fuel ashes, Residual carbon

1   Introduction

In the United Kingdom, it has become an accepted practice to add selected pulverised fuel ashes to all types of structural concretes. The publication of BS3892, Part 1 : 1982 "Specification for pulverised fuel ash for use as a Cementitious component in Structural Concrete" which standardised these pulverised fuel ashes, and BS8110 "Structural use of Concrete", which gives guidance regarding the recommended levels of addition, have been significant in increasing the use of pulverised fuel ashes in the United Kingdom.

During recent years, there has been a simultaneous growth in the use of air entrained concrete resulting from an awareness and appreciation of the benefits of air entrainment upon the freeze thaw durability of concretes.

However, when air entrainment has been specified, and contractors or ready-mixed concrete suppliers have opted to include pulverised fuel ash in the concrete, serious technical problems have occurred. Increased dosage rates of air entraining agents are invariably required, but the major problems have been associated with variation in initial air content and loss of entrained air prior to the placing of the concrete.

The effects of pulverised fuel ash upon the air entrainment of concrete have been widely reported by Larson (1964), Mielenz (1978), Berry and Malhotra (1980), Lane and Best (1982) and Uchikawa, Uchida and Ogawa (1982). They are summarised in a review paper by M J Waddicor and L Hodgkinson "A Study of some variables within Pulverised Fuel Ashes which effect the Air Entraining ability of Admixtures in Concrete" (1983). The overall conclusion of most researchers is that the most significant variable within pulverised fuel ashes which affects air entrainment is the 'Loss of Ignition' (LOI) of the ash. The material which constitutes this LOI is normally associated with carbon from the combustion of the pulverised fuel. However, the nature and composition of the LOI is undoubtedly variable and can often be unburnt coal, but can be numerous forms of carbon.

Many forms of this carbon are highly surface active and can adsorb gases or surface active organic molecules. This is the same effect as that employed when using activated charcoal in gas masks. The mechanism of adsorption in concrete is not clear, but it is likely that the surface active air entraining agent itself is adsorbed, thereby reducing its effectiveness. It is also probable that the minute air voids themselves may be adsorbed.

It was reasoned, that a polar species would be more strongly adsorbed onto an activated carbon surface than would a non-polar species. If this is so, then it is also likely that the more polar a species is, then the more likely it is to preferentially adsorb onto active sites. Based on these reasonable assumptions, it was decided to deliberately add polar compounds to improve the performance of certain air entraining agents in the presence of pulverised fuel ash. It was reasoned that the deliberately added polar species would preferentially absorb onto the carbon, thereby satisfying its surface activity, allowing the surface active air entraining agent to fulfil its normal function.

2 Experimental

Various proprietary air entraining agents were compared with respect to their ability to entrain air into normal ordinary portland cement concretes and also concretes containing pulverised fuel ashes. The mix designs used are tabulated in Appendix 1.

In air entrained concretes containing ordinary portland cement, without any pulverised fuel ash, there appeared to be no evidence of variable air content, and no problem of air degradation with any of the proprietary air entraining agents. Typical results are tabulated below.

Table 1.

Admixture Type	Admixture Dose ml/50 k cement	Initial air (%) 1 min Vibration	Air % Further 2 mins mixing	Air Loss (% of initial)	Air % Further 2 mins mixing	Air Loss (% of Initial)
Vinsol Resin (1)	30	5.2	5.1	1.9	5.2	Nil
Vinsol Resin (2)	30	5.5	5.4	1.8	5.4	1.8
Ether Sulphate	30	4.7	5.0	6.3 Gain	5.3	12.6 Gain

In air entrained concretes containing pulverised fuel ash, all of the air entraining agents required a higher than normal dosage to achieve an acceptable initial level of air entrainment. The dosage level of air entraining agent required was very dependent upon the LOI of the ash and on the source of the ash.

Of equal significance, however, was the observation that the level of air entrainment decreased with extended mixing in all the ash concretes. In ordinary portland cement mixes, no significant loss had been observed. This decrease in air content in ash concretes, occurred with all the normal air entraining agents. In the examples quoted, 30% of Pulverised Fuel Ash was incorporated in place of Ordinary portland cement. Ashes with varying levels of LOI and varying activity of LOI were deliberately selected. The effects are tabulated below:

Table 2a. Source of PFA. Didcot Power Station. (4% LOI)

Admixture Type	Admixture Dose ml/50 k cement	Initial air (%) 1 min Vibration	Air % Further 2 mins mixing	Air Loss (% of initial)	Air % Further 2 mins mixing	Air Loss (% of Initial)
Vinsol Resin (1)	200	6.1	2.8	54.0	1.7	72.0
Vinsol Resin (2)	200	5.8	2.4	58.6	1.8	69.0
Ether Sulphate	200	4.7	3.7	12.3	3.4	27.6

Table 2b. Source of PFA. Rugeley Power Station (2% LOI).

Admixture Type	Admixture Dose ml/50 k cement	Initial air (%) 1 min Vibration	Air % Further 2 mins mixing	Air Loss (% of initial)	Air % Further 2 mins mixing	Air Loss (% of Initial)
Vinsol Resin (1)	83	5.6	4.7	16.1	3.9	30.4
Vinsol Resin (2)	140	5.3	4.0	24.5	3.3	37.7
Ether Sulphate	100	9.2	7.5	18.5	6.3	31.5

Table 2c. Source of PFA. Fiddlers Ferry Power Station (4% LOI).

Admixture Type	Admixture Dose ml/50 k cement	Initial air (%) 1 min Vibration	Air % Further 2 mins mixing	Air Loss (% of initial)	Air % Further 2 mins mixing	Air Loss (% of Initial)
Vinsol Resin (1)	100	6.6	4.6	30.3	3.7	44.0
Vinsol Resin (2)	140	4.0	3.0	25.0	2.5	37.5
Ether Sulphate	132	7.0	6.4	8.6	5.6	20.0

One particular patented air entraining agent, has an active ingredient based on sodium decanoate. It also contains some sodium octanoate, as both these salts are derived from natural fatty acids. Deliberately raising the level of sodium octanoate, which does not entrain air, does not appear to have any significant effect on the initial air entraining ability of the air entraining agent in all concretes. However, it has a dramatic effect on the observed loss of air when extended mixing takes place with ash concretes.

No loss of air occurs with extended mixing. These effects are tabulated below:

Table 3.

Admixture Type	Source of PFA	Admixture dose ml/50 k cement	Initial Air (%) 1 min vibration	Air (%) further 2 mins mixing	Air Loss (% of Intal)	% Air further 2 mins mixing	Air Loss (% of Intal)
C_8/C_{10} Fatty Acid	Rugeley (2% LOI)	100	7.5	7.5	Nil	7.3	2.7
C_8/C_{10} Fatty Acid	Fiddlers Ferry (4% LOI)	100	7.5	7.6	1.3 Gain	7.5	Nil
C_8/C_{10} Fatty Acid	Didcot (4% LOI)	250	6.8	6.2	8.8	6.1	10.3

It is postulated that the sodium octanoate is more polar than the sodium deconoate and is preferentially absorbed onto active sites on the carbon, thereby satisfying the surface activity. Sodium deconoate is less polar and is not absorbed until the sodium octanoate is exhausted. It is thus free to act in a normal manner and entrains air. Similarly, the air entrained is not de-stabilised by any interaction with carbon from the pulverised fuel ash.

The deliberate addition of sodium octanoate to an air entraining agent based upon sodium decanoate is the subject of a British Patent (1989).

This air entraining agent is marketed as Cormix Stablair

3 Laboratory Trials with Pulverised Fuel Ashes Available within the United Kingdom

Two series of extensive laboratory trials were carried out using random daily samples of production pulverised fuel ash readily available in the United Kingdom.

The objective of these trials, was to evaluate the effectiveness of the air entraining agent in conditions where the ash would show typical day to day production variability.

A standard dosage was employed to entrain a target air content for the first days production sample, and thereafter, the dosage was fixed for all the samples tested. Similarly, no adjustment was made to water content throughout each test series.

The mix designs used in both series are tabulated in Appendix 2.

3.1 Ash Resources (UK) Limited

Samples of Ash Resources, High Performance Fuel Ash were collected at Little Barford during the months of September, October and November, 1987. Forty four daily samples of the PFA were collected in all.

Each sample represented a random sample of daily production from each plant.

Cormix Stablair was employed throughout at a fixed dosage rate of 150 mls/50 kg total cementitious. The mix contained 320 kg/m^3 total cementitious with an ash content of 30%.

Slump and Air content measurements were taken on all the mixes tested and the compressive strength of the concrete was measured at 7 days and 28 days.

The results of these trials are shown in the following table.

Table 4.

Date	Mix No	LOI (%)	Slump (mm)	Air (%)	Compressive Strength (N/mm^2)	
					7 Days	28 Days
1.9.87	1	3.26	100	6.4	-	-
2.9.87	2	3.71	105	6.2	25.9	35.0
3.9.87	3	4.11	100	6.4	26.2	35.5
4.9.87	4	3.11	90	6.8	25.8	35.2
7.9.87	5	2.91	75	5.6	27.3	37.0
8.9.87	6	4.04	85	5.2	29.5	39.3
9.9.87	7	3.89	75	5.2	29.7	40.0
10.9.87	8	3.29	80	5.4	27.7	38.0
14.9.87	9	3.62	70	5.7	28.3	38.8
16.9.87	10	4.21	75	5.2	28.4	38.7
17.9.87	11	3.68	65	5.0	30.7	41.0
18.9.87	12	4.20	75	5.6	27.8	38.1
21.9.87	13	3.57	70	5.6	27.1	37.0
22.9.87	14	3.31	85	6.0	26.7	36.5
23.9.87	15	3.32	85	5.8	26.6	36.8
24.9.87	16	4.19	74	5.2	30.2	40.5
28.9.87	17	3.37	70	5.4	29 9	40.0
29.9.87	18	2.91	80	6.6	24.1	34.3
30.9.87	19	3.44	75	5.6	27.5	38.0
1.10.87	20	3.13	80	5.4	24.8	35.2
2.10.87	21	3.64	70	5.2	27.6	37.8
5.10.87	22	3.25	70	5.9	28.1	38.9
6.10.87	23	3.42	65	5.8	31.3	41.0
7.10.87	24	3.50	65	5.8	32.5	43.0
8.10.87	25	3.69	50	5.2	34.6	35.2
9.10.87	26	3.63	95	5.8	28.9	39.1
12.10.87	27	3.52	70	5.6	29.8	40.0
13.10.87	28	4.26	70	5.0	32.7	43.2
14.10.87	29	3.72	65	5.5	31.5	40.8
15.10.87	30	3.78	95	6.2	28.2	39.0
16.10.87	31	3.49	70	5.5	29.2	39.1

Table 4 continued.

Date No	Mix	LOI (%)	Slump (mm)	Air (%)	Compressive Strength (N/mm²)	
					7 Days	28 Days
19.10.87	32	3.33	95	6.2	26.0	36.3
20.10.87	33	3.42	65	5.5	31.8	42.0
21.10.87	34	3.54	100	5.7	29.1	39.4
22.10.87	35	3.61	75	5.6	30.1	40.3
23.10.87	36	4.26	55	5.3	31.7	42.0
26.10.87	37	3.82	70	5.5	30.2	40.0
27.10.87	38	3.25	70	6.2	28.9	38.9
28.10.87	39	3.01	70	5.9	28.4	38.0
29.10.87	40	4.17	65	4.9	29.2	39.5
30.10.87	41	3.11	70	5.6	29.7	40.1
2.11.87	42	3.13	95	6.2	26.5	36.3
3.11.87	43	2.98	75	5.7	29.9	41.0
4.11.87	44	3.47	60	5.0	30.5	41.0

Statistical Analysis of Results

	LOI (%)	Slump (mm)	Air (%)	Compressive Strength (N/mm²)	
Average	3.55	76.5	5.66	28.8	39.7
Highest	4.28	105	6.8	34.6	43.2
Lowest	2.91	50	4.9	24.1	34.3
Standard Deviation	0.38	12.79	0.45	2.02	2.04

3.2 ARC Limited

ARC Limited collected random daily samples of pulverised fuel ash from their Didcot plant during the month of January, 1989. Air entrainment evaluations were carried out using these samples of ash in conjunction with Cormix Stablair.

All the samples were evaluated in concrete using the mix design and method specified in BS5075 Part 2. 30% of the cement content was replaced with PFA. The Cormix Stablair was dosed at a constant 150 mls/50 kg cement. The cement used was Blue Circle OPC. Slump and air content measurements were taken on all the mixes and the compressive strength of the hardened concrete was measured at 7 days and 28 days.

The results of these trials are shown in the following Table.

Table 5.

Date January 1989	LOI (%)	Slump (mm)	Air (%)	Compressive Strength (N/mm^2) 7 Day	Compressive Strength (N/mm^2) 28 Day
3	4.1	80	5.6	15.8	24.0
4	3.8	75	4.8	17.5	27.5
5	4.6	65	5.3	16.8	26.4
6	7.3	80	4.8	16.7	26.2
9	6.3	65	4.5	17.1	26.4
10	5.7	60	3.4	19.9	29.9
11	4.9	50	4.0	20.1	29.0
12	5.1	65	4.3	18.2	26.8
13	4.7	75	4.8	17.2	25.5
14	4.6	60	4.6	18.3	26.9
16	4.6	85	4.0	15.5	26.0
17	5.5	65	3.8	19.4	29.6
18	4.2	85	4.8	17.1	26.2
19	2.9	65	4.5	17.5	27.0
20	3.5	75	6.3	14.2	20.0
23	3.5	70	4.5	17.5	26.9
24	3.7	80	5.6	15.1	23.5
25	3.0	80	6.0	14.6	23.3
26	3.2	70	5.3	16.2	25.1
27	3.2	65	5.5	15.6	21.6
30	3.5	85	5.3	17.9	26.9
31	3.7	75	5.7	15.6	23.0

Statistical Analysis of Results

	LOI (%)	Slump (mm)	Air (%)	Compressive Strength (N/mm^2)	
Average	4.3	70	4.9	16.8	25.8
Highest	7.3	85	6.3	20.1	29.9
Lowest	2.9	50	3.4	14.2	21.6
Standard Deviation	1.1	9.2	0.73	1.47	2.39

3.3 Comments on the Results of the Trials

The most interesting feature which emerges from both sets of trial
mixes is the consistency of the air entrainment levels,
which were independent of the level of carbon in the ash.

In the Ash Resources mix series there is only a 2% difference
between the highest and lowest air entrainment figures from a total of
44 pulverised fuel ash samples. Again in the ARC mix series the
difference in the highest and lowest air entrainment figure was 2.9%.
These trials were done according to BS5075 Part 2 conditions and only
one result from 22 was outside the standard limits of 4.5 +/- 1.5%
air.

Conclusions

Previous work has indicated that the Loss of Ignition (LOI) of
Pulverised Fuel Ashes is the principal cause of the problems
associated with the air entrainment of concretes containing pulverised
fuel ashes. This work would support those findings.

The effects of this carbon are dependent upon the source of ash,
and the dosage rate of air entraining agent required is normally
proportional to the proportion of ash used in a concrete. With
normal air entraining agents, the dosage rate required is very
sensitive to the LOI of the ash, and because of the unavoidable
variability of this LOI on a production basis, it is extremely
difficult to produce air entrained concrete consistently with varying
batches or consignments of pulverised fuel ash, even from nominally
the same source.

Assuming the initial air content could be achieved in production,
this work would suggest that degradation of the air content would lead
to delivery problems, with loss of air in transit, where conventional
air entraining agents were employed.

Novel air entraining agents have been produced which probably
function by preferential absorption of polar species onto the active
sites on the carbon, leaving the air entraining species free to
function as normal. With such air entraining agents it has been
demonstrated that typical production variability of pulverised fuel
ashes will not cause a large variation in air content when a fixed
dosage of air entraining agent is used.

It has also been demonstrated that degradation of the air content
will not take place when using these novel air entraining agents, and
loss of air from concrete in transit should not be such a major
problem.

However, the physio-chemical nature of the carbon is so variable
between pulverised fuel ashes from different sources, it is uncertain
whether novel air entraining agents of this type will enable the use
of pulverised fuel ash from variable sources to be used without
recourse to a change in dosage.

References

BS3892, Part 1, 1982 "Specification for Pulverised fuel ash for use as Cementitious Component in Structural Concrete".

BS8110 "Structural use of concrete".

Larson, T.D, "Air entrainment and durability aspects of fly ash concrete." Proceedings ASTM Vol 64 1964, pp 866 - 886.

Mielenz, R.C, "Specifications on fly ash for use in concrete: where we are now - where should we go?" pp 978.

Berry, E.E, and Malhotra, V.M, "Fly ash for use in concrete - a critical review". ACI Journal, Mar-April 1980, pp 59 - 73.

Lane, R.O, and Best, J.F, "Properties and use of fly ash in Portland cement concrete". Concrete International, July 1982, pp 81 - 92.

Uckikawa, H, Uchida, S, and Ogawa , K, "Influence of the properties of fly ash on the fluidity and structure of fly ash cement paste". Proceedings on International Symposium, "The Use of PFA in Concrete". University of Leeds, England, March 1982.

Waddicor, M.J, and Hodgkinson, L. "A study of some variables within pulverised fuel ashes which affect the air entraining ability of admixtures in concrete". ERMCO 83, London May 22 - 26, 1983.

G.B Patent Application 8811171.1

Appendix 1 - Mix Design Used in Paragraph 2

Croxden Gravel 20 mm	8.46 k/m^3
Croxden Gravel 10 mm	3.75 k/m^3
Almington Pit Sand	5.97 k/m^3
Castle OPC	2.00 k/m^3
Pulverised Fuel Ash (Various Sources)	1.00 k/m^3
A/C Ratio	6.13
% Fines	32.5
W/C Ratio	0.58

Trial batch valume 0.01 m^3.

Appendix 2 - Mix design used in Paragraph 3

1. Ash Resources (UK) Limited trials

Coarse Aggregate	
Gravel 20 mm	350 k/m^3
15 mm	350 k/m^3
10 mm	465 k/m^3
Fine Aggregate ex RMC Canttarm	625 k/m^3
Ketton OPC	224 k/m^3
Ash Resources PFA ex Little Barford Power Station	96 k/m^3
A/C Ratio	5.59 %
% Fines	34.9

Trial batch volume 0.01 m^3

2. ARC Limited Trials

Coarse Aggregate	
Gravel 20 mm	845 k/m^3
10 mm	400 k/m^3
Fine Aggregate ex Almington Pit Sand	600 k/m^3
Blue Circle OPC	210 k/m^3
ARC Limited PFA ex Didcot Power Station	90 k/m^3
A/C Ratio	6.13
% Fines	32.5

Trial batch volume 0.01 m^3.

INFLUENCE DE LA TENEUR EN ALCALINS SOLUBLES DU CIMENT SUR LA STABILITÉ DU RÉSEAU DE BULLES D'AIR EN PRÉSENCE DE SUPERPLASTIFIANT

(Influence of soluble alkali content in cement on air void stability in the presence of superplasticizer)

P. PLANTE
Fondatec Inc., Montreal, Canada
M. PIGEON
Department of Civil Engineering, Laval University, Quebec, Canada

Résumé

Cinquante-quatre mélanges de mortier ont été préparés afin d'évaluer l'influence de la dose d'agent entraîneur d'air et de la teneur en alcalins solubles du ciment sur la stabilité du réseau de bulles d'air en présence de superplastifiant.

Deux ciments, trois agents entraîneurs d'air et deux superplastifiants (un à base de naphtalène et un à base de mélamine) ont été utilisés. Des alcalins solubles (Na_2SO_4 et K_2SO_4) ont été ajoutés à l'eau de gâchage pour certains mélanges afin d'obtenir différentes teneurs en alcalins.

De façon à évaluer correctement la stabilité du réseau de bulles d'air, le facteur d'espacement a été mesuré sur des éprouvettes fabriquées avec des échantillons de mortier prélevés à trois reprises (c'est-à-dire à 10, 30 et 60 minutes) sur une période de malaxage de minutes calculée à partir du premier contact eau-ciment.

Un indice de stabilité a été mis au point et utilisé de façon à quantifier la stabilité du facteur d'espacement. Les résultats indiquent que la stabilité du facteur d'espacement n'est pas fonction des caractéristiques initiales des mélanges (étalement, teneur en air et facteur d'espacement) ni des variations de la teneur en air.

Les résultats indiquent également que, quel que soit le type d'agent entraîneur d'air utilisé, l'addition d'un superplastifiant (15 minutes après le début du malaxage) est souvent une importante cause d'instabilité. Toutefois, il semble que la stabilité du réseau de bulles d'air s'améliore avec l'augmentation de la dose d'agent entraîneur d'air et surtout avec l'augmentation de la teneur en alcalins solubles du ciment.

Mots clés: Facteur d'espacement; indice de stabilité; agent entraîneur d'air; superplastifiant; alcalin.

Introduction

Il est maintenant bien établi que la durabilité du béton soumis à des cycles de gel-dégel est fonction des caractéristiques du réseau de bulles d'air et plus particulièrement du facteur d'espacement

(1,2,3). Le producteur de béton doit donc être en mesure de produire et de livrer au chantier de construction un béton ayant un réseau de bulles d'air correct.

La revue de la littérature scientifique (4) ne permet pas d'obtenir beaucoup d'informations sur les paramètres susceptibles d'influencer la production et la stabilité du réseau de bulles d'air. La majorité des chercheurs qui ont étudié l'entraînement de l'air ont mesuré les teneurs en air du béton plastique plutôt que les caractéristiques du réseau de bulles d'air du béton durci. De plus, les problèmes de stabilité (c'est-à-dire la variation en fonction du temps de la teneur en air et plus particulièrement du facteur d'espacement), ont été très peu étudiés (5,6). Quelques chercheurs (6,7,8) ont mentionné que l'utilisation d'un superplastifiant cause généralement une augmentation de la dimension moyenne des bulles formant le réseau de vides d'air et donc du facteur d'espacement (pour un volume d'air constant). Il a également été dit (9) que la production d'un bon réseau de vides d'air est plus difficile lorsque la teneur en alcalins (alcalins provenant du ciment ou des cendres volantes) d'un mélange augmente.

Des essais effectués à l'Université Laval ont confirmé que l'utilisation d'un superplastifiant peut affecter la stabilité du réseau de bulles d'air (10,11,12,13,14). Toutefois la stabilité du facteur d'espacement semble fonction de la combinaison des adjuvants (agent entraîneur d'air et superplastifiant) et, tout probablement, de la teneur en alcalins du ciment.

Depuis quelques années, les mesures de protection de l'environnement ont amené les cimentiers à modifier leurs usines (ex: recirculation des poussières), ce qui a causé une augmentation de la teneur en alcalins solubles des ciments. Comme les superplastifiants peuvent être une importante cause d'instabilité des vides d'air, il a été décidé d'utiliser ces adjuvants afin d'étudier l'influence des sulfates alcalins sur la stabilité du réseau de bulles d'air pour différents environnements physico-chimiques (type et dose d'agent entraîneur d'air et type de superplastifiant).

Programme d'essais

Cet article présente les résultats d'une étude comprenant 54 mélanges de mortier fabriqués en laboratoire. De façon à évaluer correctement la stabilité du facteur d'espacement en fonction du temps (le temps est calculé entre le premier contact eau-ciment et la mise en place de l'échantillon), ce facteur a été mesuré à trois reprises, soit après 10, 30 et 60 minutes. Afin d'obtenir différentes teneurs en alcalins solubles, deux ciments ont été utilisés et des alcalins solubles (sulfate de sodium et sulfate de potassium) ont été ajoutés à l'eau de gâchage pour certains mélanges. De façon à obtenir une énergie de malaxage raisonnable, les mortiers ont été malaxés pendant des périodes de 3 minutes suivies de périodes de repos de 7 minutes. Pour simuler l'ajout du superplastifiant au chantier, le superplastifiant était ajouté au mélange 20 minutes après le début de l'essai. De cette façon, le facteur d'espacement pouvait être mesuré avant et immédiatement après l'ajout du superplastifiant.

Plusieurs combinaisons type d'agent entraîneur d'air-ciment-type de superplastifiant ont été utilisées afin d'évaluer dans différents environnements physico-chimiques l'influence de la teneur en alcalins solubles des mélanges. L'objectif principal de cette recherche est d'évaluer l'influence des sulfates alcalins et non l'effet des agents entraîneurs d'air, des superplastifiants ou des autres caractéristiques des ciments.

Matériaux, caractéristiques des mélanges et procédures d'essais

Deux différents ciments Portland de type 10 (norme ACNOR A5) ont été utilisés pour la fabrication des mortiers. Les caractéristiques chimiques et physiques de ces deux ciments que les auteurs considèrent importantes pour ce qui est de l'entraînement de l'air sont présentées au tableau 1 (où les ciments sont tout simplement nommés 1 et 2).

Des alcalins solubles ont été ajoutés à l'eau de gâchage dans certains cas. Les alcalins utilisés avec le ciment 1 sont (en équivalent Na_2O): Na_2SO_4, 1 % de la masse du ciment, K_2SO_4, 1 % et $Na_2SO_4 + K_2SO_4$ 1 % + 1 %. Les alcalins utilisés avec le ciment 2 sont: Na_2SO_4, 0,5 % et 1 %, K_2SO_4, 1 % et $Na_2SO_4 + K_2SO_4$, 1 % + 1 %.

Tableau 1 : Caractéristiques des ciments

	Ciment 1	Ciment 2
Finesse (m^2/kg)	370	324
C_3A (%)	8,9	7,5
Alcalins totaux (%, Na_2O eq.)	1,0	0,7
Alcalins solubles (%, Na_2O eq.)	0,6	0,4
Filler calcaire (%)	0,0	4,5

Le sable utilisé est un sable naturel de rivière contenant principalement du gneiss granitique. Ce sable a une densité de 2,7, une absorption en eau variant entre 0,32 % et 0,54 % et un module de finesse de 2,3.

Trois différents agents entraîneurs d'air ont été utilisés: un sel sulfoné d'hydrocarbure (HS), une résine de Vinsol (RV) et un détergent synthétique (DS). Ces adjuvants ont été utilisés selon un dosage de 0,2 ml/kg de ciment pour la première série de 24 mélanges et de 0,1 ml/kg de ciment pour les 30 derniers mélanges.

Deux superplastifiants ont été utilisés, soit un produit à base de naphtalène et un autre à base de mélamine. Le naphtalène a été utilisé pour les 48 premiers mélanges selon un dosage de 6,6 ml/kg de ciment. Le mélamine a été utilisé seulement pour les 6 derniers mélanges selon un dosage de 18,2 ml/kg de ciment. Ce deuxième superplastifiant n'a été utilisé que pour confirmer la tendance obtenue avec le naphtalène. Pour tous les mélanges, le même réducteur d'eau a été utilisé (un produit à base de lignosulfonates de calcium) selon un dosage de 4,4 ml/kg de ciment.

La composition des 54 mélanges est donnée au tableau 2. Comme le montre ce tableau, de façon à pouvoir comparer systématiquement l'influence des différents paramètres, toutes les

Tableau 2: Caractéristiques des mélanges

Code	AEA (ml/kg de ciment)	SP	Alcalins (%, éq. Na$_2$O)
M-2-HS-N-0	0,2	6,6	-----------
M-2-RV-N-0	0,2	6,6	-----------
M-2-DS-N-0	0,2	6,6	-----------
M-2-HS-N-.5Na	0,2	6,6	0,5%, Na$_2$SO$_4$
M-2-HS-N-Na	0,2	6,6	1%, Na$_2$SO$_4$
M-2-HS-N-K	0,2	6,6	1%, K$_2$SO$_4$
M-2-HS-N-Na+K	0,2	6,6	2%, (Na$_2$SO$_4$+K$_2$SO$_4$)
M-2-RV-N-.5Na	0,2	6,6	0,5%, Na$_2$SO$_4$
M-2-RV-N-Na	0,2	6,6	1%, Na$_2$SO$_4$
M-2-RV-N-K	0,2	6,6	1%, K$_2$SO$_4$
M-2-RV-N-Na+K	0,2	6,6	2%, (Na$_2$SO$_4$+K$_2$SO$_4$)
M-2-DS-N-.5Na	0,2	6,6	0,5%, Na$_2$SO$_4$
M-2-DS-N-Na	0,2	6,6	1%, Na$_2$SO$_4$
M-2-DS-N-K	0,2	6,6	1%, K$_2$SO$_4$
M-2-DS-N-Na+K	0,2	6,6	2%, (Na$_2$SO$_4$+K$_2$SO$_4$)
M-1-HS-N-0	0,2	6,6	----------
M-1-HS-N-Na	0,2	6,6	1%, Na$_2$SO$_4$
M-1-HS-N-Na+K	0,2	6,6	2%, (Na$_2$SO$_4$+K$_2$SO$_4$)
M-1-RV-N-0	0,2	6,6	----------
M-1-RV-N-Na	0,2	6,6	1%, Na$_2$SO$_4$
M-1-RV-N-Na+K	0,2	6,6	2%, (Na$_2$SO$_4$+K$_2$SO$_4$)
M-1-DS-N-0	0,2	6,6	----------
M-1-DS-N-Na	0,2	6,6	1%, Na$_2$SO$_4$
M-1-DS-N-Na+K	0,2	6,6	2%, (Na$_2$SO$_4$+K$_2$SO$_4$)
M-2-.5HS-N-0	0,1	6,6	----------
M-2-.5RV-N-0	0,1	6,6	----------
M-2-.5DS-N-0	0,1	6,6	----------
M-2-.5HS-N-.5Na	0,1	6,6	0,5%, Na$_2$SO$_4$
M-2-.5HS-N-Na	0,1	6,6	1%, Na$_2$SO$_4$
M-2-.5HS-N-K	0,1	6,6	1%, K$_2$SO$_4$
M-2-.5HS-N-Na+K	0,1	6,6	2%, (Na$_2$SO$_4$+K$_2$SO$_4$)
M-2-.5RV-N-.5Na	0,1	6,6	0,5%, Na$_2$SO$_4$
M-2-.5RV-N-Na	0,1	6,6	1%, Na$_2$SO$_4$
M-2-.5RV-N-K	0,1	6,6	1%, K$_2$SO$_4$
M-2-.5RV-N-Na+K	0,1	6,6	2%, (Na$_2$SO$_4$+K$_2$SO$_4$)
M-2-.5DS-N-.5Na	0,1	6,6	0,5%, Na$_2$SO$_4$
M-2-.5DS-N-Na	0,1	6,6	1%, Na$_2$SO$_4$
M-2-.5DS-N-K	0,1	6,6	1%, K$_2$SO$_4$
M-2-.5DS-N-Na+K	0,1	6,6	2%, (Na$_2$SO$_4$+K$_2$SO$_4$)
M-1-.5HS-N-0	0,1	6,6	----------
M-1-.5HS-N-Na	0,1	6,6	1%, Na$_2$SO$_4$
M-1-.5HS-N-Na+K	0,1	6,6	2%, (Na$_2$SO$_4$+K$_2$SO$_4$)
M-1-.5RV-N-0	0,1	6,6	----------
M-1-.5RV-N-Na	0,1	6,6	1%, Na$_2$SO$_4$
M-1-.5RV-N-Na+K	0,1	6,6	2%, (Na$_2$SO$_4$+K$_2$SO$_4$)
M-1-.5DS-N-0	0,1	6,6	----------
M-1-.5DS-N-Na	0,1	6,6	1%, Na$_2$SO$_4$
M-1-.5DS-N-Na+K	0,1	6,6	2%, (Na$_2$SO$_4$+K$_2$SO$_4$)
M-2-.5HS-M-0	0,1	18,2	----------
M-2-.5HS-M-Na	0,1	18,2	1%, Na$_2$SO$_4$
M-2-.5RV-M-0	0,1	18,2	----------
M-2-.5RV-M-Na	0,1	18,2	1%, Na$_2$SO$_4$
M-2-.5DS-M-0	0,1	18,2	----------
M-2-.5DS-M-Na	0,1	18,2	1%, Na$_2$SO$_4$

combinaisons de deux ciments et de trois agents entraîneurs d'air selon deux dosages ont été utilisées avec le superplastifiant N et différents ajouts alcalins. Pour tous les mélanges, le rapport eau/ciment a été fixé à 0,45 et le rapport sable/ciment à 3. De façon à identifier correctement chaque mélange, un code contenant l'information pertinente sur les composants des mélanges a été utilisé. Ce code est présenté en annexe.

Les résultats des mesures de l'étalement du mortier à 10, 30 et 60 minutes après le premier contact eau-ciment sont présentés au tableau 3. L'étalement initial moyen (39 %, $\sigma=12$) est généralement correct. Une importante augmentation de l'étalement a évidemment été observée lors de l'ajout du superplastifiant au mélange (15 minutes). Puis, l'étalement a diminué en fonction du temps (voir les valeurs de l'étalement à 60 minutes). La diminution de l'étalement en fonction du temps peut être expliquée en partie par l'évaporation d'une petite quantité d'eau, en partie par la diminution normale de l'efficacité du superplastifiant et en partie par l'avancement de l'hydratation. Les faibles ouvrabilités obtenues à la fin de certains essais n'ont toutefois pas causé de difficultés particulières pour la mise en place puisque le mortier était vibré dans les moules.

Les agents entraîneurs d'air ont été utilisés selon deux dosages: 0,2 ml/kg de ciment pour 24 mélanges et 0,1 ml/kg de ciment pour les 30 autres. Le premier dosage a été choisi afin d'obtenir une teneur en air initiale comprise entre 10 et 15 %, c'est-à-dire un volume d'air généralement suffisant pour obtenir un facteur d'espacement initial d'environ 250 µm, et le deuxième afin de mettre en relief l'effet du dosage sur la stabilité. Les résultats de la mesure de la teneur en air du mortier plastique (selon la méthode volumétrique ASTM C173) sont présentés au tableau 4.

Les mortiers ont été fabriqués à l'aide d'un malaxeur standard à mortier d'une capacité de 0,02 m^3. Tous les matériaux étaient pesés et préparés à l'avance et la procédure d'essai était identique pour tous les mélanges. Les alcalins (lorsque utilisés) étaient dissous dans l'eau de gâchage. Le ciment et la majorité de l'eau (contenant le réducteur d'eau) étaient mélangés afin d'obtenir

Tableau 3: Etalement du mortier (ASTM C-230) (%)

Code	Temps (minutes)		
	10	30	60
M-2-HS-N-0	70	120	100
M-2-RV-N-0	45	110	60
M-2-DS-N-0	60	115	100
M-2-HS-N-.5Na	45	115	85
M-2-HS-N-Na	50	80	65
M-2-HS-N-K	40	90	65
M-2-HS-N-Na+K	40	70	50
M-2-RV-N-.5Na	45	90	70
M-2-RV-N-Na	45	80	50
M-2-RV-N-K	40	80	60
M-2-RV-N-Na+K	30	45	30
M-2-DS-N-.5Na	40	80	50
M-2-DS-N-Na	45	70	45
M-2-DS-N-K	40	90	70
M-2-DS-N-Na+K	30	55	30
M-1-HS-N-0	40	80	40
M-1-HS-N-Na	30	35	25
M-1-HS-N-Na+K	25	30	20
M-1-RV-N-0	40	85	65
M-1-RV-N-Na	55	70	55
M-1-RV-N-Na+K	30	45	20
M-1-DS-N-0	30	100	35
M-1-DS-N-Na	45	65	35
M-1-DS-N-Na+K	55	75	50
M-2-.5HS-N-0	60	135	100
M-2-.5RV-N-0	45	90	75
M-2-.5DS-N-0	50	125	85
M-2-.5HS-N-.5Na	30	80	60
M-2-.5HS-N-Na	30	55	35
M-2-.5HS-N-K	25	85	55
M-2-.5HS-N-Na+K	25	55	35
M-2-.5RV-N-.5Na	40	90	75
M-2-.5RV-N-Na	35	60	40
M-2-.5RV-N-K	25	85	40
M-2-.5RV-N-Na+K	30	60	35
M-2-.5DS-N-.5Na	45	100	60
M-2-.5DS-N-Na	40	75	40
M-2-.5DS-N-K	25	85	50
M-2-.5DS-N-Na+K	35	90	60
M-1-.5HS-N-0	20	75	40
M-1-.5HS-N-Na	20	40	25
M-1-.5HS-N-Na+K	25	65	35
M-1-.5RV-N-0	30	95	50
M-1-.5RV-N-Na	25	55	20
M-1-.5RV-N-Na+K	30	55	20
M-1-.5DS-N-0	35	85	55
M-1-.5DS-N-Na	25	45	30
M-1-.5DS-N-Na+K	30	45	20
M-2-.5HS-M-0	55	80	60
M-2-.5HS-M-Na	50	95	55
M-2-.5RV-M-0	55	105	50
M-2-.5RV-M-Na	50	85	50
M-2-.5DS-M-0	60	100	70
M-2-.5DS-M-Na	55	105	60

Tableau 4: Teneur en air (%, mortier plastique)

Code	Temps (minutes)		
	10	30	60
M-2-HS-N-0	13.3	19.1	13.4
M-2-RV-N-0	12.1	12.4	15.2
M-2-DS-N-0	13.3	23.2	20.1
M-2-HS-N-.5Na	12.2	19.4	14.5
M-2-HS-N-Na	18.1	13.6	13.8
M-2-HS-N-K	13.5	12.9	12.2
M-2-HS-N-Na+K	13.2	13.6	8.9
M-2-RV-N-.5Na	18.6	16.3	16.7
M-2-RV-N-Na	14.9	14.9	16.4
M-2-RV-N-K	12.1	15.9	13.3
M-2-RV-N-Na+K	10.0	8.6	12.7
M-2-DS-N-.5Na	12.7	13.6	17.5
M-2-DS-N-Na	11.1	15.8	18.5
M-2-DS-N-K	19.1	12.8	14.5
M-2-DS-N-Na+K	13.2	8.9	12.3
M-1-HS-N-0	11.7	11.0	13.4
M-1-HS-N-Na	6.9	12.7	12.5
M-1-HS-N-Na+K	13.9	13.9	14.6
M-1-RV-N-0	8.9	11.1	13.5
M-1-RV-N-Na	15.0	8.9	12.5
M-1-RV-N-Na+K	12.0	9.7	7.5
M-1-DS-N-0	11.0	13.2	13.7
M-1-DS-N-Na	13.0	11.8	14.0
M-1-DS-N-Na+K	12.0	10.4	12.3
M-2-.5HS-N-0	11.9	13.9	16.3
M-2-.5RV-N-0	7.4	16.4	13.7
M-2-.5DS-N-0	9.1	15.3	15.2
M-2-.5HS-N-.5Na	7.7	12.8	12.8
M-2-.5HS-N-Na	17.8	8.3	3.9
M-2-.5HS-N-K	14.7	12.1	11.1
M-2-.5HS-N-Na+K	11.8	17.9	7.8
M-2-.5RV-N-.5Na	7.9	13.2	13.3
M-2-.5RV-N-Na	15.6	6.4	14.9
M-2-.5RV-N-K	8.9	8,4	7.4
M-2-.5RV-N-Na+K	6.5	11.9	6.9
M-2-.5DS-N-.5Na	12.1	14.7	12.4
M-2-.5DS-N-Na	9.8	13.9	13.3
M-2-.5DS-N-K	10.0	14.3	14.7
M-2-.5DS-N-Na+K	13.4	12.8	12.9
M-1-.5HS-N-0	11.4	11.9	6.4
M-1-.5HS-N-Na	14.2	13.3	8.8
M-1-.5HS-N-Na+K	9.7	6.6	6.5
M-1-.5RV-N-0	11.7	6.6	8.5
M-1-.5RV-N-Na	8.1	8.6	5.9
M-1-.5RV-N-Na+K	8.3	4.5	4.3
M-1-.5DS-N-0	16.2	12.9	11.5
M-1-.5DS-N-Na	9.5	13.5	10.2
M-1-.5DS-N-Na+K	11.3	8.6	6.7
M-2-.5HS-M-0	9.6	11.8	15.9
M-2-.5HS-M-Na	12.8	12.9	14.0
M-2-.5RV-M-0	10.2	11.0	13.7
M-2-.5RV-M-Na	9.7	16.7	16.7
M-2-.5DS-M-0	13.8	15.7	15.9
M-2-.5DS-M-Na	14.9	13.8	14.1

une pâte. Après 2 minutes de malaxage, le reste de l'eau (contenant l'agent entraîneur d'air) était ajouté. Finalement, le sable était incorporé au mélange. Le superplastifiant était ajouté après 15 minutes de malaxage (après la première série de mesures et le premier échantillonage). De façon à fournir une énergie de malaxage raisonable, le mortier était mélangé pour des périodes de trois minutes suivies de périodes de repos de 7 minutes.

Les éprouvettes ont été fabriquées après 10, 30 et 60 minutes. Après une période de mûrissement suffisante, des tranches étaient sciées à partir des éprouvettes puis polies de façon à pouvoir mesurer le facteur d'espacement des bulles d'air selon la norme ASTM C 457 "modified point count method". Pour chaque mesure, nous avons examiné deux tranches de 100 mm x 100 mm (à un grossissement de 100 X). Sur chacune de ces tranches, l'analyse a été faite en utilisant 1000 points d'arrêt (soit un total de 2000 pour chaque facteur d'espacement).

Résultats

Les caractéristiques du réseau de bulles d'air mesurées sur les éprouvettes fabriquées après 10, 30 et 60 minutes sont présentées aux tableaux 5 (teneur en air), 6 (surface volumique) et 7 (facteur d'espacement).

En comparant les résultats montrés aux tableaux 4 et 5, on constate que les teneurs en air mesurées sur le mortier durci sont, en général, assez près de celles mesurées sur le mortier plastique (par la méthode volumétrique) même s'il y a dans certains cas des différences assez importantes (voir la figure 1). La valeur moyenne de la teneur en air est de 13,0 % pour les mélanges contenant une dose d'agent entraîneur d'air de 0,2 ml/kg de ciment, et de 11,2 % pour ceux contenant une dose de 0,1 ml/kg, ce qui représente une différence assez faible.

Les valeurs de la surface volumique sont données au tableau 6. La moyennne des valeurs initiales est assez faible (11,2 mm^{-1}), mais peut être considérée suffisante pour un mortier, bien qu'elle soit inférieure à 25 mm^{-1}, la valeur limite fréquemment mentionnée pour le béton. En effet, pour ces valeurs de la surface volumique, il a été possible d'obtenir des facteurs d'espacement initiaux corrects (voir le

Tableau 5: Teneur en air (%, mortier durci)

Code	Temps (minutes)		
	10	30	60
M-2-HS-N-0	16.1	17.9	18.9
M-2-RV-N-0	14.1	21.1	20.3
M-2-DS-N-0	16.6	21.2	23.3
M-2-HS-N-.5Na	14.3	16.5	17.4
M-2-HS-N-Na	14.0	23.0	18.9
M-2-HS-N-K	11.6	16.6	14.9
M-2-HS-N-Na+K	12.2	16.4	15.7
M-2-RV-N-.5Na	15.9	23.9	18.2
M-2-RV-N-Na	13.7	19.1	16.3
M-2-RV-N-K	11.0	14.0	14.8
M-2-RV-N-Na+K	11.2	13.6	13.8
M-2-DS-N-.5Na	15.6	23.2	19.1
M-2-DS-N-Na	15.9	22.5	20.0
M-2-DS-N-K	15.4	21.7	18.0
M-2-DS-N-Na+K	13.2	17.5	15.4
M-1-HS-N-0	13.0	19.7	14.2
M-1-HS-N-Na	11.1	14.4	15.3
M-1-HS-N-Na+K	10.0	13.0	10.0
M-1-RV-N-0	12.0	17.8	13.9
M-1-RV-N-Na	15.0	15.6	15.9
M-1-RV-N-Na+K	11.1	12.3	10.7
M-1-DS-N-0	12.8	15.9	12.6
M-1-DS-N-Na	15.4	17.9	15.0
M-1-DS-N-Na+K	14.6	18.2	13.4
M-2-.5HS-N-0	14.6	23.5	18.6
M-2-.5RV-N-0	12.3	22.1	20.5
M-2-.5DS-N-0	15.4	23.5	21.9
M-2-.5HS-N-.5Na	12.3	14.8	15.2
M-2-.5HS-N-Na	12.9	15.2	12.9
M-2-.5HS-N-K	11.8	18.3	14.7
M-2-.5HS-N-Na+K	9.0	14.0	12.4
M-2-.5RV-N-.5Na	12.9	18.7	17.2
M-2-.5RV-N-Na	12.9	11.9	13.4
M-2-.5RV-N-K	12.1	12.3	12.2
M-2-.5RV-N-Na+K	11.2	12.9	13.1
M-2-.5DS-N-.5Na	15.0	20.0	19.8
M-2-.5DS-N-Na	13.9	18.2	16.0
M-2-.5DS-N-K	13.2	20.8	17.1
M-2-.5DS-N-Na+K	13.4	17.1	19.0
M-1-.5HS-N-0	12.4	14.7	12.7
M-1-.5HS-N-Na	9.4	12.8	10.7
M-1-.5HS-N-Na+K	9.2	10.4	7.9
M-1-.5RV-N-0	9.5	14.6	12.4
M-1-.5RV-N-Na	8.0	11.6	10.3
M-1-.5RV-N-Na+K	7.7	9.6	7.3
M-1-.5DS-N-0	13.6	19.9	18.6
M-1-.5DS-N-Na	12.0	16.2	13.6
M-1-.5DS-N-Na+K	10.9	11.3	7.4
M-2-.5HS-M-0	12.0	13.8	12.3
M-2-.5HS-M-Na	12.3	13.7	12.6
M-2-.5RV-M-0	12.7	12.4	12.2
M-2-.5RV-M-Na	12.6	12.6	12.0
M-2-.5DS-M-0	15.5	17.5	16.5
M-2-.5DS-M-Na	15.8	16.0	15.8

Tableau 6: Surface volumique des bulles d'air (mm^{-1})

Code	Temps (minutes)		
	10	30	60
M-2-HS-N-0	12.9	9.5	13.3
M-2-RV-N-0	11.0	7.0	10.2
M-2-DS-N-0	26.3	15.5	16.5
M-2-HS-N-.5Na	10.9	8.0	13.0
M-2-HS-N-Na	9.3	10.0	15.7
M-2-HS-N-K	11.7	11.0	15.0
M-2-HS-N-Na+K	7.7	9.5	12.6
M-2-RV-N-.5Na	11.9	11.8	18.7
M-2-RV-N-Na	10.2	12.1	17.4
M-2-RV-N-K	10.2	7.7	11.8
M-2-RV-N-Na+K	8.4	13.7	13.2
M-2-DS-N-.5Na	15.7	12.4	16.7
M-2-DS-N-Na	16.1	14.7	17.2
M-2-DS-N-K	15.9	13.8	18.6
M-2-DS-N-Na+K	12.4	15.6	15.8
M-1-HS-N-0	20.7	13.8	18.4
M-1-HS-N-Na	9.6	10.5	11.0
M-1-HS-N-Na+K	9.4	8.2	9.8
M-1-RV-N-0	11.2	6.0	9.1
M-1-RV-N-Na	8.9	10.6	10.1
M-1-RV-N-Na+K	8.6	8.8	10.4
M-1-DS-N-0	14.4	8.2	11.7
M-1-DS-N-Na	15.3	18.3	21.2
M-1-DS-N-Na+K	10.8	12.9	15.2
M-2-.5HS-N-0	10.0	5.8	9.1
M-2-.5RV-N-0	10.7	5.9	7.6
M-2-.5DS-N-0	14.8	8.3	9.9
M-2-.5HS-N-.5Na	9.2	6.1	8.3
M-2-.5HS-N-Na	7.8	7.7	10.9
M-2-.5HS-N-K	8.7	5.3	8.2
M-2-.5HS-N-Na+K	8.0	6.8	8.6
M-2-.5RV-N-.5Na	9.1	5.5	8.1
M-2-.5RV-N-Na	7.7	9.2	10.7
M-2-.5RV-N-K	8.9	6.2	8.8
M-2-.5RV-N-Na+K	7.3	7.3	8.3
M-2-.5DS-N-.5Na	12.5	8.6	11.7
M-2-.5DS-N-Na	12.4	12.8	16.5
M-2-.5DS-N-K	11.8	8.7	12.4
M-2-.5DS-N-Na+K	11.4	8.6	11.3
M-1-.5HS-N-0	9.7	5.2	7.9
M-1-.5HS-N-Na	8.7	7.6	11.0
M-1-.5HS-N-Na+K	8.3	8.6	10.8
M-1-.5RV-N-0	10.3	4.2	7.1
M-1-.5RV-N-Na	9.5	8.5	11.8
M-1-.5RV-N-Na+K	8.7	8.9	11.4
M-1-.5DS-N-0	16.7	8.1	11.3
M-1-.5DS-N-Na	11.0	13.0	13.7
M-1-.5DS-N-Na+K	9.1	10.4	14.1
M-2-.5HS-M-0	12.3	7.9	9.8
M-2-.5HS-M-Na	6.9	6.3	7.4
M-2-.5RV-M-0	9.9	6.5	9.1
M-2-.5RV-M-Na	7.5	7.6	8.9
M-2-.5DS-M-0	15.4	13.2	19.0
M-2-.5DS-M-Na	11.8	12.9	13.6

Tableau 7: Facteur d'espacement (µm) et indice de stabilité

Code	Temps (minutes)			Indice de stabilité
	10	30	60	($\partial L=L1-L2$)
M-2-HS-N-0	162	197	134	-35
M-2-RV-N-0	216	215	149	1
M-2-DS-N-0	79	95	79	-16
M-2-HS-N-.5Na	224	251	146	-27
M-2-HS-N-Na	282	144	104	138
M-2-HS-N-K	268	187	156	81
M-2-HS-N-Na+K	389	214	191	175
M-2-RV-N-.5Na	191	118	102	73
M-2-RV-N-Na	254	147	113	107
M-2-RV-N-K	332	331	201	1
M-2-RV-N-Na+K	383	187	178	196
M-2-DS-N-.5Na	138	106	103	32
M-2-DS-N-Na	139	99	94	40
M-2-DS-N-K	150	113	102	37
M-2-DS-N-Na+K	223	128	145	95
M-1-HS-N-0	136	118	131	18
M-1-HS-N-Na	350	233	227	117
M-1-HS-N-Na+K	403	338	363	65
M-1-RV-N-0	255	310	269	-55
M-1-RV-N-Na	267	207	207	60
M-1-RV-N-Na+K	383	331	335	52
M-1-DS-N-0	203	276	241	-73
M-1-DS-N-Na	153	106	112	47
M-1-DS-N-Na+K	226	151	176	75
M-2-.5HS-N-0	232	223	187	9
M-2-.5RV-N-0	253	239	186	24
M-2-.5DS-N-0	153	153	146	0
M-2-.5HS-N-.5Na	317	384	270	-67
M-2-.5HS-N-Na	335	299	242	36
M-2-.5HS-N-K	336	352	298	-16
M-2-.5HS-N-Na+K	508	367	310	141
M-2-.5RV-N-.5Na	301	307	217	-6
M-2-.5RV-N-Na	347	327	239	20
M-2-.5RV-N-K	321	478	315	-157
M-2-.5RV-N-Na+K	438	373	314	65
M-2-.5DS-N-.5Na	189	194	134	-5
M-2-.5DS-N-Na	207	146	128	61
M-2-.5DS-N-K	229	179	155	50
M-2-.5DS-N-Na+K	233	237	162	-4
M-1-.5HS-N-0	308	472	359	-164
M-1-.5HS-N-Na	437	338	294	99
M-1-.5HS-N-Na+K	467	400	412	67
M-1-.5RV-N-0	378	546	398	-168
M-1-.5RV-N-Na	443	381	282	62
M-1-.5RV-N-Na+K	517	439	399	78
M-1-.5DS-N-0	155	197	149	-42
M-1-.5DS-N-Na	263	156	167	107
M-1-.5DS-N-Na+K	359	299	332	60
M-2-.5HS-M-0	255	313	322	-58
M-2-.5HS-M-Na	419	430	395	-11
M-2-.5RV-M-0	295	472	329	-177
M-2-.5RV-M-Na	386	377	336	9
M-2-.5DS-M-0	155	144	110	11
M-2-.5DS-M-Na	191	173	167	18

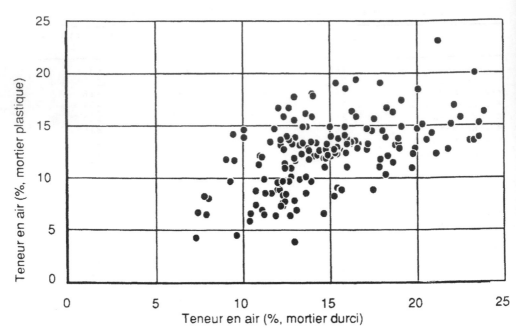

Fig.1: Relation entre la teneur en air du mortier plastique
et du mortier durci

tableau 7) avec une teneur en air raisonnable pour un mortier. Les
faibles valeurs de la surface volumique sont probablement dues aux
dosages de l'agent entraîneur d'air qui se situent à la limite
inférieure de la gamme de dosage suggérée par les producteurs.
Rappelons toutefois que pour les 24 premiers mélanges, le dosage de
l'agent entraîneur d'air a été choisi de façon à obtenir un facteur
d'espacement initial de l'ordre de 250 μm (le résultat moyen est en
fait de 242 μm). Pour les 30 derniers mélanges, la dose de l'agent
entraîneur d'air a été réduite de 50 % et nous avons alors obtenu 314
μm comme moyenne pour le facteur d'espacement initial (à 10 minutes).
Il est également possible que les caractéristiques du malaxeur à
mortier, qui développe une importante énergie de malaxage, soient la
cause de la présence d'une quantité significative de grosses bulles
d'air, amenant ainsi une importante diminution de la surface volumique.

La surface volumique des vides d'air a systématiquement
chuté après l'ajout du superplastifiant, passant de 11,2 mm^{-1} en
moyenne ($\sigma = 3,5$) à 9,5 mm^{-1} ($\sigma = 3,2$). Cette variation démontre bien

l'effet du superplastifiant sur le réseau de bulles d'air. Cet effet peut s'expliquer de deux façons: premièrement, le superplastifiant peut entraîner des grosses bulles d'air ce qui augmente la dimension moyenne des bulles et donc diminue la surface volumique; deuxièmement, le superplastifiant augmente la fluidité de la pâte ce qui facilite probablement la coalescence des bulles. Lorsqu'il y a coalescence, on ne mesure pas de variation de la teneur en air ce qui explique, du moins en partie, pourquoi la relation entre la teneur en air (mesurée sur le mortier plastique ou durci) et le facteur d'espacement ne peut être qu'approximative (voir la figure 2). Après 60 minutes de malaxage, l'effet du superplastifiant s'est atténué et le cisaillement des bulles par les pales du malaxeur tend à faire augmenter de nouveau la valeur moyenne de la surface volumique (cette valeur est de 12,2 mm^{-1} en moyenne (σ=4,1) à 60 minutes).

Les résultats de la mesure du facteur d'espacement sont présentés au tableau 7. La plupart des valeurs à 10 minutes ne sont pas très éloignées de 250 μm, la valeur visée pour les 24 premiers mélanges

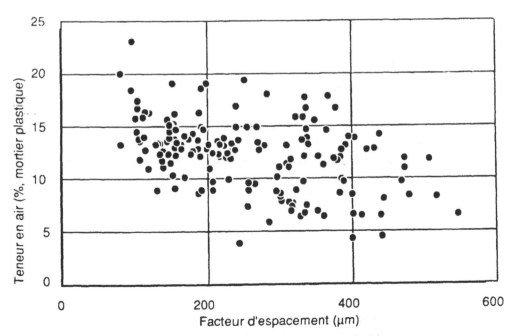

Fig.2: Relation entre le facteur d'espacement et la
teneur en air du mortier plastique

(un facteur d'espacement inférieur à 300 μm permet généralement d'éviter les problèmes d'écaillage dus au gel en présence de sels de déglaçage (15)). Les valeurs élevées du facteur d'espacement initial ont surtout été obtenues dans les mélanges avec ajout d'une forte quantité d'alcalins (voir le code des mélanges), indépendamment du dosage de l'agent entraîneur d'air.

De façon générale, excepté lors de l'ajout du superplastifiant, le facteur d'espacement tend à diminuer en fonction du temps de malaxage (tableau 7). Ce phénomène n'a pas été observé pour les mélanges de béton fabriqués dans notre laboratoire (10,11,12,13,14), mais a déjà été noté pour des mélanges de mortier (16). Il s'explique probablement par l'importante quantité d'énergie développée par le malaxeur à mortier. Toutefois, la valeur du facteur d'espacement augmente souvent immédiatement après l'addition du superplastifiant. Le superplastifiant, qui fluidifie la pâte, peut faciliter deux phénomènes qui sont susceptibles de faire augmenter le facteur d'espacement: premièrement la perte de bulles d'air lors des périodes de malaxage et de repos, et deuxièmement la coalescence des bulles d'air qui cause une diminution de la surface volumique.

Analyse des résultats

Production du réseau de bulles d'air

Comme nous l'avons montré dans d'autres publications (10,11), il semble que la production d'un bon facteur d'espacement, c'est-à-dire le résultat de la mesure à 10 minutes, soit relativement simple. Pour les 54 mélanges de mortier, le facteur d'espacement initial moyen est de 282 μm. Toutefois, les résultats sont assez variables et semblent surtout fonction de la teneur en alcalins solubles des mélanges. La valeur moyenne du facteur d'espacement initial est de 219 μm (σ=74, 21 résultats) pour les mélanges ayant une teneur en alcalins solubles inférieure à 1 % de la masse du ciment, et 322 μm (σ=104, 33 résultats) pour ceux dont le contenu en alcalins solubles est supérieur à 1 %. La loi de Student-Fischer permet de différencier ces deux échantillons avec un risque inférieur à 1 %. Cela confirme que l'ajout de sulfates alcalins nuit, dans les premières minutes de malaxage, à la production du réseau de bulles d'air.

Même si la relation entre la teneur en air et le facteur d'espacement est très imprécise (voir la figure 2), il semble qu'en général il s'agisse simplement de bien choisir la dose de l'agent entraîneur pour obtenir, après quelques minutes de malaxage, un bon facteur d'espacement avec une teneur en air raisonnable. Toutefois, des problèmes d'instabilité du facteur d'espacement, problèmes que le producteur peut difficilement régler, peuvent souvent survenir. C'est pourquoi le but principal de notre étude est l'examen des facteurs qui peuvent influencer la stabilité du facteur d'espacement.

Stabilité du réseau de bulles d'air

Afin de quantifier la stabilité du facteur d'espacement, nous avons mis au point un indice de stabilité. Cet indice est calculé en soustrayant la valeur du facteur d'espacement obtenue lors de la deuxième mesure (30 minutes) de celle obtenue lors de la première mesure (10 minutes); $\partial L = L1 - L2$. De cette façon, nous obtenons la variation du facteur d'espacement lors de l'ajout du superplastifiant. La troisième mesure du facteur d'espacement est simplement utilisée afin de vérifier l'évolution du réseau de bulles d'air sur une période de temps plus prolongée. Lors de l'ajout du superplastifiant, la variation du facteur d'espacement est généralement immédiate, c'est pourquoi l'indice de stabilité permet de bien évaluer l'influence de cet adjuvant. L'indice de stabilité peut être négatif (lorsque le facteur d'espacement augmente) ou positif (lorsqu'il diminue). Suite aux nombreuses expériences faites dans leurs laboratoires (11,12,13,14), les auteurs sont d'avis que l'indice de stabilité est valable pour une gamme assez étendue de valeurs initiales du facteur d'espacement, c'est-à-dire qu'une augmentation du facteur d'espacement de 100 µm est tout aussi significative (représente une détérioration tout aussi importante) pour une valeur initiale de 100 que pour une valeur de 250 µm.

Si le facteur d'espacement augmente de façon significative entre la deuxième et la troisième mesure, l'indice de stabilité ne représente pas la détérioration complète du réseau de bulles d'air. Le tableau 7 montre cependant que le facteur d'espacement tend généralement à diminuer entre les deux dernières mesures. Ce phénomène est probablement causé par l'importante énergie de malaxage combinée à la diminution de l'efficacité du superplastifiant.

La valeur de l'indice de stabilité pour chacun des 54 mélanges est présenté au tableau 7. Certains résultats indiquent une importante diminution du facteur d'espacement (∂L=175), d'autres une augmentation majeure (∂L= -177) et, dans certains cas, le facteur d'espacement n'a que très peu varié (-10<∂L<10).

Les résultats des essais démontrent clairement que la stabilité du facteur d'espacement n'est pas reliée aux caractéristiques initiales du mortier (étalement, teneur en air et facteur d'espacement) ni aux variations de la teneur en air. On présente à la figure 3 la relation entre la teneur en air initiale du mortier plastique et l'indice de stabilité pour les 54 mélanges. Pour les mélanges ayant une teneur en air initiale égale ou inférieure à 10 %, l'indice de stabilité moyen est de 28 (σ=83, 17 résultats), comparativement à 20 (σ=79, 37 résultats) pour les mélanges ayant une teneur en air initiale

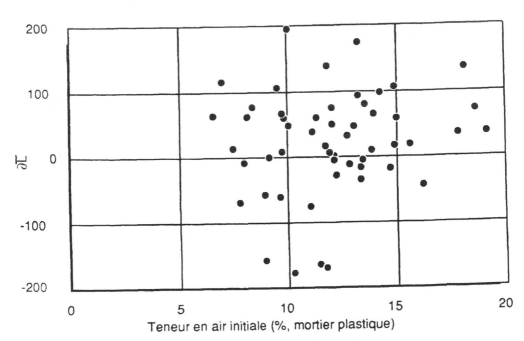

Fig.3:   Relation entre l'indice de stabilité et la
teneur en air initiale du mortier plastique

476

supérieure à 10 %. La loi de Student-Fischer ne permet pas de
différencier, même avec un risque de 50 %, ces deux échantillons. Il
n'existe donc pas de relation précise entre la teneur en air initiale
du mortier plastique et l'indice de stabilité, bien que 30 mélanges
aient été fabriqués avec une dose d'agent entraîneur d'air de seulement
0,1 mL/kg de ciment.

La figure 4 montre la relation entre le facteur
d'espacement initial et l'indice de stabilité pour les 54 mélanges de
mortier. Pour les mélanges ayant un facteur d'espacement initial égal
ou inférieur à 300 µm, l'indice de stabilité moyen est de 19 (σ=64, 31
résultats), comparativement à 29 (σ=98, 23 résultats) pour ceux ayant
un facteur d'espacement initial supérieur à 300 µm. La loi de Student-
Fischer ne permet pas, même avec un risque de 50 %, de différencier ces
deux échantillons. La stabilité du réseau de bulles d'air n'est pas
donc reliée non plus à la valeur initiale du facteur d'espacement

A la figure 5,* on présente la relation entre la variation
de la teneur en air (ΔA=%A1-%A2) et l'indice de stabilité. Cette figure
montre bien que ces deux variables ne sont pas reliées. Pour les

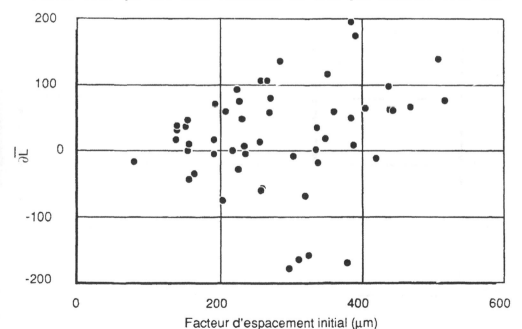

Fig.4: Relation entre le facteur d'espacement initial
et l'indice de stabilité

Not available

477

mélanges dont la teneur en air a augmenté ou est demeurée constante
($\partial A \leq 0$), l'indice de stabilité moyen est de 12 ($\sigma=78$, 31 résultats),
comparativement à 38 ($\sigma=81$, 23 résultats) pour les mélanges dont la
teneur en air a diminué. La loi de Student-Fischer ne permet pas, même
avec un risque de 20 %, de différencier ces deux échantillons.

Comme le montre la figure 6, les résultats obtenus avec le
superplastifiant N sont très variables. En effet, l'indice de stabilité
varie de -168 (ce qui indique une importante détérioration du réseau de
bulles d'air) à 196 (une importante amélioration). Avec le
superplastifiant M, l'indice de stabilité varie de -177 à 18. Il est
donc clair, bien que le produit à base de mélamine n'ait été utilisé
que pour 6 mélanges, que les deux superplastifiants, qui sont des
produits très différents, sont également susceptibles de causer de
fortes instabilités du réseau de bulles d'air.

La relation entre le type d'agent entraîneur d'air et
l'indice de stabilité est présentée à la figure 7. Les résultats
obtenus avec le DS sont moins variables qu'avec les deux autres. Il
semble donc moins risqué d'obtenir de très importantes détériorations
du réseau de bulles d'air en utilisant cet adjuvant. Toutefois les

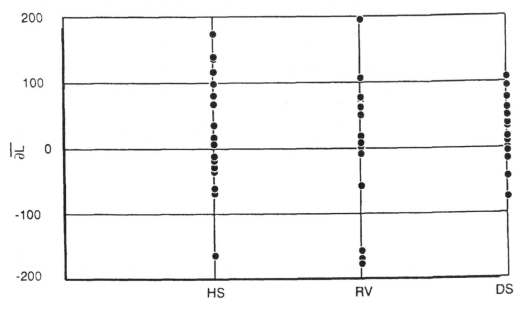

Fig. 7: Relation entre le type d'adjuvant entraîneur d'air
et l'indice de stabilité   ✳ Not available.

trois agents entraîneurs d'air ont donné, en moyenne, des résultats assez semblables et il est statistiquement impossible de les différencier. Ainsi, l'indice de stabilité moyen pour les mortiers contenant du HS est de 32 (σ=87, 18 résultats), comparativement à 10 (σ=98, 18 résultats) pour ceux contenant le RV et 27 (σ=46, 18 résultats) pour ceux contenant le DS.

La figure 8 montre la relation entre la dose d'agent entraîneur d'air et l'indice de stabilité. On constate que les très grandes détériorations (∂L<-100) du réseau de bulles d'air ont toutes été obtenues avec la faible dose d'adjuvant. Pour les mélanges fabriqués avec 0,1 ml/kg de ciment d'agent entraîneur d'air, l'indice de stabilité moyen est de 1 (σ=82, 30 résultats), comparativement à 50 (σ=68, 24 résultats) pour ceux fabriqués avec une dose de 0,2 ml/kg de ciment. La loi de Student-Fischer permet de différencier, avec un risque inférieur à 5 %, ces deux échantillons. Il existe donc une relation entre la dose d'agent entraîneur d'air et la stabilité du facteur d'espacement.

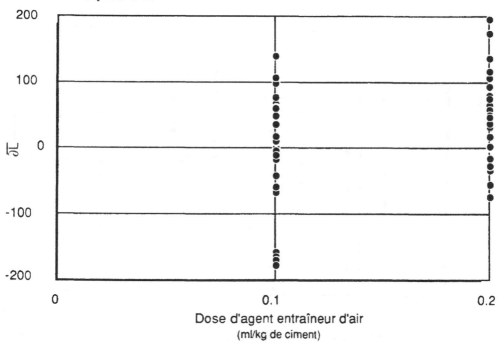

Fig.8: Relation entre la dose d'agent entraîneur d'air et l'indice de stabilité

On présente à la figure 9 la relation entre la teneur en alcalins solubles des mélanges et l'indice de stabilité du facteur d'espacement. On constate clairement une tendance à la hausse des indices de stabilité avec l'augmentation de la teneur en alcalins solubles, c'est-à-dire que les fortes détériorations du réseau de bulles d'air ont été obtenues avec les mélanges contenant peu d'alcalins. Pour les mélanges ayant une teneur en alcalins solubles égale ou inférieure à 1 % de la masse du ciment, l'indice de stabilité moyen est de -35 (σ=67, 21 résultats), comparativement à 60 (σ=64, 33 résultats) pour ceux ayant une teneur en alcalins solubles supérieure à 1 %. La loi de Student-Fischer permet de différencier, avec un risque inférieur à 1 %, ces deux échantillons. La teneur en alcalins solubles est donc un paramètre fondamental qui influence de façon relativement importante la stabilité du réseau de bulles d'air en présence de superplastifiant.

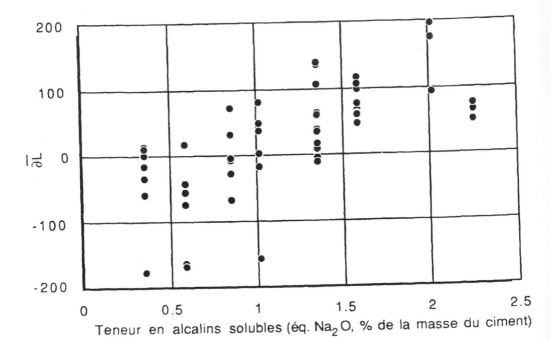

Fig.9: Relation entre la teneur en alcalin soluble des mélanges et l'indice de stabilité

Conclusions

Les résultats des essais confirment que, lors de l'utilisation de superplastifiants, les problèmes de stabilité du réseau de bulles d'air peuvent être assez fréquents. Ces résultats démontrent cependant que la stabilité du facteur d'espacement en présence de superplastifiant augmente avec la teneur en alcalins solubles du ciment. Par ailleurs, la production d'un bon facteur d'espacement semble plus difficile lorsque la teneur en alcalins est élevée.

La stabilité du facteur d'espacement en présence de superplastifiant ne semble pas influencée de façon significative par le type d'agent entraîneur d'air. Ainsi, aucun des trois adjuvants entraîneurs d'air utilisés n'a pu garantir des réseaux de bulles d'air stables indépendamment des autres variables (ajout de superplastifiant, dose d'adjuvant entraîneur d'air et teneur en alcalins solubles). Une augmentation de 100 % de la dose d'agent entraîneur d'air a cependant permis d'obtenir des réseaux de bulles d'air généralement plus stables.

Les problèmes de stabilité sont certainement plus difficiles à prévoir et donc à éviter que les problèmes de production. Les résultats indiquent que la stabilité du réseau de bulles d'air n'est pas reliée aux caractéristiques initiales du réseau de bulles d'air (teneur en air et facteur d'espacement initiaux) et que les variations du facteur d'espacement sont indépendantes des variations de la teneur en air.

Il est important de rappeller que les résultats discutés dans cet article ont été obtenus lors d'essais sur des mélanges de mortier. Il serait intéressant de vérifier ces résultats avec des mélanges de béton. Par ailleurs, d'autres recherches seront nécessaires afin de préciser l'influence des alcalins et surtout d'identifier leur mode d'action.

Bibliographie

1-POWERS, T.C., (1949), "The Air Requirement of Frost-Resistant Concrete", Proceedings of the Highway Research Board, no. 29, pp. 189-211.

2-VERBECK, G.J. and KLIEGER, P. (1957), "Studies of "Salt" Scaling of Concrete", Highway Research Board Bulletin No. 150, pp. 1-13.

3-PIGEON, M., PERRATON, D. and PLEAU, R., "Scaling Tests of Silica Fume Concrete and the Critical Spacing Factor Concept", Concrete Durability, Katharine and Bryant Mather International Conference, American Concrete Institute, publication SP-100, John M. Scanlon, Editor, April 1987, pp. 1155-1182.

4-WHITING, D. and STARK, D. (1983), "Control of Air Content in Concrete", National Highway Research Program 258 Report, 263p.

5-GEBLER, S. and KLIEGER, P., (1983), "Effect of Fly Ash on the Air-Void Stability of Concrete", Proceedings of the CANMET/ACI First International Conference on the Use of Fly Ash, Silica Fume, Slag and Other Mineral By-Products in Concrete, American Concrete Institute, Special Publication SP-79, pp. 103-120.

6-MALHOTRA, V.M., (1981), "Effect of Repeated Dosages of Superplasticizers on Slump, Strength and Freeze-Thaw Resistance of Concrete", Materials and Structures, 14 (80), pp. 78-89.

7-LITVAN, G.G., (1983), "Air Entrainment in the Presence of Superplasticizers", Journal of the American Concrete Institute, Proceedings, 80(3) pp.326-331.

8-TOGNON, G. and CANGIANO, S., (1982), "Air Contained in Superplasticized Concretes", Journal of the American Concrete Institute, Proceedings, 79(5), pp.350-354.

9-PISTILLI, M.F., (1983), "Air-Void Parameters Developped by Air-Entraining Admixtures, as Influenced by Soluble Alkalies from Fly Ash and Portland Cement", Journal of the American Concrete Institute, Proceedings, 80(3), pp. 217-222.

10-PLANTE,P., PIGEON,M. and FOY,C. "The Influence of Water-Reducers on the Production and Stability of the Air Void System in Concrete" Accepted for publication, Cement and Concrete Research, June 1988

11-PIGEON, M, PLANTE, P. and PLANTE, M. "Air Void Stability, part I: Influence of Silica Fume and Other Parameters" Accepted for publication in the ACI Materials Journal, June 1988. 76(5) Sept-Oct 1989. 482-490

12-PLANTE,P., PIGEON,M. and SAUCIER, F. "Air Void Stability, part II: Influence of Superplasticizer and Cement" submitted for publication to the ACI Materials Journal, September 1988. 86(6) Nov-Dec 1989. 581-9

13-PIGEON,M., SAUCIER,F. and PLANTE, P. "Air Void Stability part III: Field Tests of Superplasticized Concrete" submitted for publication to the ACI Materials Journal, October 1988.

14-SAUCIER,F., PIGEON,M. and PLANTE,P. "Air Void Stability part IV: Retempering" submitted for publication to the ACI Materials Journal, January 1989.

15-FOURNIER, B., BERUBE, M.A. and VEZINA, D. (1987), "Condition Survey of Concrete Structures Built with Potentially Alkali-Reactive Limestone Aggregates from the Quebec City Area (Quebec, Canada)" Concrete Durability, Katharine and Bryant Mather International Conference, American Concrete Institute, publication SP-100, John M. Scanlon, Editor, April 1987, pp. 1343-1364.

16-PLANTE, P. et PIGEON, M.(1989) "The Influence of the Soluble Alkali Content on Air Void Stability" Accepté pour présentation au troisième congrès international sur les superplastifiants et autres adjuvants chimiques, Ottawa, Canada, Octobre 1989.

ANNEXE 1

Codification des mélanges

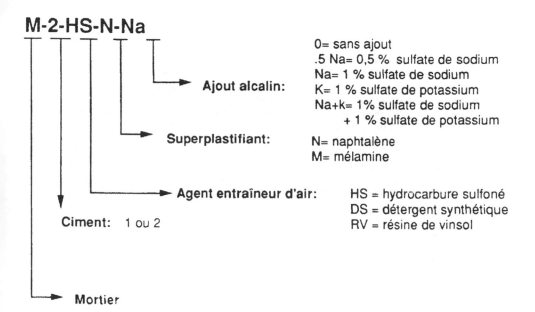

M-2-HS-N-Na

Ajout alcalin:
0= sans ajout
.5 Na= 0,5 % sulfate de sodium
Na= 1 % sulfate de sodium
K= 1 % sulfate de potassium
Na+k= 1% sulfate de sodium
+ 1 % sulfate de potassium

Superplastifiant:
N= naphtalène
M= mélamine

Agent entraîneur d'air:
HS = hydrocarbure sulfoné
DS = détergent synthétique
RV = résine de vinsol

Ciment: 1 ou 2

Mortier

IMPROVEMENT OF DRYING SHRINKAGE AND SHRINKAGE CRACKING OF CONCRETE BY SPECIAL SURFACTANTS

M. SHOYA and S. SUGITA
Department of Civil Engineering, Hachinohe Institute of Technology,
Hachinohe, Japan
T. SUGAWARA
Department of Civil Engineering, Hachinohe National College of Technology,
Hachinohe, Japan

Abstract
This paper describes the efficacies of special organic surfactants as
chemical admixtures to reduce drying shrinkage and shrinkage cracking
in air-dried concrete. These admixtures have been recently invented in
Japan and tentatively called as shrinkage reducing agents because of
the high ability reducing drying shrinkage of concrete by 20 to 40% in
their standard dosages.
  Shrinkage tests using small specimens made of cement pastes under
various humidity conditions showed that the mechanism of the reduction
of shrinkage by the use of these admixtures could be almost explained
by the theory of capillary tension.
  There was almost little lowering of mechanical properties of
concretes due to the addition of four kinds of admixtures in their
standard dosages and it was also confirmed that their additions
attained the prolongation of the time in passing-through cracking and
the decrease of the crack width in proportion to their dosages from
the restraint shrinkage cracking tests.

Key words: organic surfactants, shrinkage reducing agents, surface
           tension, capillary tension theory, drying shrinkage,
           shrinkage cracking, crack width, creep coefficient

1 Introduction

Preventing the occurrence of cracking due to drying shrinkage in
concrete structures must be remarkably effective not only to prolong
the service life and to maintain their durability but also not to
impair the beauty of the appearances.
  Shrinkage reducing admixtures composed of special organic surfactants
have been developed in our country to act so as to reduce drying
shrinkage of concrete remarkably and the efficacy has been confirmed
by Shoya et al. (1985,1987) and others. Four kinds of these
admixtures which have been now in the market of Japan are just to be
also expected to have high efficacies to prevent the cracking due to
drying shrinkage of concrete.
  Then, in order to scheme their positive utilization, the effects of
each admixture must be clarified from the view point of the changes in
the chemical compositions between these four admixtures, in their

484

amount of use and in the methods in application like the addition or impregnation under the different restraint conditions for shrinkage.
 Furthermore, the details of actual acting mechanism of these admixtures in concrete must be clarified from the scientific interest on their physico-chemical actions. In this paper, the action of shrinkage reducing admixture was mainly investigated by the shrinkage test using small prisms made of cement pastes in the various humidity conditions, and the performance for reducing shrinkage cracking of concrete using the cracking frame was also investigated relating to the change of drying shrinkage due to its addition.

2 Experimental methods

2.1 Materials used and mix proportions
(1) Materials
 Four types of shrinkage reducing agent (SRA) composed of organic surfactants are used in this study. The physical properties of each SRA were shown in Table 1. Each SRA is either soluble or dispersive in water and has the high performance to reduce the surface tension of its water solution remarkably as shown in the table.
 In case of the addition of a certain SRA, the surface tension of water solution lowered by about 50% to that of pure water in the dosage of 5%. A and B type of SRA consist of lower alcohol alkylen oxide adducts with chemical composition of $RO-(AO)n-H$ (R:alkyl radical, A: alkylen radical) with molecular weight of one hundred sixty and three hundred in round figures, respectively. C and D of SRA consist of poly propylene glycol with the chemical composition of $HO-(C_3H_6O)n-H$ and a glycol ether derivative with the composition of $R-(C_3H_6O)n-OH$ (R :alkyl radical).

Table 1. Physical properties of SRA

type of SRA	Appearance	Specific gravity	Viscosity	PH	Standard dosage	surface * tension	remarks
A	Blue colored transparent liquid	1.00 (20℃)	20cps (20℃)	6	7.5 kg/m	34.5 dyn/cm	soluble in water
B	Low viscous transparent liquid	0.98 (20℃)	16cps (20℃)	6±1	12 kg/m	47.8 dyn/cm	soluble in water
C	Colorloss or light-colored liquid	1.02±0.02 (20℃)	100±20cps (20℃)	7±1	2 to 6% to Cement content	37.0 dyn/cm	soluble in water
D	light-yellow colored liquid	1.04	---	.	1 to 3% to Cement content	38.8 dyn/cm	insoluble in water

* value in the water solution of SRA by 5% in weight

Normal portland cement was used. The physical and chemical properties of the cement used are shown in Table 2. Fine aggregate for concrete specimens was the land sand whose fineness modulus, specific gravity and water absorption were 2.75, 2.57 and 1.10%, respectively.

Table 2. Physical and chemical properties of
cement

| Specific gravity | Fineness | Time of setting (hr-min) | | | Soundness (Pat test) | Compressive Strength (MPa) (40×40×160mm mortar) | | |
	Specific surface (Blaine method) (cm²/g)	Amount of water (%)	Initial	Final		3 days	7 days	28 days
3.16	3350	28.5	3-08	4-22	OK	14.3	23.6	41.7

Chemical Composition: SiO_2=22.2%, Al_2O_3=5.6%, Fe_2O_3=3.0%, CaO=63.3%, MgO=1.8%, SO_3=2.0%,
R_2O=0.4% (K_2O=0.31%, Na_2O=0.23%)

Coarse aggregate was the crushed stone of andesite whose maximum size, fineness modulus, specific gravity and water absorption were 25mm, 6.93, 2.71 and 0.49%, respectively. Air-entraining water reducer (AEWR) whose main compound was lignin sulphonic acid and high range-water reducer (HWR) whose main compound was highly polymarized aromatic sulphonic acid was used in the part of experiments. Cement expansive additives (E) consisted mainly of anhydrous sulphate was also used in some experiments.

(2)Mix proportions
The conditions of mix proportion for cement paste and concrete are shown in table 3. The low water cement ratio of 30% for cement paste was decided to obtain the homogeneous sample.
The water cement ratio of 55% in concrete specimens containing AEWR was adopted because of intending a easy comparison of the performance of each SRA in the range of usual mix proportions.
The water reduced concrete using HWR was adopted to obtain the target slump in case of SRA D because of the effect dissipating the air bubble resulting in less air-entrainment by AE agent.

Table 3. Mix proportions used in tests

| Type of | Max. size of Aggregates (mm) | slump in cm or (FLOW in mm) | air (%) | W C+E | s/a (%) | Unit Content (kg/m³) | | | | | SRA** Additives | *** AEWR | **** HWR |
						W *	C	E	S	G			
Cement Paste	—	(190)									0	—	—
		(185~205)	—	30	—	477	1590	—	—	—	A,B,C,D 2% to cement content		
		(145~200)									A,B,C,D 6%		
Concrete	25	8±1	4.5	55	42.0	157	285	0	763	1114	A,B,C,D 0 3.75,7.5 kg/m³	393 cc/m³	600 cc/m³
		8±1	4.5	55	42.0	157	255	30	763	1114	Impregnation A 100,300 cc/m²	393 cc/m³	—

*: in containing SRA   SRA: shrinkage reducing agent
AEWR: air-entraining water reducing agent
HWR: high-performance water reducer

486

2.2 Testing method

(1) Drying shrinkage test for cement paste

The test was initiated from the age of 28 days and the measurement was continued for 60 days when the weight loss and length change were judged almost asymptotic to the constant values.

Tests were made in the dessicators kept in the required humidity conditions under the temperature of 20 °C.

The humidity conditions were 20, 45, 58, 79 and 90% R.H. arranged by the saturated water solutions of CH_3COOk, KNO_2, $NaBr2H_2O$, NH_4Cl, and $ZnSO_4.7H_2O$, respectively.

The length change of the prismatic small specimen having 1x1x16cm in size was measured using the dial gauge with minimum reading of 1/1000 mm between two gauge plugs buried at the end of specimen.

Weight loss was measured by the automatic platform scale with the minimum reading of 1 mg.

(2) Shrinkage cracking test for concrete specimen

The combination of tests for restraint shrinkage cracking using cracking frame was shown in Table 4.

Twenty-two kinds of tests were schemed with the varieties of the type of SRA, the method of use of SRA including whether or not adding the expansive additives and the cross sectional area of restraint steel frame.

Specimen geometry and testing apparatus are shown in Fig 1.

This apparatus for restraint shrinkage cracking was applied correspondingly to the draft of·J.I.S. besides the selection of various rates of restraint by outer steel frame.

The rate of restraint of shrinkage by steel frame, R, can be evaluated by the following equation using Young's modulus ratio, n.

$$R = nAs / (Ac + nAs) \qquad (1)$$

  As : The cross sectional area of steel frame.

  Ac : The cross sectional area of concrete in the centre portion.

Table 4. Schematic list for restraint craking test on concrete

	S R A		Unit con-tent of E	Outer steel frame for restraint		Number in sets of test
	addition (kg/m³)	impregna-tion(cc/m²)	(kg/m³)	(cm²)		
Non	3.75 7.5	100 300	0 30	8.8 21.7 36.5		
O			O	O O O		3
O			O	O		1
	C C		O	O		2
	B B		O	O		2
	A		O	O O O		3
	A		O	O		1
		A A	O O	O		4
	D		O	O O O		3
	D		O	O		1

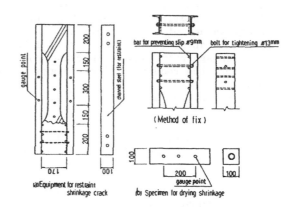

(a)Equipment for restraint shrinkage crack

(b) Specimen for drying shrinkage

bar for preventing slip ⌀9mm   bolt for tightening ⌀13mm

(Method of fix)

Fig.1. Outline of test apparatus

If n is assumed to be 15, R is calculated as about 57%, 76%, and 86% when As is $8.8cm^2$, $21.7cm^2$ and $36.5cm^2$, respectively.

The cross section of restrained concrete is 10 by 10cm in the center portion and the length of its straight portion is 30cm.
Tests were made in the controlled room where the temperature and humidity were kept at 20 ℃ and 55% R.H.
Testing subjects are as follows;
1) The strains of steel frame and concrete in the center portion
2) Time in initial cracking and passing-through cracking
3) Crack width
4) Shrinkage stress
5) Free shrinkage of prismatic specimens having the cross section of 10 by 10cm and length of 40cm corresponding to the straight portion of the cracking apparatus
6) Compressive and tensile strength at the beginning of test and at the cracking time using the cylindrical specimens
Tests were began at the age of 7 days after spray curing with sheet.
Two methods for application of SRA were examined, one was as additives and another was as impregnating agent which was spread by brush at the age of 5 days.
Length change was measured by the contact-type strain gauge and Whittmore strain gauge of 1/1000 mm reading.
Crack width was measured with both crack gauge of 1/20 mm reading and contact-type strain gauge.

3 Results and considerations

3.1 Effect of SRA on drying shrinkage of cement paste
(1) Effect of SRA on shrinkage and weight loss of cement paste.
 Table 5 shows the test results after 60 days drying obtained by 1x1x16cm prismatic bars. Shrinkage and weight loss were obtained as average values of three specimens.
 From the table, it can be recognized that the addition of SRA reduces the shrinkage but tends to increase more or less the weight loss. This means that the reduction of shrinkage can't be caused by suppressing the loss of water.
 Fig.2 shows the ratios of shrinkage of paste specimens containing SRA to that of control specimens without SRA in the humidity conditions of 20%,45%, 58% and 79% R.H.
 From the figure, the effect of SRA on reducing shrinkage became pronounced in the medium humidity conditions from 20 to 45% R.H.,but on the other hand, it weakened in the high humidity condition of 79%R.H.
 Among all the types of SRA, A was the most effective and the addition of 2% to cement content in weight ratio reduced shrinkage by about 40% and that of 6% to cement content did by a little less than 60%. D showed about the same effectiveness as A. C showed the variation in its effect by the change of humidity conditions and the amount of dosage of SRA.

Table 5. Results of shrinkage and weight loss
tests on cement paste

R. H. (%)		Control	A		B		C		D	
			2%	6%	2%	6%	2%	6%	2%	6%
2 0	1) *	278	153	120	176	155	237	140	175	126
	**	185	179	181	201	208	181	196	180	137
4 5	2) *	238	123	82	152	139	180	106	150	100
	**	121	132	151	136	156	130	157	128	113
5 8	3) *	181	125	74	144	137	191	100	156	100
	**	90	109	132	108	120	110	131	98	100
7 9	4) *	135	102	60	111	106	154	80	117	71
	**	37	40	82	52	57	50	85	54	49
9 0	5) *	80	30	29	41	35	89	49	48	25
	**	7	4	51	12	15	12	38	23	23

1) CH_3COOK 2) KNO_2 3) $NaBr·2H_2O$ 4) NH_4Cl 5) $ZnSO_4·7H_2O$
* shrinkage($\times 10^{-5}$)
** loss of water(mg/cc)

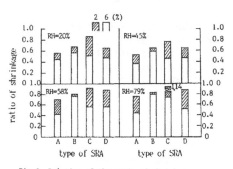

Fig.2. Relation of the ratio of shrinkage versus
type and dosage of SRA

Fig.3 shows the relation between shrinkage ε sh and the relative
humidity R.H.   Shrinkage generally tends to increase with the
lowering of humidity, however, that of control paste without SRA
tends to have almost little increase in the humidity condition less
than 45% R.H.

On the other hand, when SRA was added, the same tendency was observed
in the humidity less than 58% R.H.

This means that the region where the shrinkage showed the almost
constant value moved to the higher humidity by the use of SRA.

This fact will also indicate the change of internal mechanism of
shrinkage due to its addition.

Fig.4 showed the relation of shrinkage ε sh versus weight loss W'
per unit paste volume. From the figure, it was found that the
shrinkage of paste containing SRA showed the lower value in the same
loss of water as that of control paste and that the weight loss
increased with the lowering of humidity and continued to increase in

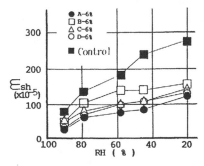

Fig.3. Relation of shrinkage in equilibrium state
versus relative humidity

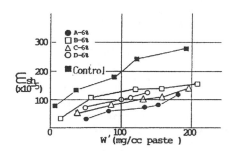

Fig.4. Relation of shrinkage in equilibrium state
versus weight loss

the humidity conditions where the shrinkage ceased its increasing tendency.
This results also suggests that the cause of shrinkage differs in above two different humidity regions.

(2) Consideration on the mechanism of reduction in shrinkage of paste containing SRA.
It is very profitable to have an information on acting mechanism of SRA in paste. Sato et al.(1983) have already reported that the reduction of shrinkage of paste containing SRA can be explained by the theory of capillary tension reflecting the reduction of the surface tension of the pore water by the effect of SRA.
According to the capillary condensation theory, the negative pressure ΔP induced is expressed as following, when the meniscus is formed whose main radii of curvature are $r1$ and $r2$ after on the liquid face having the surface tension γ recedes due to drying.

$$\Delta P = - \gamma \cdot (1 / r1 + 1 / r2) \qquad (2)$$

Assuming the thermodynamic equilibrium condition on the capillary condensation,

$$\Delta P = R \cdot T / (M \cdot Vf) \cdot \ln(P / Ps) \qquad (3)$$

where, R: gas constant T: absolute temperature (°K)
        M: molecular weight V: specific volume
   P / Ps: relative pressure

Internal stress σcp induced by the negative pressure Δp is expressed as the following equation.

$$\sigma cp = Wv \cdot \Delta P \qquad (4)$$

Where, Wv represents the residual water content by volume.

After Dr. Kondo (1957), drying shrinkage strain εsh corresponds to the value of internal stress divided by dynamic modulus of elasticity. If there is little difference in elasticity between solids and the amount of residual water in any pore size can't change by the addition of SRA, that is, if the pore volume and pore size are unchangeable by its addition, shrinkage strain becomes linearly relating to surface tension of the solution. Table 6 shows the radius of the pore calculated from Eq.(2) and (3) supposing that the

Table 6. Relation of surface tension of water solution of SRA versus the radius of capillary pore in equilibrium state

	Control	A 2%	B 2%	C 2%	D 2%
Surface tension of solution (dyn/cm)	72.8	32.6	46.3	36.0	38.4
R. H. (%)	(100)[*]	(44.8)[*]	(63.6)[*]	(49.4)[*]	(52.7)[*]
90	102.2	45.8	65.0	50.5	53.9
79	45.7	20.5	29.1	22.6	24.1
58	19.8	8.9	12.6	9.8	10.4
45	13.5	6.0	8.6	6.7	7.1
20	6.7	3.0	4.3	3.3	3.5

* percentage of the surface tension of the solution of SRA to that of pure water

surface tension of water solution where SRA is added by 2% in weight percentage to cement content would be equal to that of pore water in cement paste.

The radius of pore decreases in proportion to the lowering of the surface tension in the same humidity conditions.

Fig.5 shows the relation between the pore radius r(Å) and the amount of the residual water Wv in the case of 2% dosage of SRA to cement content.

Wv is obtained by the oven drying for 2days after the completion of shrinkage test. Wv in the control paste specimen tends to show the smallest value comparing to those in specimens containing SRA in the same pore radius. However, it can be generally judged that there is almost little difference in above relations regardless with or without SRA.

Fig.6 shows the relation of the measured shrinkage strain ε sh versus the capillary pore radius r(Å) corresponding to the humidity condition under which shrinkage was measured.

In the figures, the chain line corresponds to the calculated shrinkage value simply reduced by the lowering rate of reduction in surface tension r due to the addition of SRA. This shows that the estimated shrinkage becomes in agreement with the measured value and that the mechanism of reducing shrinkage by SRA can be explained by the theory of capillary tension in the cases of SRA A, B and D. However, in the case of C, measured shrinkage is in discord with the estimated value.

This discrepancy is suspected because the discordance in surface tension between the actual pore water in paste and the water solution whose surface tension was measured and used for calculation.

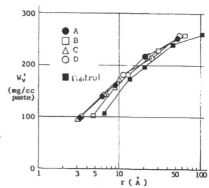

Fig.5. Relation of the amount of water remained versus the radius of capillary pore

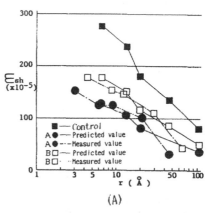

(A)

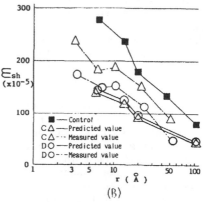

(B)

Fig.6. Relation of shrinkage of cement paste versus the radius of capillary pore

3.2 Investigations on the effect of SRA on shrinkage cracking
(1) Effect of SRA on the mechanical properties of concrete.
The effect of SRA on the mechanical properties of concrete was studied.

Fig.7 shows the test results on wet cured concretes containing all the types of SRA.

According to the figure, a little lowering of strength and modulus of elasticity was found at 7days in the case of use of SRA B and C, however, the gradual improvements were observed at the age of 28 days.

In the use of D, the strengths and elasticity were found higher than in control concrete because the air bubble couldn't be entrained by AE agent in the concrete containing D.

Then, there will be little problem on the development of their strengths and modulus of elasticity in the use of SRA here examined.

(2) Drying shrinkage of concrete containing SRA.

Fig.8 shows the results on drying shrinkage tests for concretes containing each SRA by 7.5kg/m^3. From the figure, the efficacy of SRA to reduce shrinkage of concrete was found to be the following order;

$$D \geqq A > C > B$$

This order differs from that of cement paste in that SRA D has the larger efficacy to reduce shrinkage than A.

The reversed order in A and D is suspected to be mainly related to the difference in the unit cement content between the cement paste and the concrete.

However, the rate of reduction in shrinkage of concrete by the use of SRA amounts to 40% at the maximum.

This maximum rate of reduction is in conform with that in cement paste in the same dosage of SRA by weight percentage to cement content.

The mechanism of the action of SRA to reduce shrinkage will be concluded as same as that in cement paste.

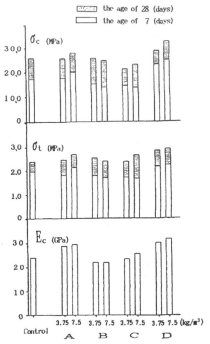

Fig.7. Effect of the addition of SRA on the mechanical properties of concrete

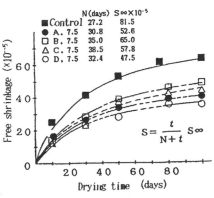

Fig.8. Relation of drying shrinkage of concrete containing SRA versus drying time

(3) The effect of SRA on shrinkage cracking of concrete.
Fig.9 shows the relation between the time of the occurrence in passing-through cracking T and the dosage of SRA.
Fig.10 does between T and the amount of spreading of SRA for impregnation.

The time T is in a clear direct proportion to the amount of use of SRA.

In the same dosages of SRA, the largest efficacy to prolong cracking can be expected in D. However, when compared by the specified standard dosages by manufacturers (A=7.5kg/m^3, B=12kg/m^3, C=4% by weight percentage to cement content, D=2% by weight percentage to cement content), there is little difference in the efficacies between A, C. and D except B.

The reason of the lower efficacy of B will be attributed to the smallest effect to reduce shrinkage among four types of SRA.

In the case of impregnation of SRA A shown in Fig.10, the pronounced efficacy to prolong cracking was found when the expansive cement concrete was selected as the base concrete.

However, when the normal concrete was examined for the base concrete, the amount of impregnation by 300cc/m^2 has the almost same efficacy with the addition by 2 to 3kg/m^3, which means a little lack of the performance in the actual use for impregnation.

Fig.11 .shows the changes of the cracking time T and the restrained strain Δε, which can be obtained by the subtraction of the strain of concrete restrained in the apparatus from the free shrinkage strain in the case of three different rates of outer restraint given by the change of the cross sectional area A s of cracking frame.

The efficacies to prolong cracking in the addition of SRA A and D by 3.75kg/m^3 are maintained regardless of the restraint rates by frame.

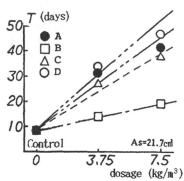
Fig.9. Relation of time in passing-through cracking versus dosage of SRA

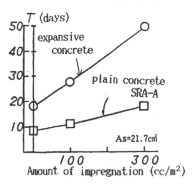
Fig.10. Relation of time in passing-through cracking versus the amount of impregnation

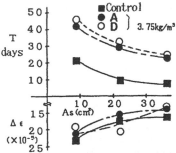
Fig.11. Effect of SRA on the cracking tendency of concrete with different cross areas of outer steel frame

493

The restrained strain $\Delta \varepsilon$ also changed with the rate of restraint.
Shoya and et al.(1985,1987) have already proposed the following
equation to estimate the time T when passing-through cracking
occurred.

$$T = \Delta \varepsilon \cdot (a + b \cdot Q) \tag{5}$$

where, a and b are experimental constants, in special, b is
dependent on the type of SRA. Q is the dosage of SRA.

It will be possible to predict the cracking time if the restrained
strain $\Delta \varepsilon$ could be expressed as the function of cross sectional
area of outer steel for restraint.

Fig.12 shows the relation between the crack width at one month after
the occurrence of the passing-through cracking and the dosage of SRA.

The crack width extends with the elapse of the time. The rate of
its extension relates to the speed of the shrinkage and the final
crack width will depend on the amount of shrinkage strain. The use
of SRA is pronouncedly effective to control the crack width and can be
expected to keep it less than 0.1mm by the choice of the type and the
dosage of SRA.

Fig.13 shows the relation of the shrinkage stress σ sh induced in
the concrete portion of the restraining apparatus versus the age.

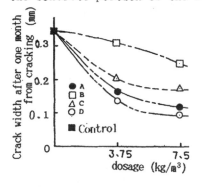

Fig.12. Relation of crack width versus type of SRA

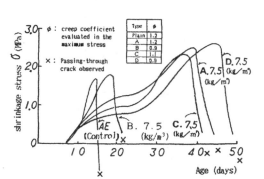

Fig.13. Effect of SRA on shrinkage stress and
creep coefficient of concrete in the
maximum shrinkage stress

Shrinkage stress σ sh was calculated by the following expression.

$$\sigma \text{ sh} = \varepsilon \text{ s} \cdot As \cdot Es / Ac \tag{6}$$

where ε sh and As are the strain and the cross sectional area of
outer steel frame, respectively. Ac is the cross sectional area of
concrete in the center portion.

Shrinkage stress can be also calculated by the next expression.

$$\sigma \text{ sh} = \Delta \varepsilon \cdot Ec / (1 + \phi) \tag{7}$$

where, ϕ is the creep coefficient. As shown in the figure,
the use of SRA tends to slow the development of shrinkage stress,
which will result in the prolongation of the cracking time.

The creep coefficients estimated from Eq. 7 at the time when the
shrinkage stress reaches the maximum value become nearly equal values
of more or less than 1.1 regardless with or without SRA.

Authors have confirmed the almost same reduction of compressive creep as shrinkage by the addition of SRA.

However, it will be concluded that the effectual relaxation of shrinkage stress by both the great reduction of shrinkage and the faster advance of creep of concrete due to the prominent effect of SRA.

4 Conclusions

Conclusions are drawn as following from the test results on paste and concrete containing four kinds of shrinkage reducing agents.

1) The efficacy of shrinkage reducing agent SRA to reduce drying shrinkage of cement paste is dependent on its type, but it showed the maximum reduction by about 40% and 60% in the dosage of 2% and 6% to cement content, respectively.
2) The cause of the reduction of shrinkage by the addition of SRA was confirmed to be not attributed to the suppression of loss of water and can be well explained by the theory of capillary tension reflecting the lowering of surface tension of pore water by SRA.
3) There was almost little difference in mechanical properties between concretes with and without SRA within the range of dosage available in practical use.
4) The efficacy of SRA to reduce drying shrinkage of concrete was judged to be similar with that of cement paste.
5) The use of SRA showed the prolongation of cracking time and the reduction of crack width in concrete.
 Three types of SRA A, C and D showed the almost same efficacies to prevent shrinkage cracking in the specified dosage by manufacturers.
 These reasons were considered being attributed to the great reduction of shrinkage and the effective relaxation of shrinkage stress by creep.

References

Kondo, M.(1969) A study on the causes of volume change of cement paste, Materiaux et Constructions, vol2, N 7 pp23-32.
Sato, K. et al.(1983) Mechanism of reducing drying shrinkage of hardened cement by some organic additives. Cement and Concrete (in Japanese), 442, pp9-19.
Shoya, M. et al.(1985) On the reducing effect of drying shrinkage of concrete due to special admixture. Proc. of 39th General Meeting, CAJ (in Japanese), PP367-379.
Shoya, M. et al.(1987) Effective use of shrinkage reducing admixture to control shrinkage cracking of concrete, Proc. of the First International RILEM Congress, vol.3, PP1346-1349.

6 TECHNOLOGY

WORKING WITH SUPERPLASTICIZERS IN CONCRETE: A WIDE FIELD APPLICATION

N. CILASON and A. H. S. ILERI
S. T. F. A. Construction Company Inc., Turkey
M. CHIRUZZI
Grace Italiana S.p.A. Construction Products, Italy

Abstract
With the present day requirements for a highly durable concrete of
incredible talents from the moment its components merge, there are
specific characteristics it has to have and definite "requirements"
with which it must meet. Off-hand, these are : workability,
placeability, cohesiveness and slump retention, in the fresh state;
and ultimate integrity, or durability, in the hardened state.

Workability, together with the other requirements in the fresh
state not to be compromised by any site engineer, must be maintained.
However, depending on the global location of mostly and largely
environmentally susceptible construction, these characteristics most
often tend, unfortunately, just to be the numerator in a fraction,
namely the "w" in w/c ratio. Therefore, the most important step to be
taken in order to reach the satisfactorily hardened state, is to
supply involved personnel with a concrete that can be placed with a
minimum of water and a maximum of "placeability". This is where the
proper choice of materials, particularly admixtures, comes in.

One of the main concerns in the use of superplasticizers (SP) is
the effect which local materials and ambient temperatures may
potentially have on fresh concrete; this was noticed and reported in
various projects in the Middle East, North Africa and in Turkey,
where we have had satisfactory results.

For the work presented in this paper, admixture limits were kept
at 0.8 - 2.0 % by wt. of cement for incrementally launched
post-tensioned pre-cast beam segments and pre-cast beams for use on
the viaducts, flyovers and bridges of the Thracian Motorway Project
and including the extremely large anchorage and foundation pours of
the Second Bosphorous Bridge Crossing.

More specifically, an SP based on naphthalene sulphonate
formaldehyde condensate (NSFC) has been used in order to achieve the
best concrete quality in terms of:
1. high workability at very low water/cement ratios
2. good slump retention
3. good cohesiveness, increased density, strength and durability.

All of these, together with a significant cost reduction obtained
by means of modification for the steam-curing cycle, resulted in a
durable and economical concrete which we call "QUALITY CONCRETE".

1   Introduction : Concrete admixtures, their effect on the properties of fresh and hardened concrete

With the onset of large scale construction in the second half of the 20th century, the need for concrete which is workable and easy to place in conjunction with tight project schedules, economy and most importantly the understanding that concrete strength and durability are directly related to proper mix design, batching and placing methods, chemical admixtures were, are, and it seems, still will be developed.

Todays admixtures offer to us great advantages in achieving the Control of Quality and the maintenance of project schedules.

The contents of this presentation will provide information compiled by the STFA (*) Quality Control department from examples of construction in the Middle East, North Africa and Turkey from ambient temperature ranging from − 10°C to 60°C, using different aggregate sources from natural Wadi sands to crushed rock fines, from igneous to sedimentary rocks, from blended to normal Portland cements and sulphate resisting to PC 500's. All the concrete mixes were designed to Fuller's ideal grading curves, for maximum density.

We have aimed to present more specifically a construction approach to research rather than a research adaptation to construction.

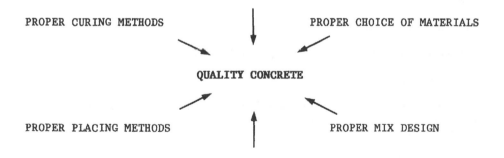

QUALITY ASSURANCE

PROPER CURING METHODS   PROPER CHOICE OF MATERIALS

QUALITY CONCRETE

PROPER PLACING METHODS   PROPER MIX DESIGN

PROPER BATCHING

* Sezai Turkes Feyzi Akkaya Construction Company Inc.

2 Properties

The wide range of applications for ADMIXTURES and, specifically superplasticizers, has defined their need for use by the fact that they produce concrete of required qualities. These can briefly be itemized by:

2.1 Fresh concrete
- compatible with the requirements of ready mixing
- low w/c ratio
- high workability
- continuity in the control of desired qualities (i.e. Slipforming)
- consistency
- self Levelling (i.e. less workmanship = economy)
- good pumping.

2.2 Hardened concrete
- high early strength gain
- enhancing the properties of hardened concrete
- maintaining tight project scheduling by specialized application
- durability.

3 Projects

A few typical projects of the sort where different types of concrete with admixtures were used:

3.1 KOMURHAN - TOHMA Combined Road and Railway Bridge
Malatya, Turkey

Method of construction
- slipformed piers
- Free Cantilever Box Beam Bridge Deck / post tensioned

Requirements

a. controlled setting times (6 - 7 hr initial setting time)
b. high early strength (300 kg/cm^2 in 72 hrs)
c. workability (150 - 200 mm slump)

3.2 BULK WATER STORAGE RESERVOIR ROOF STRUCTURE
Meccah, Saudi Arabia

Method of construction
- slipformed piers
- cable stayed roof structure

Requirements
a. controlled setting times under high temperature conditions
b. high early strength to meet tight project schedule
c. workability.

3.3 THERMAL POWER PLANT
Orhaneli, Turkey

Method of construction
- slipformed cooling towers

Requirements

a. consistency of concrete delivery from ground up to 270 meters
b. controlled setting times
c. extremely good workability.

3.4 BRIDGES AND VIADUCTS OF THE KINALI-SAKARYA MOTORWAY,
including 2nd Bosphorous Bridge Crossing,
Istanbul, Turkey

Method of construction
- slipformed piers
- incrementally launched segmental beams
- precast beams.

Requirements

a. tight project schedules necessitated extremely speedy load transfers for incremental launching and precast beams
b. high early strengths
c. workability.

4 Description of effects on some properties of concrete

4.1 Setting time
ASTM C 403, "Standard Test Method for Time of Setting of Concrete Mixtures by Penetration Resistance" defines the determination of initial and final setting time of concrete by testing that portion of mortar which will pass a #4 sieve (4.75 mm), for time at a given penetration resistance of 500 psi (35 kg/cm^2) for initial set, and time for a given penetration resistance of 4000 psi (280 kg/cm^2) for final set.

Setting time is an important factor to consider relative to the type and amount of admixture to be used. This is specifically aimed at stringent requirements for slipforming, precasting and pre or post tensioned in-situ concrete.

This type of construction necessitates the need to control, at will, the strength and setting time of concrete to be poured. In slipforming varying doses of SUPERPLASTICIZERS are called upon to maintain a smooth and speedy sliding of the formwork.

In order to determine the required setting times, the following tables for a given admixture and w/c ratio were developed and subsequently used (figures 1-5).

With respect to changing ambient temperatures or to varying admixture rates at a given temperature there are one of the most effective means of projecting pours.

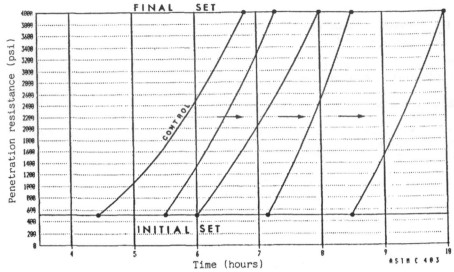

Fig. 1. Effect of a superplasticizer on the setting time of
concrete as percentage increases at a given temperature
Project: Gabes Harbour – Tunisia

Cement content: 400 kg/m³   Aggregate: El Kebir
w/c ratio: 0.48   Ambient temperature: 30°C

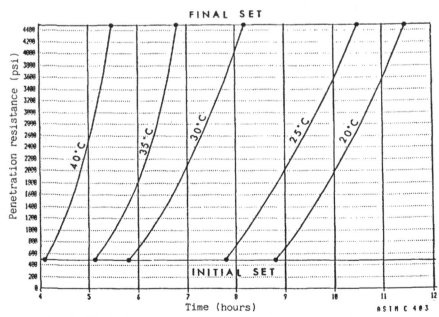

Fig. 2. Effect of increasing temperature on the setting time of
concrete at a set superplasticizer content
Project: Gabes Harbour – Tunisia

w/c ratio   : 0.48   cement content: 400 kg/m³
aggregate   : El Kebir

500

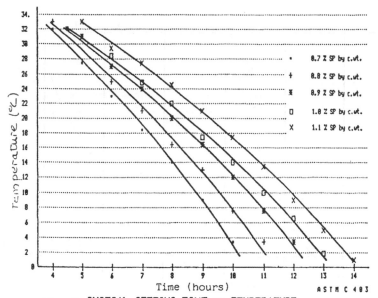

Fig. 3. INITIAL SETTING TIME vs TEMPERATURE

Project: Komurhan-Tohma
Combined Road and Railway Bridge, Malatya – Turkey

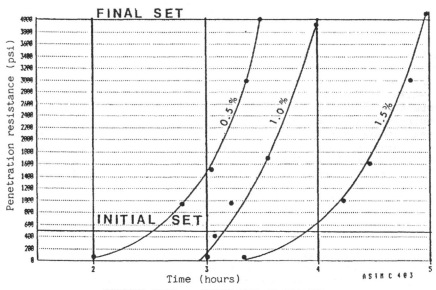

Fig. 4. SETTING TIME vs INCREASING SP DOSAGES

Project: Bulk Water Storage Reservoir, Muna – Saudi Arabia

w/c ratio   : 0.50   cement content: 360 kg/m^3
coarse aggr.: crushed rock   fine aggr.   : wadi sand
          ambient temperature : 41 – 44°C

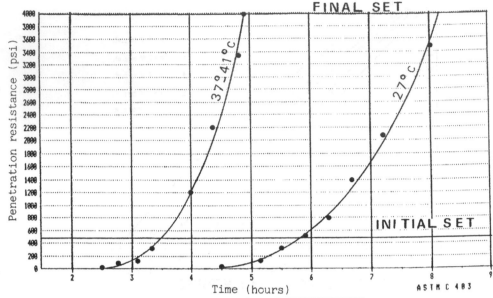

Fig. 5. SETTING TIME vs VARYING TEMPERATURES

Project: Bulk Water Storage Reservoir, Muna — Saudi Arabia
(Superplasticizer at supplier recommended dosage)

4.2 Workability
Workability is a vital property as far as the finished product is
concerned, since concrete must have a workability such that
compaction to maximum density is achieved with a reasonable amount
of work. This need arises from the relation between the degree of
compaction and the resulting compressive strength values. The
requirements for workable concrete are apparent in todays fast
paced construction with dense reinforcement, precast elements,
slipforming requisites, the use of pumping techniques and methods
for placing which require self-levelling properties.

In order to determine which percentages will yield the most
optimum results (fig. 6) to achieve these requirements, the
research into which specific admixture to use and this at which
rate to add into the mix is a matter for testing at site labs
(Table 1).

4.3 High early strength
Lowered w/c ratios in projects involving the use of many different
aggregrates, different cements and all these with varying
mix-designs have shown us that very low ratios produce very
satisfactory results in a wide range of concrete applications.

Among the most important considerations here is the work in the
precast field. Load transfer times depend on the attaining of
specified strength after very short periods.

As can be seen, (figures 7, 8 and 9) water reductions (0.33 —

502

0.50 w/c ratios) yield the strengths required for post tensioning
34% above the specified values at 50% of the specified time (Free
Cantilever Bridge at Komurhan–Tohma, Turkey).

Table 1. Relationship between superplasticizer content and slump.

SP content	0.0 %		0.8 %		1.2 %	
Class	slump (mm)	Ave. (mm)	slump (mm)	Ave. (mm)	slump (mm)	Ave. (mm)
45	36 35	35.5	125 96 85	102	165 175	170
30	5 24 4	11	30 28 48 44	37.5	90 83 109 100 97 91	95

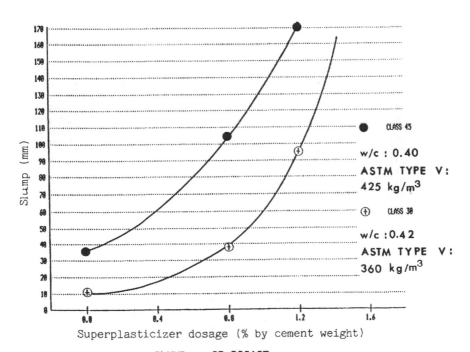

Fig. 6. SLUMP vs SP DOSAGE
Project: New Galata Bridge, Istanbul – Turkey

503

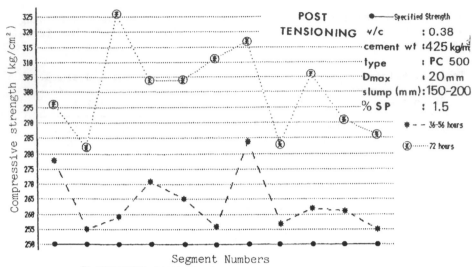

fig. 7 STRENGTH GAIN FOR BOX BEAM SEGMENTS

KOMURHAN-TOHMA FREE CANTILEVER BRIDGE

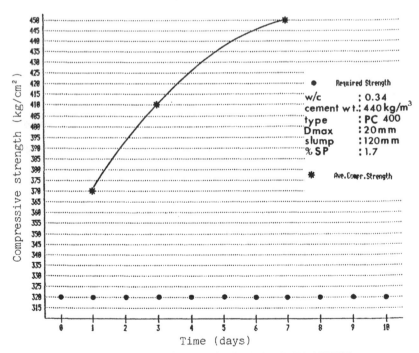

fig. 8 STRENGTH GAIN FOR PRECAST BEAM LOAD TRANSFERS

KINALI-SAKARYA MOTORWAY TURKEY

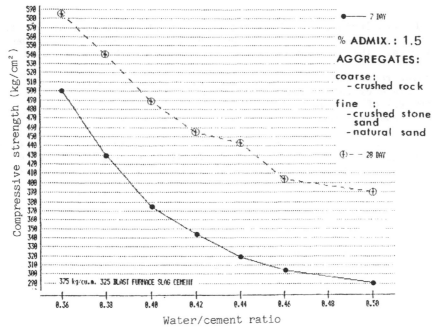

Fig. 9. COMPRESSIVE STRENGTH vs W/C RATIO

5 Conclusions

To sum up what admixtures are required to do in the field the mention of slipforming technology has to be made. This is one of the more important areas of application for superplasticizers, as one is confronted with most of the needs for proper concreting.

Slipforming necessitates:
a. gaining the compressive strength for smooth operation
b. controlling, at will, the desired setting times
c. high workability.

In order to slip the formwork satisfactory strengths must be achieved while keeping workability levels and setting times under strict control. The need for a consistent concrete is apparent here.

Two examples descriptive of this type of work are the Bulk Water Storage Reservoir in Muna, Saudi Arabia and the Free Cantilever Combined Road and Railway Bridge in Malatya, Turkey. Although only the piers of the bridge were slipformed the basic requirements for the box beam deck, also, are similar.

With the results obtained in the field, we can safely say that:
i.   an increase in superplasticizer % INCREASES WORKABILITY
ii. an increase in superplasticizer % DECREASES WATER CONTENT
iii. by the use of superplasticizer a GOOD CONTROL OF SETTING TIME
     is achieved by the proper control of concrete as affected by
     temperatures.

505

In conclusion, we find that most standard specifications for contracts require that superplasticizers meet performance criteria. The following characteristics are usually called for:

- water reduction
- consistency (at constant w/c ratio)
- control of setting time
- compressive strength.

It is important to note that performance requirements vary with different sources of concrete constituents and that laboratory or field testing in conjunction with lab testing be carried out with ultimate durability in mind.

REFERENCES

STFA Construction Company
    Quality Control Dept., Project Files, Kinali–Sakarya Motorway, Istanbul, Turkey.
STFA Construction Company
    Quality Control Dept., Project Files, B.W.S. Reservoir, Meccah, Saudi Arabia.
STFA Construction Company
    Quality Control Dept., Project files, Thermal Power Plant, Orhaneli, Turkey
STFA Construction Company
    Quality Control Dept., Project files, Road and Railway Bridge, Komurhan–Tohma, Turkey.
STFA Construction Company
    Quality Control Dept., Project Files, New Galata Bridge Construction, Istanbul, Turkey.
STFA Construction Company
    Quality Control Dept., Project Files, Harbor Construction, Gabes, Tunisia.
V.S. Ramachandran, (1984), Concrete Admixtures Handbook – Properties Science and Technology, Noyes Publications.
M.R. Rixom and M.P. Mailvaganam (1986), Chemical Admixtures for concrete, 2nd Edition, E & FN Spon.
A.M. Neville, J.J. Brooks (1987), Concrete technology Longman Scientific & Technical.
A.M. Neville (1978), Properties of Concrete, 2nd edition , Pitman Publishing Ltd.
P.C. Aitcin (1980), Concrete Construction.
B.P.J. Verkerk, Admixtures, their use and benefit in concrete, Symposium Presentation.
Relevant ASTM Standards.
A.A. Burk,Jr., J.M. Gaidis and A.M. Rosenberg, (1981), Adsorption of Naphthalene–based Superplasticizers on Different Cements, Presented at 2nd International Conference on Superplasticizers in Concrete, Ottawa, Canada.

PROPRIETES DE CERTAINS BETONS A ADJUVANTS UTILISES DANS L'INDUSTRIE DES PREFABRIQUES EN BETON
(Properties of concretes containing admixtures used in precast concrete manufacture)

I. IONESCU and I. ISPAS
Research Institute for the Construction Materials Industry, Romania

Resumé
Ce compte - rendu présente les principales propriétés des
bétons des classes Bc20 ... Bc60 avec des additifs superplastifiants a base de mélamine sulfonée et des bétons
des classes Bc30 ... Bc40 avec des additifs mixtes à base
de lignosulfonates d'ammonium.
En fonction des modifications des propriétés des bétons
frais et durcis obtenus en utilisant des additifs dans
leur composition,nous présentons succintement les domaines
d'utilisation des additifs à base de MFS et de LSA lors
de la réalisation des éléments préfabriqués en béton et
des ouvrages en béton coulé.
Mots-clef: Additifs,Ciment,Granulats,Béton,Ouvrabilité,
              Résistance

1. Introduction

L'utilisation des additifs plastifiants à base de lignosulfonate de calcium et des additifs entraîneurs d'air
à base de résine Vinsol dans les ouvrages,en Roumanie,date
depuis les années 50.Vers la fin des années 70 et le
début des années 80, parallèlement ont commencé la production et la recherche des directions d'utilisation,dans
l'exécution des ouvrajes en béton,des additifs superplastifiants à base de mélamine - formaldéhyde sulfonée,conven-
tionnellement notés MFS et des additifs à base de lignosulfonate d'ammonium,conventionnellement notés LSA,à effet
mixte de plastifiants et d'entraîneurs d'air.Le présent
ouvrage comprend certains résultats de recherches et
experimentations effectuées pour l'utilisation des additifs
MFS et LSA dans l'industrie des préfabriqués en béton.

2 Materiaux et méthodes de recherche utilisées

2.1 Matériaux utilisés pour les recherches et les essais
exécutés dans les unités de préfabriqués

Pour la préparation des bétons on a utilisé: des additifs MFS,dans une solution aqueuse avec une concentration de 19 - 21 % de substance active;des additifs LSA en solution préparée en laboratoire avec une concentration de 20% de substance active;des ciments portland sans adjuvants en fabrication de type P.40, P.45; P.50; des ciments portland avec 15 - 20 % d'adjuvants de scorie basique granulée de haut fourneau en fabrication de type Pa 35; des ciments P.40 à adjuvant de 25 - 30% de cendre volante de centrale thermoélectrique lors de la préparation du béton,notés Pac; des ciments P.40 à adjuvant de 7 - 15% de silice ultrafine lors de la préparation du béton,notés Pas; des ciments portland résistants aux sulfates de type BS 4072 notés SR ...En tant que granulats minéraux et en fonction des cas,on a utilisé des granulats naturels de rivière lavés et triés comme il suit 0-3, 0-5, 3-7, 5-10, 7-16, 10-20 mm, surtout pour les bétons appartenant aux classes Bc 50 - Bc 60,les catégories de granulats 7-16 ou 10-20 mm étant remplacées totalement en partiellement avec des granulats obtenus par le concassage de certaines roches naturelles dures,comme par exemple le granite,le basalte, la granodiorite,etc.Des granulats et des mélanges de catégories correspondant aux spécifications des normes roumaines pour le béton ont été utilisés.L'eau utilisée à la préparation des bétons en laboratoire et dans les unités de préfabriqués a été l'eau potable.

2.2 Méthodes de recherche utilisées
Ont été effectuées des recherches concernant l'influence des additifs sur la prise des ciments en utilisant de test Vicat et sur la constance de volume; des recherches sur la résistance des mortiers type de ciment en utilisant méthode ISO TC-72; des recherches sur les propriétés des bétons à l'état frais,concernant la teneur en air inclus, l'ouvrabilité déterminée par le tassement du Cône Abrams, le degré de compactage Waltz,la densité. Recherahes concernant les propriétés des bétons à l'état durci,comprenant les résistances mécaniques,la perméabilité,la résistance au gel-dé-gel répété, les retraits,le module d'élasticité statique et dynamique,L'adhérence aux armatures à profil lisse et à empreintes,etc.Tous les tests ont été effectués conformément aux spécifications roumaines de vérification des bétons,similaires aux normes utilisées dans le même but dans les autres pays,excepté les essais de gel-dégel répété.It faut mentionner qu'un cycle d'essais de gel-dégel répété comprend le gel des éprouvettes cubiques de 14,1 x 14,1 cm saturées préalablement avec de l'eau,pendant 4 heures à - 17 ± 2^0C,puis le dégel dans léau à + $2o^0$C $\pm 2^0$C Les travaux réalisés en laboratoire ont été effectués à des températures de 18-22^{0}C, et ceux réalisés dans les unités de préfabriqués à la température du milieu ambiant, variant en fonction de la saison,entre 12 ... 28^{0}C.

3 Quelques résultats obtenus

Certains résultats des recherches effectuées en laboratoire et des expérimentations réalisées dans les unités de préfabriqués sont ci-dessous présentés.

3.1 Prise des ciments à additifs MFS et LSA

Certains résultats des recherches effectuées sont présentés dans le tableau 1;leur analyse indique principalment:

Tableau 1. Influence des additifs sur la prise des ciments

Type de ciment	Additif Type	%	Eau de pre- paration de la pâte	Ouvrabili- té de la pâte	Temps de prise heures,minutes Debut	Fin	Effet obtenu	Obser- vations
1	2	3	4	5	6	7	8	9
P40	–	–	Normale	Normale	1,30	3,00	–	Témoin
P40	MFS	0,50	Réduite 12%	Egale témoin	1,00	2,35	Accélérateur	–
P40	MFS	0,60	Réduite 14%	– " –	0,50	2,05	Accélérateur	–
P40	MFS	0,80	Réduite 17%	– " –	0,40	1,30	Accélérateur	–
P40	MFS	0,50	Egale témoin	Augmentée	2,45	5,30	Retardateur	–
P40	MFS	0,60	Egale témoin	Très augmentée	4,15	7,45	Retardateur	–
Pa35	–	–	Normale	Normale	3,15	5,30	–	Témoin
Pa35	MFS	0,60	Réduite 15%	Egale témoin	1,45	3,00	Accélérateur	–
Pa35	MFS	0,60	Egale témoin	Très augmentée	4,45	7,15	Retardateur	–
Pac	–	–	Normale	Normale	2,45	4,45	–	–
Pac	MFS	0,60	Réduite 16%	Egale	2,15	4,30	Légèrement accélérateur	
Pac	MFS	0,60	Egale témoin	Très augmentée	5,30	7,15	Retardateur	
SR	–	–	Normale	Normale	2,30	3,45	–	Témoin
SR	MFS	0,60	Réduite 17%	Egale témoin	1,15	2,00	Accélérateur	–
SR	MFS	0,60	Egale témoin	Très augmentée	4,45	7,30	Retardateur	–
P40	LSA	0,25	Réduite 7%	Egale témoin	1,15	2,45	Sensiblement egal	–
P40	LSA	0,25	Egale témoin	Augmentée	2,30	4,30	Retardateur	–
Pac	LSA	0,30	Réduite 6%	Egale témoin	4,45	7,30	Retardateur	–
Pac	LSA	0,30	Egale témoin	Augmentée	6,30	9,15	Retardateur	–
SR	LSA	0,25	Egale témoin	Egale témoin	2,45	4,00	–	–
SR	LSA	0,25	Egale témoin	Très augmentée	5,15	8,30	Retardateur	–

La prise et l'ouvrabilité des pâtes de ciment sont influencées
par le type d'additif et la quantité d'eau de préparation.
Pour une quantité d'eau de préparation égale aux pâtes
témoin.,l'ouvrabilité des pâtes de ciment augmente
différemment,en fonction du type d'additif:indépendamment
du type de ciment,aussi bien le début que la fin de la
prise sont généralement retardés.Pour une ouvrabilité
égale des pâtes de ciment et de l'eau de préparation
différemment réduite en fonction du type de ciment,le type
de l'additif et le pourcentage d'additiv utilisé,l'effet de
des additifs se manifeste différemment,soit : par la
diminution significative du temps de début et fin de prise
dans le cas des pâtes de ciment portland sans adjuvants
de type P40,P45,P50 et SR et portland avec ajuvant de
scorie basique granulée de haut fourneau lors du broyage,
de type Pa35 préparés avec l'adjuvant MFS,la réduction
du temps de prise augmentant généralement en même temps
que l'augmentation du pourcentage d'eau réduit lors de
la préparation de la pâte.Dans le cas des ciments P40 avec
adjuvant de 25-30% de cendre volante lors de la préparation
du béton et de l'additif MFS, l'accélération du début de
prise est relativement modeste,la fin de prise étant
sensiblement égale à celle des pâtes témoin.Pour les pâtes
préparées avec du ciment sans adjuvants dans la fabrication de type P40,P45 et SR avec de l'additif LSA,
l'ouvrabilité égale aux pâtes témoin et de l'eau de
préparation réduite,le temps de début et de fin de prise
est sensiblement égal au temps des pâtes témoin ou bien
réduit de manière insignifiante.Pour les pâtes de ciment
P40 avec adjuvant de cendre volante lors de la préparation
du béton de l'additif LSA et de l'eau de préparation réduite
le temps de début et fin de prise augmente sensiblement,
voir tableau 1.A des temperatures de 25-28°C de préparation des pâtes de ciments sans adjuvants P40,P50 et SR,
avec des additifs MFS,ouvrabilité égale et teneur en eau
réduite,des diminutions importantes du temps de début de
prise ont été constatées(12 - 30 minutes) mais aussi de
fin de prise,45 - 60 minutes.
C'est pour cela qu'il est très important : de connaître
l'influence de l'addition d'additifs sur le temps de prise
des ciments,surtout pour les unités de préfabriquée qui
les utilisent dans de but de réduire la quantité d'eau
de préparation du béton;d'accélérer ainsi le durcissement
des bétons; de réduire le traitement thermique et la
durée de décoffrage des éléments préfabriqués.La constance
de volume des pâtes de ciment avec de l'additif MFS et
LSA a été bonne dans tous les cas.

3.2. Propriétés des bétons à additifs MFS et LSA à
     l'état frais
Les propriétés des bétons à additifs à l'état frais
subissent des modifications en fonction de la composition
du béton témoin qui aide à faire la comparaison,du type

et du pourcentage d'additif utilisé,proportion dont l'eau
de préparation du béton a été reduite à cause de
l'utilisation de l'additif.Certains resultats des recherches effectuées sont présentés dans lc tableau 2.

3.3 Propriétés des bétons à additifs MFS et LSA à l'état
     durci

Tableau 2. Influence des additifs sur les propriétés de
           certains bétons frais.

Ciment		Additif		Eau	Densité	Air inclus %	Tasse-ment cm	Gc Waltz	Obser-vations
Type	kg/m³	Type	%	l/m³	kg/m³				
P40	300	-	-	192	2362	2,2	8	1,08	Témoin
P40	300	MFS	0,50	170	2370	2,1	9	1,07	
P40	300	MFS	0,60	165	2386	2,0	8	1,08	
Pac	360	-	-	204	2347	2,4	9	1,06	Témoin
Pac	360	MFS	0,50	174	2359	2,2	8	1,07	
Pac	360	MFS	0,60	168	2375	2,0	8	1,07	
Pac	360	MFS	0,60	202	2351	3,3	14	-	
Pas	300	-	-	192	2353	2,1	8	1,08	Témoin
Pas	300	MFS	0,60	162	2370	1,9	8	1,08	
Pas	300	MFS	0,60	192	2347	2,4	15	-	
SR	300	-	-	180	2366	2,2	8	1,07	Témoin
SR	300	MFS	0,60	158	2394	2,1	9	1,06	
SR	300	MFS	0,50	180	2358	2,3	16	-	
P40	300	LSA	0,25	178	2338	4,3	9	1,06	
Pac	360	LSA	0,30	188	2328	3,6	8	1,08	
Pas	300	LSA	0,30	176	2339	3,9	9	1,06	
Pa35	350	-	-	198	2358	2,2	8	1,07	
Pa35	350	LSA	0,25	182	2346	3,1	10	-	

Les tableaux 3 et 4 présentent quelques résultats des
recherches et expérimentations effectuées sur les propriétés
des bétons à additifs à l'état durci.
L'analyse des données des tableaux 3 et 4 démontre
principalement: par l'utilisation des additifs MFS à

quantité d'eau de préparation égale et ouvrabilité du
béton frais fortement améliorée on obtient des résistances
mécaniques initiales et finales,ainsi que des résistances
au gel-dégel répété pratiquement égales aux résistances
des bétons témoins.L'adhérence de ces bétons aux armatures
en acier-béton n'en est pas affectée.En utilisant les
additifs MFS pour la préparation des bétons à ouvrabilité
égale à celle des bétons témoin on obtient une augmentation
des résistances mécaniques à 1... 90 jours;conformement
à la quantité d'eau de préparation réduite de la composition du béton,ces bétons subissent une augmentation de
la résistance au gel-dégel répété et entre certaines
limites,l'adhérence aux armatures.

Tableau 3. Résistances mécaniques pour certains bétons
          à additifs.

| Ciment | | Addi-tif | | Eau l/m³ | Résistance à la compression N/mm² au jour | | | | | Rt 28 N/ mm² | Adhé-rence aux armatu-res N/mm² | Obser-vation |
Type	kg/m³	Type	%		1	3	7	28	90			
P40	300	–	–	198	6,6	12,5	18,2	29,7	36,4	2,61	2,98	Témoin
P40	300	MFS	0,60	198	6,4	12,9	19,3	30,7	37,2	2,73	3,06	
P40	300	MFS	0,40	180	10,1	19,6	24,2	35,3	43,0	2,85	3,26	
P40	300	MFS	0,60	172	12,1	22,7	27,1	40,2	46,1	3,14	3,41	
P40	300	MFS	0,80	168	14,3	24,7	29,6	42,5	47,4	3,40	3,47	
P40	300	MFS	1,00	163	15,8	26,1	30,9	43,7	49,2	3,42	3,51	
P40	400	–	–	192	18,7	26,2	38,6	50,2	54,7	3,52	3,37	Témoin
P40	400	MFS	0,60	168	26,9	33,2	46,9	57,3	61,4	3,78	3,56	
P40	500	–	–	190	30,3	40,2	52,6	64,5	69,8	4,03	–	Témoin
P40	500	MFS	0,60	166	40,1	55,1	67,2	79,6	83,1	4,26	–	
Pac	360	–	–	204	5,8	12,3	18,6	30,9	38,2	2,78	3,01	Témoin
Pac	360	MFS	0,60	168	13,1	23,1	28,4	40,7	46,5	3,30	3,43	
Pas	330	–	–	195	7,5	13,4	21,4	35,3	41,6	3,36	3,27	Témoin
Pas	330	MFS	0,60	166	13,6	23,6	31,7	42,8	48,5	3,62	3,51	
Pas	450	–	–	192	26,1	37,2	44,6	53,4	61,7	3,98	3,60	Témoin
Pas	450	MFS	0,60	168	32,3	44,6	51,2	60,1	66,3	4,12	3,70	
Pas	500	–	–	192	30,2	41,4	53,8	66,9	72,5	–	–	
Pas	500	MFS	0,60	170	38,5	54,0	66,9	82,7	89,6	–	–	
P40	330	–	–	192	8,9	16,1	24,4	35,6	42,3	3,34	3,30	
P40	330	LSA	0,25	192	5,7	12,6	18,9	29,8	34,7	3,29	3,12	
P40	330	LSA	0,25	176	8,8	17,5	26,5	38,2	44,7	3,57	3,42	

Suivre le comportement dans de conditions différentes
d'exploitation,dans le temps,jusqu'a 10 ans,de certaines
structures de résistances et constructions réalisées avec
des éléments préfabriqués produits industriellement dans
des bétons des classes Bc 20 ... Bc 60 dont on a utilisé
des superplastifiants de type MFS dans la composition,
démontre un bon comportement pendant l'exploitation.
Dans des conditions d'utilisation des additifs MFS à ouvrabilité égale à celle du béton témoin,il en résultent les
suivants avantages techniques et économiques :
- sensible augmentation des résistances des bétons à des
dosages de ciments égaux;
- possibilité de réduction des dosages de ciment,si il
y a résistances mécaniques égales;
- possibilité de réduction de la consommation d'énergie
lors du traitement thermique des bétons,aux dosages de
ciments égaux au béton témoin etc.Les unités de préfabriqués
peuvent ainsi opter pour la meilleure variante dans des
conditions concrètes.

Tableau 4. Perméabilité et résistance au gel-dégel répété
          pour bétons à additifs

Ciment		Additif		Eau	Résis-	Perméabilité		Résis-	Obser-
Type	kg/m³	Type	%	l/m³	tance à 28 jours	Pres-sion max. d'essai Bars	Infil-tra-tion eau cm	tance aux cycles gel-dégel	vations
P40	300	–	–	198	29,7	6	9,6	50	Témoin
P40	300	MFS	0,60	198	30,7	8	8,9	50	–
P40	300	MFS	0,60	172	40,2	16	5,4	100	–
P40	300	LSA	0,25	183	30,2	16	8,9	100	–
P40	300	LSA	0,30	182	29,4	16	5,2	100	–
Pac	360	–	–	204	30,9	8	12,9	35	Témoin
Pac	360	MFS	0,60	168	40,7	12	4,7	50	–
Pac	360	LSA	0,30	186	33,7	12	2,9	75	–
Pas	330	–	–	195	35,3	12	5,2	50	Témoin
Pas	330	MFS	0,60	195	36,9	12	5,5	50	–
Pas	330	MFS	0,60	166	32,8	20	6,1	100	–
Pas	330	LSA	0,30	195	31,9	12	6,7	75	–
Pas	330	LSA	0,30	180	36,2	20	4,2	100	–

En utilisant les additifs LSA pour la préparation des bétons,à
une quantité d'eau de préparation égale et ouvrabilité
améliorée,on obtient généralement des réductions des
résistances mécaniques à 1 ... 90 jours dues à l'augmentation
de la teneur en air inclus du béton,mais en même temps
une augmentation de la résistance gel-degel de ces bétons.
En utilisant les additifs de type LSA lors de la préparation des bétons à ouvrabilité egale a celle des bétons
témoin,on obtient une réduction de la quantité d'eau de
préparation de 8-10%,en maintenant une quantité d'air inclus
suffisante pour assurer une meilleure résistance au gel-
degel répété du béton,Pour ces bétons les résistances
mécaniques à 1... 90 jours et l'adhérence aux armatures
sont pratiquement égales ou légèrement supérieures aux
résistances des bétons témoins.De bons résultats ont été
également obtenus lors des expérimentations réalisées
dans les unités de préfabriqués en utilisant des formules
différentes de traitement thermique pour le durcissement
des éléments préfabriqués en béton armé et précontraint
appartenant aux classes BC20 - BC60 avec des additifs MFS
dans la composition.
C'est à partir de ces recherches et expérimentations qu'ont
été élaborées,les prescriptions techniques roumaines
d'utilisation des additifs MFS et LSA pour la préparation
des bétons dans les unités de préfabriqués.Les additifs
de type MFS sont admis pour la préparation des bétons
utilisés à la réalisation des éléments préfabriqués en
béton simple,armé ou précontraint dont le durcissement
se fait à la température du milieu ambiant ou par traitement thermique.Les additifs de type LSA sont admis
seulement pour la préparation des bétons utilisés pour
la réalisation des éléments préfabriqués en béton simple,
béton armé et béton précontraint dont le durcissement
du béton se fait sans traitement thermique et principalement
pour la réalisation d'éléments préfabriqués soumis directement
à l'action de phénomène de gel-degel répété.

4 Conclusions

Les recherches et les expérimentations effectuées ont mis
en évidence les propriétés des bétons à additifs de type
MFS et LSA à l'état frais et durci,et permis l'élaboration
des prescriptions techniques roumaines d'utilisation des
additifs du type MFS et LSA lors de la préparation des
bétons employés à la réalisation d'éléments préfabriqués
en béton simple,béton armé et béton précontraint,avec
d'importants avantages techniques et économiques.

Bibliographie

Ionescu I. Actuels aspects de l'utilisation des superplas-
    tifiants dans la technologie du béton,et des préfa-
    briques,Revue Materiale de Construcţii nr.1/1980
    p.41-51.
Ionescu I et coll.Certains aspects des expérimentations
    effectuées sur les lignes technologiques de
    l'Entreprise des Matériaux de Construction Braşov,
    avec utilisation des superplastifiants dans la
    technologie des préfabriques,Revue Materiale de Construc-
    ţii nr.1/1980 p.51-54.
Ionescu I.,Ispas T., La pratique actuelle du béton Chap.
    3 Ed.Tehnică Bucureşti 1986.
Ionescu I et coll. Instructions techniques et technolo-
    giques de préparation des bétons utilisés pour la
    réalisation des éléments préfabriques,indicatif
    CD 137/1989,Buletinul Construcţiilor nr.2/1989

USE OF MAGNESIUM CALCIUM SILICATE ADMIXTURE FOR INSTANT-STRIPPING CONCRETE BLOCKS

K. KOHNO
Department of Civil Engineering, University of Tokushima, Japan
K. HORII
Department of Civil Engineering, Anan College of Technology, Japan

Abstract
Some concrete products are manufactured by the instant-stripping method using no-slump concrete. It is desirable to improve the compactability during vibrating and pressing compaction by the use of a suitable admixture to the concrete. Therefore, the use of an admixture composed of magnesium calcium silicate to no-slump concrete was investigated and the qualities of an instant-stripping concrete block were discussed.

The results of this investigation showed that the use of this admixture in no-slump concrete is effective for the improvement of concrete qualities such as the strength and the surface texture of a concrete block, although the water content of the concrete mixture increases from 5 to 10 kg/m³, compared with that of plain concrete, to obtain the same CF value.

Key words: Admixture for instant-stripping, No-slump concrete, CF value, T98 value, Compressive strength, Concrete block.

1 Introduction

Concrete products, such as concrete blocks, pavement blocks, plain concrete pipe, oval pipe and concrete sleepers, are manufactured by the instant-stripping method using no-slump concrete as moist earth. It is desirable to improve the compactability during vibrating and pressing compaction by the selection of an optimum mix proportion [1], especially the use of suitable admixture, as the fresh concrete has an extremely dry consistency.

Therefore, the use of an admixture composed of magnesium calcium silicate, which has been developed for instant-stripping concrete products, was investigated using no-slump concrete and the effect was tested on the qualities of concrete blocks manufactured by an instant-stripping method.

516

2 Materials used and mix proportions

2.1 Materials used

Ordinary Portland cement (specific gravity:3.15, Blaine's finess:3130cm²/g and 28-day compressive strength:40.9MPa) was used. Crushed hardstone of a maximum size of 20mm (specific gravity:2.60 and absorption:1.39%) and river sand (specific gravity:2.61, absorption:1.31% and FM:2.81) were used as coarse and fine aggregates.

An admixture composed of magnesium calcium silicate was used for no-slump concrete. The chemical composition and main physical properties are shown in Table 1. Silica fume (SiO_2:89.5%, specific gravity:2.20 and specific surface area:15.5m²/g) was used to test the water-tightness of no-slump concrete.

Table 1. Chemical compositions and physical properties of admixture

Chemical compositions (%)								Physical properies			
SiO_2	Fe_2O_3	Al_2O_3	CaO	MgO	Na_2O	K_2O	ig.loss	Mean particle diameter (μm)	10μm residual fineness	Absolute specific gravity	Bulk specific gravity
51.6	0.6	1.6	7.8	19.2	0:15	0.29	18.8	60	99	2	0.4—0.6

2.2 Mix proportions of no-slump concrete

In order to discuss the effect of mix factors, the water content, cement content, fine aggregate percentage of concrete mix and the dosage of an admixture were varied under zero-slump, as shown in Table 2. The standard dosage of the admixture for instant-stripping was one percent of cement weight.

Table 2. Mix proportions of no-slump concrete

Test Series		Water cement ratio W/C (%)	Fine aggregate percentage s/a(%)	Water content W (kg/cm³)	Cement content C (kg/cm³)	Admixture (%/cement)
s/a	Plain	31	36,40,44,48,52	110	350	0
	Admixture	33	36,40,44,48,52	115	350	1
Water	Plain	27,28,30,31,33	40	95,100,105,110,115	350	0
	Admixture	30,31,33,34,35	40	105,110,115,120,125	350	1
Cement	Admixture	46,43,40,37,34	46,43,40,37,34	115	250,300,350,400,450	1
		38,35,33,30,28	46,43,40,37,34	95,105,115,120,125	250,300,350,400,450	1
Dosage		31,33,35	40	110,115,122	350	0,1,2
Water-tight-ness	Admixture	33	40	115	350	1
	S F *	32	40	113	315+35	10

* Silicafume (SiO_2=89.5%. specific gravity=2.20, specific surface area=15.5m²/g)

3 Testing procedures

3.1 Mixing of no-slump concrete

All the concretes was mixed in a 50ℓ pan-type mixer for forced circulating mixing. The mortar was first mixed for

60 seconds, and the mixing was continued for an additional
90 seconds after the addition of crushed hardstone.

3.2 Test of fresh concrete
The CF test using a compacting factor testing apparatus and
the solidity ratio test shown in Fig. 1 were done immediately
after the mixing of the concrete. CF value was calculated
by the following equation:

$$\text{CF value} = \frac{\text{Mass of concrete obtained by an actual filling}}{\text{Mass of concrete calculated by the specified mix proportion without air-void}}$$

The T_{98} value was determined
from the relationship between
the solidity ratio of fresh
concrete and a vibrating time
of up to 120 seconds, using
the apparatus; and then the
time required for the solidity
ratio of packed concrete to
reach a level of 98 percent
(T_{98} value) was calculated
by means of this relation [2].

3.3 Fabrication of concrete
    block
Concrete was placed in an
instant-stripping mold with
an external vibrator of 180
Hz high frequency. During
compaction, the top surface was pressed and finished with a
steel block, as can be seen in Photo 1.

A wooden plate was then set on the surface and the mold
was turned upside down. Finally, the concrete block of 15x
15x54cm in size was stripped by lifting up the mold using
the handles on both sides as shown in Photo 2.

Handle →
Holding
bar ↓

Cylindrical
weight
(ϕ9.5x5.5cm)

Cylindrical guide
(ϕ9.7x7.8cm)

Cylindrical mold
for concrete
(ϕ10x20cm)

Vibrating table
(Frequency:3600vpm)
(Amplitude:0.2mm)

Fig.1. Test apparatus of solidity
ratio of no-slump concrete

Photo 1. Instant-stripping apparatus   Photo 2. Fabrication of concrete
and vibrating compaction of   block by instant-strip-
no-slump concrete   ping method

3.4 Curing and testing of concrete block

After stripping, concrete blocks were stored at a constant
room temperature of about 20°C for one day. Then the surface
texture of each block was observed, the weight and size of
the block were measured, and they were cured in a water tank
at 20°C within the given periods of time. The concrte blocks
using a different dosage of admixture were cured in a steam
chamber under the condition of presteaming+temperature rise+
isothermal curing+soking period = 2+3(15°C/h)+3(65°C)+14hours,
to compare with standard curing.

At the age of 14 days, each block was cut into three parts
by a concrete cutter. Each part was used for a compressive
strength test using a universal testing machine. Each
strength is given as the mean value of three parts.

3.5 Water permeability test of no-slump concrete

In order to investigate the water-tightness of no-slump concrete, the coefficient of water permeability was measured by
an apparatus using three closed pressure chambers, as can be
seen in Photo 3. The concrete specimen was a 15x30cm cylinder with a 2cm diameter
hole at the center and
was tested under the hy-
draulic pressure of 3MPa.
The amount of outflow
water was measured using
a glass cylinder and the
coefficient of water per-
meability was calculated
according to Darcy's law.
The test was performed at
the ages of 7 and 28 days.
The mean value of the
three specimens was taken
as the results of the set
of measurements.

Photo 3. Water permeability test
apparatus

4 Test results and considerations

4.1 Effect of mix factors on consistency of no-slump concrete

4.1.1 Effect of mix factors on CF value

The effects of varying mix factors, such as fine aggregate
percentage, water content and cement content, on the CF value
of no-slump concrete are shown in Fig. 2.

When the fine aggregate percentage increases, the CF value
of fresh no-slump concrete tends to clearly increase within
the limit of this investigation. At the same fine aggregate
percentage, the CF value of concrete using admixture is a
little lower than that of plain concrete. The CF value of
no-slump concrete increases with the increment of water content, as can be seen in Fig. 2(2), and the concrete using an
admixture for instant-stripping shows a higher water content,

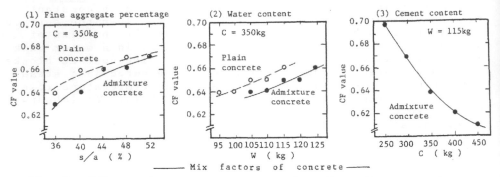

Fig. 2.　Effects of mix factors on CF value of no-slump concrete

from 5 to 10 kg/m³ compared with plain concrete, to obtain the same CF value. As the cohesiveness of no-slump concrete increases with the increment of the cement factor, the CF value clearly decreases in rich mix (see Fig. 2(3)), and so it is desirable to increase the water content to improve the compactability of no-slump concrete.

The CF value of 0.64 to 0.66 is suitable for the concrete using an admixture for instant-stripping, when the concrete block is fabricated using the mold.

4.1.2 Effect of mix factors on T_{98} value

Fig. 3 shows the effects of mix factors on the T_{98} value of no-slump concrete. As can be seen in Fig. 3(1), there is an optimum fine aggregate percentage, at which T_{98} value is minimized. This value is about 40 percent in both concretes. As

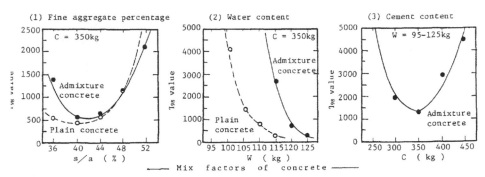

Fig. 3.　Effects of mix factors on T_{98} value of no-slump concrete

the water content increases, the T_{98} value of no-slump concrete decreases, and this means that the vibrating time to keep the required solidity ratio is able to be shortened. When the cement factor has increased from 250 to 450kg/m³, the T_{98} value reachs a minimum in the mix of 350kg/m³, as shown in Fig. 3(3).

It is considered that T_{98} value is an index to obtain the
optimum fine aggregate percentage in the mix of no-slump
concrete, although the testing method is not simple and the
testing time becomes longer compared with those of the CF
test.

4.2 Effect of mix factors on strength of concrete block
4.2.1 Effect of fine aggregate percentage, water content and
      cement content on block strength
The effects of the fine aggregate percentage, water content
and cement content on the compressive strength of concrete
block manufactured by instant-stripping method are shown in
Fig. 4(1)-(3). When the fine aggregate percentage of con-

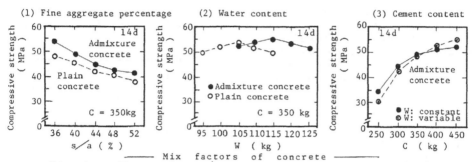

Fig. 4. Effects of mix factors on compressive strength of
         concrete block stripped immediately after molding

crete mix increases, the compressive strength of the concrete
block tends to decrease. This tendency is similar to that
found in ordinary concrete [3]. A concrete block using admixture yields slightly more strength than that yielded by a
plain concrete block at the same fine aggregate percentage.
In the relationship between the water content of no-slump
concrete and the block strength, the optimum water content,
at which the strength reaches its maximum, is obtained in
each kind of concrete. This is considered to be for the
reason that the volume of cement paste and its workability
increases and then the concrete may be well compacted. The
optimum water content of admixture concrete is found at a
point 10kg/m³ higher than that of plain concrete. The richer
the cement factor in a concrete mix, the lower the ratio of
strength development, as the compactability deteriorates of
an extremely low water-cement ratio. Therefore, the strength
is improved when the water content increases in richer mixes
as shown in Fig. 4(3).
    When the concrete mix with a slightly higher water content
and an admixture for instant-stripping is used, the surface
texture of the concrete block tends to improve as can be seen
in Photo 4. Larger air-voids are observed on the surfaces
of plain concrete blocks compared with the concrete blocks
using the admixture.

(1) Admixture concrete   (2) Plain concrete

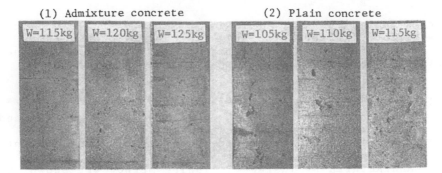

Photo 4. Examples of surface texture of concrete blocks

4.2.2 Effect of dosage of admixture on block strength
Fig. 5 shows the effect of dosage of an admixture for instant-
stripping on the compressive
strength of a concrete block.
The block strength increases a
little with the dosage of the
admixture and the solidity ratio
of the concrete block is slightly
higher. The surface texture after
stripping tend to be improved by
the addition of admixture.

In addition, the compressive
strength of a steam-cured concrete
block is a little lower than that
of a standard cured one at the age
of 14 days. However, a high
strength of about 50MPa can be
obtained in the mix of a cement
factor of 350kg/m³ and a standard
admixture dosage of one percent
per cement factor.

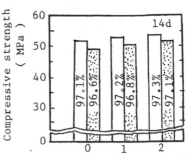

Fig.5. Effect of dosage of
admixture on block
strength

4.3 Water-tightness of no-slump concrete
The result of the water permeability test of no-slump concretes is shown in Table 3
as compared with that of
regular concrete used for
ordinary concrete products.
The coefficient of the
water permeability of no-
slump concrete becomes
extremely low at the age
of 28 days, and therefore
it is considered that this
concrete is sufficiently
water-tight. The coeffi-
cient of water permeability
of no-slump concrete using

Table 3. Test result of water-tightness

Type of concrete	Coefficient of water permeability (x10⁻¹¹cm/sec)	
	7d	28d
Admixture for instant stripping	5.56	0.97
Silica fume	5.13	0.92
Regular concrete*	–	1.54

* W/C = 53%, C = 320kg, Slump = 8cm

an admixture approches the value of silica fume concrete, which has high water-tightness [4]. The value at the age of 28 days is superior to that of the regular concrete. This means that the water-tightness of no-slump concrete may be improved by the selection of a suitable mix and the use of powerful vibrating compaction.

5 Conclusions
On the basis of this investigation, the following conclusions can be stated.

(1) The CF value of 0.64 to 0.66 is suitable for the concrete block using an admixture for instant-stripping.

(2) The T_{98} value of fresh no-slump concrete may be an index to obtain the optimum fine aggregate percentage.

(3) The water content of no-slump concrete using admixture composed of magnesium calcium silicate increases from 5 to 10kg/m^3 compared with plain concrete.

(4) The use of this admixture for no-slump concrete is effective for the improvement of compactability and the surface texture can be improved.

(5) The block strength and water-tightness of no-slump concrete is improved by the use of this admixture as the solidity ratio of concrete becomes a little higher.

References

[1] Kohno, K., Hayashi, M. and Takemura, K. (1970) On the mix proportion and strength of instant-stripping concrete. Cement Concrete No. 282, pp.31-38.

[2] Horii, K., Kohno, K. and Amo, K. (1987) Mixing method and properties of silica fume concrete for factory products. Transactions of the JCI Vol. 9, pp. 9-16.

[3] Kohno, K. and Mizuguchi, H. (1971) Effect of fine aggregate percentage on workability and compressive strength. Semento Gijutsu Nenpo Vol. 25, pp.190-193.

[4] Kohno, K. and Komatsu, H. (1986) Use of ground bottom ash and silica fume in mortar and concrete. ACI SP-91 Vol. 2, pp.1279-1292.

USE OF SUPERPLASTICIZERS FOR CONCRETE IN UNDERGROUND CONSTRUCTION WORK

F. MAÑA
School of Architecture, Technical University of Catalunya, Barcelona, Spain

SUMMARY

The proliferation of underground construction work in big cities and the durability requirement of concealed concrete construction (screens, piles etc.) require this material to respond to rigorous impermeability demands.

Comparative tests were carried out between fluid concretes and plastic concretes with added superplasticizers with ample operating time. The pouring procedure for both concretes was the same.

It has been shown that the impermeability achieved with the latter, in the circumstances studied, was at least 2.6 times greater than that of the fluid concretes.

The availability of building land in the city of Barcelona cannot,at present ,keep up with demand. The repercussion of the cost price of land on each m^2 constructed is out of all proportion and, within the restrictions on height laid down by Municipal bye-laws, maximum advantage tends to be taken of this resource.

Moreover, Barcelona is a city in which the number of cars greatly exceeds the number of parking spaces available. Municipal bye-laws oblige each building undertaking to cater for its own parking requirements within the site itself.

The logical result of this framework of requirements is that practically all newly-constructed buildings contain a large number of underground floors, giving rise to a curious type of "singular construction" in which there are as many, or more, underground floors in section as in height.

The geomorphology of the city of Barcelona, which lies between a mountain range and the sea, favours a situation that may well be a feature of many other urban agglomerations - the less the available land is put to agricultural and industrial uses as it becomes reclassified for building purposes, the lower the consumption of subsurface water by means of wells. The direct result of this is that today's phreatic levels are constantly on the increase. Basements which had always been dry now display significant areas of damp and even, in especially conflictive areas, of flooding.

We are well aware that the impermeability of inhabited space, together with structural resistance, is the demand most clearly sought after by the consumer. No wonder, then, that at the present time the flooding of basements in the environments we are describing is one of the causes leading to the highest and costliest number of suits brought against architects and builders.

Contrary to this proliferation of underground building space and the demand for impermeability of such space at all costs, we are employing procedural strategies that do not guarantee the fulfilment of this demand. Aside from other important questions that need not concern us here (such as the suitability of the solutions used in the joints between underpinnings or between the latter and the foundation slab), the concretes that result from de construction of the perimeter screens are usually excessively porous, usually displaying nucleii of segregated aggregate and an excess of " connected capillarity " which leads to an inadequate impermeability of the mass.

The construction of the screen walls prior to undertaking overall excavation of the site involves the pouring into a previously opened shaft of significant quantities of concrete (from 30 to 50 m^3) through pipes (to avoid segregation). The requirement of good circulation of the concrete mass inside the narrow piping and within the shaft, without segregations, determines the use of high fluidity slurries (18 to 22cm Abrams cone).

Traditionally, the concrete in most underground construction work suffers from this defect. The most economic way of undertaking foundation piles employed today is that based on the prior excavation of a shaft using a screw and the injection or simple deposition of concrete at the bottom of this shaft. This concrete, because of the special features of its execution, has to be fluid. If this fluidity is achieved by using a large quantity of water, the resulting material is porous and its durability in the face of eventual chemical attack from the environment must be considered low.

We know that raising the water content is not the only means of obtaining slurries of fluid consistency. For some time now, the industry has been supplying us with a whole range of products capable of fluidifying initially rigid concretes, but these products, until recently, were not suitable for pouring large masses of concrete on the job. The "operating time" of these products (roughly 20 minutes) was not sufficient to facilitate this type of operation, thus running the risk that the pouring process might break down when the concrete already poured into the shaft instantly set (figure 1.1).

The appearance of a new generation of superplasticizers, with an operating time of the same order of value as the concrete itself has enabled us to successfully tackle the use of fluid concretes with low water content.

In the laboratory tets carried out in order to study the viability of the proposal, we obtained, for an initial Abrams cone of 4cm, to which was added 1.5%

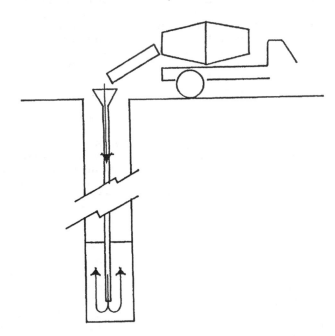

FIGURE 1   Procedure for filling an underpinning

of the weight of cement in the form of superplasticizers, a 15 cm cone,150 minutes after making the addition.

With this procedure it seemed possible to make the construction requirements (fluidity for pouring the concrete "in situ") compatible with those of a functional character (impermeability).

We present here the conclusions obtained from impermeability tests carried out on core samples cut from the concrete in the support screens of the building under construction as an extension to the headquarters of the Guild of Architects of Catalonia.

The core samples cut directly from the screens, as an alternative to studies applied to moulded test pieces, have the advantage that the different incidences of execution have accumulated in the concrete, but the drawback that if we do not want to compromise the final quality of the work, their number cannot be sufficient for correct statistical processing.

The study was limited to extracting and testing 3 10x20 core samples from one screen element constructed with a concrete of plastic consistency with an added superplasticizer, which were compared with three others, taken from the same depth, from a nearby underpinning constructed with fluid concrete (photograph 2).

The permeability tests were carried out using the L'Ecole des Ponts et Chaussées de Paris procedure.

The permeability values obtained were as follows:

	level 1	level 2	level 3	mean
additive	50	0	3	17.67
fluid	130	39	8.6	59.2

(all the values must be multiplied by 10^{-7})

It can be seen from these results that:

a)   In the vertical pouring of large masses of concrete an accumulation of water occurs at the top, which reduces the resistances and increases the permeability.

b)   The permeability achieved in the concretes with additives was of the order of 2.6 times lower than that

527

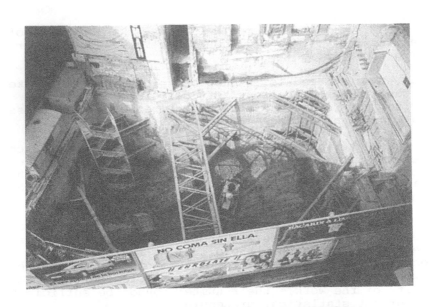

for fluid concretes if we compare individual results, and of the order of 3.35 if we compare their mean values.

Although a detailed economic study was not carried out, it can be noted that:

The weight of cement saved was significant (more so if we consider that the benefits of the concrete with additives were somewhat higher than those of the fluid concrete).

The pouring of the concrete with additives proved quicker because its Abrams cone was greater and the slurry, despite its fluidity, displayed such cohesion that it was not necessary to take special precautions with regard to its possible segregation.

SPECIFIC DATA OF THE WORK

1.   Volume   of   concrete   from   a   specimen underpinning .... 30 m^3.
(5 cement mixers of 6 m^3 were required).

         The pouring was carried out in a shaft full of water with bentonite, which is a technique normally used to keep the shaft open.

2. Demand parameters of the concretes.

For the fluid concrete:

Fck .... 175 kp/cm[1]
Consistency ....fluid (18-20 cm)
Maximum aggregate size......... 20 mm

For the concrete with additives:

Fck .... 175 kp/cm^2
Consistency ....plastic (3-5 cm)
Maximum aggregate size .... 20 mm

3. Characteristics of the concrete supplied. Values obtained from the reception check tests and from the concrete's own manufacturer.

For the fluid concrete;

Type C-175-liquid-20

Proportioning per m^3

```
Cement.....................................320kg
River sand (0-6 mm).......................932  "
Crushed aggregate 1 (6-12 mm).............416  "
Crushed aggregate 2 (12-20 mm)...........416  "
Water.....................................240 litres
Plasticizer for transportation...........0.5 litres

Mean cone.................................18 cm
```

Plastic concrete for additives:

Type C-175-plastic-20

Proportioning per m^3

```
P-450 ARI Cement.........................263 kg
River sand (0-6 mm)......................1011  "
Crushed aggregate 1 (6-12 mm)............438  "
Crushed aggregate 2 (12-20 mm)...........438  "
Water....................................205 litres
Plasticizer for transportation...........0.45  "
Mean cone.................................5 cm
```

Given their importance for the sealing of the slurries and, therefore, for the impermeability, the granulometric curves for the three aggregates employed are included in the figures attached (figure 2).

4. On-site operations

The quantity of additive employed in each mixing (cement mixer) was 1.7% of the weight of cement.

The mean age of the concrete at the time of addition was 45 minutes.

Ambient temperature at the time of addition and pouring was 14 degrees Centigrade.

The mean cone obtained, at 50% of emptying of the truck, was 24 cm (ranging from 19 to 25). This measurement was carried out between 5 and 15 minutes after adding the superplasticizer.

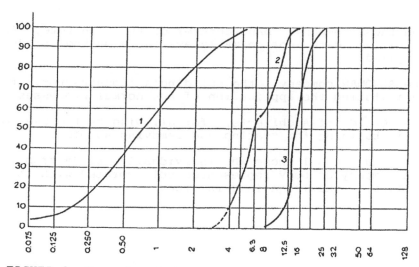

FIGURE 2   Granulometric graphs of the aggregates

   1. sand, with 4.7% clay

   2. crushed aggregate of dry course pebbles 1
      (6-12)

   3. crushed aggregate of dry course pebbles 2
      (12 -20)

5.   Resistances achieved after 7 and 10 days, measured
on moulded test pieces.

fluid concrete........199 kp/cm² (7d); 254 kp/cm²(28d)

plastic concrete without additives....208 kp/cm²(7d) ;
262 kp/cm²(28d)

Acknowledgements

The Materials Testing Laboratory of the Escuela Tecnica
Superior de Caminos of the UPC for its precise handling
of the permeability tests.

The   SECOTEC company ,for extending its normal checking
procedures.

The   GRACE company,for supplying products,   advising on
their application and the prior testing carried out.

The LACSA company, manufacturer of the concrete, for supplying data not normally available on the product supplied.

The AGROMAN company, the building's constructor, for facilitating the application of non-normal processes.

The Guild of Architects, for allowing us the use of their building for testing and removing samples.

... and to all those who individually, aside from the company or body represented, collaborated with their particular knowledge in this study.

IMPROVING QUALITY WITH A NEW ROLLER COMPACTED DAM (RCD) CONCRETE ADMIXTURE

M. TANABE, M. TAKADA and K. UMEZAWA
Nisso Master Builders Co. Ltd, Japan

Abstract
In recent dam constructions projects, the compacting of dry concrete with a vibratory roller i.e. the RCD (Roller Compacted Dam–Concrete) method, has been used widely. However, with the reduction in quality that results from the aggregate and fly ash that is used in this concrete, the unit water content is sometimes increased. As well, due to the increased structural variety of dams, higher design strength is demanded.

In order to cope with this situation, a new admixture was developed for use with RCD concrete. This new admixture is able to achieve greater compacting effect than presently used admixtures at the same unit water content. As well, this admixture is able to reduce 5-10% unit water content at the same compacting effect.

This study outlines the action of this new admixture, and examines its strength development and its compacting performance when it is used in RCD concrete.
Key words: RCD Concrete, Admixture, Compacting effect

1 Introduction

With RCD concrete, the unit cement content must be reduced as much as possible, to reduce the heat of hydration. Also, concrete compaction with a vibration roller over a large surface area requires ultra stiff consistency concrete with a long workability.

The quality of aggregate used in recent dam construction projects has been steadily reduced, due to available sources and other limitations. At the same time, the low quality of the available fly ash has necessitated increases in the unit water content to ensure consistency required for compaction.

For the above reasons, it has become desirable to develop a high performance admixture with even better water reducing capability than current admixtures.

The authors thereupon developed an admixture for use with RCD concrete which exhibits a strong dispersing action and a strong wetting action, and which greatly reduces the unit water content, increases strength, and improves placeability.

2 Chemical Composition and Analysis of our New Admixture

Our new admixture, developed for use with RCD (Roller Compacted Dams), is composed of a combination of a cement dispersing agent and a non-ionic surface activation agent.

The cement dispersion agent is lignin sulfonate, and the non-ion surfactant is polyoxyethylene nonylphenylether.

The effect of the new admixture on improving the quality of RCD concrete can be considered in terms of the following:

Concrete is produced from the hydration product of the reaction of cement with water combining with aggregate to form a single body. In this reaction of the cement components with water, the two must make uniform contact, a prerequisite to which is the uniform wetting of the cement particles.

As well, a necessary condition to the dispersing agent being effective on the cement particles is uniform distribution of the solution of the dispersing agent around the entire circumference of all of the particles. Usually, however, dispersion agents (water reducing agents), used as surface activation agents, do not show much surface expansion (wetting action) or micelle production.

Moreover, in the RCD concrete mix-design developed in our research, the unit water content is a fair amount lower than that of normal concrete, making it difficult to distribute uniformly all over the surface of the cement particles. Therefore, an admixture for use with RCD concrete cannot be merely a water reducing agent, but also a wetting agent, in order to achieve maximum cement dispersion effect. A combination of the two enables greater water reducing effect and consistency, attaining .greater strength development through even cement hydration.

Based on the above concept, a new admixture for RCD concrete was developed by combining a dispersing agent, which shows high water reducing properties, with a non-ionic surfactant, which has the relatively high HLB (10-14) which is required due to the relatively high hydrophilic properties of concrete, to attain good wetting properties.

Table 1 shows the RDA cement dispersion and wetting effect in terms of the measured zeta potential and contact angle.

Table 1. Dispersing and wetting effect of RDA

Type of admixture	Zeta potential (mV)	Surface tension (dyn/cm)	Contact angle (θ)
P L A I N	+4.0	72.8	65
A E W R A	-11.7	61.1	50
R D A	-15.9	34.4	50

AEWRA : AE water reducing agent
RDA : New admixture

3 Outline of the Experiment

3.1 Concrete mix conditions

Currently, control of the rise in concrete temperature and reduction of the early age heat of hydration in cement for use with RCD concrete is generally managed by the use of moderate heat portland cement with fly ash cement to which 20–30% fly ash has been added. With this, the use of a retarding type AE water reducing agent (lignin type) is common.

In this test, the performance of RDA was evaluated in comparison to a retarding type AE water reducing agent, which was used with moderate heat portland cement and fly ash (substitution rate: F/C+F = 30%).

The concrete mix-design is shown in Table 2. The unit binding material contents were 130, 120, and 110 kg/m^3.

Table 2. Concrete Mix Conditions

Unit binder content(kg/m^3) (C+F)		130,120,110	
Rate of Fly Ash(%) F/(C+F)		30	
Target VC value(Sec.)		20±5	
Admixture	Type	AEWRA	RDA
	Dosage	(C+F)x0.25%	(C+F)x1.0,2.0%
Temperature (℃)		20	

3.2 Materials

Tables 3–6 show the physical properties of the moderate heat portland cement, fly ash, fine aggregate (crushed sand), and coarse aggregate (crushed stone) used in the testing.

Table 3. Physical Properties of the Moderate Portland Cement

Specific gravity	Specific surface (cm^2/g)	Setting time			Sound-ness (Pat test)	Comp.strength(kgf/cm^2)			Ig·loss (%)	MgO (%)	SO$_3$ (%)
		Water (%)	Initial (h-m)	Final (h-m)		3 days	7 days	28 days			
3.20	3250	28.1	3-09	4-32	good	130	185	339	0.6	1.2	1.9

Table 4. Physical Properties of the fly ash

Specific gravity	Specific surface (cm^2/g)	Unit water content ratio(%)	Comp.Strength(kgf/cm^2)	
			28 days	91 days
2.21	2790	98	74	96

Table 5. Physical Properties of the fine aggregate

Specific gravity	Absorption (%)	Washing loss (%)	Durability (%)	Unit weight (kg/ℓ)	Solid volume percentage (%)	Fineness modulus
2.65	1.40	4.0	2.6	1.75	68.5	2.73

Table 6. Physical Properties of the Coarse aggregate

Max. size (mm)	Specific gravity	Absorption (%)	Washing loss (%)	Durability (%)	Unit weight (kg/ℓ)	Solid volume percentage (%)	Fineness modulus
150	2.71	0.40	0.35	3.2	1.83	67.8	8.43

3.3 Test Procedures

(a) Concrete mixing
All materials were mixed for 3 min. in a 150 liter capacity tilt mixer.

(b) Consistency
The consistency of the RCD concrete was determined by measuring the vibrating compaction (VC) value by the vibrating consistency test. Fig. 1 is a diagram of the VC value tester. The change in consistency over time was measured at prescribed times (1.0, 2.5, 4.0hr) after mixing.

(c) Preparation of specimens for compressive strength testing The material was wet screened through a 40 mm sieve, and was placed in 3 layers in a ϕ15x30cm form. Using the surface vibrator shown in Fig. 2, each layer was vibrated until the mortar rose to the surface.

The specimens were water cured at 20±3°C to the prescribed ages (7, 28, 91 days).

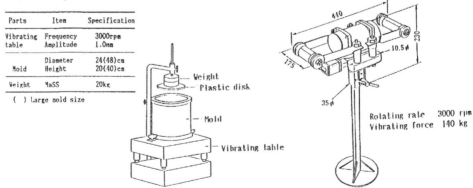

Parts	Item	Specification
Vibrating table	Frequency	3000rpm
	Amplitude	1.0mm
Mold	Diameter	24(48)cm
	Height	20(40)cm
Weight	Mass	20kg

( ) Large mold size

Weight
Plastic disk
Mold
Vibrating table

Rotating rate 3000 rpm
Vibrating force 140 kg

Fig.1. VC tester   Fig.2. Surface vibration

4 Test Result and Discussion

4.1 RDA water reducing effect
With the combination of a dispersing agent and a wetting agent described above, RDA acts to disperse the cement particles.

Fig. 3 shows the effect of RDA on the unit water content. RDA exhibited excellent water reducing effects. To obtain the target VC value (20±5 sec), the AE water reducing agent required a unit water content of 95kg/m^3, whereas RDA required 90kg/m^3 at the dosage of 1.0% and 85kg/m^3 at the dosage of 2.0%. If the same unit water content is used as with the AE water reducing agent, the VC value decreases. When the surface vibrator is used for compaction, the greater the flowability of the concrete components (cement paste and mortar) the earlier the necessary degree of compaction is reached. Therefore, the water reduction resulting from the use of RDA enables a reduction in compaction time together with a strength increase.

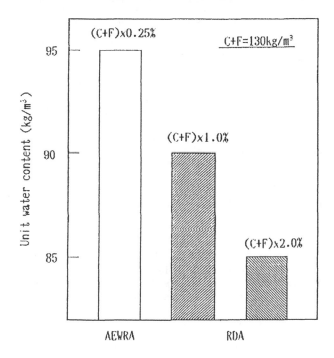

Fig.3. Water reducing effect of RDA

4.2 RDA Increase in strength.
Table 7 shows the results of concrete compressive strength under various mix conditions. With its reduction of the unit water content, RDA, compared to the AE water reducing agent, increased the concrete strength by 15% at the dosage of 1%, and by 30% at the dosage of 2%.

Table 7. Concrete test results

Admixture		Unit binder content C+P (kg/m³)	Water-binder ratio W/(C+F) (%)	Sand-aggregate ratio S/a (%)	Unit water content W (kg/m³)	VC value (Sec.)		Compressive strength (kgf/cm²)		
Type	Dosage (C+F)x%					Small	Large	7 days	28 days	91 days
AEWRA	0.25	130	73.1	32	95	26.0	58.3	61.2	105	164
		120	79.2		95	27.9	66.0	50.4	84.5	145
	1.0	130	69.2		90	24.5	65.7	70.2	121	185
		120	75.0		90	26.7	69.3	64.0	100	162
RDA	2.0	130	63.1	31	82	36.4	115	87.4	146	220
			65.4		85	23.1	62.0	80.4	139	203
			69.2		90	15.4	27.7	65.3	131	194
		120	70.8		85	29.9	73.0	77.7	114	183
		110	81.8	33	90	17.8	69.0	53.0	95.2	155

Fig. 4 shows the relationship between the unit binding material content and the compressive strength. From the figure, it can be seen that, at the same unit binding material content, RDA develops significantly greater compressive strength.

Fig. 5 shows the relationship between the binding material/water ratio and the compressive strength in RDA concrete. A directly proportional relationship was observed in which increases in the binding material/water ratio was accompanied by increases in compressive strength. At the same binding material/water ratio, RDA developed slightly greater strength than did the AE water reducing agent.

4.3 Change in Consistency over Time. In the RCD method using ultra stiff consistency lean mix concrete, generally one lift is divided into some number of layers, each of which is leveled by a bulldozer, after which the surface of the lift is compacted with a vibration roller.

Therefore, there is often a fairly long period of time between concrete mixing and compaction. In such instance, as little change in consistency over time as possible is best for placing.

Table 8 shows the relationship between sitting time and the VC value for concrete using RDA and an AE water reducing agent. As the table shows, the use of RDA instead of the water reducing agent enables a $10kg/m^3$ reduction in the unit water content with little change in the VC value over time, greatly improving concrete compaction.

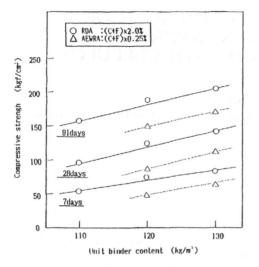

Fig.4. Relationship between the unit binder content and the compressive strengh

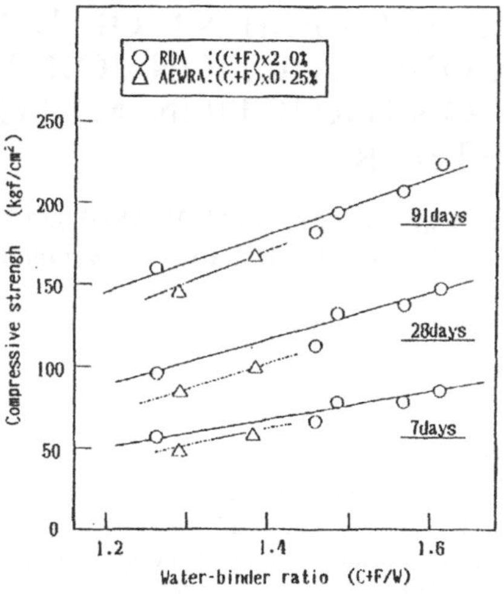

Fig.5. Relationship between the water-binder ratio and the compressive strengh

Table 8. Test results of consistency over time

Admixture		Rate of Fly Ash	Unit binder content	Water-binder ratio	Sand-aggregete ratio	Unit water content	VC value (Sec.)			
Type	Dosage	F/(C+F)	C+F	W/C+F	S/a	W				
	(C+F)x%	(%)	(kg/m³)	(%)	(%)	(kg/m3)	0 hr	1.0 hr	2.5 hr	4.0 hr
A E W R A	0.25	30	130	73.1	32	95	21.2	29.8	35.4	41.3
R D A	2.0	30	130	65.4	31	85	20.7	25.2	28.9	33.4

5 Conclusion

The results of the comparison of the newly developed admixture for use with RCD concrete with the existing AE water reducing agent are as follows:

(1) RDA, at the dosage of 2%, enables a 10kg/m³ reduction in the unit water content over the use of the AE water reducing agent.

(2) The reduction of the unit water content led to a 30% increase in the compressive strength.

(3) The change in consistency of RDA concrete over time was lower, and placeability was improved.

IMPROVEMENT OF BOND STRENGTH OF CONSTRUCTION JOINTS IN INVERTED CONSTRUCTION METHOD WITH CELLULOSE ETHER

Y. TAZAWA, K. MOTOHASHI and T. OHNO
Civil Engineering Department, Kajima Institute of Construction Technology, Japan

Abstract
In the case of using inverted construction method, the construction joint is located under the existing concrete, and thus has insufficient bond strength because of bleeding water at the top or settlement of the newly placed concrete.

Addition of a cellulose ether (Hydroxy Propyl Methyl Cellulose, abbreviated by HPMC) to concrete as a bleeding reducing admixture was examined for the purpose of the improvement of bond strength for such joints. The following observations were obtained in the present investigation:
(a) A concrete using HPMC shows no detectable amount of bleeding.
(b) When an air detraining agent (tribtylphosphate) is added in suitable quantity, the air content of the HPMC concrete can be controlled within the range of 3 to 4%.
(c) The structural properties of the HPMC concrete can be made comparable to those of conventional concrete.
(d) When the HPMC concrete is used as the newly placed concrete in inverted construction method, the bond strength of the joint measured by flexural tests can be taken up to 60% of that of unjointed specimens.
(e) Aluminum powder is added to compensate the volume changes. It also serves to improve the bond strength of the joint.
Key words: Cellulose ether, Water soluble polymer, Bleeding, Bond strength, Construction joint, Inverted construction method, Aluminum powder

1 Introduction

In the case of using inverted construction method, the construction joint is located under the existing concrete, and generally, it has an insufficient bond strength because of the bleeding water between upper existing concrete and lower newly placed concrete.

There are very few papers dealing with the structural properties of the inverted joints in this special construction method. Suzuki (1968) suggested strengthening this inverted construction joint using prepacked mortar between the old (upper) and the new (lower) concrete surfaces. He reported a bond strength (from shear tests) of 10 kgf/cm^2 between the upper concrete and the prepacked mortar.

On the other hand, the previous work by Hansen (1952) suggests the possibility of using some types of water soluble polymers such as cellulose ether and acrylic polymer in concrete because these polymers prevent the rapid loss of water in cement grout.

The research has been performed in recent years at the Kajima Institute of Construction Technology, to investigate this problem and some of the important results have already been published by Kotani et al (1983, 1984).

Since the addition of cellulose ethers (like HPMC used in this study) to concrete is known to reduce the bleeding and laitance, the possibility of using such cellulose added concrete in the inverted construction method was examined in this study.

Two series of experiments were carried out to clarify the effects of cellulose ether addition to concrete and to determine:
1. Basic properties of fresh and hardened concrete and
2. Applicability of such concrete to inverted construction method

2 Properties of fresh and hardened concrete with HPMC

Some experiments are conducted to examine the effect of the dosage of cellulose ether on properties of fresh and hardened concrete in case of concretes having varied unit water content and water-cement ratio. The experiments are designed using orthogonal array table (L27) including these experimental factors.

Cellulose ether used in this experiment is hydroxy propyl methyl cellulose (HPMC) whose viscosity at 2% solution is about 22000 cps at 20°C.

2.1 Experimental methods
Experimental factors and levels employed are unit water content, water-cement ratio, and dosages of HPMC, and superplasticizer, as shown in Table 1.

Table 1. Variables for the experiment

Factor	Level		
	1	2	3
A. Unit water content (kg/m^3)	155	165	175
B. Water cement ratio (%)	45	55	65
C. Dosage of HPMC (W×wt.%)	0	0.3	0.6
D. Dosage of DA-X (W×Vol.%)	0	0.05	0.10
E. Dosage of super-plasticizer (cc/C=100kg)	0	400	800

The addition of HPMC to fresh concrete causes increased air content, though the problem can be controlled by using suitable doses of a detraining agent and therefore the dosage of this agent is treated as one of the variables in the experiment, as shown in Table 1. In this series of experiments, the detraining agent used contained mostly tributylphosphate and is designated as DA-X.

The physical properties of materials used for casting specimens in all experiments are: Ordinary portland cement (specific gravity: 3.16), river gravel (Gmax: 25mm, specific gravity: 2.65), river sand (FM: 2.96, specific gravity: 2.62), AE water-reducing agent (lignosulphonic acid) and superplasticizer (high-condensation-triagin compound).

The mixing and the tests for fresh concrete were conducted in a chamber with controlled temperature (20°C) and RH (80%).

Bleeding, slump and air content were measured in fresh concrete and the compressive strength (at 7, 28 and 91 days), splitting tensile strength (28 days) and Young's modulus (secant modulus at the stress level equivalent to 1/3 of compressive strength, at 28 days), were determined for the hardened concrete.

2.2 Experimental results

(1) Properties of fresh concrete
The effect of various factors on the rate of bleeding is shown in Fig.1. Water content, water-cement ratio, dosage of HPMC and the interaction between water-cement ratio and dosage of HPMC were judged to be "significant" for 5% level of significance. The effect of dosage of HPMC on the rate of bleeding was exceedingly greater than that of other factors. In fact, in case of 0.6% of unit quantity of water by weight for HPMC dosage, the rate of bleeding was too small to be detected regardless of other factors.

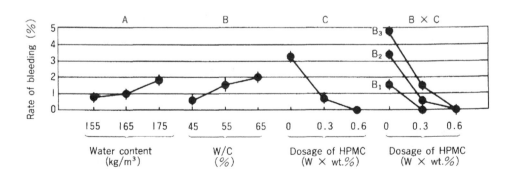

Fig.1. The effect of various factors on rate of bleeding

The factorial effect on air content is shown in Fig.2. The air content increased with increase in the amount of HPMC and conversely decreased with the increase of DA-X. The air content of concrete became more than 10% when using more than 0.3% of unit water content by weight of HPMC, without any DA-X addition. It can be concluded that the increased viscosity of HPMC concrete causes air entrappment and leads to this excessive air content in fresh concrete. On the other hand, when using 0.05 or 0.10% of unit quantity of water by volume of DA-X in HPMC concrete, the air content decreased to the range of 3 to 4%. This shows the effectiveness of DA-X as an air detraining agent.

The changes in the recorded values of slump is shown in Fig.3. All factors and the interaction between unit water content and dosage of HPMC were judged to be " significant" for 5% level of significance on slump. Effect of unit water content was particularly greater than that of other factors. It was observed that slump increased with the addition of HPMC and decreased with the addition of DA-X. In spite of the increased slump, the HPMC concrete was found to be less workable and difficult to place. This shows that the increased air content in HPMC concrete interferes with the direct interpretation of results of the slump test.

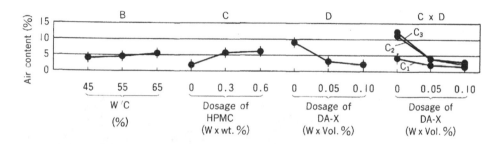

Fig.2. Variation of air content

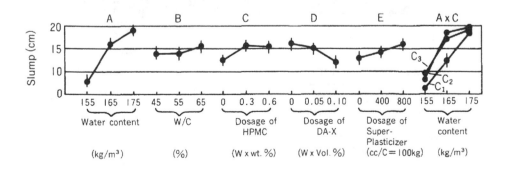

Fig.3. Variation of slump

(2) Properties of hardened concrete

The diagram of factorial effect on compressive strength is shown in Fig.4. From this figure, it can be seen that the effect of water-cement ratio and dosage of DA-X was greater than that of other factors and the sum of coefficient of determination on these experimental factors became about 85%. Therefore, it could be concluded that the compressive strength of hardened concrete is related to cement-void ratio calculated from water, cement and air content of fresh concrete.

Fig.5 shows the variation of compressive strength at 28 days with cement-void ratio. According to this diagram, the coefficient of correlation between both factors was 0.943 and the compressive strength at 28 days obviously increases with increases in the cement-void ratio. No significant difference could be identified between the concretes with and without HPMC as far as this relation between cement-void ratio and compressive strength at 28 days is concerned. Compressive strength tests at 7 days and 91 days also yielded similar results.

Also, from the data plotted in Figures 6 and 7, showing the variation of tensile strength and Young's modulus with compressive strength, it can be seen that by suitable addition of air detraining agent the structural properties of HPMC concrete (compressive and tensile strengths and Young's modulus) can be made comparable to those of conventional concrete.

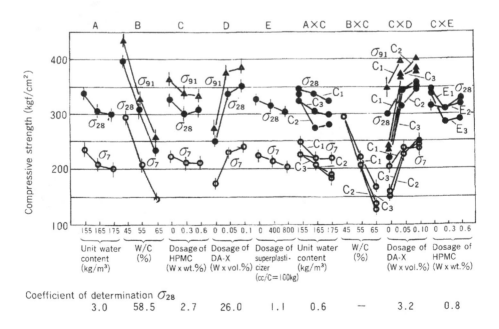

Fig.4. Variation of compressive strength

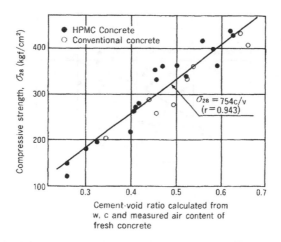

Fig.5. Relationship between cement-void ratio
and compressive strength at 28 days

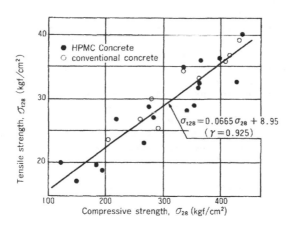

Fig.6. Relationship between compressive strength
and tensile strength

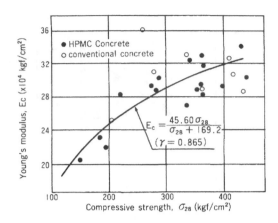

Fig.7. Relationship between compressive strength
and Young's modulus

3 Applicability of HPMC concrete to inverted construction method

In the case of casting of walls in underground tanks, etc. it becomes
necessary to cast the concrete from below, leading to the joint
formation under the existing concrete. In such cases, because of the
accumulation of bleeding water at the top of the new concrete layer,
the bond strength of the joint would not be sufficient. Now, since
the HPMC concrete was found, in the previous section, to have reduced
bleeding and increased viscosity, it might be a good alternative to
conventional concrete for use in inverted construction.
    The specimens cast were tested in flexure to check the integrity
of the joint. In addition, the HPMC concrete specimens were tested
for drying shrinkage and resistance to freezing and thawing since it
was thought that these properties were of significance from the point
of view of concrete durability.

3.1 Properties of bottom concrete and bond
      strength of jointed specimens

(1) Experimental methods
This experiment was also designed based on the orthogonal array table
L9. The factors and levels adopted are shown in Table 2. Because it
was thought that the bond strength of construction joint would be
influenced by consistency, bleeding and early volume change of fresh
concrete, slump, dosage of HPMC and aluminum powder are considered as
factors in this series of experiments.
    A slump of 18 cm was accomplished by adding a superplasticizer to
concrete having slump of 12 cm. When using HPMC, 0.1% DA-X by volume
of unit quantity of water was used. Different doses of aluminum
powder were used to compensate early volume change of concrete.

Table 2. Variables for the experiment

Factors	Level		
	1	2	3
A. Slump (cm)	12 + 2	18 + 2	12 + 2
B. Dosage of HPMC (W×wt.%)	0	0.3	0.6
C. Dosage of aluminum powder (C×wt.%)	0	0.005	0.010

The mix proportions of concrete were decided through trial mixing using water-cement ratio of 0.55 and the same weight of coarse aggregate per unit volume of concrete.

Bleeding, slump, air content, early volume change and compressive strength (using cylinders of 10cm diameter and 20cm height) were measured. Also, flexural strength was determined for jointed and unjointed specimens using 15 x 15 x 53 cm prisms.

Preparation of jointed specimens and the testing method for flexural strength using 4-point loading are shown in Fig.8 and Photos 1 and 2. After the joint surface of upper-existing concrete was brushed away at 7 days, the new concrete was placed and compacted sufficiently with internal vibrator. All specimens were stored under a moist cloth in a chamber with a temperature of 20°C and 80% R.H..

Volume change in concrete was measured using cylinderical specimens of diameter 150mm and height 300mm. The change in the height of the specimen was recorded using a dial gauge beginning 30 minutes after adding water to the mixer. The reading at the end of 24 hours is taken as a measure of the early volume change of the specimen and is expressed as a percentage of the initial volume.

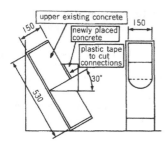

(a) Preparation of specimens having a joint

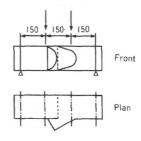

(b) Testing method of the flexural strength

Fig.8. Preparation of specimens having a joint and testing method for the strength

Photo 1.

Casting of jointed specimens

Photo 2.

View of flexural strength test

(2) Experimental results

The temperature of mixed concrete were within the range of 17.0–21.0°C. The air content of concrete with and without HPMC was 3±1% and 4±1%, respectively. Slump in both cases was kept at 12±2cm and 18±2cm.

a) The properties of newly placed concrete

Fig.9 shows the diagrams of the factorial effect judged to be "significant" for 5% level of significance on some properties of newly placed concrete.

According to this diagram, bleeding could not be detected in case of more than 0.3% of unit quantity of water by weight for HPMC dosage.

Early volume change of concrete without aluminum powder was about minus 0.4%. When using 0.005 and 0.01% of aluminum powder by weight of cement, the volume change was found to become plus 0.8 and 2.1%, respectively, because of the induced hydrogen gas evolution.

In this case, the compressive strength decreased to about 85% and 75% of the strength for the concrete without aluminum powder.

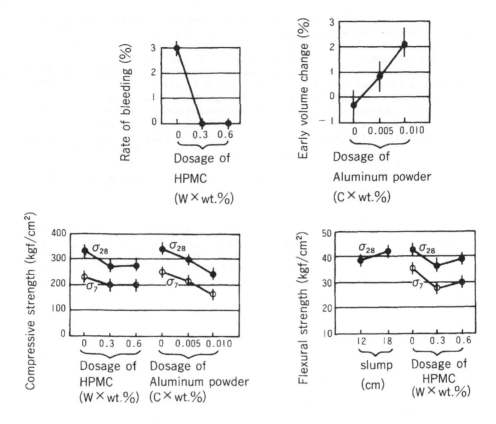

Fig.9. Effects of each factors on the properties of concrete

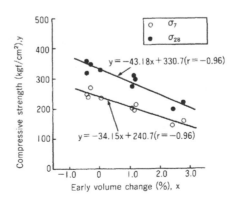

Fig.10. Relationship between early volume change
and compressive strength

Fig.10 shows the relationship between the early volume change and the compressive strength. Regardless of the dosage of HPMC and aluminum powder, the compressive strength decreased with increase of early volume change and the coefficient of correlation between both factors was about 0.96. And the compressive strength of HPMC concrete decreased about 15% as compared with that of non-HPMC concrete. Because the HPMC provided high viscosity to concrete, the early volume change caused by hydrogen evolution (induced by aluminum powder) was larger than that of non-HPMC concrete. The decrease in the compressive strength of HPMC concrete can thus be attributed to "expansion" due to hydrogen evolution, which leads to a decreased density of the matrix.

The ratio of the flexural strength to the compressive strength at 28 days were within the range of 1/8-1/7 regardless of HPMC content.

b) Flexural strength of specimens
Fig.11 shows the diagrams of factorial effect on the flexural strength of specimens jointed invertedly and the flexural strength ratio i.e. the ratio of the flexural strength of specimens jointed invertedly to that of unjointed specimens. The effect of adding different doses of HPMC and aluminum powder on the flexural strength of specimens jointed invertedly can be seen to be very significant (level of significance; 0.05).

The flexural strength of inverted joint specimens was highly influenced by HPMC content. Addition of aluminum powder to concrete also contributed to some improvement in the bond strength of the joint.

Half of the jointed specimens that were made by inverted construction using conventional concrete broke immediately upon removing the form or during handling before flexural testing. Even those specimens which could be tested in flexure, gave a strength of only 1 kgf/cm^2.

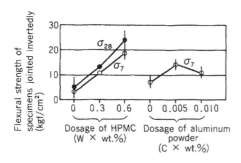 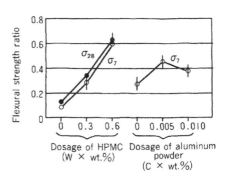

Fig.11. Effects of each factor on flexural strength
of specimen jointed invertedly

In the case of adding 0.010% of aluminum powder by weight of cement only, without any HPMC, the flexural strength of the joint was found to be about 8 kgf/cm^2. On the other hand, the flexural strength of specimens jointed invertedly was about 5, 12 and 24 kgf/cm^2, at 28 days in case of 0, 0.3 and 0.6% of the unit quantity of water by weight for HPMC dosage, respectively.

The bond strength in flexure of the unjointed specimen was observed to be about 39 kgf/cm^2. It can thus be seen that by using 0.6% HPMC by weight of the unit water content, the flexural strength can be taken up to about 60% of that of unjointed specimens.

The ruptured surface of the newly placed concrete for the specimen without HPMC and aluminum powder is shown in Photo 3 and that of the specimen using 0.010 wt.% of aluminum powder and 0.6 wt.% of HPMC is shown in Photo 4. Regardless of aluminum powder content, in case without HPMC the specimens broke cleanly at the joint. In the case of specimens cast with HPMC, the fracture surface is much more ragged with mortar from the original concrete sticking to the ruptured surface as shown in Photo 4.

Therefore, it can be thought that the bond strength of joint in inverted construction method is obviously improved by using HPMC as compared with the conventional concrete or the concrete using only aluminum powder.

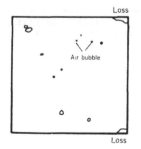
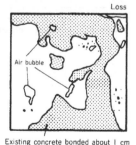

Photo 3. Surface of rupture

$\left(\begin{array}{l}\text{HPMC dosage:W} \times \text{0wt.\%} \\ \text{Aluminum powder dosage:C} \times \text{0wt.\%}\end{array}\right)$

Photo 4. Surface of rupture

$\left(\begin{array}{l}\text{HPMC dosage:W} \times \text{0.6wt.\%} \\ \text{Aluminum powder dosage:W} \times \text{0.010wt.\%}\end{array}\right)$

3.2 Drying shrinkage and resistance to freezing and thawing

(1) Experimental methods
The mix proportions and relevant properties of fresh and hardened
concrete are shown in Tables 3 and 4.
Drying shrinkage-reducing agent (SRA), whose main ingredient is
alkylene-oxide, was used to reduce drying shrinkage of HPMC concrete.
The specimens for drying shrinkage test were 10x10x40cm. The
change of length of specimens were measured in the chamber (20°C,
60% RH) after curing in water for 7 days. The specimen dimensions
for freezing and thawing test were the same and the tests were
conducted according to the method specified in ASTM C 666 (1984).
The dynamic modulus of elasticity was taken as a measure of the
resistance to freezing and thawing.

Table 3. Mix proportions

| No. | Dosage of HPMC (W×wt.%) | Dosage of Aluminum powder (C×wt.%) | Dosage* of SRA (kg/m^3) | Gmax (mm) | W/C (%) | s/a (%) | Unit content (kg/m^3) | | | | |
							W	C	AE** agent	DA-X	SP*** agent
1	–	–	–	25	55	43	158	287	0.72	0.08	2.1
2	0.6	–	–	25	55	43	158	287	0.72	0.08	2.1
3	0.6	0.007	–	25	55	43	158	287	0.72	0.08	2.1
4	0.6	0.007	7.5	25	55	43	158	287	0.72	0.08	2.1

  * SRA was added to mixer as part of unit water content
  ** AE water reducing agent
  *** Superplasticizer

Table 4. Concrete properties

No.	Slump (cm)	Air content (%)	Rate of bleeding (%)	Early volume change (%)	Compressive Strength $\sigma 28$ (kgf/cm^2)	Durability index (%)	Spacing factor (μ)
1	19.5	3.7	2.2	-0.27	365	97	231
2	21.0	4.6	0.0	-0.56	351	14	223
3	21.5	4.4	0.0	1.29	269	91	187
4	19.0	2.1	0.0	1.57	301	–	–

(2) Experimental results

a) Drying shrinkage
Time history of drying shrinkage is shown in Fig.12 for the different
concretes. Drying shrinkage of concrete using HPMC, HPMC and
aluminum powder and conventional concrete without HPMC are about 860,

890, and 760 micros respectively at 12 months. Though the drying
shrinkage of HPMC concrete (860 micros) is greater than that of
conventional concrete (760 micros), it was found that addition of SRA
(710 micros) helps to control this excessive shrinkage.

b) Resistance to freezing and thawing
The results of freezing and thawing test are shown in Fig.13. From
the data, the relative dynamic modulus of elasticity and the
durability index were calculated by the method specified in ASTM C
666 (1984). The durability index on the basis of 300 cycles of
repetition of freezing and thawing was about 97% for conventional
concrete, about 14% for HPMC concrete and about 91% for HPMC concrete
using aluminum powder.

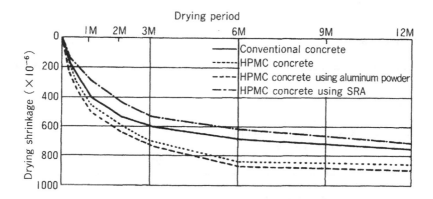

Fig.12. Time history of drying shrinkage

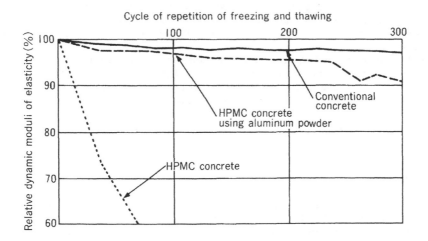

Fig.13. Results of freezing and thawing test

Spacing factor, as defined in ASTM C 457 (1982), was also determined for the different concretes, and the values are listed in Table 4. It can be seen that in the case of HPMC concrete, even though there is no change in the spacing factor compared to the conventional concrete, there is a drastic reduction in the durability index of the concrete. The reason for this could be that HPMC addition increases the viscosity of the capillary water and thus restricts its movement during freezing, thereby causing accelerated deterioration.

In the case of adding aluminum powder to the HPMC concrete the spacing factor reduces and the resistance to freezing and thawing is almost the same as that of conventional concrete.

It can be hence concluded that, caution needs to be exercised when using HPMC concrete in environments likely to be subjected to freezing and thawing cycles.

4 Conclusions

The following conclusions can be described based on the present experimental results:

(a) The HPMC concrete introduced herein shows no detectable amount of bleeding.

(b) When an air detraining agent (tribtylphosphate) is added in suitable quantity, the air content of the HPMC concrete can be controlled within the range of 3 to 4%.

(c) It was found within the limit of present experiments that proper dosage of HPMC was 0.6% by weight of unit water content and that of the air detraining agent was 0.1% by volume of unit water content. When these agents are added, the structural properties, such as, compressive and tensile strength and Young's modulus, of the HPMC concrete become almost the same as those of conventional concrete.

(d) When applying HPMC concrete as a newly placed concrete in inverted construction method, the bond strength of the joint by flexural test can be taken up to 60% of that of the unjointed specimens.

(e) Aluminum powder is added to compensate the volume changes and it also leads to some improvement in the bond strength of the joint.

(f) In HPMC concrete, the drying shrinkage itself becomes larger and the resistance to freezing and thawing lower when compared to conventional concrete. However, this problem can be taken care of by proper additions of drying shrinkage-reducing agents (e.g. alkylene-oxide) and aluminum powder.

Acknowledgement

This work was carried out at the Kajima Institute of Construction Technology as part of an on-going research effort to improve the technique for concrete construction. The authors are very grateful to Mr. K. Kotani and Mr. Y. Nakazato for their valuable suggestions and support during the course of the work.

References

ASTM C 457 (1982) Standard practice for Microscopical determination of air-void content and parameters of the air-void system in hardened concrete.

ASTM C 666 (1984) Standard test method for Resistance of concrete to rapid freezing and thawing.

Hansen, W.C. (1952) Oil Well Cement, Proceedings of the Third International Symposium on the Chemistry of Cement, Cement and Concrete Association, London, September.

Kotani, K., Nakazato, Y. and Motohashi, K. (1983) Experimental study on bond strength of lift joints in inverted concrete pouring method. Annual Report of Kajima Institute of Construction Technology, 31, 1-8.

Kotani, K., Nakazato, Y., Motohashi, K. and Ohno, T. (1984) Some properties of concrete containing water-soluble polymer as a bleeding-reducing admixture. Annual Report of Kajima Institute of Construction Technology, 32, 1-6.

Suzuki, T. and Karasuda, S. (1968) The experimental studies on the joints caused by placing of concrete beneath the hardened concrete. Reports of the research laboratory of the Shimizu Construction Co. Ltd., 11, 9-22.

LE Ca-ACETATE COMME ADJUVANT POUR MORTIERS ET BETONS
(Calcium acetate as an admixture for mortars and concretes)

G. USAI

Department of Chemical Engineering and Materials, University of Cagliari, Italy

Resumé

On a étudié l'influence du Ca-acetate sur les réactions de prise et d' endurcissement de mortiers et bétons des Ciments Portland et Pouzzolanique. En particulier,on a examiné mortiers avec rapport eau-ciment variable et contenant le Ca-acetate en proportions de 0.053 mg à 0.320mg de sel pour gramme de ciment. Sur les épreuves on a étudié la prolongation des temps de prise,en fonction du rapport eau-ciment aussi,et les différences enregistrées regard à l'endurcissement pour maturations jusqu'à duoze mois. Cettes données experimentales peuvent être utiles pour une contribution à l'application d'adjuvants rétardeurs contenant le Ca-acetate.

Key-words:Adjuvants,Ca-acetate,Mortiers,Bétons.

1   Introduction

La consommation d'adjuvants pour ciments et bétons est en augmentation continuelle en Europe -en 1985 le béton preparé avec adjuvants a été 60% en Allemagne,30% en Italie et 20% en France,Rixom 1985- car la préparation de mortiers et bétons de qualité superiuere,à prices contenus, est désormais une necessité universellement reconnue,soit dans les projects,soit dans l'execution des oeuvres architectoniques.
La plupart des études concernant les adjuvants a été dédié aux "accélerants",grace à interets commerciaux ,Collepardi,1980,tandis que les adjuvants "retardeurs" ont connu un mineur interet entre techniciens et rechercheurs du secteur des liants.
Dans cette note -qui est l'accomplissement d'une précédente investigation,Usai 1982,- on a examiné l'effet du Ca-acetate sur les proprietés de mortiers des ciments Portland et Pouzzolanique.
L'auteur espère vivement que les resultats de cette recherche contribueront à fixer les characteristiques du Ca-acetate comme adjuvant retardeur-fluidifiant pour mortiers et bétons.

2 Matériaux et Méthodes

Dans cette investigation on a utilisé les matériaux suivants:
1) Ciment Portland normal (Norme Italiane),production Italcementi SpA (Bergamo) ;
2) Ciment Pouzzolanique (mélange de Ciment Portland (1),70%,et de pouz-zolane de Segni (Rome),30% en poids;
3) Ca-acetate hydraté (réactif pur Merck GmbH);
4) sable normal (Norme Italiane D.M. 3-6-68).
La composition des matériaux est en tableau 1.
On a preparé,ensuite,des mortiers contenant les ciments (1) et (2) dans les proportions: ciment 1 : sable 3 ,en poids.
Le rapport eau-ciment (e/c) a été rigoureusement mantenu constant comme 0.35 ;toutefois,une autre classe d'échantillons comprenait mor-tiers avec rapport e/c variable entre 0.25 et 0.55.
Sur les mortiers frais on a executé les "tests" suivants:
a)temps de prise. Car on ne disposait pas de l'appareillage normal (Aiguille de Vicat,Turriziani 1979),on a utilisé un dispositif exper-imental (pénétromè tre) constitué d'une aiguille d'acier (13 cm) munie d'un poids de 200 g,voir fig. 1.
Le retard de prise -et la prolongation du temps de prise aussi- en fon-ction du contenu du sel Ca-acetate,a été mis en rapport à la pénétra-tion de l'aiguille dans un volume cylindrique de mortier,diamètre 3cm, hauteur 4cm. Voir fig. 2,3 et 4.
b)rhéologie des mortiers. Dans le bût de controler l'effet de l'adju-vant sur l '"ouvrabilité" (Greger,1955;Joisel,1973) du mortier(facili-té d'écoulement du mortier sous l'action d'une force contante)on a ob-

Tableau 1. Composition des Matériaux, % en poids.

	Ciment Portland	Pouzzolane de Segni (Rome)
Al_2O_3	5.10	17.10
Fe_2O_3	4.80	9.25
SiO_2	25.30	45.60
CaO	60.20	9.70
p.p.c.	4.50	4.01
n.det.	- --	14.34

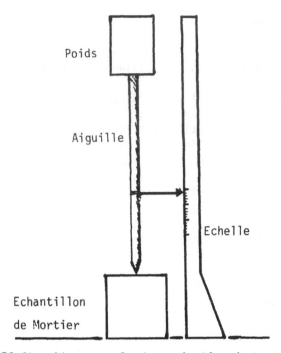

Poids

Aiguille

Echelle

Echantillon

de Mortier

Fig. 1. Pénétromètre pour la determination du temps de prise

Fig.1:schème de l'appareillage dit "pénétromètre" utilisé pour la détermination (non-normalisée) de la fin et de la prolongation de la prise.

tenu des courbes tension-écoulement par un viscosimètre Haake,modification du viscosimètre de Couette.
Voir fig.5 et tableau 3.
On a ensuite preparé des échantillons de mortier des deux ciments,purs et avec adjuvant. Les échantillons ont été laissés endurcir -sous sable et eau- jusq'à 12 mois,temperature 20°C.
Sur les échantillons endurcis on a executé les "tests" suivants:
c)surfaces spécifiques. Les surfaces spécifiques selon B.E.T.(Collepardi 1967) ont été determinées par un appareil "Adsorptomat"Aminco avec registration automatique des isothèrmes d'absorption de l'azote à -196°C,1 atm. Les determination ont regardé mortiers à 15 jours d'endurcissement. Voir tableau 2.
d)résistances à la compression. On a utilisé une presse hydraulique.
Voir fig.6 et 7.
e)porosité des mortiers. On a utilisé un porosimètre Aminco à pénétration forcée de mercure jusqu'à 500 atm.
Voir fig. de 8 à 11 et tableau 4.

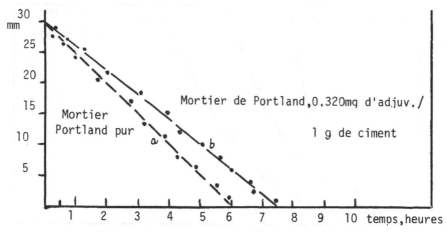

Fig.2. Prolongation du temps de prise.

Fig.2:la ligne droite "a" est rapportée à la pénétration de l'aiguille (fig.1) dans le mortier pur de ciment Portland,rapport e/c=0,35,avec fin de la prise à 6 heures;la ligne droite "b" est rapportée à la pénétration dans le même mortier mais avec contenu de Ca-acetate égal à 0.320 mg/1g de ciment.Fin de la prise :7,5 heures;

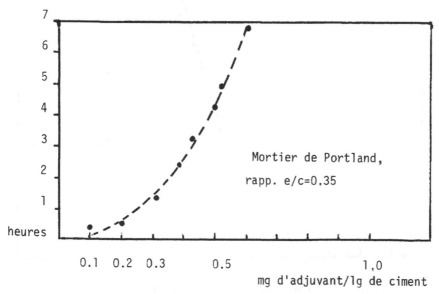

Fig.3.Prolongation du temps de prise.

Fig.3:se rapporte à la prolongation des temps de prise obtenue pour mortiers de Portland (rapport e/c=0.35),en fonction du contenu de sel;

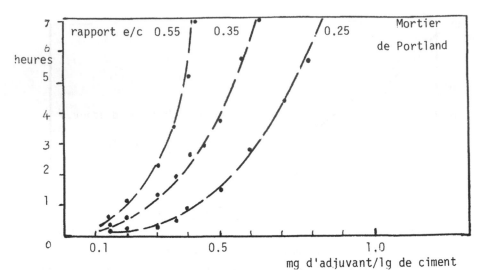

Fig. 4. Prolongation du temps de prise.

Fig.4:se rapporte à la prolongation des temps de prise pour mortiers de Portland,en fonction du contenu d'adjuvant,avec variation du rapport eau-ciment de 0.25 à 0.55;

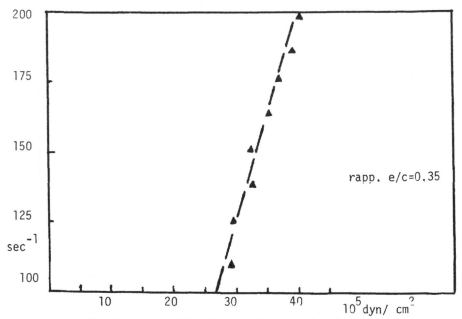

Fig.5.Courbe d'écoulement,Mortier Portland pur.

Fig.5:est la "courbe d'écoulement" d'un mortier de Portland,rapport eau -ciment 0.35; on obtient un valeur de "viscosité apparente" par le rap- port abscisses-ordonnées,unités ċP (centiPoises);

Tableau 2. Surfaces spécifiques selon B.E.T. des Mortiers ,15 jours d'
endurcissement. Rapport eau/ciment =0.35,contenu d'adjuvant
0.053 mg de sel / 1g de ciment.

Mortier de Portland pur	53.5 m^2/g	id. avec adjuvant 55.9 m^2/g
Mortier de Pouzzolane pur 46.8 "		id. " " 47.2 "

Tableau 3.Viscosités apparentes des Mortiers de ciment, unités cP 10^3

Portland pur	(a)	(b)	(c)	(d)
17.0	16.5	15.0	14.6	14.0

Pouzz. pur				
18.8	17.0	16.6	15.0	14.5

D maximum =195 sec^{-1} rapp. e/c=0.35

Contenu d'adjuvant: (a) 0.053 (b) o.100 (c) 0.180 (d) 0.320 mg/1 g

Tableau 4.Porosité des Mortiers,six mois d'endurcissement.
Volume de mercure pénétré dans 1g de mortier, unités cmc 10^{-3}.

Portland pur	(a)	(b)	(c)
18.0	15.0	12.0	10.0

Pouzz. pur			
10.0	9.0	7.0	5.5

Rapp.E/C =0.35 .Contenu d'adjuvant: (a) 0.053 (b)0.180 (c)0.320

mg /1g de ciment

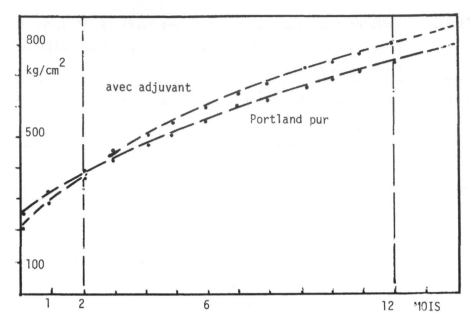

Fig.6.Résistances à la compression,Mortiers Portland.

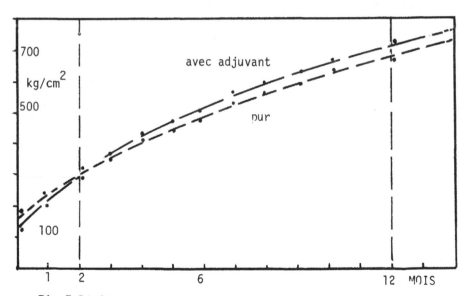

Fig.7.Résistances à la compression,Mortiers Pouzzolane.

Fig.6:(et fig.7):sont ici rapportées les croissances des résistances à la compression des mortiers endurcis,à 6 mois,rapport eau-ciment 0.35, purs et avec adjuvant (0.320 mg de sel/1g de ciment);

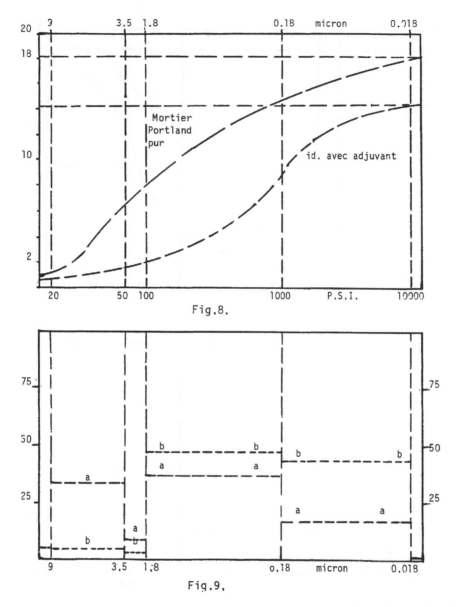

Fig.8.

Fig.9.

Fig. 8 et 10:"courbes de pénétration" rélatives à mortiers de Portland, pur et avec adjuvant ,et à mortiers pouzzolaniques. Rapport e/c=0.35, contenu d'adjuvant 0.320 mg/1g de ciment.
Sur l'axe supérieur on a indiqué le diamètre moyen des pores,unités micron,et sur les abscisses on a indiqué les pressions de pénétration en unités P.S.I.. En ordonnées on rapporte le volume de mercure pénétré dans 1g d'échantillon,unités millièmes de cmc.;

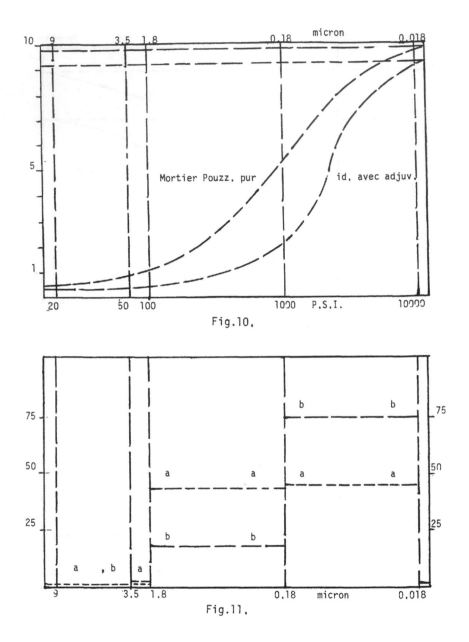

Fig.10.

Fig.11.

Fig.9 et 11:diagrammes de distribution du volume des pores en fonction du diamètre moyen;en ordonnées le pourcentage de volume.
a--a:mortier pur;b--b:mortier avec adjuvant. 14,7PSI=1atm;1 micron= 10 exp-6 mètres.

3 Discussion

Lorsque le ciment commence son hydratation,les sels de Calcium-dont il
est composé- subissent hydrolise et libèrent,ainsi,une grande quantité
d'ions de Calcium (Nakshatra,1984). En presence d'une solution de Ca-acetate,la réaction d'hydrolise peut être empechée,avec un ralentisse-
ment de la réaction générale d'hydratation.
Une théorie propose,en outre,(van Olphen 1963)que le grain de ciment en
hydratation peut adsorber,sur sa surface,des ions acetate avec charge
négative. Le nuage négatif entourant les grains donne répulsion mutuel-
le (action "déflocculante") ,et ça rend plus "ouverte" la structure des
grains en réaction,délivrant des molecules d'eau qui fluidifient toute
l'ensemble. Cettes deux actions du sel Ca-acetate -l'une rétardante et
l'autre fluidifiante- ont été objet de l'investigation ici contenue; cependant le même adjuvant -selon ses proportions- peut agir avec effets
opposés,en regard à l'hydratation des ciments (Collepardi 1980).
On a partagé la discussion sur les données experimentales selon les proprietés étudiées.

a) Temps de prise
L'action rétardante du sel est très évidente dès fig.2,3 et 4.
A' conditions experimentales égales,la prise des mortiers de ciment Portland est majeurement rétardée en regard à la prise des mortiers de Pouzzolane;ça car dans le Portland le pourcentage des sels de Calcium,notam-
ment C3S et C3A,est plus élévé et,pourtant,l'action de l'adjuvant résul-
te favorisée (Bogue 1955). La confrontation parmi le pourcentage d'adjuvant -exprimé comme mg de sel par 1g de ciment- et le rapport eau-ciment
est très interessante,voir fig.4. On observe une action concomitante
parmi l'effet de l'adjuvant et la diluition du liant (augmentation du
rapport e/c de 0.25 à 0.55),Powers 1964.

b) Rhéologie des mortiers
En fig.5 est répresentée la "courbe d'écoulement" d'un mortier de ciment
Portland. En ordonnées on donne le degré d'écoulement (unités sec exp-1)
et en abscisses on donne les correspondentes tensions d'écoulement (unités dynes cm exp-2). Par les courbes d'écoulement on obtient un valeur
de la "viscosité apparente" du mélange solide-eau,pour un valeur du degré d'écoulement,par le rapport tension-degré en unités centiPoises cP
(Heinz 1968). En tableau 3 sont reportés les valeurs de viscosité apparente des mortiers en étude. Les valeurs,ordre 10.000 - 20.000 cP,sont
élévés,consideré le pourcentage de solide (ciment et sable)des mortiers,
presque 90% de la masse totale.
Les mortiers de ciment pouzzolanique montrent viscosités plus élévées
que les mortiers de Portland,avec égal rapport e/c.
Ça car dans le ciment pouzzolanique est remarquable la presence de la

pouzzolane,un matériau avec elevée activité de surface,surement majeure que laquelle du ciment à égale finesse ,Bastianoni 1970.
Dans les mélanges les frottements parmi les grains de solide sont évidemment importants pour la rhéologie du système:grains de solide avec élévée surface active,comme les pouzzolanes,donneront mélanges avec viscosités élévées,Roy 1963. L'action fluidifiante du sel est bien évidente, voir les données en tableau 3.

c) Surfaces spécifiques
On a évalué,selon la méthode B.E.T.,les surfaces spécifiques des mortiers (Portland et Pouzzolaniques),à la maturation de 15 jours. Voir tableau 2. Les mortiers conténant le sel Ca-acetate montrent une augmentation des surfaces. Le developpement des réactions d'hydratation des gels du ciment peut être suivi par l'évaluation de l'augmentation des surfaces (Czernin 1960). La grande dispersion des particules,dérivante de l'action de l'adjuvant,favorise une réaction d'hydratation plus complète. L'augmentation des surfaces,pourtant,est due à l'action de l'adjuvant Ca-acetate (Lea 1970).

d) Résistances à la compression
En fig.6 et 7 on a reporté la croissance des résistances mécaniques des mortiers,jusqu'à 12 mois. Jusq'à deux mois,les mortiers purs ont résistances majeures,mais procedant l'hydratation des ciments,les mortiers avec adjuvant atteignent et supèrent les résistances des mortiers purs. Lorsque la concentration du sel Ca-acetate est diminuée dans l'eau,l'hydratation des ciments recommence plus vite,grace à l'action de l'adjuvant sur les surfaces des gels -comme déja dit- et les résistances resultent améliorées,Collepardi 1973.
Ces effets sont moins marqués pour les ciments pouzzolaniques;à cet propos aussi,il faut rappeler que la "diluition" des composés C3S et C3A- dans les liants de pouzzolane- en est responsable.

e) Porosités
Les figures (de 8 à 11)montrent les "courbes de pénétration" (volume de mercure pénétré dans 1 g de mortier,vieillissement 6 mois,en fonction du diamètre moyen des pores et de la pression de pénétration,Ritter 1945), et les diagrammes de distribution du volume des vides (volume des pores en fonction du diamètre moyen,Winslow 1959). En tableau 4,on peut noter comme le volume des vides diminue dans les mortiers contenant le Ca-acetate,à conditions égales.
Les mortiers avec adjuvant resultent,au total,poins poureux que les mortiers purs,car la réaction d'hydratation a procedé plus completement grace à la dispersion des grains de ciment en hydratation operée par le sel, avec les effets déja examinés. En particulier,il faut noter que -avec égale porosité totale- souvent a diminué le pourcentage des pores "grands"

au profit des pores plus petits.
Dans un matériau solide,à conditions égales,un certain volume de pores
"grands" -en comparaison avec un égal volume de pores "petits"- rends
le solide moins résistant mécaniquement (Massidda 1980):on explique
ainsi l'améliorement des résistances à la compression dans les mortiers
avec Ca-acetate. Enfin,en tout cas,les mortiers de ciment pouzzolanique
montrent porosités mineures que les mortiers de ciment Portland,proba-
blement car les ciments pouzzolaniques hydratés possèdent structures
d'hydratation moins "ouvertes" (Collepardi 1972).

Conclusions
Les résultats experimentales obtenus dans cette récherche permettent
d'aboutir aux conclusions suivantes:
1)entre les proportions experimentées,le sel agit comme rétardeur de la
prise et,avec égal rapport eau-ciment,accroit la fluidité des mélanges
et,donc,la "ouvrabilité" de mortiers et bétons;
2)par un dosage de l'adjuvant attentivement controlé,il est possible
adopter des rapports eau-ciment mineurs (diminution de l'eau de mélange),avec améliorement des résistances mécaniques par la mineure porosité des coulées endurcies;
3)l'effet du sel sur les réactions d'hydratation des composants du ciment est bien évident dès surfaces spécifiques des mortiers endurcis.
Cettes surfaces resultent augmentées,grace à l'action du sel sur l'hydratation des gels;
4)les mortiers de ciment Portland montrent un effet de l'action du Ca-
acetate plus marqué que les mortiers de ciment pouzzolanique,probable-
ment car la proportion des composés C3S et C3A est majeure et,donc,ma-
jeure l'action de l'adjuvant sur leure hydratation.

Bibliographie.

Rixom,M.R. (1986)Chemical Admixtures for Concrete,Spon Ltd London.
Collepardi,M. (1980)Scienza e Tecnica del Calcestruzzo,Milano,pp.324.
Usai,G. (1982) Atti Facoltà Ingegneria Univ.Cagliari,18,pp.157-166.
Turriziani,R. (1979) I Leganti,Roma,pp.371.
Greger,N.H. (1955) J.AM.Ceram.Soc.,,39,pp.98-106.
Joisel,A. (1973)Les Adjuvants du Ciment,Paris,pp.21-30.
Collepardi,M. (1967) Rendic. Facoltà Scienze Univ.Cagliari,37,pp.603-10.
Nakshatra,B.S. (1984)Il Cemento,81,pp.21-29.
van Olphen,H. (1963)Clay Colloid Chemistry,New York,pp.111 .
Collepardi,M. (1980)Scienza e Tecnologia del Calcestruzzo,Milano,p.336.
Bogue,R.H. (1955)The Chemistry of Portland Cement,New York,pp.480.
Powers,T.C. (1964)The Chemistry of Cement,London,pp.391-400.
Heinz,W. (1968)Reologia e Reometria,Firenze,pp.16-20.
Bastianoni,M.(1970)Il Cemento,67,pp.123-130.

Roy,H. (1965) Trans. Indian Ceram. Soc.,24,pp.147-149.

Czernin,W. (1960)La Chimica del Cemento,Milano,pp.59-70.

Lea,F.M. (1970)The Chemistry of Cement and Concrete,Glasgow,pp.305-306.

Collepardi,M.(1973)Il Cemento,69,pp.3-14.

Ritter,L.C.(1945) Industr. Eng. Chem.,17,pp.782.

Winslow,N.M. (1959)ASTM Bulletin.

Massidda,L.(1980) Atti Facoltà Ingegneria Univ.Cagliari,14,pp.215-231.

Collepardi,M; (1972)Il Cemento,68,pp.143-150.

METHOD OF DETERMINATION OF OPTIMAL QUANTITY OF CEMENT IN CONCRETE, BY USING SUPERPLASTICIZERS WHEN CEMENT-STONE DENSITY IS GIVEN IN ADVANCE

T. R. VASOVIC and S. P. MANIC
Technical Division TKK Srpenica, Yugoslavia

ABSTRACT

In technology of concrete, by using superplasticizers, fresh concrete obtains different rheological characteristics in comparing with starting concrete. Some new features have been appearing in form of cement emulsion density, cement-stone density, value of cohesion of fresh concrete, permanent volume of cement emulsion, characteristic water-cement ratio, and maximal fresh concrete cohesion.

On the basis of the above mentioned features of fresh concrete, the possibility of their numerical determining was reached. Accordingly, numerical determination of the optimal quantity of cement in concrete is determined.

Key words: Superplasticizers, Cement, Cement emulsion, Cement-stone, Cement-stone density, Fresh concrete cohesion, Optimal quantity of cement.

INTRODUCTION

In technology of concrete, results of numerous researches show that, by using the superplasticizers, fresh concrete obtains different rheological characteristics in comparing with starting concrete. Those changes of rheological characteristics, as we have noticed, are connected with cement emulsion density in starting concrete. Also, the dosage of superplasticizer has direct influence on cement emulsion density.

The cement emulsion density, by concrete hardening, transforms itself into the cement-stone density. It makes possible to find the answer on many questions about fresh and hardened concrete. This is possible only for those concretes into which superplasti er is added.

In this paper, the authors have done a modest step in development of procedure of determination optimal quantity of cement in concrete, by using cement emulsion density in fresh concrete as a parameter, when cement-stone density is given in advance.

In this very moment, there are not acceptable and clear definitions of different rheological features of fresh concrete, and their connection with hardened concrete as well. For that reason, it is useful some definitions of used terms to be given.

THE VOLUME OF CEMENT EMULSION (Vce)

The volume of cement emulsion can be determined as a sum of volumes of cement and water, as ingredients of fresh concrete.

$$Vce=C(1/gc+W/C) \tag{1}$$

Vce Volume of cement emulsion in l
C   Quantity of cement in kg.
gc   Specific density of cement

THE DENSITY OF CEMENT EMULSION (Gce)

Cement emulsion density is determined as a concentration of cement particles in an available volume of cement and water solution.

$$Gce=\frac{1/gc}{1/gc+W/C} = \frac{0,32}{0,32+W/C} \tag{2}$$

THE DENSITY OF CEMENT-STONE (Gck)

Cement-stone density can be determined as a concentration of cement hydration products in an available volume.

$$Gck=\frac{1/gc+0,23(1-0,254)+0,19}{1/gc+W/C} = \frac{0,68}{0,32+W/C} \tag{3}$$

THE COHESION OF FRESH CONCRETE (Vk)

The cohesion of fresh concrete is determined as attractional force between particles of the fresh concrete as a body, by which it resists fresh concrete mass spreading, while vibrating on vibration desk.

$$Vk=D0/Ds \tag{4}$$

Vk Value of cohesion
D0 Diameter of the Abrams cone base (20 cm)
Ds Mean diameter of spreading concrete, made
    by Abrams cone.

Until now, cohesion of fresh concrete, as a criterion of prediction of hardened concrete features, did not have adequate treatment. Namely, features of the hardened concrete can be recognized even on the diagram of the values of the cohesion of fresh concrete.

An example of this relation is shown on figure 1. The compressive strengh of hardened concrete, as we can see, depands on the cohesion of fresh concrete.

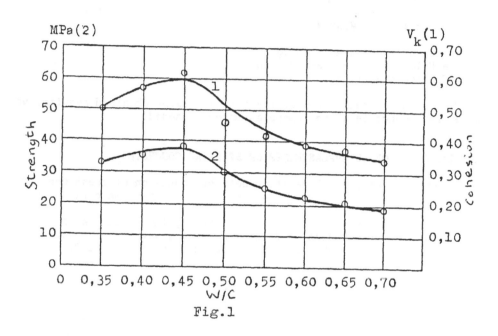

Fig.1

It is obvious, also, that by the constant aggregate (grain-size distribution curve), and the constant quantity of cement, the maximal cohesion of the fresh concrete corresponds to maximal compressive strength of the hardened concrete.

The same water-cement ratio corresponds to extreme values of the compressive strength and cohesion. This water-cement ratio is called "characteristic".

Characteristic water-cement ratio and the constant quantity of cement make possible, by using formula (1), to find the optimal quantity of cement emulsion in the fresh concrete, for aggregate given in advance.

Therefore, optimal volume of the cement emulsion in starting concrete is that one by which cohesion of the fresh concrete and compressive strength of the hardened concrete are maximal.

This optimal volume will be constant for applied aggregate. From that, one can see that for different aggregate, we get different optimal volume of cement emulsion. We can easily prove that by experiment, or using formula (4).

The importance of optimal volume of the cement emulsion is that using that value, and the value of water-cement ratio, calculated by formula (3), one can find optimal quantity of cement by using formula (5). Of course, the desired value of cement-stone density is given in advance.

$$C = \frac{Vce}{1/gc + W/C} = \frac{Vce}{0,32 + W/C}$$ (5)

C   Quantity of cement in kg
Vce Volume of cement emulsin in l
W/C Water-cement ratio

This can be useful in practice only if, using obtained results, we make fresh concrete where superplasticizer is added.

THE INFLUENCE OF SUPERPLASTICIZER ON STARTING CONCRETE

Starting concrete, which relation between cohesion of fresh concrete and compressive strength is shown on figure 1, was made with super-plasticizer. The relation between the dosage of superplasticizer and cement emulsion density we adjust so that by increasing cement emul-sion density, the cohesion of fresh concrete always rises. With res-pect to that rule, we have obtained results shown on figure 2.

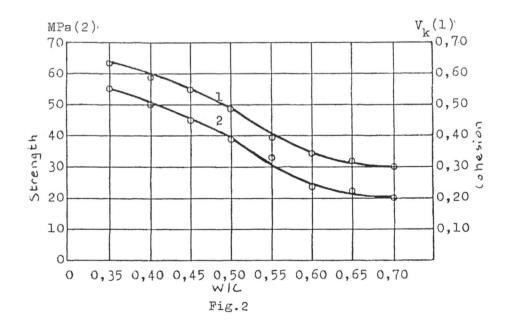

Fig.2

Comparing diagrams on figure 1 and figure 2, we can conclude at once that those relations between cement emulsion density in fresh concrete, and cement-stone density in hardened concrete, have their full meaning only if we make fresh concrete where superplasticizer is added. The dosage of superplasticizer we determine on the base of the rule mentioned above (increasing cement emulsion density implies rise of the fresh concrete cohesion).

The cement-stone without capillary porosity is one of the main conditions of high-quality and high-durability concrete.

Hardened concrete, without capillary pores, according to (3), has density value equal to one (eventual porosity is the result of a bad tamping and vibrating of fresh concrete; we can avoid that by using superplasticizer).

Adjusting the density of cement emulsion in fresh concrete, one can control the cement-stone density of hardened concrete in advance, influencing all relevant parameters of its mechanical and other characteristics. To obtain the cement stone without capillary porosity, the cement-stone density must not be less than 0,47 (in formula (2) W/C=0,36).

SIMILAR IDEAS KNOWN IN TECHNICAL LITERATURE

There are numerous empirical methods of testing workability of fresh concrete. Most of them are different only in some details. Almost all of them simulate some phases of concreting without intention to determine relation between fresh and hardened concrete.

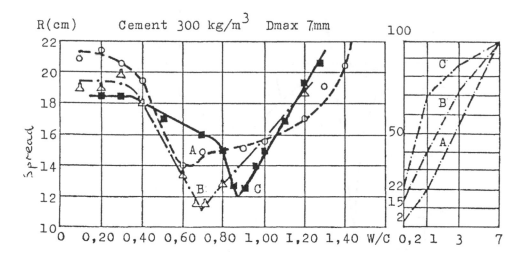

Fig.3

Some authors tried to find some of those relations, using princi-
pls of the fresh concrete rneology:

Tassios (3) has, on the base of numerous tests of concrete consistency, by increasing the quantity of water in fresh concrete, with constant cement and aggregate suggested "the measure of plasticity". On
figure 3, one can see diagrams of spread test for fresh concrete, with
constant quantity of cement, and three different aggregates.

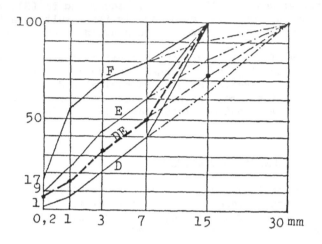

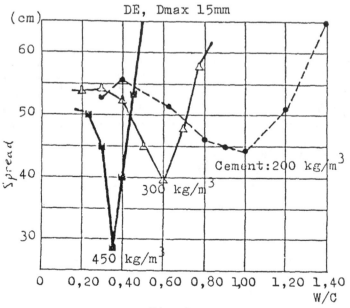

Fig.4

Figure 4 represents also the results of the spread test, now with constant aggregate, and three different quantities of cement in fresh concrete.

It is obvious that two concretes with equal spread, and different cement quantity or different aggregates, have a very different plasticity. Suggested "modulus of plasticity" is not the measure of plasticity of a certain concrete mixture in rheological sense, but the measure of possibility to create concrete of better fluidity and plasticity.

Losinger(4) has, by examining consistency of fresh concrete using Vebe method, found that the influence of internal friction of the large particles can be eliminated by vibrating; the cohesion acts as the only cause of internal resistance.

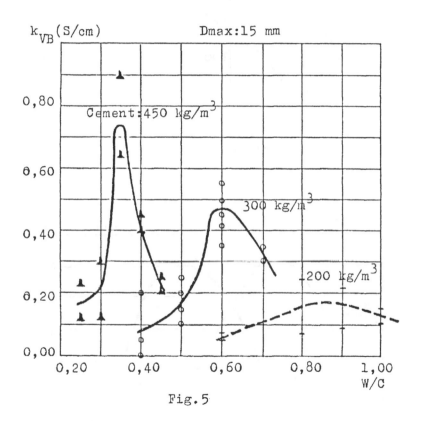

Fig.5

The sample of concrete slumps by gravitation, and the disk slump rate depands on value of cohesion. He suggested that cohesion can be expressed by numerical coefficient Kvb, which expresses disk slump rate in s/cm. On figure 5 are given the diagrams of the coefficient Kvb, for constant aggregate, and three different quantities of cement.

Newman (2) has examined the shear strength of fresh concrete. His results are shown on figure 6. Diagrams present the shear strength of
fresh concrete before (01A), during (OB) and after vibrating (01C).
   By vibrating of concrete, the direction 01A will move into the curve OB, and when vibrating stops, it will move into direction 01C, parallel with flat part of the curve OB. By vibrating, the angle of the
internal friction increases, because of concrete tamping.

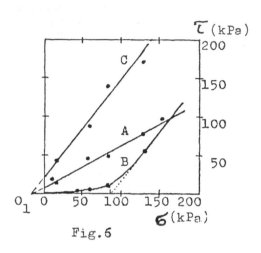

Fig.6

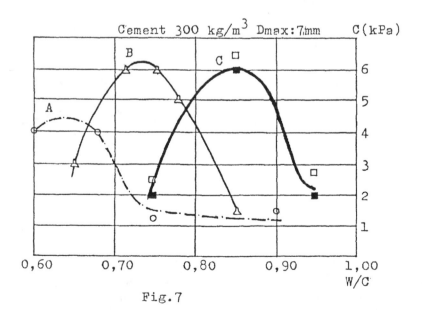

Fig.7

On figure 7 are shown diagrams of cohesion, by constant quantity of cement and three different aggregates.

In Bibliography, we can see that mentioned authors and many others, have felt that in cohesion of fresh concrete the real rheological relation between fresh and hardened concrete exists.

CONCLUSION

The suggested procedure for determination of optimal quantity of cement in concrete represents a step forward in comparison with present knowledge in that area. It takes into account aggregates, quality of future hardened concrete by cement-stone density, cement-stone density by cement emulsion density, and plasticity and workability by the value of cohesion of fresh concrete. Adding of superplasticizer in adequate dosage, enables the fabrication of concrete with different qualities, with great density and high mechanical performances.

References

1. A. M. Neville, Svojstva betona, Gradjevinska knjiga, Beograd, 1976.
2. K. Newman, Concrete Systems, Composite Materials, Elsevier Publishing Company, 1966.
3. T. P. Tassios, Plasticity and Cohesiveness of Fresh Concrete, Proceedings of a RILEM Seminar, Important Properties and their Measurment, Leeds, 1973.
4. R. Losinger, Die Messung von Verarbeitbarkeit von Frishbeton, Roesch-Vogt, Bern, 1956.
5. T. R. Vasović, Contribution to the study of a rheological properties of young and hardened concrete with addition of superplasticizers, ACI-RILEM, Monterey, 1985.
6. T. R. Vasović, Kohezija svežeg betona, merenje veličine kohezije i njen uticaj na mehaničke karakteristike očvrslog betona, JUDIMK, Portorož, 1986.
7. T. R. Vasović, Gustina cementnog kamena – osnovni ključ za rešenje problema trajnosti betonskih konstrukcija, savez IT, Beograd, 1988.
8. T. R. Vasović, Zaštitni sloj betona do armature u betonskoj konstrukciji – kvalitet i predlog kriterijuma za ocenu kvaliteta, SDGKJ, Dubrovnik, 1989.
9. V. Ukrainčik, Prilog optimizaciji obradivosti svežeg betona – Disertacija. Gradjevinski institut. Zagreb, 1979.

A STUDY OF SOME ASPECTS OF MICROAIR AS AIR ENTRAINING ADMIXTURE IN FLYASH CONCRETE

T. A. WEIGEL, J. P. MOHSEN and D. J. HAGERTY
Department of Civil Engineering, University of Louisville, USA

Abstract
In the past decade, several new air entraining admixtures for concrete have been introduced. With increasing use of these products, some field problems have been reported. Specifically, it has been reported that: air contents significantly exceeded target air contents; traditional methods of measuring air content could not be relied upon; low compressive strength; and large variations between the measured air content of fresh concrete and that of the hardened concrete. To investigate the reported problems, a testing program was initiated. Parameters investigated include the variation of air content with mixing time, the variation of strength versus mixing time and the degree of agreement between plastic air content and the air content of hardened concrete.
Key words: Admixture, air content, compressive strength.

1 Introduction

In the past decade several new air entraining admixtures for concrete have been introduced. Some claims of superiority of these new products over traditional vinsol resin admixtures include: more stable air content with mixing; increased resistance to damage from freeze/thaw cycles; reduced permeability; reduced segregation and bleeding and improved plasticity and workability.

With increasing use of these products, some field problems have been reported. Amon (1987) reported the following problems for one such admixture (MicroAir): final air contents which significantly exceed target values; traditional methods of measuring air content could not be relied upon to to accurately determine true air content; low compressive strengths resulting from excessive air contents; and large variations between the measured air content of fresh concrete and that of the hardened concrete.

To investigate both claims for and reported problems with these new admixtures, a testing program was initiated at the University of Louisville. Because of specific reported problems, MicroAir was chosen as the air entraining admixture for the initial study. MicroAir is a product of Master Builders of Cleveland, Ohio and is a multi-component mixture of fatty acids, salts of sulfonic acid and stabilizing agents.

2 Testing Program

Mix data for the testing program were:

Target air content	7%
Design compressive strength	3500 psi
Target compressive strength	4700 psi
Admixture dosage	43 ml
Water	24.5 lb
Flyash	11.5 lb
Cement	46.2 lb
Fine Aggregate	98.3 lb
Coarse Aggregate	121.7 lb
Water cement ratio	0.42

The mixing procedure for each batch was as follows. The air entraining admixture was added directly to the mixing water. After all ingredients were added to the mixer, the mass was mixed for three minutes, rested for three minutes, and then mixed to the final time. Time zero occurred after all the components were added to the mixer.

The first set of tests consisted of a total of six batches. At time t = 15 minutes, the following tests were conducted: air content (pressure meter), slump, unit weight and temperature. If the air content did not fall in the range 6.5% to 7.5% or if the slump did not fall in the range 3.5 in to 4.5 in, the batch was rejected. Those batches which were not rejected were mixed to the final time (15, 30, 45, 60, 75 or 90 minutes - see Tables 1 and 2) at which time the same set of tests was performed. In addition, three strength test cylinders and three petrographic samples were made. Test results are summarized in Table 1 and Figures 1 and 2.

Table 1. Test Results for Set 1

Batch	Initial Values			Final Values (At "Batch" Time)	
	Time	Air Content	Slump	Air Content	Slump
(min)	(min)	(%)	(in)	(%)	(in)
15	15	6.5	4.5	Same as initial values	
30	15	5.0	3.5	6.5	3.0
45	15	6.6	3.5	7.6	2.0
60	15	7.0	4.0	9.0	2.5
75	15	6.6	4.0	6.4	2.0
90	15	7.5	4.5	6.8	2.0

Batch	Compressive Strength @ 28 days
(min)	(psi)
15	4500
30	4800
45	3800
60	3500
75	5350
90	4300

A great deal of difficulty was encountered in achieving target air contents for the first set of tests. Before testing was begun, the research team fully expected the batches would be at full air content at the time the initial tests were done (15 minutes). This expectation was not realized, as results from the second set of tests demonstrate. Because at the time it was felt that target air contents were not being achieved, many batches were rejected and the amount of admixture used was frequently adjusted in an effort to achieve the proper air content. As a result, final air contents varied widely, as Table 1 shows.

After completion of the first set of tests, it was suspected that it may take longer than 15 minutes for MicroAir to become fully involved in the mix to produce the desired air content. For this reason, during the second set of tests, air contents were sampled at both 8 minutes and 20 minutes. Air pressures measured at 20 minutes (using a fixed

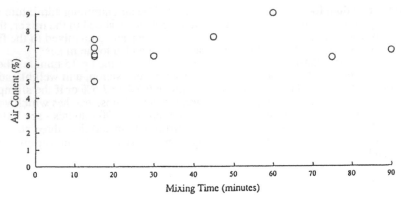

Figure 1. Air Content vs. Mixing Time (Set 1)

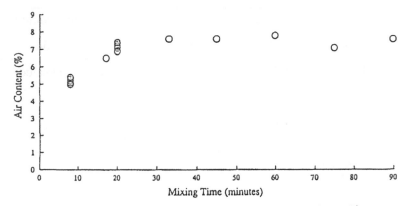

Figure 2. Air Content vs. Mixing Time (Set 2)

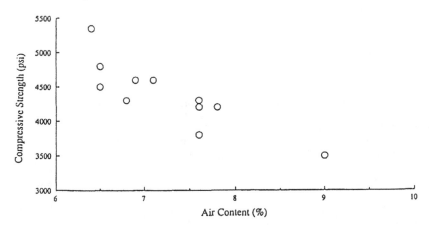

Figure 3. Compressive Strength vs. Final Air Content (Both Sets)

amount of admixture) were very consistent, as examination of Table 2 will show. The amount of admixture shown in the mixture proportions (43 ml) is the amount used for the second set of tests.

For the second set of tests (also consisting of six batches), the following testing procedure was used. At time t = 8 minutes a pressure meter reading was taken. At time t = 20 minutes the following data were collected: temperature, air content (pressure meter), slump and unit weight. If the air content was not within the range 6.5% to 7.5%, the batch was rejected. If the measured slump was not within the range 3.5 in to 4.5 in, the batch was rejected.

After these tests were made, the mixer was never stopped until the final mixing time was reached, and at that time the mix was deposited into a wheelbarrow and the same tests were repeated. In addition, three petrographic and three compressive strength specimens were prepared. Additional water was never added to the mix and fresh concrete was never returned to the mixer after a test was performed. Test results are shown in Table 2 and Figures 2 and 3.

Table 2. Test Results for Set 2

Batch	Initial Values			Final Values (At "Batch" Time)	
	Time	Air Content	Slump	Air Content	Slump
(min)	(min)	(%)	(in)	(%)	(in)
20	20	6.9	3.5	Same as initial values	
33	17	6.5	4.0	7.6	3.5
45	20	7.4	3.5	7.6	3.0
60	20	7.3	4.0	7.8	3.0
75	20	7.1	4.0	7.1	2.5
90	20	7.2	4.0	7.6	2.5

Batch	Air Content at 8 Minutes	Compressive Strength @ 28 days
(min)	(%)	(psi)
20	5.3	4600
33	5.0	4200
45	5.4	4200
60	5.2	4200
75	5.1	4600
90	5.1	4300

3. Conclusions

The tests performed in this study show that air content is well correlated with mixing time. For MicroAir, the air content increased rapidly during the first 20 minutes of mixing and then remained relatively stable. The air content increased from 5% at 8 minutes to 7% at 20 minutes in all of the batches. This explains the variation in the results for the first set of tests, where the initial air contents were measured at 15 minutes. For MicroAir, initial testing time appears to be critical. A test made after mixing for 20 minutes appears to be more representative of the final air content than does a test made at 15 minutes. This may explain some of the reported field problems.

Petrographic inspection results are not presented in this paper. Air content measurements made on the petrographic samples correlated extremely well with the corresponding pressure meter readings. This indicates that the pressure meter used was operating properly and it was properly used. Further, this result seems to refute field observations that traditional methods of measuring air content cannot be used reliably on concrete incorporating the new admixtures (or at least for MicroAir).

Another observation is in order. Batches of concrete made with the admixture tested seem to be very sensitive to water content. This sensitivity seems to be related to use of flyash in the mix, because this sensitivity was not observed in batches made without flyash (not reported in this paper). Batches of concrete made using the admixture tested appear to yield very reproducible properties if the mix proportions of water and aggregates are adjusted to compensate for changes in aggregate moisture content.

Reference

Amon, J. (1987), Law Engineering, Nashville, Tennessee, Private Communication.

Units Equivalence

To convert	to	Multiply by
pound force per square inch (psi)	kilopascal (kPa)	6.89
inch (in)	millimeter (mm)	25.4
pound force (lb)	newton (N)	4.45

INDEX

This index has been compiled using the keywords provided by the authors of the individual papers, with some editing and additions to ensure consistency. The numbers refer to the first page of the relevant paper.

Absorption   63, 142, 307, 317, 360
Accelerated testing   241, 289, 317, 382
Accelerators   106, 142, 197, 251
Acid resistance   156
Adsorption   158
Aggregates   51, 524
Aggregate-cement interface   392
Aggressivity   360
Air content   94
Air-detraining agents   540
Air-entraining agents   1, 20, 360, 429, 449, 460, 507, 578
Alkyl alkoxy silane   317
Aluminium   540
Amino alcohol derivatives   317

Bleeding   429, 540
Bond strength   540

Calcium acetate   540
Calcium chloride   106, 251, 375
Calcium lignosulphonate   135, 289, 507
Calcium nitrate   106
Calcium nitrite   251, 299, 317, 382
Calorimetry   120, 142, 402
Capillary absorption   307, 325, 360
Carbonation   10, 219, 229, 241, 269, 289, 317
Cellulose ether   540
Cement paste   569
Chemical resistance   156
Chlorides   251, 279, 299, 317, 346, 375, 382
Chloride-free accelerators   197, 251
Cohesion   569
Compaction   533
Compressive strength   94, 142, 156, 168, 175, 183, 197, 209, 229, 317, 325, 402, 433, 440, 507, 533, 540, 556, 578
Construction joints   540
Corrosion   219, 251, 269, 289, 299, 346, 375, 382
Corrosion inhibitors   219, 251, 279, 299, 375, 382
Cracks and cracking   279, 484
Creep   209, 402, 484
Curing   183, 197

Density   569
Dispersing agent   533
Dosage   63, 120, 496, 516
Drying shrinkage, *see* Shrinkage
Durability   229, 241, 307, 317, 346, 507
Dynamic viscosity   20

Electrochemical methods   142, 219, 251, 289, 299, 375, 382
Expansion   360, 433

Ferrocement   360
Fillers   80
Finishability   34
Flash setting   135
Flexural strength   142, 540
Flow curves   20
Flow table test   1, 569
Flowing concrete   51, 429, 524
Fluorogypsum   135
Fly ash   142, 168, 183, 197, 219, 251, 269, 325, 449, 533, 578
 Formwork   51
Freeze-thaw resistance   10, 307, 360, 429, 507, 540

High rise structures   94
High strength concrete   63, 94
Hot weather concreting   120
Hydration   106, 142
Hydroxide   106

Inverted construction method   540

Joints   540

Magnesium calcium silicate   516
Mathematical models   209
Melamine formaldehyde sulphonates   34, 63, 168, 175, 229, 289, 507
Microsilica, *see* silica fume
Modulus of elasticity   402, 540
Molybdates   299
Moulds   516

Naphthalene formaldehyde sulphonates   63, 168, 175, 229, 289, 392
Nitrites   219, 279

No-slump concrete   516
Non-destructive testing   360

Particle size distribution   1, 440
Permeability   10, 20, 183, 229, 251, 325, 346, 360, 392,
    507, 516, 524
Plasticizers   80, 142, 289, 360, 507
Pore water pressure   20
Porosity   241, 307, 325, 392, 556
Pozzolanas   142, 219, 269, 325, 392, 556
Precast concrete   516
Pulverized-fuel ash, see Fly ash

Ready-mixed concrete   94
Reinforcement   51, 279
Resonance   360
Retarders   120, 142, 360, 440
Rheology   20, 34, 80, 556
Roller compacted concrete   533

Seawater   279, 360, 375
Segregation   51, 440
Segregation-controlling admixtures   20, 34, 51, 440
Setting time   120, 135, 142, 197, 496, 556
Settlement   540
Shrinkage   10, 209, 325, 402, 484, 540
Shrinkage reducing agents   484
Silica fume   1, 63, 156, 175, 229, 241, 307, 346, 402,
    516
Site measurements   94, 496, 524
Size effects   183
Slag   142, 219, 440
Slipforming   496
Slump loss   10, 440
Slump retention   94
Slump testing   10, 34
Sodium molybdate   299
Sodium thiocyanate   106
Solution chemistry   106
Specific surface   1, 556
Spectrophotometry   142
Steam curing   197, 375
Sulphate resistance   325, 346
Sulphates   106, 325, 346
Superplasticizers   1, 10, 20, 34, 63, 94, 168, 175,
    183, 197, 209, 241, 269, 325, 392, 402, 429, 440,
    460, 496, 524, 540, 569
Surface activation agent   533

Temperature effects   120, 197, 402, 496
Tensile strength   402

Underwater concrete   440, 524
Ultrasonic pulse velocity testing   360

Vacuum dewatering   51
Vibration   20
Viscosity   20, 34, 51, 80, 556

Water/cement ratio   229, 392
Water penetration   183, 241, 325, 360, 516
Water-reducing admixtures   120, 135, 156, 229, 346, 533
Waterproofers   229
Wave propagation   20
Wetting and drying tests   382
Workability   10, 20, 516, 540
Workability retention   63, 94

X-ray diffraction analysis   106

Zinc oxide   375

Printed and bound by CPI Group (UK) Ltd, Croydon, CR0 4YY

01/11/2024

01782621-0014